TTL PIN CONFIGURATIONS

74123

Pin	Signal		Signal	Pin
1	$\overline{A}_1$		V_{CC}	16
2	B_1		$R_{ext}/C_{ext\,1}$	15
3	R_{D1}		$C_{ext\,1}$	14
4	$\overline{Q}_1$		Q_1	13
5	Q_2		Q_2	12
6	C_{ext2}		R_{D2}	11
7	R_{ext}/C_{ext2}		B_2	10
8	GND		$\overline{A}_2$	9

74132

(quad NAND Schmitt trigger, pins 1–14, V_{CC} = 14, GND = 7)

74138

Pin	Signal		Signal	Pin
1	A_0		V_{CC}	16
2	A_1		$\overline{0}$	15
3	A_2		$\overline{1}$	14
4	$\overline{E}_1$		$\overline{2}$	13
5	$\overline{E}_2$		$\overline{3}$	12
6	E_3		$\overline{4}$	11
7	$\overline{7}$		$\overline{5}$	10
8	GND		$\overline{6}$	9

74139

Pin	Signal		Signal	Pin
1	$\overline{E}_a$		V_{CC}	16
2	A_{0a}		$\overline{E}_b$	15
3	A_{1a}		A_{0b}	14
4	$\overline{0}_a$		A_{1b}	13
5	$\overline{1}_a$		$\overline{0}_b$	12
6	$\overline{2}_a$		$\overline{1}_b$	11
7	$\overline{3}_a$		$\overline{2}_b$	10
8	GND		$\overline{3}_b$	9

74147

Pin	Signal		Signal	Pin
1	$\overline{I}_4$		V_{CC}	16
2	$\overline{I}_5$			15
3	$\overline{I}_6$		$\overline{A}_3$	14
4	$\overline{I}_7$		$\overline{I}_3$	13
5	$\overline{I}_8$		$\overline{I}_2$	12
6	$\overline{A}_2$		$\overline{I}_1$	11
7	$\overline{A}_1$		$\overline{I}_9$	10
8	GND		$\overline{A}_0$	9

74148

Pin	Signal		Signal	Pin
1	$\overline{I}_4$		V_{CC}	16
2	$\overline{I}_5$		$\overline{EO}$	15
3	$\overline{I}_6$		$\overline{GS}$	14
4	$\overline{I}_7$		$\overline{I}_3$	13
5	$\overline{EI}$		$\overline{I}_2$	12
6	$\overline{A}_2$		$\overline{I}_1$	11
7	$\overline{A}_1$		$\overline{I}_0$	10
8	GND		$\overline{A}_0$	9

74150

Pin	Signal		Signal	Pin
1	D_7		V_{CC}	24
2	D_6		D_8	23
3	D_5		D_9	22
4	D_4		D_{10}	21
5	D_3		D_{11}	20
6	D_2		D_{12}	19
7	D_1		D_{13}	18
8	D_0		D_{14}	17
9	E		D_{15}	16
10	$\overline{Y}$		S_0	15
11	S_3		S_1	14
12	GND		S_2	13

74151

Pin	Signal		Signal	Pin
1	I_3		V_{CC}	16
2	I_2		I_4	15
3	I_1		I_5	14
4	I_0		I_6	13
5	Y		I_7	12
6	$\overline{Y}$		S_0	11
7	$\overline{E}$		S_1	10
8	GND		S_2	9

74154

Pin	Signal		Signal	Pin
1	$\overline{0}$		V_{CC}	24
2	$\overline{1}$		A_0	23
3	$\overline{2}$		A_1	22
4	$\overline{3}$		A_2	21
5	$\overline{4}$		A_3	20
6	$\overline{5}$		E_1	19
7	$\overline{6}$		E_0	18
8	$\overline{7}$		$\overline{15}$	17
9	$\overline{8}$		$\overline{14}$	16
10	$\overline{9}$		$\overline{13}$	15
11	$\overline{10}$		$\overline{12}$	14
12	GND		$\overline{11}$	13

74163

Pin	Signal		Signal	Pin
1	$\overline{SR}$		V_{CC}	16
2	CP		TC	15
3	D_0		Q_0	14
4	D_1		Q_1	13
5	D_2		Q_2	12
6	D_3		Q_3	11
7	CEP		CET	10
8	GND		$\overline{PE}$	9

74164

Pin	Signal		Signal	Pin
1	D_{sa}		V_{CC}	14
2	D_{sb}		Q_7	13
3	Q_0		Q_6	12
4	Q_1		Q_5	11
5	Q_2		Q_4	10
6	Q_3		$\overline{MR}$	9
7	GND		CP	8

74165

Pin	Signal		Signal	Pin
1	$\overline{PL}$		V_{CC}	16
2	CP		$\overline{CE}$	15
3	D_4		D_3	14
4	D_5		D_2	13
5	D_6		D_1	12
6	D_7		D_0	11
7	$\overline{Q}_7$		D_S	10
8	GND		Q_7	9

74181

Pin	Signal		Signal	Pin
1	$\overline{B}_0$		V_{CC}	24
2	$\overline{A}_0$		$\overline{A}_1$	23
3	S_3		$\overline{B}_1$	22
4	S_2		$\overline{A}_2$	21
5	S_1		$\overline{B}_2$	20
6	S_0		$\overline{A}_3$	19
7	C_n		$\overline{B}_3$	18
8	M		$\overline{G}$	17
9	$\overline{F}_0$		$C_n + 4$	16
10	$\overline{F}_1$		$\overline{P}$	15
11	$\overline{F}_2$		$A = B$	14
12	GND		$\overline{F}_3$	13

74190

Pin	Signal		Signal	Pin
1	D_1		V_{CC}	16
2	Q_1		D_0	15
3	Q_0		CP	14
4	$\overline{CE}$		$\overline{RC}$	13
5	$\overline{U/D}$		TC	12
6	Q_2		$\overline{PL}$	11
7	Q_3		D_2	10
8	GND		D_3	9

74191

Pin	Signal		Signal	Pin
1	D_1		V_{CC}	16
2	Q_1		D_0	15
3	Q_0		CP	14
4	$\overline{CE}$		$\overline{RC}$	13
5	$\overline{U/D}$		TC	12
6	Q_2		$\overline{PL}$	11
7	Q_3		D_2	10
8	GND		D_3	9

74192

Pin	Signal		Signal	Pin
1	D_1		V_{CC}	16
2	Q_1		D_0	15
3	Q_0		MR	14
4	CP_D		$\overline{TC}_D$	13
5	CP_U		$\overline{TC}_U$	12
6	Q_2		$\overline{PL}$	11
7	Q_3		D_2	10
8	GND		D_3	9

74193

Pin	Signal		Signal	Pin
1	D_1		V_{CC}	16
2	Q_1		D_0	15
3	Q_0		MR	14
4	CP_D		$\overline{TC}_D$	13
5	CP_U		$\overline{TC}_U$	12
6	Q_2		$\overline{PL}$	11
7	Q_3		D_2	10
8	GND		D_3	9

74194

Pin	Signal		Signal	Pin
1	MR		V_{CC}	16
2	D_{SR}		Q_0	15
3	D_0		Q_1	14
4	D_1		Q_2	13
5	D_2		Q_3	12
6	D_3		CP	11
7	D_{SL}		S_1	10
8	GND		S_0	9

74244

Pin	Signal		Signal	Pin
1	$\overline{OE}_a$		V_{CC}	20
2	I_{a0}		$\overline{OE}_b$	19
3	Y_{b0}		Y_{a0}	18
4	I_{a1}		I_{b0}	17
5	Y_{b1}		Y_{a1}	16
6	I_{a2}		I_{b1}	15
7	Y_{b2}		Y_{a2}	14
8	I_{a3}		I_{b2}	13
9	Y_{b3}		Y_{a3}	12
10	GND		I_{b3}	11

74245

Pin	Signal		Signal	Pin
1	$S/\overline{R}$		V_{CC}	20
2	A_0		$\overline{CE}$	19
3	A_1		B_0	18
4	A_2		B_1	17
5	A_3		B_2	16
6	A_4		B_3	15
7	A_5		B_4	14
8	A_6		B_5	13
9	A_7		B_6	12
10	GND		B_7	11

74280

Pin	Signal		Signal	Pin
1	I_6		V_{CC}	14
2	I_7		I_5	13
3			I_4	12
4	I_8		I_3	11
5	Σ_E		I_2	10
6	Σ_O		I_1	9
7	GND		I_0	8

74283

Pin	Signal		Signal	Pin
1	Σ_2		V_{CC}	16
2	B_2		B_3	15
3	A_2		A_3	14
4	Σ_1		Σ_3	13
5	A_1		A_4	12
6	B_1		B_4	11
7	C_{IN}		Σ_4	10
8	GND		C_{out}	9

74373

Pin	Signal		Signal	Pin
1	$\overline{OE}$		V_{CC}	20
2	Q_0		Q_7	19
3	D_0		D_7	18
4	D_1		D_6	17
5	Q_1		Q_6	16
6	Q_2		Q_5	15
7	D_2		D_5	14
8	D_3		D_4	13
9	Q_3		Q_4	12
10	GND		E	11

74374

Pin	Signal		Signal	Pin
1	$\overline{OE}$		V_{CC}	20
2	Q_0		Q_7	19
3	D_0		D_7	18
4	D_1		D_6	17
5	Q_1		Q_6	16
6	Q_2		Q_5	15
7	D_2		D_5	14
8	D_3		D_4	13
9	Q_3		Q_4	12
10	GND		CP	11

74395

Pin	Signal		Signal	Pin
1	$\overline{MR}$		V_{CC}	16
2	D_S		Q_0	15
3	D_0		Q_1	14
4	D_1		Q_2	13
5	D_2		Q_3	12
6	D_3		$\overline{Q}_3$	11
7	PE		$\overline{CP}$	10
8	GND		$\overline{OE}$	9

Christopher Charron
267 Rounseville Rd.
Rochester, ma 02770
(508) 763-2619

Christopher Charron
267 Rounseville Rd.
Rochester, ma 02770
(508) 763-2619

Digital
Electronics

FOURTH EDITION

Digital Electronics

A Practical Approach

William Kleitz

Tompkins Cortland Community College

PRENTICE HALL
Upper Saddle River, New Jersey 07458

Library of Congress Cataloging-in-Publication Data
Kleitz, William.
    Digital electronics : a practical approach / William Kleitz.—
4th ed.
      p.   cm.
    Includes bibliographical references (p.   ) and indexes.
    ISBN 0-13-352188-5
    1. Digital electronics.   I. Title.
TK7868.D5K55   1996
621.39′5—dc20

          95-1578
          CIP

Cover photo: Bruce Peterson/H. Armstrong Roberts
Editor: Dave Garza
Developmental Editor: Carol Hinklin Robison
Production Editor: Louise N. Sette
Copy Editor: Linda Thompson
Text Designer: Rebecca M. Bobb
Cover Designer: Brian Deep
Production Manager: Deidra M. Schwartz
Marketing Manager: Debbie Yarnell
Illustrations: Steve Botts and Jane Lopez

This book was set in Times Roman by The Clarinda Company and was printed and bound by Von Hoffman Press, Inc. The cover was printed by Phoenix Color Corp.

© 1996 by Prentice-Hall, Inc.
Simon & Schuster / A Viacom Company
Upper Saddle River, New Jersey 07458

Earlier editions © 1993, 1990, 1987 by REGENTS/PRENTICE HALL.

Photo Credit: Figure 4–38(a) on p. 96 courtesy of Jon Reis Photography.

Printed in the United States of America

10 9 8 7 6 5 4 3 2

ISBN: 0-13-352188-5

Prentice-Hall International (UK) Limited, *London*
Prentice-Hall of Australia Pty. Limited, *Sydney*
Prentice-Hall of Canada, Inc., *Toronto*
Prentice-Hall Hispanoamericana, S. A., *Mexico*
Prentice-Hall of India Private Limited, *New Delhi*
Prentice-Hall of Japan, Inc., *Tokyo*
Simon & Schuster Asia Pte. Ltd., *Singapore*
Editora Prentice-Hall do Brasil, Ltda., *Rio de Janeiro*

To my wife, *Leeann,*
for typing the original manuscript and providing encouragement and
understanding throughout the preparation of this text;
and to my daughters, *Shirelle and Hayley.*

Preface

To the Instructor

It is time to reevaluate the way that digital electronics is taught. At first, digital electronics was a theoretical science, but now it is a down-to-earth, workable technology that can be taught using a practical approach.

Digital Electronics: A Practical Approach, Fourth Edition, places emphasis on analytical reasoning and basic digital design using the standard integrated circuits that are used in industry today. Throughout the text, actual ICs are used, and reference is made to the appropriate manufacturers' data sheets. Because of this approach, students become proficient at using the terminology and timing diagrams that are the standard in manufacturers' data manuals and industrial settings.

A strong effort was made to make the text easy to read and understand so that motivated students can teach themselves topics that require extra work without the constant attention of the instructor. Several digital system design applications and troubleshooting exercises are included. Also, there are ample illustrations, examples, and review questions to help students reach a point where they can reason out the end of chapter problems on their own. After all, that is the main goal of this book—to help students think and reason on their own.

This book can be used for a one- or two-semester course in digital electronics and is intended for students of technology, computer science, or engineering programs. Although not mandatory, it is helpful if students using this text have an understanding of, or are concurrently enrolled in, a basic electricity course. A laboratory component to provide hands-on reinforcement of the material presented in this book can be very helpful. Laboratory exercises can be developed by building, testing, debugging, and analyzing the operation of any of the examples or system design applications that are provided within the text.

A Practical Approach to Learning Digital Electronics

What is it that students really need to learn to function in the modern digital electronics field? That is the question that I ask myself every time I teach a digital course from my textbook or when I visit an electronics facility. The answer is that students need to learn the practical skills required to design and troubleshoot actual digital circuitry that they will see on the job. *Practical* means that the circuits that they study must be made up of *actual* integrated circuits and the specifications that they learn about are taken from *actual* manufacturers' data sheets. Practical also means that the students must be taught how to think on their own and how to reason out new concepts as they come up on the job. They must also be able to teach themselves new material as new developments in the field arise.

To address these needs, this book makes a strong effort to use actual ICs and data sheets in its examples. The text covers the basic fundamentals of any circuit design so that the students, knowing the basic building blocks, can teach themselves the newest technology when faced with it on the job. Also, material has been added to help reduce the anxiety that students feel when they first start a new job and are faced with schematics of large systems to analyze. To get students used to these large-scale circuit diagrams, I have included four "real-world" schematics that contain several of the ICs and circuits described in this text. By working the Schematic Interpretation problems at the end of each chapter, and referencing the schematics in Appendix G, students are forced to dig into the complex diagrams a little bit deeper in each successive chapter, so that by the end of the text they will have covered almost all of the complete circuits.

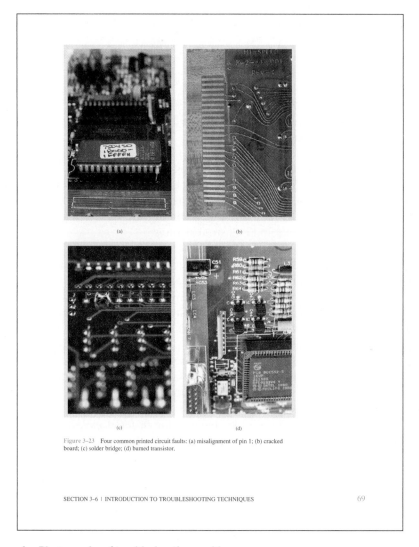

(a)　　　　　(b)

(c)　　　　　(d)

Figure 3–23　Four common printed circuit faults: (a) misalignment of pin 1; (b) cracked board; (c) solder bridge; (d) burned transistor.

SECTION 3–6 | INTRODUCTION TO TROUBLESHOOTING TECHNIQUES　　69

Figure 1　Photographs of troubleshooting problems.

Problem Sets

A key part of learning any technical subject matter is for the student to have practice solving problems of varying difficulty. The problems at the end of each chapter are grouped together by section number. Within each section there are several basic problems designed to get the student to solve a problem using the fundamental information presented in the chapter. Besides the basic problems, there are four other problem types:

D (Design) Problems designated with the letter *D* ask the students to modify an existing circuit or to design an original circuit to perform a specific task. This type of exercise stimulates creative thinking and instills a feeling of accomplishment upon successful completion of a circuit design.

T (Troubleshooting) Problems designated with the letter *T* present the student with a malfunctioning circuit to be diagnosed or ask for a procedure to follow to test for proper circuit operation. This develops the student's analytical skills and prepares him or her for troubleshooting tasks that would typically be faced on the job.

C (Challenging) Problems designated with the letter *C* are the most challenging to solve. They require a thorough understanding of the material covered and go a step beyond, by requiring the student to develop some of his or her own strategies to solve a problem that is different from the examples presented in the chapter. This also expands the student's analytical skills and develops critical thinking techniques.

S (Schematic interpretation) Problems designated with the letter *S* are designed to give the student experience interpreting circuits and ICs in complete system schematic diagrams. The student is asked to identify certain components in the diagram, describe their operation, modify circuit elements, and design new circuit interfaces. This gives the student experience working with real-world large-scale schematics like the ones that he or she will see on the job.

Margin Annotation Icons

Several annotations are given in the page margins throughout the text. These are intended to highlight particular points that were made on the page. They can be used as the catalyst to develop a rapport between the instructor and the students and initiate team discussions among the students. Three different icons are used to distinguish between the annotations.

Common Misconception: These annotations point out areas of digital electronics that have typically been stumbling blocks for students and need careful attention. Pointing out these potential problem areas helps students avoid making that mistake.

Team Discussion: These annotations are questions that tend to initiate a discussion about a particular topic. The instructor can use them as means to develop cooperative learning by encouraging student interaction.

Helpful Hint: These annotations offer suggestions for circuit analysis and highlight critical topics presented in that area of the text. Students use these tips to gain insights into important concepts.

Teaching and Learning Digital Electronics

I would like to share with you some teaching strategies that I've developed from using this text for the past 9 years. Needless to say, students have become very excited about learning digital electronics due to the increasing popularity of the digital computer and the expanding job opportunities for digital technicians and engineers. Students are also attracted to the subject area because of the availability of inexpensive digital ICs,

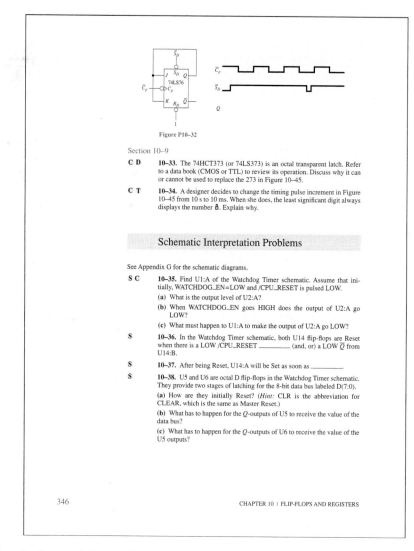

Figure P10–32

Section 10–9

C D **10–33.** The 74HCT373 (or 74LS373) is an octal transparent latch. Refer to a data book (CMOS or TTL) to review its operation. Discuss why it can or cannot be used to replace the 273 in Figure 10–45.

C T **10–34.** A designer decides to change the timing pulse increment in Figure 10–45 from 10 s to 10 ms. When she does, the least significant digit always displays the number 8. Explain why.

Schematic Interpretation Problems

See Appendix G for the schematic diagrams.

S C **10–35.** Find U1:A of the Watchdog Timer schematic. Assume that initially, WATCHDOG_EN=LOW and /CPU_RESET is pulsed LOW.

(a) What is the output level of U2:A?

(b) When WATCHDOG_EN goes HIGH does the output of U2:A go LOW?

(c) What must happen to U1:A to make the output of U2:A go LOW?

S **10–36.** In the Watchdog Timer schematic, both U14 flip-flops are Reset when there is a LOW /CPU_RESET _____ (and, or) a LOW $\overline{Q}$ from U14:B.

S **10–37.** After being Reset, U14:A will be Set as soon as _____.

S **10–38.** U5 and U6 are octal D flip-flops in the Watchdog Timer schematic. They provide two stages of latching for the 8-bit data bus labeled D(7:0).

(a) How are they initially Reset? (*Hint:* CLR is the abbreviation for CLEAR, which is the same as Master Reset.)

(b) What has to happen for the Q-outputs of U5 to receive the value of the data bus?

(c) What has to happen for the Q-outputs of U6 to receive the value of the U5 outputs?

CHAPTER 10 I FLIP-FLOPS AND REGISTERS

Figure 2 System Schematic Interpretation

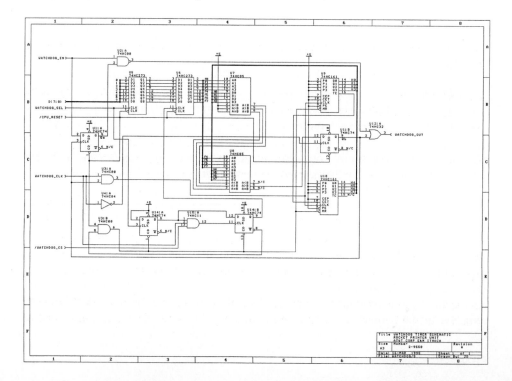

To study the operation of the circuit in more detail, let's first review some basic electronics. An *NPN* transistor is basically two diodes; a *P* to *N* from base to emitter and another *P* to *N* from base to collector, as shown in Figure 9–2. The base-to-emitter diode is forward biased by applying a positive voltage on the base with respect to the emitter. A forward-biased base-to-emitter diode will have 0.7 V across it and will cause the collector-to-emitter junction to become almost a short circuit with approximately 0.3 V across it.

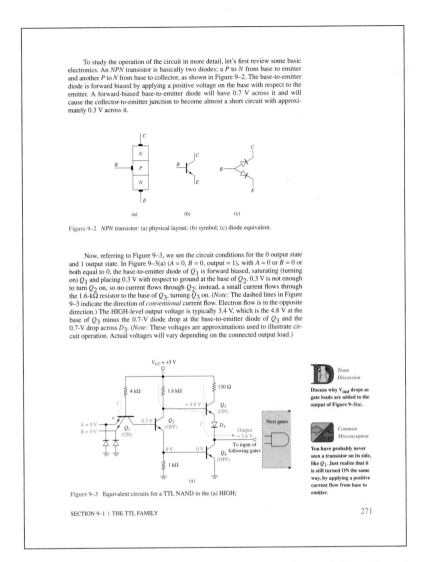

Figure 9–2 *NPN* transistor: (a) physical layout; (b) symbol; (c) diode equivalent.

Now, referring to Figure 9–3, we see the circuit conditions for the 0 output state and 1 output state. In Figure 9–3(a) ($A = 0$, $B = 0$, output = 1), with $A = 0$ or $B = 0$ or both equal to 0, the base-to-emitter diode of Q_1 is forward biased, saturating (turning on) Q_1 and placing 0.3 V with respect to ground at the base of Q_2. 0.3 V is not enough to turn Q_2 on, so no current flows through Q_2; instead, a small current flows through the 1.6-kΩ resistor to the base of Q_3, turning Q_3 on. (*Note:* The dashed lines in Figure 9–3 indicate the direction of *conventional* current flow. Electron flow is in the opposite direction.) The HIGH-level output voltage is typically 3.4 V, which is the 4.8 V at the base of Q_3 minus the 0.7-V diode drop at the base-to-emitter diode of Q_3 and the 0.7-V drop across D_3. (*Note:* These voltages are approximations used to illustrate circuit operation. Actual voltages will vary depending on the connected output load.)

Team Discussion

Discuss why V_{out} drops as gate loads are added to the output of Figure 9–3(a).

Common Misconception

You have probably never seen a transistor on its side, like Q_1. Just realize that it is still turned ON the same way, by applying a positive current flow from base to emitter.

Figure 9–3 Equivalent circuits for a TTL NAND in the (a) HIGH;

SECTION 9–1 I THE TTL FAMILY

271

Figure 3 Margin annotations point out Common Misconceptions and Team Discussions.

which have enabled them to construct useful digital circuits in the lab or at home at a minimal cost.

Student Projects: I always encourage the students to build some of the fundamental building-block circuits that are presented in this text. The circuits that I recommend are the 5-V power supply in Figure 11–39, the 60-Hz pulse generator in Figure 11–40, the cross-NAND switch debouncer in Figure 11–37, and the seven-segment LED display in Figure 12–42. Having these circuits provides a starting point for the student to test many of the other circuits in the text at his or her own pace, at home.

Team Discussions: As early as possible in the course I take advantage of the Team Discussion margin annotations. These are cooperative learning exercises where the students are allowed to form teams, discuss the problem, and present their conclusion to the class. These activities give them a sense of team cooperation and creates a student network connection that will carry on throughout the rest of their studies.

Laboratory Component: Giving the students the opportunity for hands-on laboratory experience is a very useful component of any digital course. An important feature of this text is that there is enough information given for any of the circuits that they can be built and tested in lab and you can be certain that they will give the same response as shown in the text.

Circuit Illustrations: Almost every topic in the text has an illustration associated with it. Because of the extensive art program, I normally lecture directly from illustration to illustration. To do this I have an overhead transparency made of every figure in the text. The most critical figures are available in the transparency master package provided for instructors adopting the text.

Testing: Rather than let a long period of time elapse between tests, I try to give a half-hour quiz each week. Besides the daily homework, this forces the students to study at least once per week. I also believe that it is appropriate to allow them to have a formula sheet for the quiz or test (along with a TTL or CMOS databook). This sheet can have anything they want to write on it. Making up the formula sheet is a good way for them to study and eliminates a lot of routine memorization that they would not normally have to do on the job.

The Learning Process: The student's knowledge is generally developed by learning the theory and the tools required to understand a particular topic, working through the examples provided, answering the review questions at the end of each section, and, finally, solving the problems at the end of the chapter. I always encourage the students to rework the solutions given in the examples without looking at the solutions in the book until they are done. This gives them extra practice and a secure feeling knowing the detailed solution is right there at their disposal.

Chapter Organization

Basically, the text can be divided into two halves: Chapters 1 to 8 cover basic digital logic and combinational logic, and Chapters 9 to 17 cover sequential logic and digital systems. *Chapters 1 and 2* provide the procedures for converting between the various number systems and introduce the student to the electronic signals and switches used in digital circuitry. *Chapters 3 and 4* cover the basic logic gates and introduce the student to timing analysis and troubleshooting techniques. *Chapter 5* shows how several of the basic gates can be connected together to form combinational logic. Boolean algebra, De Morgan's theorem and Karnaugh mapping are used to reduce the logic to its simplest form. *Chapters 6, 7 and 8* discuss combinational logic used to provide more advanced functions like parity checking, arithmetic operations and code converting.

The second half of this book begins with a discussion of the operating characteristics and specifications of the TTL and CMOS logic families *(Chapter 9)*. *Chapter 10* introduces flip-flops and the concept of sequential timing analysis. *Chapter 11* makes the reader aware of the practical limitations of digital ICs and some common circuits that are used in later chapters to facilitate the use of medium-scale ICs. *Chapters 12 and 13* expose the student to the operation and use of several common medium-scale ICs used to implement counter and shift register systems. *Chapter 14* deals with oscillator and timing circuits built with digital ICs and with the 555 timer IC. *Chapter 15* teaches the theory behind analog and digital conversion schemes and the practical implementation of ADC and DAC IC converters. *Chapter 16* covers memory and microprocessor bus concepts and then uses memory ICs and programmable logic to implement several system designs. *Chapter 17* introduces microprocessor hardware and software to form a bridge between digital electronics and a follow-up course in microprocessors. The book concludes with several Appendices used to supplement the chapter material.

If time constraints only allow for a single-semester course, then the following sections should be covered to provide a coherent overview of digital electronics:

Sections 1.1–1.5, 1.8–1.13

Sections 2.1–2.2

Sections 3.1–3.3, 3.5–3.6

Sections 4.1–4.3, 4.5–4.6

Sections 5.1–5.4

Sections 6.1–6.2

Sections 9.1–9.2

Sections 10.1–10.8

Sections 12.1–12.6

Sections 13.1–13.6

Sections 15.1, 15.5, 15.6, 15.10

Sections 16.1, 16.2, 16.4

Also, if the course is intended for nonelectrical technology students, then the following sections could be omitted to eliminate any basic electricity requirements:

Sections 2.6–2.8

Sections 9.1–9.3, 9.8

Sections 11.3–11.6

Sections 14.2–14.4

Sections 15.2–15.4, 15.12

Unique Learning Tools

Special features included in this textbook to enhance the learning and comprehension process are:

- Performance-based *objectives* at the beginning of each chapter outline the goals to be achieved.

- A *summary* at the end of each chapter provides a review of the topics covered.

- Over 200 *examples* are worked out step by step to clarify problems that are normally stumbling blocks.

- A *four-color format* provides a visual organization to the various parts of each section.

- *Troubleshooting applications* and problems are used throughout the text to teach testing and debugging procedures.

- *Review questions* summarize each section and are answered to see that each learning objective is met.

- More than 900 *problems* and *questions* are provided to enhance problem-solving skills. A complete range of problems, from straightforward to very challenging, is included.

- A *glossary* at the end of each chapter serves as a summary of the terminology just presented.

- Over 1000 detailed *illustrations* give visual explanations and serve as the basis for all discussions.

- *Color operational notes* are included on several of the illustrations to describe the operation of a particular part of the figure.

- Reference to *manufacturers' data sheets* throughout the book provides a valuable experience with real-world problem solving.

- A *supplementary index of ICs* provides a quick way to locate a particular IC by number.

- *Timing waveforms* are used throughout the text to illustrate the timing analysis techniques used in industry and give a graphical picture of the sequential operations of digital ICs.

- Several *tables of commercially used ICs* provide a source for state of the art circuit design.

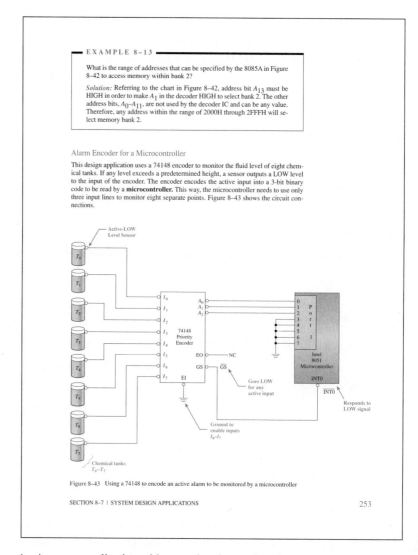

Figure 4 A system application with operational notes in color.

- Several photographs are included to illustrate specific devices and circuits discussed in the text.

Extensive Supplements Package

An extensive package of supplementary material is available to aid in the teaching and learning process.

- *Instructor's Solutions Manual* containing solutions and answers to in-text problems.

- 150 transparency masters of textbook illustrations and 50 four-color transparencies.

- *Lab Manual: A Troubleshooting Approach,* authored by Michael Wiesner (Heald Technical College), provides hands-on laboratory experience to reinforce the material presented in the textbook.

- *Solutions Manual* to *Lab Manual: A Troubleshooting Approach.*

- *Lab Manual: A Design Approach,* by David M. Perkins (University of Southern Colorado), presents experiments from a design-oriented approach. It includes objectives, equipment lists, and safety tips, and urges students to build with only a basic direction in mind.

- *Solutions Manual* to *Lab Manual: A Design Approach.*

- A *Test Item File,* authored by Les Taylor (DeVry Institute of Technology), contains over 1000 additional multiple-choice questions that can be used to develop weekly quizzes, tests, or final exams. It is available in manual or disk form (IBM compatible).

- *Test Manager 20,* IBM version, contains over 1000 test questions.

- A *Student Study Guide,* authored by David Bechtel (DeVry Institute of Technology), contains an overview of each chapter and several additional review questions.

- An *Electronics Workbench® Circuits File Disk,* (an IBM compatible disk), containing selected examples from the text.

- *Electronics Workbench®* software available upon adoption. Ask your Prentice Hall representative for details.

- *Digital Electronics* videos from Bergwall Video Productions available upon adoption. Ask your Prentice Hall representative for details.

- There is a bulletin board available through America Online® if you have questions or comments about this text.

Changes in the Fourth Edition

The first, second, and third editions were developed from an accumulation of 12 years of class notes. Having taught from the third edition for the past 3 years has given me the opportunity to review several suggestions from my students and other faculty regarding such things as: ways to improve a circuit diagram, clarifying an explanation, and redesigning an application to make it easier to duplicate in lab.

More than 120 schools have adopted the third edition. To write the fourth edition, I have taken advantage of the comments from these schools as well as my own experience and market research to develop an even more practical and easier to learn from textbook. Besides rewriting several of the examples and applications based on my classroom experience, the following material has been added:

- Chapter-end summaries

- A four-color format for the text, illustrations, and photographs

- A new section on surface-mount devices (SMDs)

- A new section on low-voltage (74LV) integrated circuits

- 200 margin annotations describing common misconceptions, helpful hints, and team discussion questions

- 80 chapter-end problems that refer to real-world schematic diagrams of actual digital instruments and systems

- Several new circuit operational notes in the figures

- Several new IC data sheets for reference in the appendix

- 30 color photographs of circuits and instruments

To the Student

Digital electronics is the foundation of computers and microprocessor-based systems found in automobiles, industrial control systems, and home entertainment systems. You are beginning your study of digital electronics at a good time. Technological advances made in the past 20 years have provided us with integrated circuits that can perform complex tasks with a minimum amount of abstract theory and complicated circuitry. Before you are through with this book, you'll be developing exciting designs that you've always wondered about, but now can experience firsthand.

The study of digital electronics also provides the prerequisite background for your future studies in microprocessors and microcomputer interfacing. It also provides the job skills to become a computer service technician, production test technician, digital design technician, or a multitude of other jobs related to computer and microprocessor-based systems.

This book is written as a learning tool, not just as a reference. The concept and theory of each topic is presented first. Then an explanation of its operation is given. This is followed by several worked-out examples and, in some cases, a system design application. The review questions at the end of each chapter will force you to dig back into the reading to see that you have met the learning objectives given at the beginning of the chapter. The problems at the end of each chapter will require more analytical reasoning, but the procedures for their solutions were already given to you in the examples. One good way to prepare for homework problems and tests is to cover up the solutions to the examples and try to work them out yourself. If you get stuck, you've got the answer and an explanation for the answer right there.

I also suggest that you take advantage of your study guide. The more practice you get, the easier the course will be. I wish you the best of luck in your studies and future employment.

William Kleitz

Acknowledgments

Thanks are due to the following professors for reviewing the manuscript and providing numerous valuable suggestions which have contributed toward bringing *Digital Electronics* to its fourth edition:

Scott Boldwyn, Missouri Technical School

Henry Baskerville, Heald Institute

Darrell Boucher, Jr., High Plains Institute of Technology

Steven R. Coe, DeVry Institute of Technology

Terry Collett, Lake Michigan College

Mike Durran, Indiana Vocational Technical College

Donald P. Hill, RETS Electronic Institute

Nazar Karzay, Indiana Vocational Technical College

Charles L. Laye, United Electronics Institute

Lew D. Mathias, Indiana Vocational Technical College

Serge Mnatzakanian, Computer Learning Center

Chrys A. Panayiotou, Brevard Community College

Richard Parett, ITT Technical Institute

Bob Redler, Southeast Community College

Ron Scott, Northeastern University

Edward Small, Southeast College of Technology

Ron L. Syth, ITT Technical Institute

Edward Troyan, LeHigh Carbon Community College

Vance Venable, Heald Institute of Technology

Ken Wilson, San Diego City College

A special thank you is extended to Scott Wager and Mitch Wiedemann of Tompkins Cortland Community College for their fine work on new schematics in the fourth edition.

I am grateful to Russell Hunt from Simco Company, Kevin White and Scott Heffron from Bob Dean Corporation, Dick Quaif from DQ Systems, Alan Szary and Paul Constantini from Precision Filters, Inc., and Jim Delsignore from AT&T Corporation for their technical assistance, and to Signetics (Philips) Corporation, Intel Corporation, Texas Instruments, Inc., Hewlett-Packard Company, and Advanced Micro Devices, Inc., for providing the data sheets used in this book. Also, thanks to my students of the past 16 years who have helped me to develop better teaching strategies and have provided suggestions for clarifying several of the explanations contained in this book; and also to Carol Robison, Louise Sette, and Dave Garza from the editorial and production staff at Prentice Hall.

Contents

CHAPTER 1

Number Systems and Codes 1

	Outline 1
	Objectives 1
	Introduction 1
1–1	Digital Representations of Analog Quantities 2
1–2	Digital Versus Analog 4
1–3	Decimal Numbering System (Base 10) 4
1–4	Binary Numbering System (Base 2) 5
1–5	Decimal-to-Binary Conversion 7
1–6	Octal Numbering System (Base 8) 9
1–7	Octal Conversions 9
1–8	Hexadecimal Numbering System (Base 16) 11
1–9	Hexadecimal Conversions 12
1–10	Binary-Coded-Decimal System 14
1–11	Comparison of Numbering Systems 15
1–12	The ASCII Code 15
1–13	Applications of the Number Systems 17
	Summary 20
	Glossary 20
	Problems 21
	Schematic Interpretation Problems 23
	Answers to Review Questions 23

CHAPTER 2

Digital Electronic Signals and Switches 24

	Outline 24
	Objectives 25
	Introduction 25

2–1   Digital Signals   26
2–2   Clock Waveform Timing   26
2–3   Serial Representation   28
2–4   Parallel Representation   29
2–5   Switches in Electronic Circuits   32
2–6   A Relay as a Switch   33
2–7   A Diode as a Switch   36
2–8   A Transistor as a Switch   39
2–9   The TTL Integrated Circuit   43
2–10   The CMOS Integrated Circuit   46
2–11   Surface-Mount Devices   47
       Summary   48
       Glossary   49
       Problems   50
       Schematic Interpretation Problems   53
       Answers to Review Questions   53

CHAPTER 3

Basic Logic Gates   56

       Outline   56
       Objectives   57
       Introduction   57
3–1   The AND Gate   58
3–2   The OR Gate   60
3–3   Timing Analysis   62
3–4   Enable and Disable Functions   64
3–5   Using Integrated-Circuit Logic Gates   66
3–6   Introduction to Troubleshooting Techniques   67
       Summary   72
       Glossary   73
       Problems   73
       Schematic Interpretation Problems   78
       Answers to Review Questions   78

CHAPTER 4

Inverting Logic Gates   80

       Outline   80
       Objectives   81
       Introduction   81
4–1   The Inverter   82
4–2   The NAND Gate   82
4–3   The NOR Gate   85
4–4   Logic Gate Waveform Generation   87
4–5   Using Integrated-Circuit Logic Gates   93
4–6   Summary of the Basic Logic Gates and IEEE/IEC Standard Logic Symbols   95
       Summary   98
       Glossary   98
       Problems   99
       Schematic Interpretation Problems   105
       Answers to Review Questions   105

CHAPTER 5

Boolean Algebra and Reduction Techniques 106

Outline 106
Objectives 107
Introduction 107
5–1 Combinational Logic 108
5–2 Boolean Algebra Laws and Rules 110
5–3 Simplification of Combinational Logic Circuits Using Boolean Algebra 115
5–4 De Morgan's Theorem 119
5–5 The Universal Capability of NAND and NOR Gates 130
5–6 AND–OR–INVERT Gates for Implementing Sum-of-Products Expressions 135
5–7 Karnaugh Mapping 139
5–8 System Design Applications 146
Summary 149
Glossary 149
Problems 150
Schematic Interpretation Problems 162
Answers to Review Questions 163

CHAPTER 6

Exclusive-OR and Exclusive-NOR Gates 164

Outline 164
Objectives 165
Introduction 165
6–1 The Exclusive-OR Gate 166
6–2 The Exclusive-NOR Gate 167
6–3 Parity Generator/Checker 170
6–4 System Design Applications 173
Summary 176
Glossary 176
Problems 176
Schematic Interpretation Problems 179
Answers to Review Questions 179

CHAPTER 7

Arithmetic Operations and Circuits 180

Outline 180
Objectives 181
Introduction 181
7–1 Binary Arithmetic 182
7–2 Two's-Complement Representation 188
7–3 Two's-Complement Arithmetic 191
7–4 Hexadecimal Arithmetic 192
7–5 BCD Arithmetic 195
7–6 Arithmetic Circuits 197
7–7 Four-Bit Full-Adder ICs 201
7–8 System Design Applications 204
7–9 Arithmetic/Logic Units 207
Summary 210
Glossary 211
Problems 212
Schematic Interpretation Problems 216
Answers to Review Questions 216

CHAPTER 8

Code Converters, Multiplexers, and Demultiplexers   218

Outline 218
Objectives 219
Introduction 219
8–1 Comparators 220
8–2 Decoding 222
8–3 Encoding 230
8–4 Code Converters 235
8–5 Multiplexers 243
8–6 Demultiplexers 248
8–7 System Design Applications 252
Summary 258
Glossary 258
Problems 259
Schematic Interpretation Problems 266
Answers to Review Questions 267

CHAPTER 9

Logic Families and Their Characteristics   268

Outline 268
Objectives 269
Introduction 269
9–1 The TTL Family 270
9–2 TTL Voltage and Current Ratings 272
9–3 Other TTL Considerations 278
9–4 Improved TTL Series 283
9–5 The CMOS Family 285
9–6 Emitter-Coupled Logic 290
9–7 Comparing Logic Families 291
9–8 Interfacing Logic Families 293
Summary 300
Glossary 301
Problems 302
Schematic Interpretation Problems 306
Answers to Review Questions 307

CHAPTER 10

Flip-Flops and Registers   308

Outline 308
Objectives 309
Introduction 309
10–1 *S-R* Flip-Flop 310
10–2 Gated *S-R* Flip-Flop 314
10–3 Gated *D* Flip-Flop 316
10–4 Integrated-Circuit *D* Latch (7475) 316
10–5 Integrated-Circuit *D* Flip-Flop (7474) 318
10–6 Master-Slave *J-K* Flip-Flop 322
10–7 Edge-Triggered *J-K* Flip-Flop 326
10–8 Integrated-Circuit *J-K* Flip-Flop (7476, 74LS76) 328
10–9 Using an Octal *D* Flip-Flop in a
Microcontroller Application 335
Summary 327

Glossary   327
Problems   339
Schematic Interpretation Problems   346
Answers to Review Questions   347

CHAPTER 11
Practical Considerations for Digital Design   348

Outline   348
Objectives   349
Introduction   349
11–1   Flip-Flop Time Parameters   350
11–2   Automatic Reset   365
11–3   Schmitt Trigger ICs   367
11–4   Switch Debouncing   372
11–5   Sizing Pull-Up Resistors   375
11–6   Practical Input and Output Considerations   376
Summary   381
Glossary   382
Problems   383
Schematic Interpretation Problems   388
Answers to Review Questions   389

CHAPTER 12
Counter Circuits and Applications   390

Outline   390
Objectives   391
Introduction   391
12–1   Analysis of Sequential Circuits   392
12–2   Ripple Counters   396
12–3   Design of Divide-by-N Counters   400
12–4   Ripple Counter Integrated Circuits   406
12–5   System Design Applications   412
12–6   Seven-Segment LED Display Decoders   418
12–7   Synchronous Counters   424
12–8   Synchronous Up/Down-Counter ICs   429
12–9   Applications of Synchronous Counter ICs   437
Summary   441
Glossary   441
Problems   442
Schematic Interpretation Problems   448
Answers to Review Questions   449

CHAPTER 13
Shift Registers   450

Outline   450
Objectives   451
Introduction   451
13–1   Shift Register Basics   452
13–2   Parallel-to-Serial Conversion   454
13–3   Recirculating Register   454
13–4   Serial-to-Parallel Conversion   456
13–5   Ring Shift Counter and Johnson Shift Counter   456

13–6   Shift Register ICs   460
13–7   System Design Applications for Shift Registers   468
13–8   Driving a Stepper Motor with a Shift Register   473
13–9   Three-State Buffers, Latches, and Transceivers   476
       Summary   481
       Glossary   482
       Problems   483
       Schematic Interpretation Problems   489
       Answers to Review Questions   489

CHAPTER 14

Multivibrators and the 555 Timer   492

       Outline   492
       Objectives   493
       Introduction   493
14–1   Multivibrators   494
14–2   Capacitor Charge and Discharge Rates   494
14–3   Astable Multivibrators   498
14–4   Monostable Multivibrators   500
14–5   IC Monostable Multivibrators   503
14–6   Retriggerable Monostable Multivibrators   507
14–7   Astable Operation of the 555 IC Timer   510
14–8   Monostable Operation of the 555 IC Timer   515
14–9   Crystal Oscillators   518
       Summary   520
       Glossary   520
       Problems   521
       Schematic Interpretation Problems   524
       Answers to Review Questions   525

CHAPTER 15

Interfacing to the Analog World   526

       Outline   526
       Objectives   527
       Introduction   527
15–1   Digital and Analog Representations   528
15–2   Operational Amplifier Basics   529
15–3   Binary-Weighted Digital-to-Analog Converters   530
15–4 $R/2R$ Ladder Digital-to-Analog Converters   531
15–5   Integrated-Circuit Digital-to-Analog Converters   533
15–6   IC Data Converter Specifications   536
15–7   Parallel-Encoded Analog-to-Digital Converters   537
15–8   Counter-Ramp Analog-to-Digital Converters   539
15–9   Successive-Approximation Analog-to-Digital Conversion   540
15–10   Integrated-Circuit Analog-to-Digital Converters   542
15–11   Data Acquisition System Application   547
15–12   Transducers and Signal Conditioning   550
       Summary   554
       Glossary   555
       Problems   557
       Schematic Interpretation Problems   560
       Answers to Review Questions   560

CHAPTER 16

Semiconductor Memory and Programmable Arrays 562

Outline 562
Objectives 563
Introduction 563
16–1 Memory Concepts 564
16–2 Static RAMs 567
16–3 Dynamic RAMs 574
16–4 Read-Only Memories 578
16–5 Memory Expansion and Address Decoding Applications 582
16–6 Programmable Logic Devices 587
Summary 595
Glossary 596
Problems 597
Schematic Interpretation Problems 600
Answers to Review Questions 601

CHAPTER 17

Microprocessor Fundamentals 602

Outline 602
Objectives 603
Introduction 603
17–1 Introduction to System Components and Buses 604
17–2 Software Control of Microprocessor Systems 607
17–3 Internal Architecture of a Microprocessor 607
17–4 Instruction Execution Within a Microprocessor 609
17–5 Hardware Requirements for Basic I/O Programming 612
17–6 Writing Assembly Language and Machine Language Programs 614
17–7 Survey of Microprocessors and Manufacturers 617
Summary 619
Glossary 620
Problems 621
Schematic Interpretation Problems 623
Answers to Review Questions 624

APPENDIX A   BIBLIOGRAPHY   625

APPENDIX B   MANUFACTURERS' DATA SHEETS   626

APPENDIX C   EXPLANATION OF THE IEEE/IEC STANDARD FOR LOGIC SYMBOLS (DEPENDENCY NOTATION)   665

APPENDIX D   ANSWERS TO ODD-NUMBERED PROBLEMS   670

APPENDIX E   DESIGNING WITH PLD SOFTWARE   690

APPENDIX F   REVIEW OF BASIC ELECTRICITY PRINCIPLES   703

APPENDIX G   SCHEMATIC DIAGRAMS   710

INDEX   719

SUPPLEMENTARY INDEX OF INTEGRATED CIRCUITS   725

1 Number Systems and Codes

OUTLINE

1–1 Digital Representations of Analog Quantities

1–2 Digital Versus Analog

1–3 Decimal Numbering System (Base 10)

1–4 Binary Numbering System (Base 2)

1–5 Decimal-to-Binary Conversion

1–6 Octal Numbering System (Base 8)

1–7 Octal Conversions

1–8 Hexadecimal Numbering System (Base 16)

1–9 Hexadecimal Conversions

1–10 Binary-Coded-Decimal System

1–11 Comparison of Numbering Systems

1–12 The ASCII Code

1–13 Applications of the Number Systems

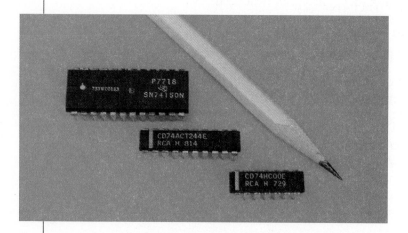

Objectives

Upon completion of this chapter, you should be able to

- Determine the weighting factor for each digit position in the decimal, binary, octal, and hexadecimal numbering systems.

- Convert any number in one of the four number systems (decimal, binary, octal, or hexadecimal) to its equivalent value in any of the remaining three numbering systems.

- Describe the format and use of binary-coded-decimal (BCD) numbers.

- Determine the ASCII code for any alphanumeric data by using the ASCII code translation table.

Introduction

Digital circuitry is the foundation of digital computers and many automated control systems. In a modern home, digital circuitry controls the appliances, alarm systems, and heating systems. Under the control of digital circuitry and microprocessors, newer automobiles have added safety features, are more energy efficient, and are easier to diagnose and correct when malfunctions arise.

Other uses of digital circuitry include the areas of automated machine control, energy monitoring and control, inventory management, medical electronics, and music. For example, the numerically controlled (NC) milling machine can be programmed by a production engineer to mill a piece of stock material to prespecified dimensions with very accurate repeatability, within 0.01% accuracy. Another use is energy monitoring and control. With the high cost of energy it is very important for large industrial and commercial users to monitor the energy flows within their buildings. Effective control of heating, ventilating, and air-conditioning can reduce energy bills significantly. More and more grocery stores are using the universal product code (UPC) to check out and total the sale of grocery orders, as well as to control inventory and replenish stock automatically. The area of medical electronics uses digital thermometers, life-support systems, and monitors. We have also seen more use of digital electronics in the reproduction of music. Digital reproduction is less susceptible to electrostatic noise and therefore can reproduce music with greater fidelity.

Digital electronics evolved from the principle that transistor circuitry could easily be fabricated and designed to output one of two voltage levels based on the levels placed at its inputs. The two distinct levels (usually +5 volts (V) and 0 V) are HIGH and LOW and can be represented by 1 and 0.

The binary numbering system is made up of only 1's and 0's and is therefore used extensively in digital electronics. The other numbering systems and codes covered in this chapter represent groups of binary digits and therefore are also widely used.

1–1 Digital Representations of Analog Quantities

Most naturally occurring physical quantities in our world are **analog** in nature. An analog signal is a continuously variable electrical or physical quantity. Think about a mercury-filled tube thermometer; as the temperature rises, the mercury expands in analog fashion and makes a smooth, continuous motion relative to a scale measured in degrees. A baseball player swings a bat in an analog motion. The velocity and force with which a musician strikes a piano key are analog in nature. Even the resulting vibration of the piano string is an analog, sinusoidal vibration.

So why do we need to use **digital** representations in a world that is naturally analog? The answer is that if we want an electronic machine to interpret, communicate, and store analog information it is much easier for the machine to handle it if we first convert the information to a digital format. A digital value is represented by a combination of ON and OFF voltage levels that are written as a string of 1's and 0's.

For example, an analog thermometer that registers 72 degrees can be represented in a digital circuit as a series of ON and OFF voltage levels. (We'll learn later that the number 72 converted to digital levels is 0100 1000.) The convenient feature of using ON/OFF voltage levels is that the circuitry used to generate, manipulate, and store them is very simple. Instead of dealing with the infinite span and intervals of analog voltage levels, all we need to use is ON or OFF voltages (usually +5 V = ON and 0 V = OFF).

A good example of the use of a digital representation of an analog quantity is the audio recording of music. Compact disks (CDs) and digital audio tapes (DATs) are becoming commonplace and are proving to be superior means of recording and playing back music. Musical instruments and the human voice produce analog signals, and the human ear naturally responds to analog signals. So, where does the digital format fit in? Although the process requires what appears to be extra work, the recording industries convert analog signals to a digital format and then store the information on a CD or DAT. The CD or DAT player then converts the digital levels back to their corresponding analog signals before playing them back for the human ear.

To accurately represent a complex musical signal as a digital string (a series of 1's and 0's), several samples of an analog signal must be taken, as shown in Figure 1–1(a). The first conversion illustrated is at a point on the rising portion of the analog signal. At that point, the analog voltage is 2 V. Two volts is converted to the digital string 0000 0010, as shown in Figure 1–1(b). The next conversion is taken as the analog signal in Figure 1–1(a) is still rising, and the third is taken at its highest level. This process will continue throughout the entire piece of music to be recorded. To play back the music, the process is reversed. Digital-to-analog conversions are made to recreate the original analog signal. If a high enough number of samples are taken of the original analog signal, an almost exact reproduction of the original music can be made.

It certainly is extra work, but digital recordings have virtually eliminated problems such as record wear and the magnetic tape hiss associated with earlier methods of audio recording. These problems have been eradicated because, when imperfections are introduced to a digital signal, the slight variation in the digital level does not change an ON level to an OFF level, whereas a slight change in an analog level is easily picked up by the human ear.

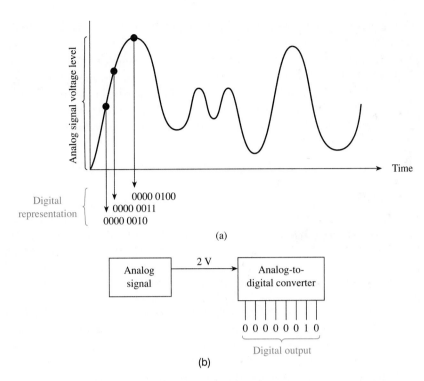

(a)

(b)

<figure>
<figcaption>Figure 1–1 (a) Digital representation of three data points on an analog waveform; (b) Converting a 2-V analog voltage into a digital output string.</figcaption>
</figure>

Helpful Hint

One of the more interesting uses of analog-to-digital (A-to-D) and digital-to-analog (D-to-A) conversion is in compact disk (CD) audio systems. Also, several A-to-D and D-to-A examples are given in Chapter 15.

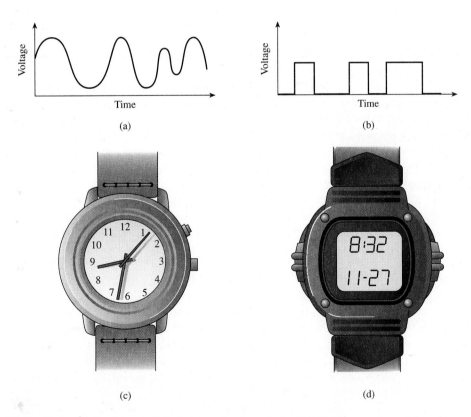

(a)

(b)

(c)

(d)

<figure>
<figcaption>Figure 1–2 Analog versus digital: (a) analog waveform; (b) digital waveform; (c) analog watch; (d) digital watch.</figcaption>
</figure>

1–2 Digital Versus Analog

Digital systems operate on discrete digits that represent numbers, letters, or symbols. They deal strictly with ON and OFF states, which we can represent by 0's and 1's. *Analog* systems measure and respond to continuously varying electrical or physical magnitudes. Analog devices are integrated electronically into systems to continuously monitor and control such quantities as temperature, pressure, velocity, and position and to provide automated control based on the levels of these quantities. Figure 1–2 shows some examples of digital and analog quantities.

Review Questions*

1–1. List three examples of *analog* quantities.

1–2. Why do computer systems deal with *digital* quantities instead of *analog* quantities?

1–3 Decimal Numbering System (Base 10)

In the **decimal** numbering system, each position contains 10 different possible digits. These digits are 0, 1, 2, 3, 4, 5, 6, 7, 8, and 9. Each position in a multidigit number will have a weighting factor based on a power of 10.

EXAMPLE 1–1

In a four-digit decimal number the least significant position (rightmost) has a weighting factor of 10^0; the most significant position (leftmost) has a weighting factor of 10^3:

$$10^3 \qquad 10^2 \qquad 10^1 \qquad 10^0$$

where $10^3 = 1000$
$10^2 = 100$
$10^1 = 10$
$10^0 = 1$

To evaluate the decimal number 4623, the digit in each position is multiplied by the appropriate weighting factor:

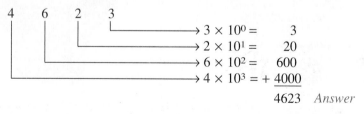

$$
\begin{aligned}
3 \times 10^0 &= 3 \\
2 \times 10^1 &= 20 \\
6 \times 10^2 &= 600 \\
4 \times 10^3 &= +\underline{4000} \\
&4623 \quad \textit{Answer}
\end{aligned}
$$

CHAPTER 1 | NUMBER SYSTEMS AND CODES

Example 1–1 illustrates the procedure used to convert from some number system to its decimal (base 10) equivalent. (In the example we converted a base 10 number to a base 10 answer.) Now let's look at base 2 (binary), base 8 (octal), and base 16 (hexadecimal).

1–4   Binary Numbering System (Base 2)

Digital electronics use the **binary** numbering system because it uses only the digits 0 and 1, which can be represented simply in a digital system by two distinct voltage levels, such as +5 V = 1 and 0 V = 0.

The weighting factors for binary positions are the powers of 2 shown in Table 1–1.

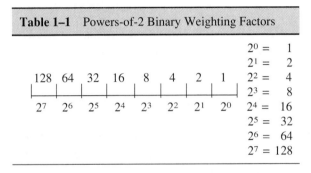

Table 1–1   Powers-of-2 Binary Weighting Factors

$$2^0 = 1$$
$$2^1 = 2$$
$$2^2 = 4$$
$$2^3 = 8$$
$$2^4 = 16$$
$$2^5 = 32$$
$$2^6 = 64$$
$$2^7 = 128$$

E X A M P L E 1 – 2

Convert the binary number 01010110_2 to decimal. (Notice the subscript 2 used to indicate that 01010110 is a base 2 number.)

Solution: Multiply each binary digit by the appropriate weight factor and total the results.

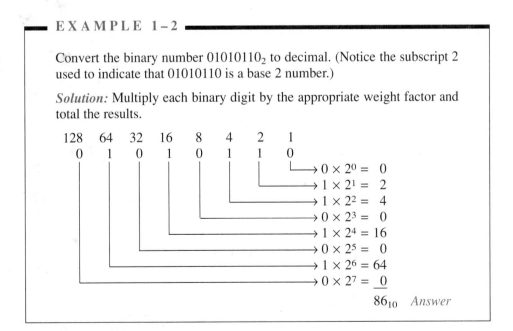

$$0 \times 2^0 = 0$$
$$1 \times 2^1 = 2$$
$$1 \times 2^2 = 4$$
$$0 \times 2^3 = 0$$
$$1 \times 2^4 = 16$$
$$0 \times 2^5 = 0$$
$$1 \times 2^6 = 64$$
$$0 \times 2^7 = \underline{0}$$

86_{10} *Answer*

Although seldom used in digital systems, binary weighting for values less than 1 is possible (fractional binary numbers). These factors are developed by successively dividing the weighting factor by 2 for each decrease in the power of 2. This is also useful to illustrate why 2^0 is equal to 1, not zero (see Figure 1–3).

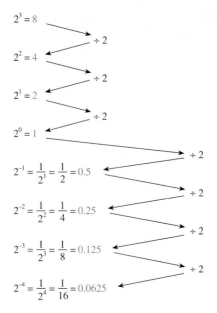

Figure 1-3 Successive division by 2 to develop fractional binary weighting factors and show that 2^0 is equal to 1.

EXAMPLE 1-3

Convert the fractional binary number 1011.1010_2 to decimal.

Solution: Multiply each binary digit by the appropriate weighting factor given in Figure 1-3 and total the results. (We will skip the multiplication for the binary digit 0 because it does not contribute to the total.)

$$
\begin{array}{cccccccc}
1 & 0 & 1 & 1 & . & 1 & 0 & 1 & 0
\end{array}
$$

$$
\begin{aligned}
1 \times 2^{-3} &= 0.125 \\
1 \times 2^{-1} &= 0.500 \\
1 \times 2^{0} &= 1 \\
1 \times 2^{1} &= 2 \\
1 \times 2^{3} &= 8 \\
\hline
11.625_{10} & \quad Answer
\end{aligned}
$$

Review Questions

1-3. Why is the binary numbering system commonly used in digital electronics?

1-4. How are the weighting factors determined for each binary position in a base 2 number?

1-5. Convert $0110\ 1100_2$ to decimal.

1-6. Convert 1101.0110_2 to decimal.

1–5 Decimal-to-Binary Conversion

The conversion from binary to decimal is usually performed by the digital computer for ease of interpretation by the person reading the number. On the other hand, when a person enters a decimal number into a digital computer, that number must be converted to binary before it can be operated on. Let's look at *decimal-to-binary* conversion.

E X A M P L E 1–4

Convert 133_{10} to binary.

Solution: Referring to Table 1–1, we can see that the largest power of 2 that will fit into 133 is 2^7 ($2^7 = 128$). But that will still leave the value 5 ($133 - 128 = 5$) to be accounted for. Five can be taken care of by 2^2 and 2^0 ($2^2 = 4$, $2^0 = 1$). So the process looks like this:

$$
\begin{array}{l}
133 \\
\underline{-128} \rightarrow 2^7 \\
\quad 5 \\
\underline{-\ 4} \rightarrow 2^2 \\
\quad 1 \\
\underline{-\ 1} \rightarrow 2^0 \\
\quad 0
\end{array}
$$

1	0	0	0	0	1	0	1
2^7	2^6	2^5	2^4	2^3	2^2	2^1	2^0

Answer: 10000101_2

Note: The powers of 2 needed to give the number 133 were first determined. Then all other positions were filled with zeros.

E X A M P L E 1–5

Convert 122_{10} to binary.

Solution:

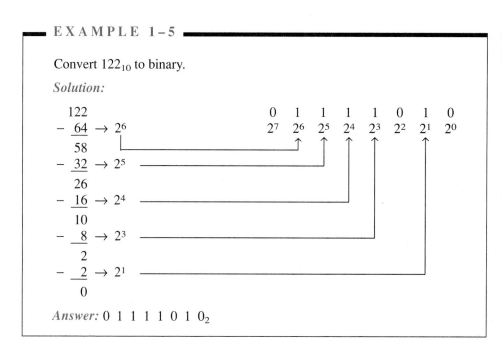

Answer: $0\ 1\ 1\ 1\ 1\ 0\ 1\ 0_2$

Helpful Hint

This is a good time to realize that a useful way to learn new material like this is to re-solve the examples with the solutions covered up. That way, when you have a problem, you can uncover the solution and see the correct procedure.

Another method of converting decimal to binary is by *successive division*. Successive division involves dividing repeatedly by the number of the base to which you are converting. For example, to convert 122_{10} to base 2, use the following procedure:

$$122 \div 2 = 61 \quad \text{with a remainder of 0} \quad \text{(LSB)}$$
$$61 \div 2 = 30 \quad \text{with a remainder of 1}$$
$$30 \div 2 = 15 \quad \text{with a remainder of 0}$$
$$15 \div 2 = 7 \quad \text{with a remainder of 1}$$
$$7 \div 2 = 3 \quad \text{with a remainder of 1}$$
$$3 \div 2 = 1 \quad \text{with a remainder of 1}$$
$$1 \div 2 = 0 \quad \text{with a remainder of 1} \quad \text{(MSB)}$$

The first remainder, 0, is the **least significant bit (LSB)** of the answer; the last remainder, 1, is the **most significant bit (MSB)** of the answer; therefore, the answer is

$$1\ 1\ 1\ 1\ 0\ 1\ 0_2$$

However, because most computers or digital systems deal with groups of 4, 8, 16, or 32 **bits** (Binary digITs), we should keep all our answers in that form. Adding a leading zero to the number $1\ 1\ 1\ 1\ 0\ 1\ 0_2$ will not change its numeric value; therefore, the 8-bit answer is

$$1\ 1\ 1\ 1\ 0\ 1\ 0_2 = 0\ 1\ 1\ 1\ 1\ 0\ 1\ 0_2$$

Common Misconception

Remember not to reverse the LSB and MSB when listing the binary answer.

EXAMPLE 1–6

Convert 152_{10} to binary using successive division.

Solution:

$$152 \div 2 = 76 \quad \text{remainder 0} \quad \text{(LSB)}$$
$$76 \div 2 = 38 \quad \text{remainder 0}$$
$$38 \div 2 = 19 \quad \text{remainder 0}$$
$$19 \div 2 = 9 \quad \text{remainder 1}$$
$$9 \div 2 = 4 \quad \text{remainder 1}$$
$$4 \div 2 = 2 \quad \text{remainder 0}$$
$$2 \div 2 = 1 \quad \text{remainder 0}$$
$$1 \div 2 = 0 \quad \text{remainder 1} \quad \text{(MSB)}$$

Answer: $1\ 0\ 0\ 1\ 1\ 0\ 0\ 0_2$

Review Questions

1–7. Convert 43_{10} to binary.

1–8. Convert 170_{10} to binary.

CHAPTER 1 I NUMBER SYSTEMS AND CODES

1–6   Octal Numbering System (Base 8)

The **octal** numbering system is a method of grouping binary numbers in groups of three. The eight allowable digits are 0, 1, 2, 3, 4, 5, 6, and 7.

The octal numbering system is used by manufacturers of computers that utilize 3-bit codes to indicate instructions or operations to be performed. By using the octal representation instead of binary, the user can simplify the task of entering or reading computer instructions and thus save time.

In Table 1–2 we see that when the octal number exceeds 7 the least significant octal position resets to zero, and the next most significant position increases by 1.

Table 1–2 Octal Numbering System		
Decimal	Binary	Octal
0	000	0
1	001	1
2	010	2
3	011	3
4	100	4
5	101	5
6	110	6
7	111-	7
8	1000	10
9	1001	11
10	1010	12

1–7   Octal Conversions

Converting from *binary to octal* is simply a matter of grouping the binary positions in groups of three (starting at the least significant position) and writing down the octal equivalent.

E X A M P L E   1 – 7

Convert $0\ 1\ 1\ 1\ 0\ 1_2$ to octal.

Solution:

$$0\ 1\ 1 \qquad 1\ 0\ 1$$
$$3 \qquad\quad 5 \qquad = 35_8 \quad Answer$$

E X A M P L E   1 – 8

Convert $1\ 0\ 1\ 1\ 1\ 0\ 0\ 1_2$ to octal.

Solution:

$$1\ 0 \qquad 1\ 1\ 1 \qquad 0\ 0\ 1_2$$

add a leading zero

$$0\ 1\ 0$$
$$2 \qquad\quad 7 \qquad\quad 1 \qquad = 271_8 \quad Answer$$

To convert *octal to binary*, you reverse the process.

EXAMPLE 1–9

Convert 6 2 4$_8$ to binary.

Solution:

$$
\underbrace{6}_{1\ 1\ 0}\quad \underbrace{2}_{0\ 1\ 0}\quad \underbrace{4}_{1\ 0\ 0} = 1\ 1\ 0\ 0\ 1\ 0\ 1\ 0\ 0_2 \quad \textit{Answer}
$$

To convert from *octal to decimal*, follow a process similar to that in Section 1–3 (multiply by weighting factors).

Helpful Hint

When converting from octal to decimal, some students find it easier to convert to binary first and then convert binary to decimal.

EXAMPLE 1–10

Convert 3 2 6$_8$ to decimal.

Solution:

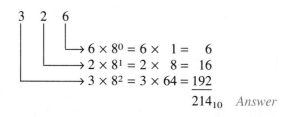

$$
\begin{array}{l}
3\quad 2\quad 6 \\
\quad\quad\quad \rightarrow 6 \times 8^0 = 6 \times \ 1 = \ \ \ 6 \\
\quad\quad \rightarrow 2 \times 8^1 = 2 \times \ 8 = \ \ 16 \\
\quad \rightarrow 3 \times 8^2 = 3 \times 64 = \underline{192} \\
\quad\quad\quad\quad\quad\quad\quad\quad\quad\quad\quad 214_{10} \quad \textit{Answer}
\end{array}
$$

To convert from *decimal to octal*, the successive-division procedure can be used.

EXAMPLE 1–11

Convert 4 8 6$_{10}$ to octal.

Solution:

$$
\left.
\begin{array}{l}
486 \div 8 = \ 60 \quad \text{remainder} \quad 6 \\
\ 60 \div 8 = \ \ \ 7 \quad \text{remainder} \quad 4 \\
\ \ \ 7 \div 8 = \ \ \ 0 \quad \text{remainder} \quad 7
\end{array}
\right\} 746_8
$$

$$
486_{10} = 746_8 \quad \textit{Answer}
$$

Check:

$$
\begin{array}{l}
7\quad 4\quad 6 \\
\quad\quad\quad \rightarrow 6 \times 8^0 = \quad \ \ 6 \\
\quad\quad \rightarrow 4 \times 8^1 = \quad 32 \\
\quad \rightarrow 7 \times 8^2 = \underline{448} \\
\quad\quad\quad\quad\quad\quad\quad\quad\quad \overline{486} \quad \checkmark
\end{array}
$$

1–9. The only digits allowed in the octal numbering system are 0 to 8. True or false?

1–10. Convert 111011_2 to octal.

1–11. Convert 263_8 to binary.

1–12. Convert 614_8 to decimal.

1–13. Convert 90_{10} to octal.

1–8 Hexadecimal Numbering System (Base 16)

The **hexadecimal** numbering system, like the octal system, is a method of grouping bits to simplify entering and reading the instructions or data present in digital computer systems. Hexadecimal uses 4-bit groupings; therefore, instructions or data used in 8-, 16-, or 32-bit computer systems can be represented as a two-, four-, or eight-digit hexadecimal code instead of using a long string of binary digits (see Table 1–3).

Table 1–3 Hexadecimal Numbering System

Decimal	Binary	Hexadecimal
0	0000 0000	0 0
1	0000 0001	0 1
2	0000 0010	0 2
3	0000 0011	0 3
4	0000 0100	0 4
5	0000 0101	0 5
6	0000 0110	0 6
7	0000 0111	0 7
8	0000 1000	0 8
9	0000 1001	0 9
10	0000 1010	0 A
11	0000 1011	0 B
12	0000 1100	0 C
13	0000 1101	0 D
14	0000 1110	0 E
15	0000 1111	0 F
16	0001 0000	1 0
17	0001 0001	1 1
18	0001 0010	1 2
19	0001 0011	1 3
20	0001 0100	1 4

Hexadecimal (hex) uses 16 different digits and is a method of grouping binary numbers in groups of four. Because hex digits must be represented by a single character, letters are chosen to represent values greater than 9. The 16 allowable hex digits are 0, 1, 2, 3, 4, 5, 6, 7, 8, 9, A, B, C, D, E, and F.

To signify a hex number, a subscript 16 or the letter H is used (that is, $A7_{16}$ or A7H). Two hex digits are used to represent 8 bits (also known as a *byte*). Four bits (one hex digit) are sometimes called a *nibble*.

1–9 Hexadecimal Conversions

To convert from *binary to hexadecimal*, group the binary number in groups of four (starting in the least significant position) and write down the equivalent hex digit.

E X A M P L E   1 – 1 2

Convert $0\ 1\ 1\ 0\ 1\ 1\ 0\ 1_2$ to hex.

Solution:

$$\underbrace{0\ 1\ 1\ 0}_{6}\quad \underbrace{1\ 1\ 0\ 1_2}_{D}\quad = 6D_{16}\quad \textit{Answer}$$

To convert *hexadecimal to binary,* use the reverse process.

E X A M P L E   1 – 1 3

Convert $A9_{16}$ to binary.

Solution:

$$\begin{array}{cc} A & 9 \\ 1\ 0\ 1\ 0 & 1\ 0\ 0\ 1 \end{array} = 1\ 0\ 1\ 0\ 1\ 0\ 0\ 1_2\quad \textit{Answer}$$

To convert *hexadecimal to decimal,* use a process similar to that in Section 1–3.

E X A M P L E   1 – 1 4

Convert $2\ A\ 6_{16}$ to decimal.

Solution:

$$
\begin{aligned}
2\quad A\quad 6 \\
6 \times 16^0 &= 6 \times 1 = 6 \\
A \times 16^1 &= 10 \times 16 = 160 \\
2 \times 16^2 &= 2 \times 256 = \underline{512} \\
&\qquad\qquad\qquad 678_{10}\quad \textit{Answer}
\end{aligned}
$$

E X A M P L E   1 – 1 5

Redo Example 1–14 by converting first to binary and then to decimal.

Solution:

$$
\begin{array}{ccc}
2 & A & 6 \\
\underbrace{0010} & \underbrace{1010} & \underbrace{0110}
\end{array} = 2 + 4 + 32 + 128 + 512 = 678_{10}\quad \textit{Answer}
$$

To convert from *decimal to hexadecimal,* use successive division.

EXAMPLE 1-16

Convert 151_{10} to hex.

Solution:

$$151 \div 16 = 9 \quad \text{remainder} \quad 7 \quad \text{(LSD)}$$
$$9 \div 16 = 0 \quad \text{remainder} \quad 9 \quad \text{(MSD)}$$
$$151_{10} = 97_{16} \quad \textit{Answer}$$

Check:

97_{16}
$\longrightarrow 7 \times 16^0 = \quad 7$
$\longrightarrow 9 \times 16^1 = \underline{144}$
$\overline{151} \quad \checkmark$

---

*Helpful
Hint*

At this point, you may be asking if you can use your hex calculator key instead of the hand procedure to perform these conversions. It is important to master these conversion procedures before depending on your calculator so that you understand the concepts involved.

EXAMPLE 1-17

Convert 498_{10} to hex.

Solution:

$$498 \div 16 = 31 \quad \text{remainder} \quad 2 \quad \text{(LSD)}$$
$$31 \div 16 = 1 \quad \text{remainder} \quad 15 \quad \text{(=F)}$$
$$1 \div 16 = 0 \quad \text{remainder} \quad 1 \quad \text{(MSD)}$$
$$498_{10} = 1 \; F \; 2_{16} \quad \textit{Answer}$$

Check:

$1 \; F \; 2_{16}$

$$2 \times 16^0 = 2 \times \quad 1 = \quad 2$$
$$F \times 16^1 = 15 \times \quad 16 = 240$$
$$1 \times 16^2 = 1 \times 256 = \underline{256}$$
$$\overline{498} \quad \checkmark$$

*Team
Discussion*

Which is the largest number—142_8, 142_{10}, or 142_{16}?

Review Questions

1–14. Why is hexadecimal used instead of the octal numbering system when working with 8- and 16-bit digital computers?

1–15. The *successive-division* method can be used whenever converting from base 10 to any other base numbering system. True or false?

1–16. Convert $0110\ 1011_2$ to hex.

1–17. Convert $E7_{16}$ to binary.

1–18. Convert $16C_{16}$ to decimal.

1–19. Convert 300_{10} to hex.

1–10 Binary-Coded-Decimal System

The binary-coded-decimal (**BCD**) system is used to represent each of the 10 decimal digits as a 4-bit binary code. This code is useful for outputting to displays that are always numeric (0 to 9), such as those found in digital clocks or digital voltmeters.

To form a BCD number, simply convert each decimal digit to its 4-bit binary code.

EXAMPLE 1–18

Convert $4\ 9\ 6_{10}$ to BCD.

Solution:

$$
\begin{array}{ccc}
4 & 9 & 6 \\
\overbrace{} & \overbrace{} & \overbrace{} \\
0100 & 1001 & 0110
\end{array}
= 0100\ \ 1001\ \ 0110_{BCD} \quad Answer
$$

To convert *BCD to decimal,* just reverse the process.

EXAMPLE 1–19

Convert $0111\ \ 0101\ \ 1000_{BCD}$ to decimal.

Solution:

$$
\begin{array}{ccc}
\underbrace{0111} & \underbrace{0101} & \underbrace{1000} \\
7 & 5 & 8
\end{array}
= 758_{10} \quad Answer
$$

EXAMPLE 1–20

Convert $0110\ \ 0100\ \ 1011_{BCD}$ to decimal.

Solution:

$$
\begin{array}{ccc}
0110 & 0100 & 1011 \\
6 & 4 & *
\end{array}
$$

---

*This conversion is impossible because 1011 is not a valid binary-coded decimal. It is not in the range 0 to 9.

1–11 Comparison of Numbering Systems

Table 1–4 compares numbers written in the five number systems commonly used in digital electronics and computer systems.

Table 1–4 Comparison of Numbering Systems				
Decimal	Binary	Octal	Hexadecimal	BCD
00	0000 0000	0 0	0 0	0000 0000
01	0000 0001	0 1	0 1	0000 0001
02	0000 0010	0 2	0 2	0000 0010
03	0000 0011	0 3	0 3	0000 0011
04	0000 0100	0 4	0 4	0000 0100
05	0000 0101	0 5	0 5	0000 0101
06	0000 0110	0 6	0 6	0000 0110
07	0000 0111	0 7	0 7	0000 0111
08	0000 1000	1 0	0 8	0000 1000
09	0000 1001	1 1	0 9	0000 1001
10	0000 1010	1 2	0 A	0001 0000
11	0000 1011	1 3	0 B	0001 0001
12	0000 1100	1 4	0 C	0001 0010
13	0000 1101	1 5	0 D	0001 0011
14	0000 1110	1 6	0 E	0001 0100
15	0000 1111	1 7	0 F	0001 0101
16	0001 0000	2 0	1 0	0001 0110
17	0001 0001	2 1	1 1	0001 0111
18	0001 0010	2 2	1 2	0001 1000
19	0001 0011	2 3	1 3	0001 1001
20	0001 0100	2 4	1 4	0010 0000

1–12 The ASCII Code

To get information into and out of a computer, we need more than just numeric representations; we also have to take care of all the letters and symbols used in day-to-day processing. Information such as names, addresses, and item descriptions must be input and output in a readable format. But remember that a digital system can deal only with 1's and 0's. Therefore, we need a special code to represent all **alphanumeric** data (letters, symbols, and numbers).

Most industry has settled on an input/output (I/O) code called the American Standard Code for Information Interchange (ASCII). The **ASCII code** uses 7 bits to represent all the alphanumeric data used in computer I/O. Seven bits will yield 128 different code combinations, as listed in Table 1–5.

Each time a key is depressed on an ASCII keyboard, that key is converted into its ASCII code and processed by the computer. Then, before outputting the computer contents to a display terminal or printer, all information is converted from ASCII into standard English.

To use the table, place the 4-bit group in the least significant positions and the 3-bit group in the most significant positions.

Have you ever tried displaying non-ASCII data to your PC screen using a disk utility program?

If you were to read a file created by the IRS for your tax return, which fields would be ASCII?

Table 1–5   American Standard Code for Information Interchange

LSB \ MSB	000	001	010	011	100	101	110	111
0000	NUL	DLE	SP	0	@	P		p
0001	SOH	DC₁	!	1	A	Q	a	q
0010	STX	DC₂	"	2	B	R	b	r
0011	ETX	DC₃	#	3	C	S	c	s
0100	EOT	DC₄	$	4	D	T	d	t
0101	ENQ	NAK	%	5	E	U	e	u
0110	ACK	SYN	&	6	F	V	f	v
0111	BEL	ETB	'	7	G	W	g	w
1000	BS	CAN	(	8	H	X	h	x
1001	HT	EM	)	9	I	Y	i	y
1010	LF	SUB	*	:	J	Z	j	z
1011	VT	ESC	+	;	K	[	k	{
1100	FF	FS	,	<	L	\	l	|
1101	CR	GS	-	=	M	]	m	}
1110	SO	RS	.	>	N	↑	n	~
1111	SI	US	/	?	O	—	o	DEL

Definitions of control abbreviations:

ACK	Acknowledge	FF	Form feed
BEL	Bell	FS	Form separator
BS	Backspace	GS	Group separator
CAN	Cancel	HT	Horizontal tab
CR	Carriage return	LF	Line feed
DC₁–DC₄	Direct control	NAK	Negative acknowledge
DEL	Delete idle	NUL	Null
DLE	Data link escape	RS	Record separator
EM	End of medium	SI	Shift in
ENQ	Enquiry	SO	Shift out
EOT	End of transmission	SOH	Start of heading
ESC	Escape	SP	Space
ETB	End of transmission block	STX	Start text
ETX	End text	SUB	Substitute
		SYN	Synchronous idle
		US	Unit separator
		VT	Vertical tab

EXAMPLE 1–21

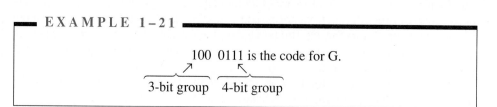

100 0111 is the code for G.

3-bit group   4-bit group

EXAMPLE 1–22

Using Table 1–5, determine the ASCII code for the lowercase letter *p*.

Solution: 1110000

1–20. How does BCD differ from the base 2 binary numbering system?

1–21. Why is ASCII code required by digital computer systems?

1–22. Convert 947_{10} to BCD.

1–23. Convert $1000\ 0110\ 0111_{BCD}$ to decimal.

1–24. Determine the ASCII code for the letter E.

1–13 Applications of the Number Systems

Because digital systems work mainly with 1's and 0's, we have spent considerable time working with the various number systems. Which system is used depends on how the data were developed and how they are to be used. In this section we will work with several applications that depend on the translation and interpretation of these digital representations.

APPLICATION 1–1

The ABC Corporation chemical processing plant uses a computer to monitor the temperature and pressure of four chemical tanks, as shown in Figure 1–4(a). Whenever a temperature or a pressure exceeds the danger limit, an internal tank sensor applies a 1 to its corresponding output to the computer. If all conditions are OK, then all outputs are 0.

Helpful Hint

This and the following five applications illustrate the answer to the common student question, Why are we learning this stuff?

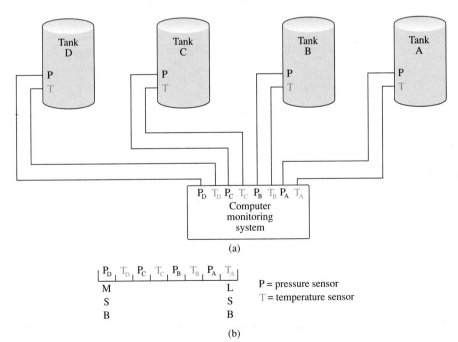

P = pressure sensor
T = temperature sensor

Figure 1–4   (a) Circuit connections for chemical temperature and pressure monitors at the ABC Corporation chemical processing plant; (b) Layout of binary data read by the computer monitoring system.

(a) If the computer reads the binary string 0010 1000, what problems exist?

Solution: Entering that binary string into the chart of Figure 1–4(b) shows us that the pressure in tanks C and B is dangerously high.

(b) What problems exist if the computer is reading 55H (55 hex)?

Solution: 55H = 0101 0101, meaning that all temperatures are too high

(c) What hexadecimal number is read by the computer if the temperature and pressure in both tanks D and B are high?

Solution: CCH   (1100 1100 = CCH)

(d) Tanks A and B are taken out of use and their sensor outputs are connected to 1's. A computer programmer must write a program to ignore these new circuit conditions. The computer program must check that the value read is always less than what decimal equivalent when no problem exists?

Solution: $<31_{10}$, because, with the 4 low-order bits HIGH, if TC goes HIGH, then the binary string will be 0001 1111, which is equal to 31_{10}.

(e) In another area of the plant, only three tanks (A, B, and C) have to be monitored. What octal number is read if tank B has a high temperature and pressure?

Solution: 14_8 ($001\ 100_2 = 14_8$)

APPLICATION 1–2

A particular brand of compact disk (CD) player has the capability of converting 12-bit signals from a CD into their equivalent analog values.

(a) What are the largest and smallest hex values that can be used in this CD system?

Solution: Largest: FFF_{16}; smallest: 000_{16}

(b) How many different analog values can be represented by this system?

Solution: FFF_{16} is equivalent to 4095 in decimal. Including 0, this is a total of 4096 unique representations.

APPLICATION 1–3

Typically, digital thermometers use BCD to drive their digit displays.

(a) How many BCD bits are required to drive a 3-digit thermometer display?

Solution: 12; 4 bits for each digit

(b) What 12 bits are sent to the display for a temperature of 147 degrees?

Solution: 0001 0100 0111

APPLICATION 1-4

Most PC-compatible computer systems use a 20-bit address code to identify each of over 1 million memory locations.

(a) How many hex characters are required to identify the address of each memory location?

Solution: Five (Each hex digit represents 4 bits.)

(b) What is the 5-digit hex address of the 200th memory location?

Solution: 000C7H ($200_{10} = C8H$; but the first memory location is 00000H, so we have to subtract 1).

(c) If 50 memory locations are used for data storage starting at location 000C8H, what is the location of the last data item?

Solution: 000F9H ($000C8H = 200_{10}$, $200 + 50 = 250_{10}$, $250 - 1 = 249_{10}$, $249_{10} = F9H$ [We had to subtract 1 because location C8H (200_{10}) received the first data item, so we needed only 49 more memory spaces.]

APPLICATION 1-5

If the part number 651-M is stored in ASCII in a computer memory, list the binary contents of its memory locations.

Solution:

$$
\begin{aligned}
6 &= 011\ 0110 \\
5 &= 011\ 0101 \\
1 &= 011\ 0001 \\
- &= 010\ 1101 \\
M &= 100\ 1101
\end{aligned}
$$

Because most computer memory locations are formed by groups of 8 bits, let's add a zero to the leftmost position to fill each 8-bit memory location. (The leftmost position is sometimes filled by a parity bit, which is discussed in Chapter 6.)

Therefore, the serial number, if strung out in five memory locations, would look like

0011 0110   0011 0101   0011 0001   0010 1101   0100 1101

If you look at these memory locations in hexadecimal, they will read

36   35   31   2D   4D

APPLICATION 1-6

To look for an error in a BASIC program, a computer programmer uses a debugging utility to display the ASCII codes of a particular part of her program. The codes are displayed in hex as 474F5430203930. Assume that the leftmost bit of each ASCII string is padded with a 0.

(a) Translate the program segment that is displayed.

Solution: GOT0 90.

(b) If you know anything about programming in BASIC, try to determine what the error is.

Solution: Apparently a number zero was typed in the GOTO statement instead of the letter O. Change it and the error should go away.

Summary

In this chapter we have learned that

1. Numerical quantities occur naturally in analog form but must be converted to digital form to be used by computers or digital circuitry.

2. The binary numbering system is used in digital systems because the 1's and 0's are easily represented by ON or OFF transistors, which output 0 V for 0 and 5 V for 1.

3. Any number system can be converted to decimal by multiplying each digit by its weighting factor.

4. The weighting factor of the least significant digit in any numbering system is always 1.

5. Binary numbers can be converted to octal by forming groups of 3 bits and to hexadecimal by forming groups of 4 bits.

6. The successive-division procedure can be used to convert from decimal to either binary, octal, or hexadecimal.

7. The binary-coded-decimal system uses groups of 4 bits to drive decimal displays such as those in a calculator.

8. ASCII is used by computers to represent all letters, numbers, and symbols in digital form.

Helpful Hint

Skimming through the glossary terms is a good way to review the chapter. You should also feel that you have a good understanding of all the topics listed in the objectives at the beginning of the chapter.

Glossary

Alphanumeric: Characters that contain alphabet letters as well as numbers and symbols.

Analog: A system that deals with continuously varying physical quantities such as voltage, temperature, pressure, or velocity. Most quantities in nature occur in analog, yielding an infinite number of different levels.

ASCII Code: American Standard Code for Information Interchange. ASCII is a 7-bit code used in digital systems to represent all letters, symbols, and numbers to be input or output to the outside world.

BCD: Binary-coded decimal. A 4-bit code used to represent the 10 decimal digits 0 to 9.

Binary: The base 2 numbering system. Binary numbers are made up of 1's and 0's, each position being equal to a different power of 2 (2^3, 2^2, 2^1, 2^0, and so on).

Bit: A single binary digit. The binary number 1101 is a 4-bit number.

Decimal: The base 10 numbering system. The 10 decimal digits are 0, 1, 2, 3, 4, 5, 6, 7, 8, and 9. Each decimal position is a different power of 10 (10^3, 10^2, 10^1, 10^0, and so on).

Digital: A system that deals with discrete digits or quantities. Digital electronics deals exclusively with 1's and 0's, or ONs and OFFs. Digital codes (such as ASCII) are then used to convert the 1's and 0's to a meaningful number, letter, or symbol for some output display.

Hexadecimal: The base 16 numbering system. The 16 hexadecimal digits are 0, 1, 2, 3, 4, 5, 6, 7, 8, 9, A, B, C, D, E, and F. Each hexadecimal position represents a different power of 16 (16^3, 16^2, 16^1, 16^0, and so on).

Least Significant Bit (LSB): The bit having the least significance in a binary string. The LSB will be in the position of the lowest power of 2 within the binary number.

Most Significant Bit (MSB): The bit having the most significance in a binary string. The MSB will be in the position of the highest power of 2 within the binary number.

Octal: The base 8 numbering system. The eight octal numbers are 0, 1, 2, 3, 4, 5, 6, and 7. Each octal position represents a different power of 8 (8^3, 8^2, 8^1, 8^0, and so on).

Problems

Section 1–4

1–1. Convert the following binary numbers to decimal.

(a) 0110   (b) 1011   (c) 1001   (d) 0111

(e) 1100   (f) 0100 1011   (g) 0011 0111

(h) 1011 0101   (i) 1010 0111   (j) 0111 0110

Section 1–5

1–2. Convert the following decimal numbers to 8-bit binary.

(a) 186_{10} (b) 214_{10} (c) 27_{10} (d) 251_{10} (e) 146_{10}

Sections 1–6 and 1–7

1–3. Convert the following binary numbers to octal.

(a) 011001   (b) 11101   (c) 1011100

(d) 01011001   (e) 1101101

1–4. Convert the following octal numbers to binary.

(a) 46_8 (b) 74_8 (c) 61_8 (d) 32_8 (e) 57_8

1–5. Convert the following octal numbers to decimal.

(a) 27_8 (b) 37_8 (c) 14_8 (d) 72_8 (e) 51_8

1–6. Convert the following decimal numbers to octal.

(a) 126_{10} (b) 49_{10} (c) 87_{10} (d) 94_{10} (e) 108_{10}

1–7. Convert the following binary numbers to hexadecimal.

(a) 1011 1001   (b) 1101 1100   (c) 0111 0100

(d) 1111 1011   (e) 1100 0110

1–8. Convert the following hexadecimal numbers to binary.

(a) $C5_{16}$ (b) FA_{16} (c) $D6_{16}$ (d) $A94_{16}$ (e) 62_{16}

1–9. Convert the following hexadecimal numbers to decimal.

(a) 86_{16} (b) $F4_{16}$ (c) 92_{16} (d) AB_{16} (e) $3C5_{16}$

1–10. Convert the following decimal numbers to hexadecimal.

(a) 127_{10} (b) 68_{10} (c) 107_{10} (d) 61_{10} (e) 29_{10}

Section 1–10

1–11. Convert the following BCD numbers to decimal.

(a) $1001\ 1000_{BCD}$ (b) $0110\ 1001_{BCD}$

(c) $0111\ 0100_{BCD}$ (d) $0011\ 0110_{BCD}$

(e) $1000\ 0001_{BCD}$

1–12. Convert the following decimal numbers to BCD.

(a) 87_{10} (b) 142_{10} (c) 94_{10} (d) 61_{10} (e) 44_{10}

Section 1–12

1–13. Use Table 1–5 to convert the following letters, symbols, and numbers to ASCII.

(a) %   (b) $14   (c) N-6   (d) CPU   (e) Pg

1–14. Insert a zero in the MSB of your answers to Problem 1–13 and list your answers in hexadecimal.

Section 1–13

C*   **1–15.** The computer monitoring system at the ABC Corporation shown in Figure 1–4 is receiving the following warning codes. Determine the problems that exist for each code (H stands for hex).

(a) $0010\ 0001_2$ (b) $C0_{16}$ (c) 88H   (d) 024_8 (e) 48_{10}

C   **1–16.** What is the BCD representation that is sent to a three-digit display on a voltmeter that is measuring 120 V?

C   **1–17.** A computer programmer observes the following hex string when looking at a particular section of computer memory: 736B753433.

(a) Assume that the memory contents are ASCII codes with leading zeros and translate this string into its alphanumeric equivalent.

(b) The programmer realizes that the program recognizes only capital (uppercase) letters. Convert all letters in the alphanumeric equivalent to capital letters and determine the new hex string.

---

*The letter C signifies problems that are more Challenging and thought provoking.

Schematic Interpretation Problems

(Note: Appendix G contains four schematic diagrams of actual digital systems. At the end of each chapter you will have the opportunity to work with these diagrams to gain experience with real-world circuitry and observe the application of digital logic that was presented in the chapter.)

S*   **1–18.** Locate the HC11D0 master board schematic in Appendix G. Determine the component name and grid coordinates of the following components. (Example: Q3 is a 2N2907 located at A3.)

     (a) U1   **(b)** U16   **(c)** Q1   **(d)** P2

S   **1–19.** Find the date and revision number for the HC11D0 master board schematic.

S   **1–20.** Find the quantity of the following ICs that are used on the watchdog timer schematic.

     (a) 74HC85   **(b)** 74HC08   **(c)** 74HC74   **(d)** 74HC32

Answers to Review Questions

1–1. Temperature, pressure, velocity

1–2. Because digital quantities are easier for a computer system to store and interpret

1–3. Because it uses only two digits, 0 and 1, which can be represented by using two distinct voltage levels

1–4. By powers of 2

1–5. 108_{10}

1–6. 13.375_{10}

1–7. $0010\ 1011_2$

1–8. $1010\ 1010_2$

1–9. False

1–10. 73_8

1–11. $010\ 110\ 011_2$ or $1011\ 0011_2$

1–12. 396_{10}

1–13. 132_8

1–14. Because hexadecimal uses 4-bit groupings

1–15. True

1–16. $6B_{16}$

1–17. $1110\ 0111_2$

1–18. 364_{10}

1–19. $12C_{16}$

1–20. BCD is used only to represent decimal digits 0 to 9 in 4-bit groupings.

1–21. To get alphanumeric data into and out of a computer

1–22. $1001\ 0100\ 0111_{BCD}$

1–23. 867_{10}

1–24. $0100\ 0101_{ASCII}$

---

*The letter S designates a Schematic interpretation problem.

2 Digital Electronic Signals and Switches

OUTLINE

2–1 Digital Signals
2–2 Clock Waveform Timing
2–3 Serial Representation
2–4 Parallel Representation
2–5 Switches in Electronic Circuits
2–6 A Relay as a Switch
2–7 A Diode as a Switch
2–8 A Transistor as a Switch
2–9 The TTL Integrated Circuit
2–10 The CMOS Integrated Circuit
2–11 Surface-Mount Devices

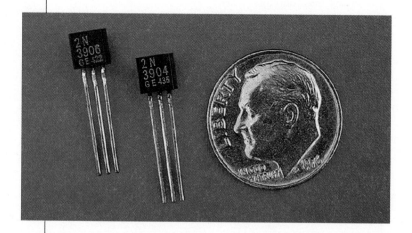

Objectives

Upon completion of this chapter, you should be able to

- Describe the parameters associated with digital voltage-versus-time waveforms.

- Convert between frequency and period for a periodic clock waveform.

- Sketch the timing waveform for any binary string in either the serial or parallel representation.

- Discuss the application of manual switches and electromechanical relays in electric circuits.

- Explain the basic characteristics of diodes and transistors when they are forward biased and reverse biased.

- Calculate the output voltage in an electric circuit containing diodes or transistors operating as digital switches.

- Perform input/output timing analysis in electric circuits containing electromechanical relays or transistors.

- Explain the operation of a common-emitter transistor circuit used as a digital inverter switch.

Introduction

As mentioned in Chapter 1, digital electronics deals with 1's and 0's. These logic states will typically be represented by a high and a low voltage level (usually $1 = 5$ V and $0 = 0$ V).

In this chapter we see how these logic states can be represented by means of a timing diagram and how electronic switches are used to generate meaningful digital signals.

2–1   Digital Signals

A digital signal is made up of a series of 1's and 0's that represent numbers, letters, symbols, or control signals. Figure 2–1 shows the **timing diagram** of a typical digital signal. Timing diagrams are used to show the HIGH and LOW (1 and 0) levels of a digital signal as it changes relative to time. In other words, it is a plot of *voltage versus time*. The *y* axis of the plot displays the voltage level and the *x* axis, the time. Digital systems respond to the digital state (0 or 1), not the actual voltage levels. For example, if the voltage levels in Figure 2–1(a) were 0.3 V and 4.0 V, the digital circuitry would still interpret it as the 0 state and 1 state and respond identically. The actual voltage level standards of the various logic families are discussed in detail in Chapter 9.

Figure 2–1(a) is a timing diagram showing the bit configuration 1 0 1 0 as it would appear on an **oscilloscope.** Notice in the figure that the LSB comes first in time. In this case, the LSB is transmitted first. The MSB could have been transmitted first as long as the system on the receiving end knows which method is used.

Figure 2–1(b) is a photograph of an oscilloscope, which is a very important test instrument for making accurate voltage and time measurements.

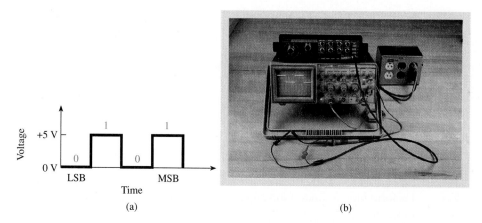

Figure 2–1   (a) Typical digital signal; (b) An oscilloscope displaying the digital waveform from a clock generator instrument.

2–2   Clock Waveform Timing

Most digital signals require precise timing. Special clock and timing circuits are used to produce clock waveforms to trigger the digital signals at precise intervals (timing circuit design is covered in Chapter 14).

Figure 2–2 shows a typical *periodic clock waveform* as it would appear on an oscilloscope displaying voltage versus time. The term *periodic* means that the waveform is repetitive, at a specific time interval, with each successive pulse identical to the previous one.

Figure 2–2 shows eight clock pulses, which we label 0, 1, 2, 3, 4, 5, 6, and 7. The **period** of the clock waveform is defined as the length of time from the falling edge of one pulse to the falling edge of the next pulse (or rising edge to rising edge) and is abbreviated t_p in Figure 2–2. The **frequency** of the clock waveform is defined as the reciprocal of the clock period. Written as a formula,

$$f = \frac{1}{t_p} \quad \text{and} \quad t_p = \frac{1}{f}$$

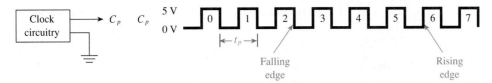

Figure 2–2 Periodic clock waveform as seen on an oscilloscope displaying voltage versus time.

The basic unit for frequency is *hertz* (Hz) and the basic unit for period is *seconds* (s). Frequency is often referred to as cycles per second (cps) or pulses per second (pps).

EXAMPLE 2–1

What is the frequency of a clock waveform whose period is 2 microseconds (μs)?

Solution:

$$f = \frac{1}{t_p} = \frac{1}{2\ \mu s} = 0.5 \text{ megahertz} \quad (0.5 \text{ MHz})$$

Team Discussion

An interesting exercise is to sketch the waveform from a 10-cps clock that is allowed to run for 1 s. How long did it take to complete 1 cycle? How did you find that time? Next, repeat for a 1-MHz clock.

Hint: To review scientific notation, see Table 2–1.

Table 2–1 Common Scientific Prefixes

Prefix	Abbreviation	Power of 10
giga	G	10^9
mega	M	10^6
kilo	k	10^3
milli	m	10^{-3}
micro	μ	10^{-6}
nano	n	10^{-9}
pico	p	10^{-12}

Helpful Hint

Frequency and time calculations can often be made without a calculator if you realize some of the common reciprocal relationships (e.g., 1/milli = kilo, 1/micro = mega). When using a calculator, if the result is not a power of 3, 6, 9, or 12, then the answer must be converted to one of these common scientific prefixes using algebra or, if available, the *ENG* key on your calculator.

EXAMPLE 2–2

If the frequency of a waveform is 4.17 MHz, what is its period?

Solution:

$$t_p = \frac{1}{f} = \frac{1}{4.17 \text{ MHz}} = 0.240\ \mu s$$

For those students who have a PC: Do you know (or could you find out) at what frequency your internal microprocessor operates?

EXAMPLE 2–3

Determine the frequency of the waveform in Figure 2–3.

Solution: The time period is 3.6 μs. Thus the frequency is

$$f = \frac{1}{t_p} = \frac{1}{3.6\,\mu s} = 278 \text{ kHz}$$

Figure 2–3 Waveform for Example 2–3.

Common Misconception

The period is labeled from rising edge to rising edge (or falling edge to falling edge) and is not just the positive pulse.

EXAMPLE 2–4

Sketch and label the x and y axes of a periodic clock waveform whose frequency is 4 MHz and voltage levels are 0.2 V and 4.0 V.

Solution:

$$t_p = \frac{1}{f} = \frac{1}{4 \text{ MHz}} = 0.25\,\mu s$$

Figure 2–4 Solution to Example 2–4.

The waveform is shown in Figure 2–4.

Review Questions

2–1. What are the labels on the x axis and y axis of a digital signal measured on an oscilloscope?

2–2. What is the relationship between clock frequency and clock period?

2–3. What is the time period from the rising edge of one pulse to the rising edge of the next pulse on a waveform whose frequency is 8 MHz?

2–4. What is the frequency of a periodic waveform having a period of 50 ns?

2–3 Serial Representation

Binary information to be transmitted from one location to another will be in either **serial** or **parallel** format. The serial format uses a single electrical conductor (and a

CHAPTER 2 I DIGITAL ELECTRONIC SIGNALS AND SWITCHES

common ground) for the data to travel on. The serial format is inexpensive because it requires only a single line, but it is slow because each bit transmitted exists for one clock period. This technique is used by computers to transmit data over telephone lines or from one computer to another (see Figure 2–5). The RS232 communications standard is a very common scheme used for this purpose.

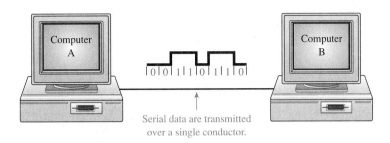

Serial data are transmitted over a single conductor.

Figure 2–5   Serial communication between computers.

Let's use Figure 2–6 to illustrate the serial representation of the binary number 0 1 1 0 1 1 0 0. The serial representation (S_o) is shown with respect to some clock waveform (C_p), and its LSB is drawn first. Each bit from the original binary number occupies a separate clock period with the change from one bit to the next occurring at each *falling* edge of C_p (C_p is drawn just as a reference).

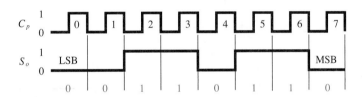

Figure 2–6   Serial representation of a binary number.

2–4   Parallel Representation

The parallel format uses a separate electrical conductor for each bit to be transmitted (and a common ground). For example, if the digital system is using 8-bit numbers, eight lines are required (see Figure 2–7). This tends to be expensive, but the entire 8-bit number can be transmitted in one clock period, making it very fast.

Inside a computer, binary data are almost always transmitted on parallel channels (collectively called the *data bus*). Two parallel data techniques commonly used by computers to communicate to external devices are the Centronics printer interface and the IEEE-488 instrumentation interface.

Figure 2–8 illustrates the same binary number that was used in Figure 2–6 (01101100), this time in the parallel representation.

If the clock period were 2 μs, it would take 2 μs × 8 periods = 16 μs to transmit the number in serial and only 2 μs × 1 period = 2 μs to transmit the same 8-bit number in parallel. Thus you can see that when speed is important, parallel transmission is preferred over serial transmission.

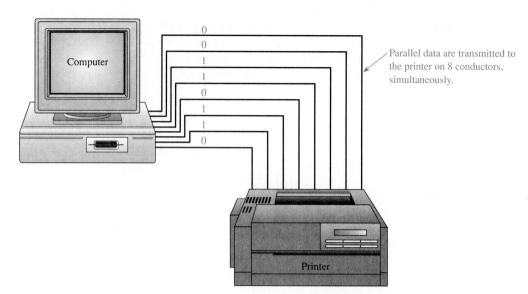

Figure 2–7   Parallel communication between a computer and a printer.

Figure 2–8   Parallel representation of a binary number.

The following examples further illustrate the use of serial and parallel representations.

EXAMPLE 2–5

Sketch the serial and parallel representations of the 4-bit number 0 1 1 1. If the clock frequency is 5 MHz, find the time to transmit using each method.

Solution: Figure 2–9 shows the representation of the 4-bit number 0 1 1 1.

CHAPTER 2 | DIGITAL ELECTRONIC SIGNALS AND SWITCHES

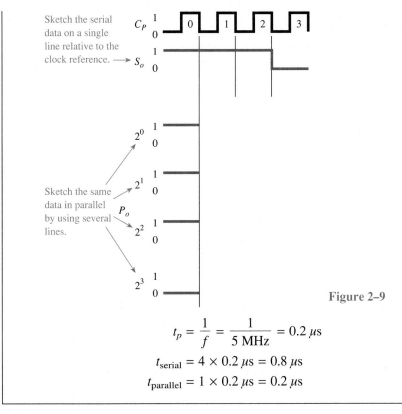

Sketch the serial data on a single line relative to the clock reference. → S_o

Sketch the same data in parallel by using several lines. → P_o

Figure 2–9

$$t_p = \frac{1}{f} = \frac{1}{5 \text{ MHz}} = 0.2 \ \mu s$$

$$t_{\text{serial}} = 4 \times 0.2 \ \mu s = 0.8 \ \mu s$$

$$t_{\text{parallel}} = 1 \times 0.2 \ \mu s = 0.2 \ \mu s$$

EXAMPLE 2–6

Sketch the serial and parallel representations (least significant digit first) of the hexadecimal number 4A. (Assume a 4-bit parallel system and a clock frequency of 4 kHz.) Also, what is the state (1 or 0) of the serial line 1.2 ms into the transmission?

Solution: $4A_{16} = 0\ 1\ 0\ 0\ 1\ 0\ 1\ 0_2$.

$$t_p = \frac{1}{f} = \frac{1}{4 \text{ kHz}} = 0.25 \text{ ms}$$

Therefore, the increment of time at each falling edge increases by 0.25 ms. Because each period is 0.25 ms, 1.2 ms will occur within the 0 period of the number 4, which, on the S_o line, is a *0* **logic state** (see Figure 2–10).

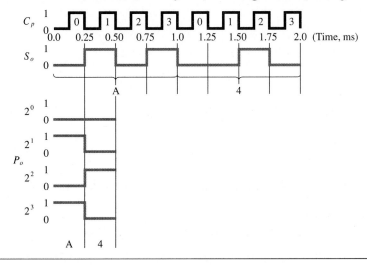

Figure 2–10

2–5. What advantage does parallel have over serial in the transmission of digital signals?

2–6. Which system requires more electrical conductors and circuitry, serial or parallel?

2–7. How long will it take to transmit three 8-bit binary strings in serial if the clock frequency is 5 MHz?

2–8. Repeat Question 2–7 for an 8-bit parallel system.

2–5   Switches in Electronic Circuits

The transitions between 0 and 1 digital levels are caused by switching from one voltage level to another (usually 0 V to +5 V). One way that switching is accomplished is to make and break a connection between two electrical conductors by way of a manual switch or an electromechanical relay. Another way to switch digital levels is by use of semiconductor devices such as **diodes** and **transistors.**

Manual switches and relays have almost *ideal* ON and OFF resistances in that when their contacts are closed (ON) the resistance (measured by an ohmmeter) is 0 ohms (Ω), and when their contacts are open (OFF), the resistance is infinite. Figure 2–11 shows the manual switch. When used in a digital circuit, a single-pole, double-throw manual switch can produce 0 and 1 states at some output terminal, as shown in Figures 2–12 and 2–13, by moving the switch (SW) to the up or down position.

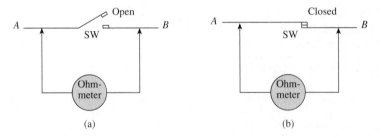

Figure 2–11   Manual switch: (a) switch open, $R = \infty$ ohms; (b) switch closed, $R = 0$ ohms.

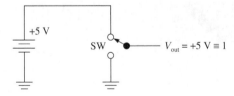

Figure 2–12   1-Level output.

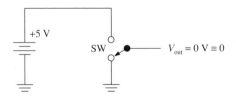

Figure 2–13   0-Level output.

2–6 A Relay as a Switch*

An electromechanical **relay** has contacts like a manual switch, but it is controlled by external voltage instead of being operated manually. Figure 2–14 shows the physical layout of an electromechanical relay. In Figure 2–14(a) the magnetic coil is energized by placing a voltage at terminals C_1–C_2; this will cause the lower contact to bend downward, opening the contact between X_1 and X_2. This relay is called *normally closed* (NC) because, at rest, the contacts are touching, or closed. In Figure 2–14(b), when the coil is energized, the upper contact will be attracted downward, making a connection between X_1 and X_2. This is called a *normally open* (NO) relay.

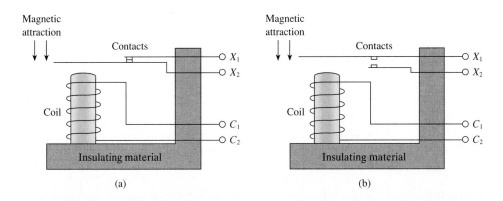

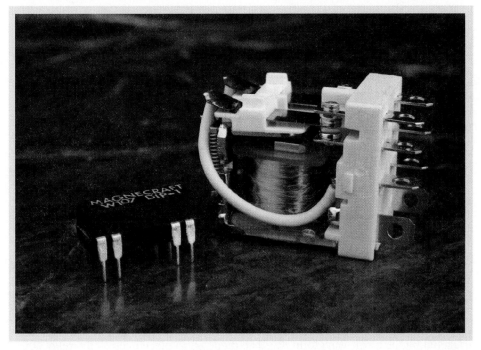

(c)

Figure 2–14 Physical representation of an electromechanical relay: (a) normally closed (NC) relay; (b) normally open (NO) relay; (c) photograph of actual relays.

---

*Systems requiring complex relay switching schemes are generally implemented using programmable logic controllers (PLC®s). PLC®s are microprocessor-based systems that are programmed to perform complex logic operations, usually to control electrical processes in manufacturing and industrial facilities. They use a programming technique called *ladder logic* to monitor and control several processes, eliminating the need for individually wired relays. PLC® is a registered trademark of Allen-Bradley Corporation.

A relay provides total isolation between the triggering source applied to C_1–C_2 and the output X_1–X_2. This total isolation is important in many digital applications, and it is a feature that certain semiconductor switches (such as transistors, diodes, and integrated circuits) cannot provide. Also, the contacts are normally rated for currents much higher than the current rating of semiconductor switches.

There are several disadvantages, however, of using a relay in electronic circuits. To energize the relay coil, the triggering device must supply several milliamperes, whereas a semiconductor requires only a few microamperes to operate. A relay is also much slower than a semiconductor. It will take several milliseconds to switch, compared to microseconds (or nanoseconds) for a semiconductor switch.

In Figure 2–15 a relay is used as a shorting switch in an electric circuit.* The +5-V source is used to energize the coil and the +12-V source is supplying the external electric circuit. When the switch (SW) in Figure 2–15(a) is closed, the relay coil will become energized, causing the relay contacts to open, which will make V_{out} change from 0 V to 6 V with respect to ground. The voltage-divider equation (see Appendix F) is used to calculate V_{out}:

$$V_{out} = \frac{12 \text{ V} \times 5 \text{ k}\Omega}{5 \text{ k}\Omega + 5 \text{ k}\Omega} = 6 \text{ V}$$

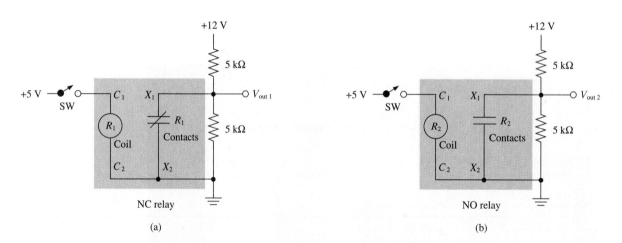

Figure 2–15 Symbolic representation of an electromechanical relay: (a) NC relay used in a circuit; (b) NO relay used in a circuit.

When the switch in Figure 2–15(b) is closed, the relay coil becomes an **energized relay coil,** causing the relay contacts to close, changing V_{out2} from 6 V to 0 V.

Now, let's go a step further and replace the 5-V battery and switch with a clock oscillator and use a timing diagram to analyze the results. In Figure 2–16 the relay is triggered by the clock waveform, C_p. The diode D_1 is placed across the relay coil to protect it from arcing each time the coil is de-energized. Timing diagrams are very useful for comparing one waveform to another because the waveform changes states (1 or 0) relative to time. The timing diagram in Figure 2–17 shows that when the clock goes HIGH (1) the relay is energized, causing V_{out3} to go LOW (0). When C_p goes LOW (0), the relay is de-energized, causing V_{out3} to go to +5 V (using the voltage-divider equation, $V_{out} = (10 \text{ V} \times 5 \text{ k}\Omega)/(5 \text{ k}\Omega + 5 \text{ k}\Omega) = 5 \text{ V}$).

---

*The principles of basic electricity required for the remainder of this chapter are given as a review in Appendix F.

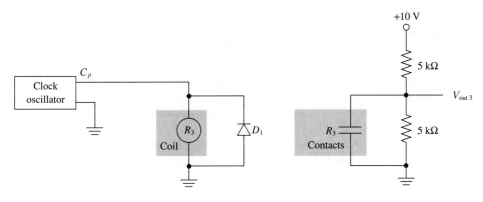

Figure 2–16   Relay used in a digital circuit.

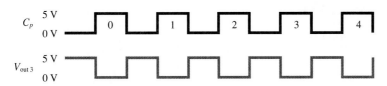

Figure 2–17   Timing diagram for Figure 2–16.

The following examples illustrate electronic switching and will help to prepare you for more complex timing analysis in subsequent chapters.

E X A M P L E   2 – 7

Draw a timing diagram for the circuit shown in Figure 2–18, given the C_p waveform in Figure 2–19.

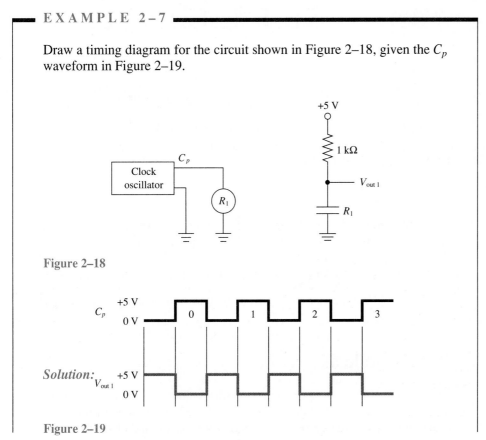

Figure 2–18

Figure 2–19

Explanation: When C_p is LOW, the R_1 coil is de-energized, the R_1 contacts are open, $I_{1 k\Omega} = 0$ A, $V_{drop\ 1 k\Omega} = I \times R = 0$ V, and $V_{out1} = 5$ V $- 0_{V\ drop} = 5$ V. When C_p is HIGH, the R_1 coil is energized, the R_1 contacts are closed, and $V_{out1} = 0$ V.

Helpful Hint

Remember that V_{out} is the voltage measured from the point in question to ground.

E X A M P L E   2 – 8

Draw a timing diagram for the circuit shown in Figure 2–20, given the C_p waveform in Figure 2–21.

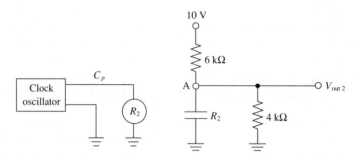

Figure 2–20

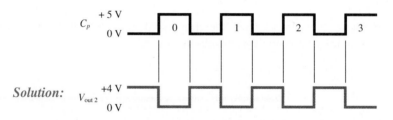

Solution:

Figure 2–21

Explanation: When the R_2 contacts are closed (R_2 is energized), the voltage at point A is 0 V, making V_{out2} equal to 0 V. When the R_2 contacts are open (R_2 is de-energized), the voltage at point A is $V_A = \dfrac{10 \text{ V} \times 4 \text{ k}\Omega}{6 \text{ k}\Omega + 4 \text{ k}\Omega} =$ 4 V and $V_{out2} = V_A = 4$ V.

Review Questions

2–9. Describe the operation of a relay coil and relay contacts.

2–10. How does a normally open relay differ from a normally closed relay?

2–7   A Diode as a Switch

Manual switches and electromechanical relays have limited application in today's digital electronic circuits. Most digital systems are based on semiconductor technology, which uses diodes and transistors. In Chapter 9 we discuss in detail the formation of digital circuits using transistors and diodes. Most electronics students should also take a separate course in electronic devices to cover the in-depth theory of the operation of diodes and transistors. However, without getting into a lot of detail, let's look at how a diode and a transistor can operate as a simple ON/OFF switch.

A diode is a semiconductor device that allows current to flow in one direction but not the other. Figure 2–22 shows a diode in both the conducting and nonconducting states. The term *forward biased* refers to a diode whose anode voltage is *more positive* than its cathode, thus allowing current flow in the direction of the arrow. (**Bias** is the voltage necessary to cause a semiconductor device to conduct or cut off current flow.) A reverse-biased diode will not allow current flow because its anode voltage is *equal to* or is *more negative* than its cathode. A diode is analogous to a check valve in a water system (see Figure 2–23).

A diode is not a perfect short in the forward-biased condition, however. The voltage-versus-current curve shown in Figure 2–24 shows the characteristics of a diode. Notice in the figure that for the reverse-biased condition, as V_{rev} becomes more negative, there is still practically zero current flow.

In the forward-biased condition, as V_{forw} becomes more positive, no current flows until a 0.7-V cut-in voltage is reached.* After that point, the voltage across the

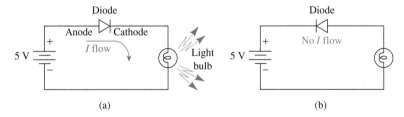

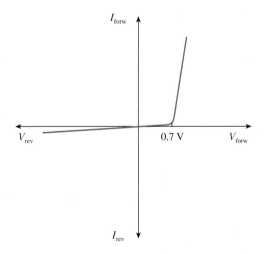

Figure 2–22   Diode in a series circuit: (a) forward biased; (b) reverse biased.

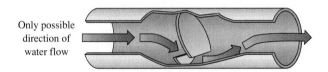

Figure 2–23   Water system check valve.

Figure 2–24   Diode voltage versus current characteristic curve.

*0.7 V is the typical cut-in voltage of a silicon diode, whereas 0.3 V is typical for a germanium diode. We will use the silicon diode because it is most commonly used in industrial applications.

diode (V_{forw}) will remain at approximately 0.7 V, and I_{forw} will flow, limited only by the external resistance of the circuit and the 0.7-V internal voltage drop.

What this means is that current will flow only if the anode is more positive than the cathode, and under those conditions the diode acts like a short circuit except for the 0.7 V across its terminals. This fact is better illustrated in Figure 2–25.

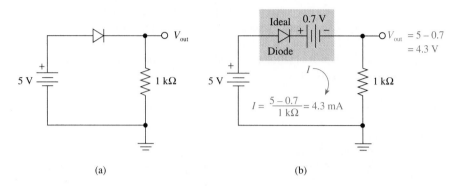

(a)   (b)

Figure 2–25 Forward-biased diode in an electric circuit: (a) original circuit; (b) equivalent circuit showing the diode voltage drop and $V_{\text{out}} = 5 - 0.7 = 4.3$ V.

The following examples and the problems at the end of the chapter demonstrate the effect that diodes have on electric circuits.

EXAMPLE 2–9

Determine if the diodes shown in Figure 2–26 are forward biased or reverse biased.

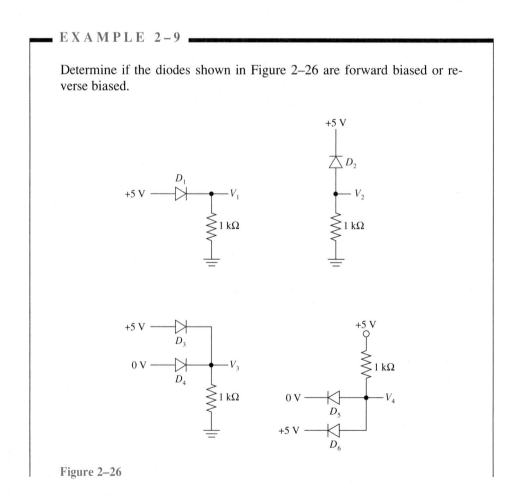

Figure 2–26

CHAPTER 2 | DIGITAL ELECTRONIC SIGNALS AND SWITCHES

Solution:

D_1 is forward biased.

D_2 is reverse biased.

D_3 is forward biased.

D_4 is reverse biased.

D_5 is forward biased.

D_6 is reverse biased.

EXAMPLE 2–10

Determine V_1, V_2, V_3, and V_4 (with respect to ground) for the circuits in Example 2–9.

Solution: V_1: D_1 is forward biased, dropping 0.7 V across its terminals. Therefore, $V_1 = 4.3$ V $(5.0 - 0.7)$.

V_2: D_2 is reverse biased. No current will flow through the 1-kΩ resistor, so $V_2 = 0$ V.

V_3: Because D_4 is reverse biased (open), it has no effect on the circuit. D_3 is forward biased, dropping 0.7 V, making $V_3 = 4.3$ V.

V_4: D_6 is reverse biased (open), so it has no effect on the circuit. D_5 is forward biased, so it has +0.7 V on its anode side, which is +0.7 above the 0-V ground level, making $V_4 = +0.7$ V.

Review Questions

2–11. To forward bias a diode, the anode is made more _____ (positive, negative) than the cathode.

2–12. A forward-biased diode has how many volts across its terminals?

2–8 A Transistor as a Switch

The bipolar transistor is a very commonly used switch in digital electronic circuits. It is a three-terminal semiconductor component that allows an input signal at one of its terminals to cause the other two terminals to become a short or an open circuit. The transistor is most commonly made of silicon that has been altered into *N*-type material and *P*-type material.

Three distinct regions make up a bipolar transistor: *emitter, base,* and *collector.* They can be a combination of *N-P-N*-type material or *P-N-P*-type material bonded together as a three-terminal device. Figure 2–27 shows the physical layout and symbol for an *NPN* transistor. (In a *PNP* transistor, the emitter arrow points the other way.)

In an electronic circuit, the input signal (1 or 0) is usually applied to the base of the transistor, which causes the collector–emitter junction to become a short or an open circuit. The rules of transistor switching are as follows:

1. In an *NPN* transistor, applying a positive voltage from base to emitter causes the collector-to-emitter junction to short (this is called "turning the transistor

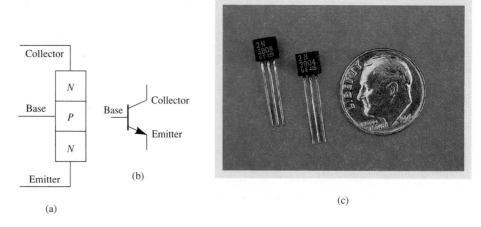

Figure 2–27   The *NPN* bipolar transistor: (a) physical layout; (b) symbol; (c) photograph.

ON"). Applying a negative voltage or 0 V from base to emitter causes the collector-to-emitter junction to open (this is called "turning the transistor OFF").

2. In a *PNP** transistor, applying a negative voltage from base to emitter turns it ON. Applying a positive voltage or 0 V from base to emitter turns it OFF.

Figure 2–28 shows how an *NPN* transistor functions as a switch in an electronic circuit. In the figure, resistors R_B and R_C are used to limit the base current and the collector current. In Figure 2–28(a) the transistor is turned ON because the base is more positive than the emitter (input signal = +2 V). This causes the collector-to-emitter junction to short, placing ground potential at V_{out} ($V_{out} = 0$ V).

In Figure 2–28(b), the input signal is removed, making the base-to-emitter junction 0 V, turning the transistor OFF. With the transistor OFF, there is no current through R_C, so $V_{out} = 5$ V $- (0$ A $\times R_C) = 5$V.

Digital input signals are usually brought in at the base of the transistor, and the output is taken off the collector or emitter. The following examples use timing analysis to compare the input and output waveforms.

Common Misconception

Students often think that the input signal to the base of a transistor must somehow be part of the output at the collector or emitter, but it is not. Once you determine if the *C*-to-*E* is a short or an open, you can ignore the base circuit altogether.

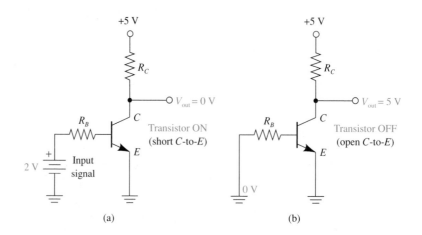

Figure 2–28   *NPN* transistor switch: (a) transistor ON; (b) transistor OFF.

**PNP* transistor circuits are analyzed in the same way as *NPN* circuits except that all voltage and current polarities are reversed. *NPN* circuits are much more common in industry and will be used most often in this book.

EXAMPLE 2–11

Sketch the waveform at V_{out} in the circuit shown in Figure 2–29, given the input signal C_p in Figure 2–30.

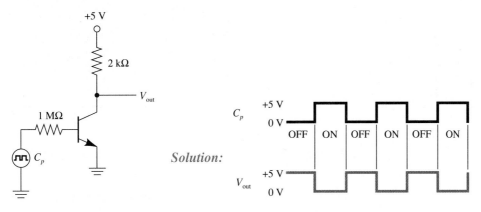

Solution:

Figure 2–29

Figure 2–30

Explanation: When $C_p = 0$ V, the transistor is OFF and the equivalent circuit is as shown in Figure 2–31(a).

$$I_C = 0 \text{ A}$$

Therefore,

$$V_C = 5 \text{ V} - (0 \text{ A} \times 2 \text{ k}\Omega) = 5 \text{ V}$$

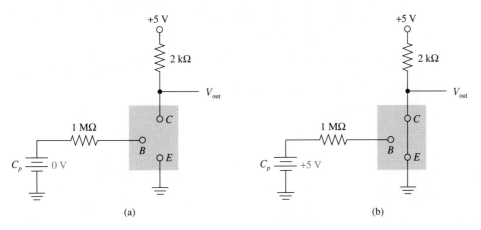

(a)   (b)

Figure 2–31

When $C_p = +5$ V, the transistor is ON and the equivalent circuit is as shown in Figure 2–31(b). The collector is shorted directly to ground; therefore, $V_{out} = 0$ V.

EXAMPLE 2–12

Sketch the waveform at V_{out} in the circuit shown in Figure 2–32, given the input signal C_p in Figure 2–33.

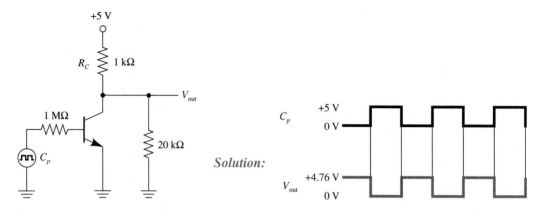

Figure 2–32

Figure 2–33

Solution:

Explanation: When $C_p = 0$ V, the transistor is OFF and the equivalent circuit is as shown in Figure 2–34(a). From the voltage-divider equation,

$$V_{out} = \frac{5\ \text{V} \times 20\ \text{k}\Omega}{20\ \text{k}\Omega + 1\ \text{k}\Omega} = 4.76\ \text{V}$$

Next, when $C_p = +5$ V, the transistor is ON and the equivalent circuit is as shown in Figure 2–34(b). Now the collector is shorted to ground, making $V_{out} = 0$ V. Notice the difference in V_{out} as compared to Example 2–11, which had no load resistor connected to V_{out}.

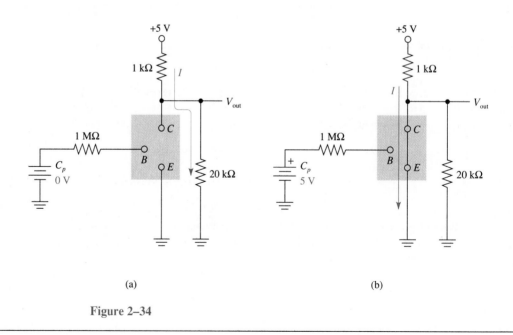

(a)

(b)

Figure 2–34

2–13. Name the three pins on a transistor.

2–14. To turn ON an *NPN* transistor, a _____ (positive, negative) voltage is applied to the base.

2–15. When a transistor is turned ON, its collector-to-emitter becomes a _____ (short, open).

2–9   The TTL Integrated Circuit

Transistor–transistor logic (**TTL**) is one of the most widely used integrated-circuit technologies. TTL integrated circuits use a combination of several transistors, diodes, and resistors integrated together in a single package.

One basic function of a TTL integrated circuit is as a complementing switch, or **inverter.** The inverter is used to take a digital level at its input and complement it to the opposite state at its output (1 becomes 0, 0 becomes 1). Figure 2–35 shows how a common-emitter-connected transistor switch can be used to perform the same function.

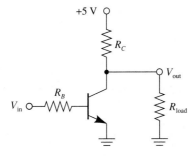

Figure 2–35   Common-emitter transistor circuit operating as an inverter.

When V_{in} equals 1 (+5 V), the transistor is turned on (called **saturation**) and V_{out} equals 0 (0 V). When V_{in} equals 0 (0 V), the transistor is turned off (called **cutoff**) and V_{out} equals 1 (approximately 5 V), assuming that R_L is much greater than R_C ($R_L \gg R_C$).

EXAMPLE 2–13

Let's assume that $R_C = 1$ kΩ, $R_L = 10$ kΩ, and $V_{in} = 0$ in Figure 2–35. V_{out} will equal 4.55 V:

$$\frac{5\text{ V} \times 10\text{ k}\Omega}{1\text{ k}\Omega + 10\text{ k}\Omega} = 4.55\text{ V}$$

But if R_L decreases to 1 kΩ by adding more loads in parallel with it, V_{out} will drop to 2.5 V:

$$\frac{5\text{ V} \times 1\text{ k}\Omega}{1\text{ k}\Omega + 1\text{ k}\Omega} = 2.5\text{ V}$$

We can see from Example 2–13 that the 1-level output of the inverter is very dependent on the size of the load resistor (R_L), which can typically vary by a factor of 10. So right away you might say, "Let's keep R_C very small so that R_L is always much greater than R_C" $(R_L \gg R_C)$. Well, that's fine for the case when the transistor is cut off $(V_{out} = 1)$, but when the transistor is saturated $(V_{out} = 0)$, the transistor collector current will be excessive if R_C is very small $(I_C = 5 \text{ V}/R_C$; see Figure 2–36).

Helpful
Hint

If you understand the idea that V_{out} varies depending on the size of the connected load, it will help you understand why gate outputs in the upcoming chapters are not exactly 0 V and 5 V. We discuss TTL and CMOS input/output characteristics in Chapter 9.

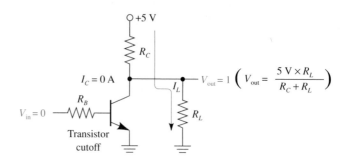

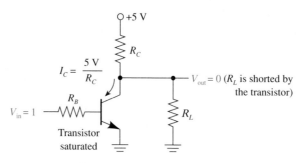

Figure 2–36   Common-emitter calculations.

Therefore, it seems that when the transistor is cut off $(V_{out} = 1)$, we want R_C to be small to ensure that V_{out} is close to 5 V, but when the transistor is saturated, we want R_C to be large to avoid excessive collector current.

This idea of needing a variable R_C resistance is accommodated by the TTL **integrated circuit** (Figure 2–37). It uses another transistor (Q_4) in place of R_C to act like a varying resistance. Q_4 is cut off (acts like a high R_C) when the output transistor (Q_3) is saturated, and then Q_4 is saturated (acts like a low R_C) when Q_3 is cut off. (In other words, when one transistor is ON, the other one is OFF.) This combination of Q_3 and Q_4 is referred to as the **totem-pole** arrangement.

Transistor Q_1 is the input transistor used to drive Q_2, which is used to control Q_3 and Q_4. Diode D_1 is used to protect Q_1 from negative voltages that might inadvertently be placed at the input. D_2 is used to ensure that when Q_3 is saturated Q_4 will be cut off totally. V_{CC} is the abbreviation used to signify the power supply to the integrated circuit.

TTL is a very popular family of integrated circuits. It is much more widely used than RTL (resistor–transistor logic) or DTL (diode–transistor logic) circuits, which were the forerunners of TTL. Details on the operation and specifications of TTL ICs are given in Chapter 9. In that chapter, you will learn why V_{out} is not exactly 0 V and 5 V (it is more typically 0.2 V and 3.4 V).

A single TTL integrated-circuit (IC) package such as the 7404 has six complete logic circuits fabricated into a single silicon **chip,** each logic circuit being the equivalent of Figure 2–37. The 7404 has 14 metallic pins connected to the outside of a plastic case containing the silicon chip. The 14 pins, arranged 7 on a side, are aligned on 14 holes of a printed-circuit board, where they are then soldered. The 7404 is called a

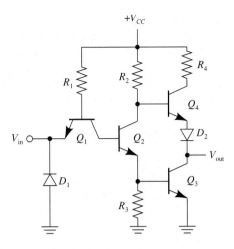

Figure 2–37 Schematic of a TTL circuit.

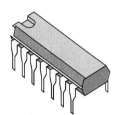

Figure 2–38 A 7404 TTL IC chip.

14-pin **DIP** (dual-in-line package) and costs less than 24 cents. Figure 2–38 shows a sketch of a 14-pin DIP IC. In subsequent chapters we see how to use ICs in actual digital circuitry.

ICs are configured as dual-in-line packages (DIPs) to ensure that the mechanical stress exerted on the pins when being inserted into a socket is equally distributed and that, although most of these pins serve as conductors to either the gates' inputs or outputs, some simply provide structural support and are simply anchored to the IC casing. These latter pins are denoted by the letters NC, meaning that they are *not* physically or electrically *connected* to an internal component.

The pin configuration of the 7404 is shown in Figure 2–39. The power supply connections to the IC are made to pin 14 (+5 V) and pin 7 (ground), which supplies power to all six logic circuits. In the case of the 7404, the logic circuits are called *inverters*. The symbol for each inverter is a triangle with a circle at the output. The circle is used to indicate the inversion function. Although *never* shown in the pin configuration top view of digital ICs, each gate is electrically tied internally to both V_{CC} and ground.

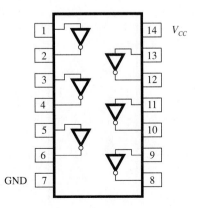

Figure 2–39 A 7404 hex inverter pin configuration.

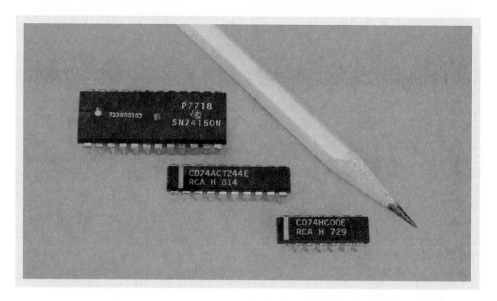

Figure 2–40   Photograph of three commonly used ICs: the 7400, 74LS244, and 74150.

Figure 2–40 shows three different ICs next to a pencil to give you an idea of their size.

2–10   The CMOS Integrated Circuit

Another common integrated-circuit technology used in digital logic is the **CMOS** (complementary metal oxide semiconductor). CMOS uses a complementary pair of metal oxide semiconductor field-effect transistors (MOSFETs) instead of the bipolar transistors used in TTL chips. (Complete coverage of TTL and CMOS is given in Chapter 9.)

The major advantage of using CMOS is its low power consumption. Because of that, it is commonly used in battery-powered devices such as hand-held calculators and digital thermometers. The disadvantage of using CMOS is that generally its switching speed is slower than TTL and it is susceptible to burnout due to electrostatic charges if not handled properly. Figure 2–41 shows the pin configuration for a 4049 CMOS **hex inverter.**

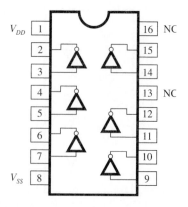

Figure 2–41   A 4049 CMOS hex inverter pin configuration.

CHAPTER 2 | DIGITAL ELECTRONIC SIGNALS AND SWITCHES

2–11 Surface-Mount Devices

The future of modern electronics depends on the ability to manufacture smaller, more dense components and systems. **Surface-mount devices (SMDs)** have fulfilled this need. They have reduced the size of DIP-style logic by as much as 70% and reduced their weight by as much as 90%. To illustrate the size difference, a 7400 IC in the DIP style measures 19.23 mm by 6.48 mm, whereas the equivalent 7400 SMD is only 8.75 mm by 6.20 mm.

SMDs have also significantly lowered the cost of manufacturing printed circuit boards. This reduction occurs because SMDs are soldered directly to a metalized footprint on the surface of a PC board, whereas holes must be drilled for each leg of a DIP. Also, SMDs can use the faster pick-and-place machines instead of the auto-insertion machines required for "through-hole" mounting of DIP ICs. (Removal of defective SMDs from PC boards is more difficult, however. Special desoldering tools and techniques are required because of the SMD's small size.)

Complete system densities can increase using SMDs because they can be placed closer together and can be mounted to both sides of a printed circuit board. This also tends to decrease the capacitive and inductive problems that occur in digital systems operating at higher frequencies. (This topic is discussed further in Chapter 9.)

The most popular SMD package styles are the SO (small outline) and the PLCC (plastic leaded chip carrier) shown in Figure 2–42. The SO is a dual-in-line plastic

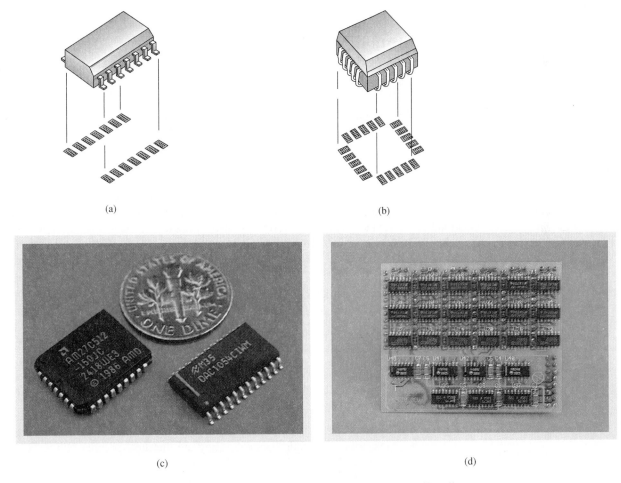

(a)   (b)

(c)   (d)

Figure 2–42   Typical surface-mount devices (SMDs) and their footprints: (a) small outline (SO); (b) plastic leaded chip carrier (PLCC); (c) photograph of actual SMDs; (d) photograph of SMDs mounted on a printed circuit board.

package with leads spaced 0.050 in. apart and bent down and out in a gull-wing format. The PLCC is the most common SMD for ICs requiring a higher pin count (those having more than 28 pins). The PLCC is square, with leads on all four sides. They are bent down and under in a J-bend configuration. They, too, are soldered directly to the metalized footprint on the surface of the circuit board.

The SO package is available for the most popular lower-complexity TTL and CMOS digital logic and analog IC devices. PLCCs are available to implement more complex logic, such as microprocessors, microcontrollers, and large memories.

Review Questions

2–16. In a common-emitter transistor circuit, when V_{out} is 0, R_C should be _____ (small, large), and when V_{out} is 1, R_C should be _____ (small, large).

2–17. Which transistor in the schematic of the TTL circuit in Figure 2–37 serves as a variable R_C resistance?

Summary

In this chapter we have learned that

1. The digital level for 1 is commonly represented by a voltage of 5 V in digital systems. A voltage of 0 V is used for the 0 level.

2. An oscilloscope can be used to observe the rapidly changing voltage-versus-time waveform in digital systems.

3. The frequency of a clock waveform is equal to the reciprocal of the waveform's period.

4. The transmission of binary data in the serial format requires only a single conductor with a ground reference. The parallel format requires several conductors but is much faster than serial.

5. Electromechanical relays are capable of forming shorts and opens in circuits requiring high current values but not high speed.

6. Diodes are used in digital circuitry whenever there is a requirement for current to flow in one direction but not the other.

7. The transistor is the basic building block of the modern digital integrated circuit. It can be switched on or off by applying the appropriate voltage at its base connection.

8. TTL and CMOS integrated circuits are formed by integrating thousands of transistors in a single package. They are the most popular ICs used in digital circuitry today.

9. SMD-style ICs are gaining popularity over the through-hole style DIP ICs because of their smaller size and reduced manufacturing costs.

Glossary

Bias: The voltage necessary to cause a semiconductor device to conduct or cut off current flow. A device can be forward biased or reverse biased, depending on what action is desired.

Chip: The term given to an integrated circuit. It comes from the fact that each integrated circuit comes from a single chip of silicon crystal.

CMOS: Complementary metal oxide semiconductor. A family of integrated circuits used to perform logic functions in digital circuits. The CMOS is noted for its low power consumption but sometimes slow speed.

Cutoff: A term used in transistor switching that signifies that the collector-to-emitter junction is turned off, or is not allowing current flow.

Diode: A semiconductor device used to allow current flow in one direction but not the other. As an electronic switch, it acts like a short in the forward-biased condition and like an open in the reverse-biased condition.

DIP: Dual-in-line packages. The most common pin layout for integrated circuits. The pins are aligned in two straight lines, one on each side of the IC.

Energized Relay Coil: By applying a voltage to the relay coil, a magnetic force is induced within it; this is used to attract the relay contacts away from their resting positions.

Frequency: A measure of the number of cycles or pulses occurring each second. Its unit is the hertz (Hz) and it is the reciprocal of the period.

Hex Inverter: An integrated circuit containing six inverters on a single DIP package.

Integrated Circuit: The fabrication of several semiconductor and electronic devices (transistors, diodes, and resistors) onto a single piece of silicon crystal. Integrated circuits are increasingly being used to perform the functions that used to require several hundred discrete semiconductors.

Inverter: A logic circuit that changes its input into the opposite logic state at its output (0 to 1 and 1 to 0).

Logic State: A 1 or 0 digital level.

Oscilloscope: An electronic measuring device used in design and troubleshooting to display a picture of waveform magnitude (y axis) versus time (x axis).

Parallel: A digital signal representation that uses several lines or channels to transmit binary information. The parallel lines allow for the transmission of an entire 4-bit (or more) number with each clock pulse.

Period: The measurement of time from the beginning of one periodic cycle or clock pulse to the beginning of the next. Its unit is the second(s), and it is the reciprocal of frequency.

Relay: An electric device containing an electromagnetic coil and normally open or normally closed contacts. It is useful because, by supplying a small triggering current to its coil, the contacts will open or close, switching a higher current on or off.

Saturation: A term used in transistor switching that signifies that the collector-to-emitter junction is turned on, or conducting current heavily.

Serial: A digital signal representation that uses one line or channel to transmit binary information. The binary logic states are transmitted 1 bit at a time, with the LSB first.

Surface-Mounted Device (SMD): The newest style of integrated circuit, soldered directly to the surface of a printed circuit board. They are much smaller and lighter than the equivalent logic constructed in the DIP through-hole-style logic.

Timing Diagram: A diagram used to display the precise relationship between two or more digital waveforms as they vary relative to time.

Totem Pole: The term used to describe the output stage of most TTL integrated circuits. The totem-pole stage consists of one transistor in series with another, configured in such a way that when one transistor is saturated, the other is cut off.

Transistor: A semiconductor device that can be used as an electronic switch in digital circuitry. By applying an appropriate voltage at the base, the collector-to-emitter junction will act like an open or a shorted switch.

TTL: Transistor–transistor logic. The most common integrated circuit used in digital electronics today. A large family of different TTL ICs is used to perform all the logic functions necessary in a complete digital system.

Problems

Sections 2–1 and 2–2

2–1. Determine the period of a clock waveform whose frequency is

(a) 2 MHz   **(b)** 500 kHz   **(c)** 4.27 MHz   **(d)** 17 MHz

Determine the frequency of a clock waveform whose period is

(e) 2 μs   **(f)** 100 μs   **(g)** 0.75 ms   **(h)** 1.5 μs

Sections 2–3 and 2–4

2–2. Sketch the serial and parallel representations (similar to Figure 2–10) of the following numbers and calculate how long they will take (clock frequency = 2 MHz).

(a) $45B_{16}$ **(b)** $A3C_{16}$

2–3. **(a)** How long will it take to transmit the number 33_{10} in serial if the clock frequency is 3.7 MHz? (Transmit the number as an 8-bit binary number.)

(b) Is the serial line HIGH or LOW at 1.21 μs?

2–4. **(a)** How long will it take to transmit the three ASCII-coded characters \$14 in 8-bit parallel if the clock frequency is 8 MHz?

(b) Repeat for \$78.18 at 4.17 MHz.

Sections 2–5 and 2–6

C   **2–5.** Draw the timing diagram for V_{out1}, V_{out2}, and V_{out3} in Figure P2–5.

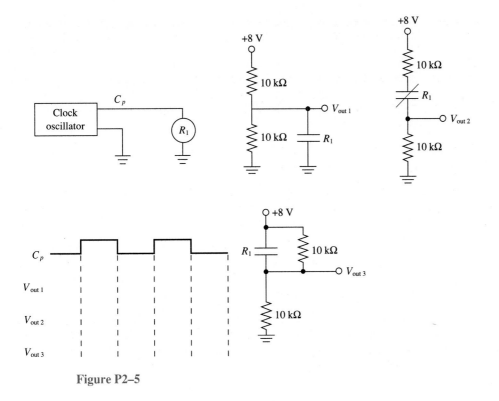

Figure P2–5

Section 2–7

2–6. Determine if the diodes in Figure P2–6 are reverse or forward biased.

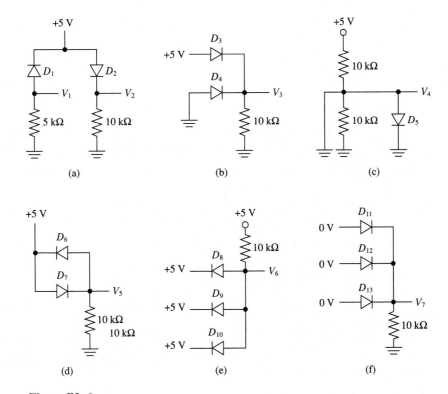

Figure P2–6

C **2–7.** Determine V_1, V_2, V_3, V_4, V_5, V_6, and V_7 in the circuits of Figure P2–6.

2–8. In Figure P2–6, if the cathode of any one of the diodes D_8, D_9, or D_{10} is connected to 0 V instead of +5 V, what happens to V_6?

2–9. In Figure P2–6, if the anode of any of the diodes D_{11}, D_{12}, or D_{13} is connected to +5 V instead of 0 V, what happens to V_7?

Section 2–8

2–10. Find V_{out1} and V_{out2} for the circuits of Figure P2–10.

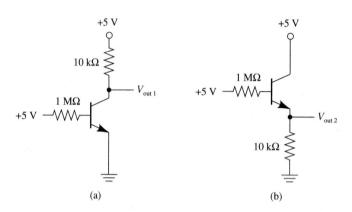

(a)

(b)

Figure P2–10

2–11. Sketch the waveforms at V_{out} in the circuit of Figure 2–32 using $R_C = 6\ \text{k}\Omega$.

Section 2–9

2–12. To use a common-emitter transistor circuit as an inverter, the input signal is connected to the _____ (base, collector, or emitter) and the output signal is taken from the _____ (base, collector, or emitter).

C **2–13.** Determine V_{out} for the common-emitter transistor inverter circuit of Figure 2–35 using $V_{in} = 0$ V, $R_B = 1$ MΩ, $R_C = 330\ \Omega$, and $R_{load} = 1$ MΩ.

C **2–14.** If the load resistor (R_{load}) used in Problem 2–13 is changed to 470 Ω, describe what happens to V_{out}.

C **2–15.** In the circuit of Figure 2–35 with $V_{in} = 0$ V, V_{out} will be almost 5 V as long as R_{load} is much greater than R_C. Why not make R_C very small to ensure that the circuit will work for all values of R_{load}?

C **2–16.** In Figure 2–35, if $R_C = 100\ \Omega$, find the collector current when $V_{in} = +5$ V.

C **2–17.** Describe how the totem-pole output arrangement in a TTL circuit overcomes the problems faced when using the older common-emitter transistor inverter circuit.

2–18. Sketch the waveform at C_p and V_{out} for Figure P2–18.

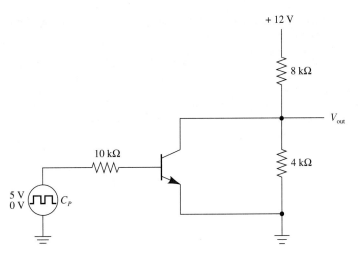

$+ 12$ V

8 kΩ

V_{out}

10 kΩ

4 kΩ

$\begin{smallmatrix} 5 \text{ V} \\ 0 \text{ V} \end{smallmatrix}$ C_P

Figure P2–18

Schematic Interpretation Problems

See Appendix G for the schematic diagrams.

S   **2–19.** Y1 in the 4096/4196 control card schematic sheet 1 is a crystal used to generate a very specific frequency. **(a)** What is its rated frequency? **(b)** What time period does that create?

S   **2–20.** Repeat Problem 2–19 for the crystal X1 in the HC11D0 master board schematic.

S   **2–21.** The circuit on the HC11D0 schematic is capable of parallel as well as serial communication via connectors P_3 and P_2. Which is parallel and which is serial? (*Hint:* TX stands for transmit, RX stands for receive.)

S   **2–22.** Is diode D_1 of the HC11D0 schematic forward or reverse biased? (*Hint:* $V_{CC} = 5$ V.)

S   **2–23.** The transistor Q_1 in the HC11D0 schematic is turned ON and OFF by the level of pin 2 on U3:A. At what level must pin 2 be to turn Q_1 ON, and what will happen to the level on the line labeled RESET B when that happens?

Answers to Review Questions

2–1. *x* axis, time; *y* axis, voltage

2–2. The clock frequency is the reciprocal of the clock period.

2–3. 125 ns

2–4. 20 MHz

2–5. It is faster.

2–6. Parallel

2–7. 4.80 μs

2–8. 600 ns

2–9. The relay coil is energized by placing a voltage at its terminals. The contacts will either make a connection (NO relay) or break a connection (NC relay) when the coil is energized.

2–10. An NO relay makes connection when energized. An NC relay breaks connection when energized.

2–11. Positive

2–12. Approximately 0.7 V

2–13. Emitter, base, collector

2–14. Positive

2–15. Short

2–16. Large, small

2–17. Q_4

3 Basic Logic Gates

OUTLINE

3–1 The AND Gate
3–2 The OR Gate
3–3 Timing Analysis
3–4 Enable and Disable Functions
3–5 Using Integrated-Circuit Logic Gates
3–6 Introduction to Troubleshooting Techniques

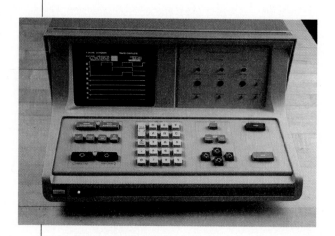

Objectives

Upon completion of this chapter, you should be able to

- Describe the operation and use of AND gates and OR gates.

- Construct truth tables for two-, three-, and four-input AND and OR gates.

- Draw timing diagrams for AND and OR gates.

- Describe the operation, using timing analysis, of an ENABLE function.

- Sketch the external connections to integrated-circuit chips to implement AND and OR logic circuits.

- Explain how to use a logic pulser and a logic probe to troubleshoot digital integrated circuits.

Introduction

Logic **gates** are the basic building blocks for forming digital electronic circuitry. A logic gate has one output terminal and one or more input terminals. Its output will be HIGH (1) or LOW (0) depending on the digital level(s) at the input terminal(s). Through the use of logic gates, we can design digital systems that will evaluate digital input levels and produce a specific output response based on that particular logic circuit design. The seven logic gates are AND, OR, NAND, NOR, INVERTER, exclusive-OR, and exclusive-NOR. The AND and OR are discussed in this chapter.

3–1   The AND Gate

Let's start by looking at the two-input AND gate shown in Figure 3–1. The operation of the AND gate is simple and is defined as follows: *The output, X, is HIGH if input A AND input B are both HIGH.* In other words, if *A* = 1 *AND B* = 1, then *X* = 1. If either *A* or *B* or both are LOW, the output will be LOW.

Figure 3–1   Two-input AND gate.

The best way to illustrate how the output level of a gate responds to all the possible input-level combinations is with a **truth table.** Table 3–1 is a truth table for a two-input AND gate. On the left side of the truth table, all possible input-level combinations are listed, and on the right side the resultant output is listed.

Table 3–1 Truth Table for a Two-Input AND Gate		
Inputs		
A	*B*	Output *X*
0	0	0
0	1	0
1	0	0
1	1	1

From the truth table we can see that the output at *X* is HIGH *only* when *both A AND B* are HIGH. If this AND gate is a TTL integrated circuit, HIGH means +5 V and LOW means 0 V (that is, 1 is defined as +5 V and 0 is defined as 0 V).

One example of how an AND gate might be used is in a bank burglar alarm system. The output of the AND gate will go HIGH to turn on the alarm if the alarm activation key is in the ON position *AND* the front door is opened. This setup is illustrated in Figure 3–2.

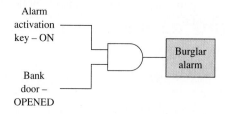

Figure 3–2   AND gate used to activate a burglar alarm.

Another way to illustrate the operation of an AND gate is by use of a series electric circuit. In Figure 3–3, using manual and transistor switches, the output at *X* is HIGH if *both* switches *A AND B* are HIGH (1).

Figure 3–3 also shows what is known as the **Boolean equation** for the AND function, *X = A* and *B,* which can be thought of as *X equals 1 if A AND B both equal 1.*

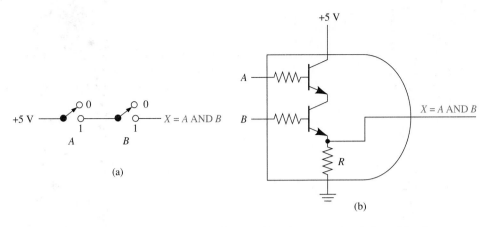

Figure 3–3 Electrical analogy for an AND gate: (a) using manual switches; (b) using transistor switches.

The Boolean equation for the AND function can more simply be written as $X = A \cdot B$ or just $X = AB$. Boolean equations will be used throughout the rest of the book to depict algebraically the operation of a logic gate or a combination of logic gates.

AND gates can have more than two inputs. Figure 3–4 shows a four-input and an eight-input AND gate. The truth table for an AND gate with four inputs is shown in

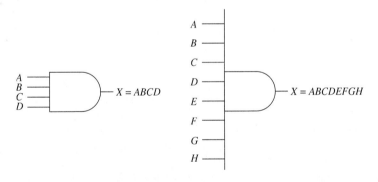

Figure 3–4 Multiple-input AND gate symbols.

Table 3–2. To determine the total number of different combinations to be listed in the truth table, use the equation

$$\text{number of combinations} = 2^N, \qquad \text{where } N = \text{number of inputs} \qquad (3\text{–}1)$$

Therefore, in the case of a four-input AND gate, the number of possible input combinations is $2^4 = 16$.

When building the truth table, be sure to list all 16 *different* combinations of input levels. One easy way to ensure that you do not inadvertently overlook a combination of these variables or duplicate a combination is to list the inputs in the order of a binary counter (0000, 0001, 0010, . . . , 1111). Also notice in Table 3–2 that the A column lists eight 0's, then eight 1's; the B column lists four 0's, four 1's, four 0's, four 1's; the C column lists two 0's, two 1's, two 0's, two 1's, and so on; and the D column lists one 0, one 1, one 0, one 1, and so on.

Table 3–2 Truth Table for a Four-Input AND Gate				
A	B	C	D	X
0	0	0	0	0
0	0	0	1	0
0	0	1	0	0
0	0	1	1	0
0	1	0	0	0
0	1	0	1	0
0	1	1	0	0
0	1	1	1	0
1	0	0	0	0
1	0	0	1	0
1	0	1	0	0
1	0	1	1	0
1	1	0	0	0
1	1	0	1	0
1	1	1	0	0
1	1	1	1	1

3–2   The OR Gate

The OR gate also has two or more inputs and a single output. The symbol for a two-input OR gate is shown in Figure 3–5. The operation of the two-input OR gate is defined as follows: *The output at X will be HIGH whenever input A OR input B is HIGH or both are HIGH*. As a Boolean equation, this can be written $X = A + B$. Notice the use of the + symbol to represent the OR function.

Figure 3–5   Two-input OR gate.

The truth table for a two-input OR gate is shown in Table 3–3.

Table 3–3 Truth Table for a Two-Input OR Gate		
Inputs		
A	B	Output X
0	0	0
0	1	1
1	0	1
1	1	1

From the truth table you can see that *X* is 1 whenever *A OR B* is 1 or if *both A* and *B* are 1. Using manual or transistor switches in an electric circuit, as shown in Figure 3–6, we can observe the electrical analogy to an OR gate. From the figure we see that the output at *X* will be 1 if *A or B,* or *both,* are HIGH (1).

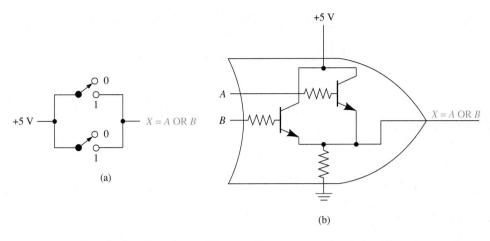

Figure 3–6   Electrical analogy for an OR gate: (a) using manual switches; (b) using transistor switches.

OR gates can also have more than two inputs. Figure 3–7 shows a three-input OR gate and Figure 3–8 shows an eight-input OR gate. The truth table for the three-input OR gate will have eight entries ($2^3 = 8$), and the eight-input OR gate will have 256 entries ($2^8 = 256$).

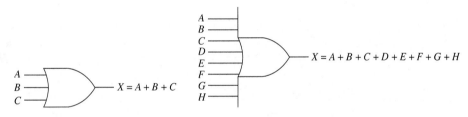

Figure 3–7   Three-input OR gate symbol.

Figure 3–8   Eight-input OR gate symbol.

Let's build a truth table for the three-input OR gate.

The truth table of Table 3–4 is built by first using Equation 3–1 to determine that there will be eight entries, then listing the eight combinations of inputs in the order of a binary counter (000 to 111), and then filling in the output column *(X)* by realizing that X will always be HIGH as long as at least one of the inputs is HIGH. When you look at the completed truth table you can see that the only time the output is LOW is when *all* the inputs are LOW.

Table 3–4 Truth Table for a Three-Input OR Gate			
A	*B*	*C*	*X*
0	0	0	0
0	0	1	1
0	1	0	1
0	1	1	1
1	0	0	1
1	0	1	1
1	1	0	1
1	1	1	1

EXAMPLE 3–1

Determine the output at *W, X, Y,* and *Z* in Figure 3–9.

Figure 3–9   Basic AND and OR gate operation.

Solution:

$$W = 0 \qquad (0 \text{ OR } 0 = 0)$$
$$X = 1 \qquad (0 \text{ OR } 1 = 1)$$
$$Y = 0 \qquad (1 \text{ AND } 0 = 0)$$
$$Z = 1 \qquad (1 \text{ AND } 1 = 1)$$

Review Questions

3–1. All inputs to an AND gate must be HIGH for it to output a HIGH. True or false?

3–2. What is the purpose of a truth table?

3–3. What is the purpose of a Boolean equation?

3–4. What input conditions must be satisfied for the output of an OR gate to be LOW?

3–3   Timing Analysis

Another useful means of analyzing the output response of a gate to varying input-level changes is by means of a *timing diagram*. A timing diagram, as described in Chapter 2, is used to illustrate graphically how the output levels change in response to input-level changes.

The timing diagram in Figure 3–10 shows the two input waveforms (*A* and *B*) that are applied to a two-input AND gate and the *X* output that results from the AND operation. (For TTL and most CMOS logic gates, 1 = +5 V and 0 = 0 V.) As you can see, timing analysis is very useful for visually illustrating the level at the output for varying input-level changes.

Timing waveforms are observed on an *oscilloscope* or a *logic analyzer.* A dual-trace oscilloscope is capable of displaying *two* voltage-versus-time waveforms on the same *x* axis. That is ideal for comparing the relationship of one waveform relative to another. The other timing analysis tool is the logic analyzer. Among other things, it can

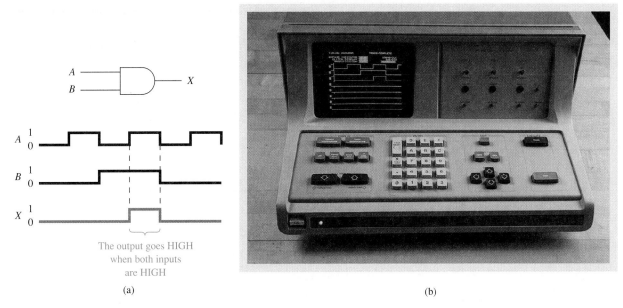

Figure 3–10   Timing analysis of an AND gate: (a) waveform sketch; (b) actual logic analyzer display.

display 8 or 16 voltage-versus-time waveforms on the same x axis [see Figure 3–10(b)]. It can also display the levels of the digital signals in a *state table,* which lists the binary levels of all the waveforms, at predefined intervals, in binary, hexadecimal, or octal. Timing analysis of 8 or 16 channels concurrently is very important when analyzing advanced digital and microprocessor systems in which the interrelationship of several digital signals is critical for proper circuit operation.

EXAMPLE 3–2

Sketch the output waveform at X for the two-input OR gate shown in Figure 3–11, with the given A and B input waveforms in Figure 3–12.

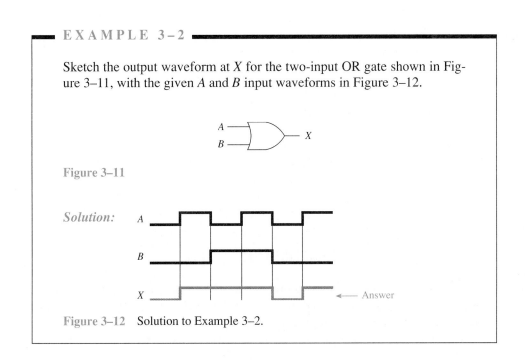

Figure 3–11

Solution:

Figure 3–12   Solution to Example 3–2.

EXAMPLE 3–3

Sketch the output waveform at X for the three-input AND gate shown in Figure 3–13, with the given A, B, and C input waveforms in Figure 3–14.

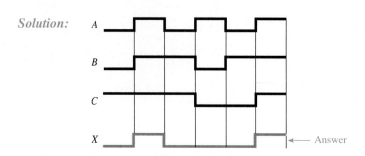

Figure 3–13

Solution:

Figure 3–14 Solution to Example 3–3.

EXAMPLE 3–4

The input waveform at A and the output waveform at X are given for the AND gate in Figure 3–15. Sketch the input waveform that is required at B to produce the output at X in Figure 3–16.

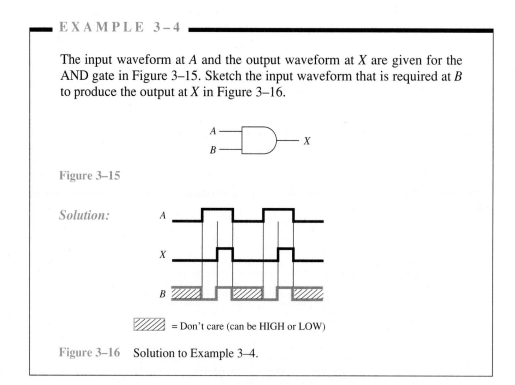

Figure 3–15

Solution:

= Don't care (can be HIGH or LOW)

Figure 3–16 Solution to Example 3–4.

3–4 Enable and Disable Functions

AND and OR gates can be used to **enable** or **disable** a waveform from being transmitted from one point to another. For example, let's say that you wanted a 1-MHz clock oscillator to transmit only *four* pulses to some receiving device. You would want to *enable* four clock pulses to be transmitted and then *disable* the transmission from then on.

CHAPTER 3 I BASIC LOGIC GATES

The clock frequency of 1 MHz converts to 1 μs (1/1 MHz) for each clock period. Therefore, to transmit four clock pulses, we have to provide an *enable* signal for 4 μs. Figure 3–17 shows the circuit and waveforms to *enable* four clock pulses. For the HIGH clock pulses to get through the AND gate to point X, the second input to the AND gate (enable signal input) must be HIGH; otherwise, the output of the AND gate will be LOW. Therefore, when the enable signal is HIGH for 4 μs, four clock pulses pass through the AND gate. When the enable signal goes LOW, the AND gate *disables* any further clock pulses from reaching the receiving device.

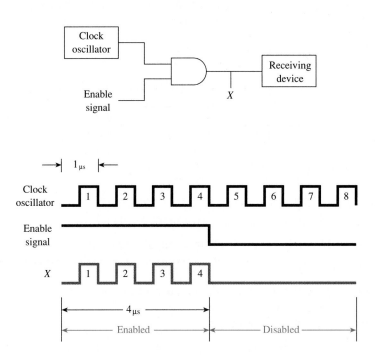

Figure 3–17   Using an AND gate to enable/disable a clock oscillator.

An OR gate can also be used to disable a function. The difference is that the enable signal input is made HIGH to disable, and the output of the OR gate goes HIGH when it is disabled, as shown in Figure 3–18.

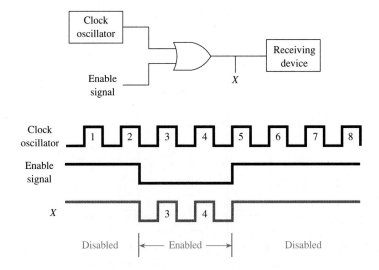

Figure 3–18   Using an OR gate to enable-disable a clock oscillator.

Review Questions

3–5. Describe the purpose of a *timing diagram*.

3–6. Under what circumstances would diagonal "don't care" hash marks be used in a timing diagram?

3–7. A _____ (HIGH, LOW) level is required at the input to an AND gate to *enable* the signal at the other input to pass to the output.

3–5   Using Integrated-Circuit Logic Gates

AND and OR gates are available as integrated circuits (ICs). The IC pin layout, logic gate type, and technical specifications are all contained in the logic data manual supplied by the manufacturer of the IC. For example, referring to a TTL or a CMOS logic data manual, we can see that there are several AND and OR gate ICs. To list just a few:

1. The 7408 (74HC08) is a quad two-input AND gate.

2. The 7411 (74HC11) is a triple three-input AND gate.

3. The 7421 (74HC21) is a dual four-input AND gate.

4. The 7432 (74HC32) is a quad two-input OR gate.

In each case, the HC stands for high-speed CMOS. For example, the 7408 is a TTL AND gate, and the 74HC08 is the equivalent CMOS AND gate. The terms *quad* (four), *triple* (three), and *dual* (two) refer to the number of separate gates on a single IC.

Let's look in more detail at one of these ICs, the 7408 (Figure 3–19). The 7408 is a 14-pin dual-in-line package (DIP) IC. The power supply connections are made to pins 7 and 14. This supplies the operating voltage for all four AND gates on the IC. Pin 1 is identified by a small indented circle next to it or by a notch cut out between pin 1 and 14 (see Figure 3–19). Let's make the external connections to the IC to form a clock oscillator enable circuit similar to Figure 3–17.

Common Misconception

Students often think that a gate output receives its HIGH or LOW voltage level from its input pin. You need to be reminded that each gate has its own totem-pole output arrangement and receives its voltage from V_{CC} or ground.

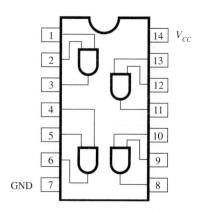

Figure 3–19   The 7408 quad two-input AND gate IC pin configuration.

CHAPTER 3 ∣ BASIC LOGIC GATES

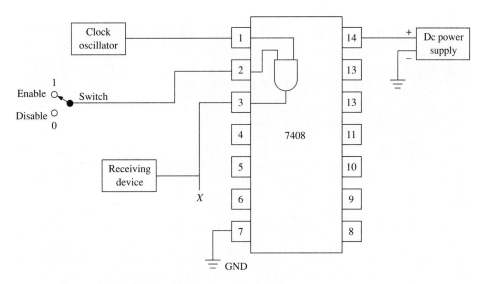

Figure 3–20 Using the 7408 TTL IC in a clock enable circuit.

In Figure 3–20, the first AND gate in the IC was used and the other three are ignored. The IC is powered by connecting pin 14 to the positive power supply and pin 7 to ground. The other connections are made by following the original design from Figure 3–17. The clock oscillator signal passes on to the receiving device when the switch is in the *enable* (1) position, and it stops when in the *disable* (0) position.

The pin configurations for some other logic gates are shown in Figure 3–21.

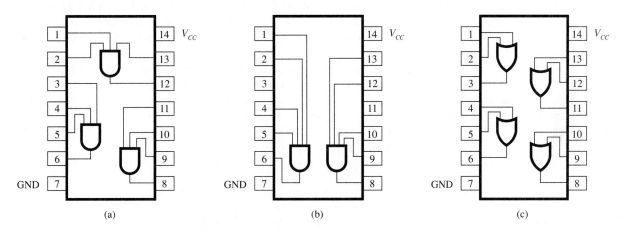

Figure 3–21 Pin configurations for other popular TTL and CMOS AND and OR gate ICs: (a) 7411 (74HC11); (b) 7421 (74HC21); (c) 7432 (74HC32).

3–6   Introduction to Troubleshooting Techniques

Like any other electronic device, integrated circuits and digital electronic circuits can go bad. **Troubleshooting** is the term given to the procedure used to find the **fault,** or *trouble,* in the circuits.

To be a good troubleshooter, you must first *understand the theory and operation* of the circuit, devices, and ICs that are suspected to be bad. If you understand how a

Figure 3–22 Logic pulser and logic probe.

particular IC is *supposed* to operate, it is a simple task to put the IC through a test or to exercise its functions to see if it operates as you expect.

There are two simple tools that we will start with to test the ICs and digital circuits. They are the logic pulser and logic probe (Figure 3–22). The **logic probe** has a metal tip that is placed on the IC pin, printed circuit board trace, or device lead that you want to test. It also has an indicator lamp that glows, telling you the digital level at that point. If the level is HIGH (1), the lamp glows brightly. If the level is LOW (0), the lamp goes out. If the level is **floating** (open circuit, neither HIGH nor LOW), the lamp is dimly lit. Table 3–5 summarizes the states of the logic probe.

Table 3–5	Logic Probe States
Logic Level	Indicator Lamp
HIGH (1)	On
LOW (0)	Off
Float	Dim

The **logic pulser** is used to provide digital pulses to a circuit being tested. By applying a pulse to a circuit and simultaneously observing a logic probe, you can tell if the pulse signal is getting through the IC or device as you would expect. As you become more and more experienced at troubleshooting, you will find that most IC and device faults are due to an open or short at the input or output terminals.

(a) (b)

(c) (d)

Figure 3–23 Four common printed circuit faults: (a) misalignment of pin 14; (b) cracked board; (c) solder bridge; (d) burned transistor.

Figure 3–23 shows four common problems that you will find on printed circuit boards that will cause opens or shorts. Figure 3–23(a) shows an IC that was inserted into its socket carelessly, causing pin 14 to miss its hole and act like an open. In Figure 3–23(b), the printed circuit board is obviously cracked, which causes an open circuit across each of the copper traces that used to cross over the crack. Poor soldering results in the *solder bridge* evident in Figure 3–23(c). In the center of this photo, you can see where too much solder was used, causing an electrical bridge between two adjacent IC pins and making them a short. Experienced troubleshooters will also visually inspect printed circuit boards for components that may appear to be darkened from excessive heat. Notice the four transistors in the middle of Figure 3–23(d). The one on the upper left looks charred and is probably burned out, thus acting like an open.

The following troubleshooting examples will illustrate some basic troubleshooting techniques using the logic probe and pulser.

Helpful Hint

You should be aware that these troubleshooting examples assume that the IC is removed from the circuit board. In-circuit testing will often give false readings because of the external circuitry connected to the IC. In that case the circuit schematic must be studied to determine how the other ICs may be affecting the readings.

EXAMPLE 3–5

The integrated-circuit AND gate in Figure 3–24 is suspected of having a fault and you want to test it. What procedure should you follow?

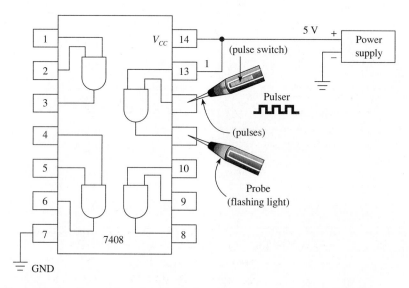

Figure 3–24   Connections for troubleshooting one gate of a quad AND IC.

Solution: First you apply power to V_{CC} (pin 14) and GND (pin 7). Next you want to check each AND gate with the pulser/probe. Because it takes a HIGH (1) on *both* inputs to an AND gate to make the output go HIGH, if we put a HIGH (+5 V) on one input and pulse the other, we would expect to get pulses at the output of the gate. Figure 3–24 shows the connections to test one of the gates of a quad AND IC. When the pulser is put on pin 12, the light in the end of the probe flashes at the same speed as the pulser, indicating that the AND gate is passing the pulses through the gate (similar in operation to the clock enable circuit of Figure 3–17).

The next check is to reverse the connections to pins 12 and 13 and check the probe. If the probe still flashes, that gate is okay. Proceed to the other three gates and follow the same procedure. When one of the gate outputs does not flash, you have found the fault.

As mentioned earlier, *the key to troubleshooting an IC is understanding how the IC works.*

E X A M P L E   3 – 6

Sketch the connections for troubleshooting the first gate of a 7421 dual AND gate.

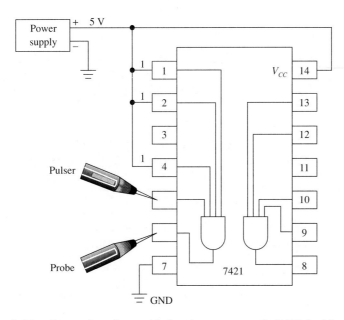

Figure 3–25   Connections for troubleshooting one gate of a 7421 dual four-input AND gate.

Solution: The connections are shown in Figure 3–25. The probe should be flashing if the gate is good. Check each of the four inputs with the pulser by keeping three inputs high and pulsing the fourth while you look at the probe. In any case, if the probe does not flash, you have found a bad gate.

EXAMPLE 3–7

Sketch the connections for troubleshooting the first gate of a 7432 quad OR gate.

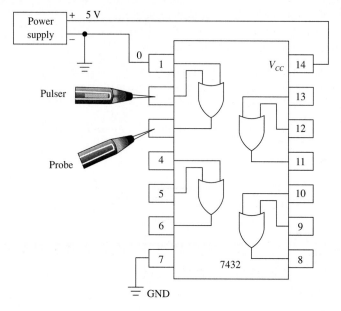

Figure 3–26 Connections for troubleshooting one OR gate of a 7432 IC.

Solution: The connections are shown in Figure 3–26. The probe should be flashing if the gate is good. Notice that the second input to the OR gate being checked is connected to a LOW (0) instead of a HIGH. The reason for this is that the output would *always* be HIGH if one input were connected HIGH. Since one input is connected LOW instead, the output will flash together with the pulses from the logic pulser if the gate is good.

Review Questions

3–8. Which pins on the 7408 AND IC are used for power supply connections, and what voltage levels are placed on those pins?

3–9. How is a *logic probe* used to troubleshoot digital ICs?

3–10. How is a *logic pulser* used to troubleshoot digital ICs?

Summary

In this chapter we have learned that

1. The AND gate requires that all inputs are HIGH in order to get a HIGH output.

2. The OR gate outputs a HIGH if any of its inputs are HIGH.

3. An effective way to measure the precise timing relationships of digital waveforms is with an oscilloscope or a logic analyzer.

4. Besides providing the basic logic functions, AND and OR gates can also be used to enable or disable a signal to pass from one point to another.

5. There are several integrated circuits available in both TTL and CMOS that provide the basic logic functions.

6. Two important troubleshooting tools are the logic pulser and the logic probe. The pulser is used to inject pulses into a circuit under test. The probe reads the level at a point in a circuit to determine if it is HIGH, LOW, or floating.

Glossary

Boolean Equation: An algebraic expression that illustrates the functional operation of a logic gate or combination of logic gates.

Disable: To disallow or deactivate a function or circuit.

Enable: To allow or activate a function or circuit.

Fault: The problem in a nonfunctioning electrical circuit. It is usually due to an open circuit, short circuit, or defective component.

Float: A logic level in a digital circuit that is neither HIGH nor LOW. It acts like an open circuit to anything connected to it.

Gate: The basic building block of digital electronics. The basic logic gate has one or more inputs and one output and is used to perform one of the following logic functions: AND, OR, NOR, NAND, INVERT, exclusive-OR, or exclusive-NOR.

Logic Probe: An electronic tool used in the troubleshooting procedure to indicate a HIGH, LOW, or float level at a particular point in a circuit.

Logic Pulser: An electronic tool used in the troubleshooting procedure to inject a pulse or pulses into a particular point in a circuit.

Troubleshooting: The work that is done to find the problem in a faulty electrical circuit.

Truth Table: A tabular listing that is used to illustrate all the possible combinations of digital input levels to a gate and the output that will result.

Problems

Section 3–1

3–1. Build the truth table for a three-input AND gate.

3–2. Build the truth table for a four-input AND gate.

3–3. If we were to build a truth table for an eight-input AND gate, how many different combinations of inputs would we have?

3–4. Describe in words the operation of an AND gate.

Section 3–2

3–5. Describe in words the operation of an OR gate.

3–6. Write the Boolean equation for

(a) A three-input AND gate

(b) A four-input AND gate

(c) A three-input OR gate

Section 3–3

3–7. Sketch the output waveform at X for the two-input AND gates shown in Figure P3–7.

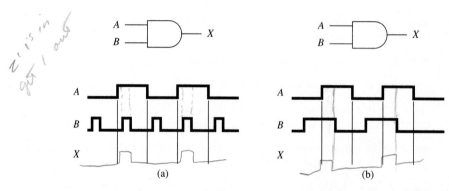

(a)　　　　　　　　(b)

Figure P3–7

3–8. Sketch the output waveform at X for the two-input OR gates shown in Figure P3–8.

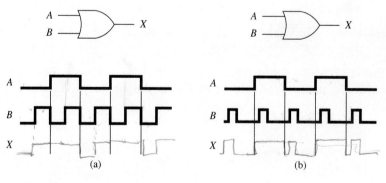

(a)　　　　　　　　(b)

Figure P3–8

3–9. Sketch the output waveform at X for the three-input AND gates shown in Figure P3–9.

CHAPTER 3 I BASIC LOGIC GATES

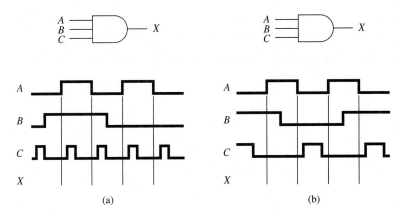

(a)   (b)

Figure P3–9

3–10. The input waveform at *A* is given for the two-input AND gates shown in Figure P3–10. Sketch the input waveform at *B* that will produce the output at *X*.

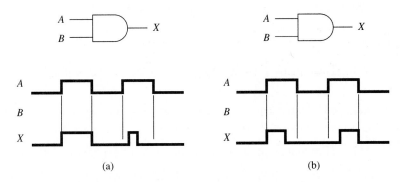

(a)   (b)

Figure P3–10

3–11. Repeat Problem 3–10 for the two-input OR gates shown in Figure P3–11.

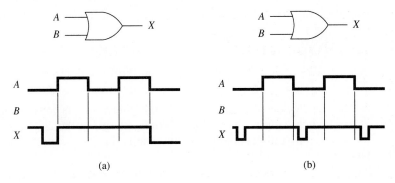

(a)   (b)

Figure P3–11

3–12. Using Figure P3–12, sketch the waveform for the *enable signal* that will allow pulses 2, 3 and 6, 7 to get through to the receiving device.

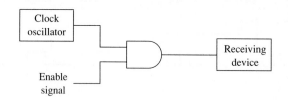

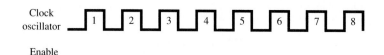

Figure P3–12

3–13. Repeat Problem 3–12, but this time sketch the waveform that will allow only the even pulses (2, 4, 6, 8) to get through.

Section 3–5

3–14. How many separate OR gates are contained within the 7432 TTL IC?

3–15. Sketch the actual pin connections to a 7432 quad two-input OR TTL IC to implement the circuit of Figure 3–18.

3–16. How many inputs are there on each AND gate of a 7421 TTL IC?

3–17. The 7421 IC is a 14-pin dual-in-line package (DIP). How many of the pins are *not* used for anything?

Section 3–6

T*   **3–18.** What are the three logic levels that can be indicated by a logic probe?

T   **3–19.** What is the function of the logic pulser?

T   **3–20.** When troubleshooting an OR gate such as the 7432, when the pulser is applied to one input, should the other input be connected HIGH or LOW? Why?

T   **3–21.** When troubleshooting an AND gate such as the 7408, when the pulser is connected to one input, should the other input be connected HIGH or LOW? Why?

C T   **3–22.** The clock enable circuit shown in Figure P3–22 is not working. The enable switch is up in the enable position. A logic probe is placed on the following pins and gets the following results. Find the cause of the problem.

---

*The letter **T** designates a problem that involves **T**roubleshooting.

CHAPTER 3 | BASIC LOGIC GATES

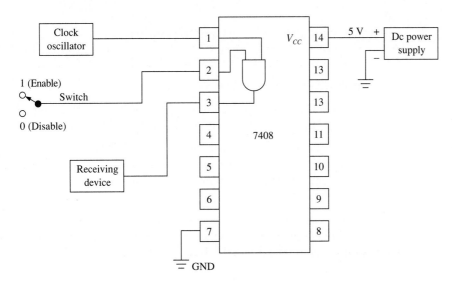

Figure P3–22

Probe on Pin	Indicator Lamp
1	Flashing
2	On
3	Off
7	Off
14	On

C T   **3–23.** Repeat Problem 3–22 for the following troubleshooting results.

Probe on Pin	Indicator Lamp
1	Flashing
2	Off
3	Off
7	Off
14	On

C T   **3–24.** Repeat Problem 3–22 for the following troubleshooting results.

Probe on Pin	Indicator Lamp
1	Flashing
2	On
3	Off
7	Dim
14	On

Schematic Interpretation Problems

See Appendix G for the schematic diagrams.

S   **3–25.** What are the component name and grid location of the 2-input AND gate and the 2-input OR gate in the Watchdog Timer schematic?

S   **3–26.** A logic probe is used to check the operation of the 2-input AND and OR gates in the Watchdog Timer circuit. If the probe indicator is ON for pin 2 of both gates and flashing on pin 1, what will pin 3 be for (a) the AND gate and (b) the OR gate?

S   **3–27.** If you wanted to check the power supply connections for the 8031 IC (U8) on the 4096/4196 circuit, which pins would you check, and what level should they be?

Answers to Review Questions

3–1. True

3–2. To illustrate how the output level of a gate responds to all possible input-level combinations

3–3. To depict algebraically the operation of a logic gate

3–4. All inputs must be LOW.

3–5. To illustrate graphically how the output levels change in response to input-level changes

3–6. When the level of an input signal will have no effect on the output

3–7. HIGH

3–8. Positive power supply of 5 V to pin 14, ground at 0 V to pin 7

3–9. It uses an indicator lamp to tell you the digital level whenever it is placed in a circuit.

3–10. It provides digital pulses to the circuit being tested, which can be observed using a logic probe.

4 Inverting Logic Gates

OUTLINE

4–1 The Inverter
4–2 The NAND Gate
4–3 The NOR Gate
4–4 Logic Gate Waveform Generation
4–5 Using Integrated-Circuit Logic Gates
4–6 Summary of the Basic Logic Gates and IEEE/IEC Standard Logic Symbols

Objectives

Upon completion of this chapter, you should be able to

- Describe the operation and use of inverter, NAND, and NOR gates.

- Construct truth tables for two-, three-, and four-input NAND and NOR gates.

- Draw timing diagrams for inverter, NAND, and NOR gates.

- Use the outputs of a Johnson shift counter to generate specialized waveforms utilizing various combinations of the five basic gates.

- Develop a comparison of the Boolean equations and truth tables for the five basic gates.

Introduction

Inverting logic gates are used like the basic AND and OR logic gates, except the inverting gates have *complemented* (inverted) outputs. Basically, there are three inverting logic gates: the *inverter,* the *NAND* (NOT–AND), and the *NOR* (NOT–OR). These gates are explained in this chapter and later combined with AND and OR gates to form the combinational logic used to provide the functional operations of complete digital systems.

4–1   The Inverter

The inverter is used to complement, or invert, a digital signal. It has a single input and a single output. If a HIGH level (1) comes in, it produces a LOW-level (0) output. If a LOW level (0) comes in, it produces a HIGH-level (1) output. The symbol and truth table for the inverter gate are shown in Figure 4–1.

The operation of the inverter is very simple and can be illustrated further by studying the timing diagram of Figure 4–2. The timing diagram graphically shows us the operation of the inverter. When the input is HIGH, the output is LOW, and when the input is LOW, the output is HIGH. The output waveform is, therefore, the exact complement of the input.

The Boolean equation for an inverter is written $X = \overline{A}$ *(X = NOT A)*. The *bar* over the *A* is an **inversion bar,** used to signify the **complement.** The inverter is sometimes referred to as the NOT gate.

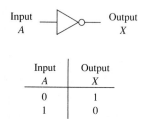

Input	Output
A	X
0	1
1	0

Figure 4–1   Inverter symbol and truth table.

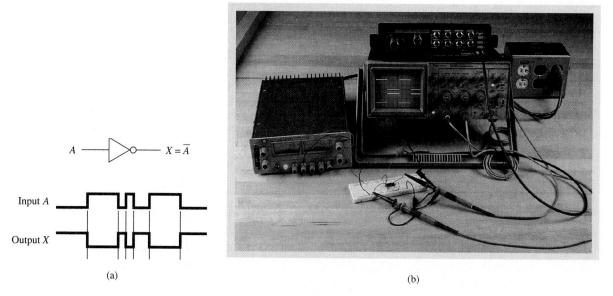

Figure 4–2   Timing analysis of an inverter gate: (a) waveform sketch; (b) oscilloscope display.

4–2   The NAND Gate

The operation of the NAND gate is the same as the AND gate except that its output is inverted. You can think of a NAND gate as an AND gate with an inverter at its output. The symbol for a NAND gate is made from an AND gate with a small circle (bubble) at its output, as shown in Figure 4–3.

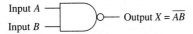

Figure 4–3 Symbol for a NAND gate.

In digital circuit diagrams, you will find the small circle used whenever complementary action (inversion) is to be indicated. The circle at the output acts just like an inverter, so a NAND gate can be drawn symbolically as an AND gate with an inverter connected to its output, as shown in Figure 4–4.

Figure 4–4 AND–INVERT equivalent of a NAND gate with $A = 1, B = 1$.

The Boolean equation for the NAND gate is written $X = \overline{AB}$. The inversion bar is drawn over (A and B), meaning that the output of the NAND is the complement of (A and B) [**NOT** (A and B)]. Because we are inverting the output, the truth table outputs in Table 4–1 will be the complement of the AND gate truth table outputs. The easy way to construct the truth table is to think of how an AND gate would respond to the inputs and then invert your answer. From Table 4–1 we can see that the output is LOW when *both* inputs *A and B* are HIGH (just the opposite of an AND gate). Also, the output is HIGH whenever either input is LOW.

Table 4–1 Two-Input NAND Gate Truth Table		
A	*B*	*X*
0	0	1
0	1	1
1	0	1
1	1	0

Helpful Hint

Some students find it easier to analyze a NAND gate by solving it as an AND gate and then inverting the result.

NAND gates can also have more than two inputs. Figure 4–5 shows three- and eight-input NAND gate symbols. The truth table for a three-input NAND gate (Table 4–2) shows that the output is always HIGH unless *all* inputs go HIGH.

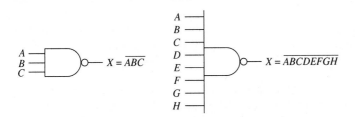

Figure 4–5 Symbols for three- and eight-input NAND gates.

| Table 4–2 | Truth Table for a | | |
| Three-Input NAND Gate | | | |
A	B	C	X
0	0	0	1
0	0	1	1
0	1	0	1
0	1	1	1
1	0	0	1
1	0	1	1
1	1	0	1
1	1	1	0

Timing analysis can also be used to illustrate the operation of NAND gates. The following examples will contribute to your understanding.

EXAMPLE 4–1

Sketch the output waveform at X for the NAND gate shown in Figure 4–6, with the given input waveforms in Figure 4–7.

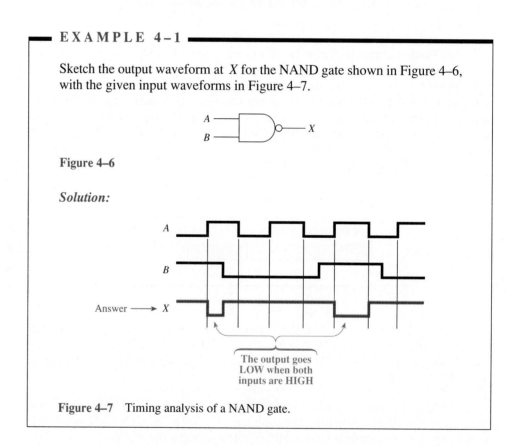

Figure 4–6

Solution:

Answer ⟶ X

The output goes LOW when both inputs are HIGH

Figure 4–7 Timing analysis of a NAND gate.

EXAMPLE 4–2

Sketch the output waveform at X for the NAND gate shown in Figure 4–8, with the given input waveforms at A, B, and Control.

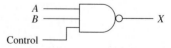

Figure 4–8

CHAPTER 4 | INVERTING LOGIC GATES

Solution: In Figure 4–9 the Control input waveform is used to *enable/disable* the NAND gate. When it is LOW, the output is stuck HIGH. When it goes HIGH, the output will respond LOW when *A* and *B* go HIGH.

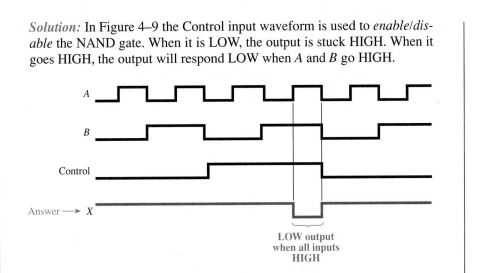

Figure 4–9 Timing analysis of a NAND gate with a Control input.

4–3   The NOR Gate

The operation of the NOR gate is the same as that of the OR gate except that its output is inverted. You can think of a NOR gate as an OR gate with an inverter at its output. The symbol for a NOR gate and its equivalent OR–INVERT symbol are shown in Figure 4–10.

Figure 4–10   NOR gate symbol and its OR–INVERT equivalent with $A = 0, B = 0$.

The Boolean equation for the NOR function is $X = \overline{A + B}$. The equation is stated "*X* equals *not* (*A* or *B*)." In other words, *X* is LOW if *A* or *B* is HIGH. The truth table for a NOR gate is given in Table 4–3. Notice that the output column is the complement of the OR gate truth table output column.

Table 4–3 Truth Table for a NOR Gate		
A	*B*	$X = \overline{A + B}$
0	0	1
0	1	0
1	0	0
1	1	0

Now let's study some timing analysis examples to get a better grasp of NOR gate operation.

Helpful Hint

To solve a timing analysis problem, it is useful to look at the gate's truth table to see what the *unique* occurrence is for that gate. In the case of the NOR, the odd occurrence is when the output goes HIGH due to all LOW inputs.

EXAMPLE 4–3

Sketch the output waveform at X for the NOR gate shown in Figure 4–11, with the given input waveforms in Figure 4–12.

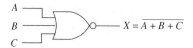

Figure 4–11

Solution:

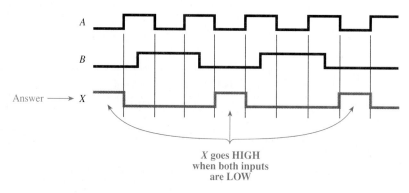

X goes HIGH
when both inputs
are LOW

Figure 4–12   NOR gate timing analysis.

EXAMPLE 4–4

Sketch the output waveform at X for the NOR gate shown in Figure 4–13 with the given input waveforms in Figure 4–14.

$$X = \overline{A + B + C}$$

Figure 4–13

Solution:

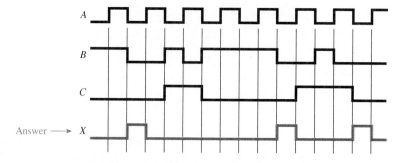

Figure 4–14   Three-input NOR gate timing analysis.

Sketch the waveform at the *B* input of the gate shown in Figure 4–15 that will produce the output waveform shown in Figure 4–16 for *X*.

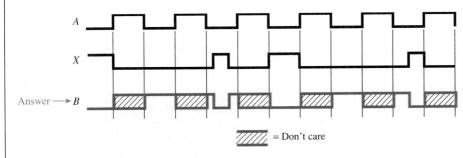

$$X = \overline{A + B}$$

Figure 4–15

Solution:

Answer $\longrightarrow$ *B*

$\diagdown\!\!\!\diagdown\!\!\!\diagdown$ = Don't care

Figure 4–16   Input waveform requirement to produce a specific output.

Review Questions

4–1. What is the purpose of an inverter in a digital circuit?

4–2. How does a NAND gate differ from an AND gate?

4–3. The output of a NAND gate is always HIGH unless *all* inputs are made _____ (HIGH, LOW).

4–4. Write the Boolean equation for a three-input NOR gate.

4–5. The output of a two-input NAND gate is _____ (HIGH, LOW) if $A = 1, B = 0$.

4–6. The output of a two-input NOR gate is _____ (HIGH, LOW) if $A = 0, B = 1$.

4–4   Logic Gate Waveform Generation

Using the basic gates, a clock oscillator, and a **repetitive waveform** generator circuit, we can create specialized waveforms to be used in digital control and sequencing circuits. A popular general-purpose repetitive **waveform generator** is the **Johnson shift counter,** whose operation is explained in detail in Chapter 13. For now, all we need are the output waveforms from it so that we may use them to create our own specialized waveforms.

The Johnson shift counter that we will use outputs eight separate repetitive waveforms: A, B, C, D and their complements, $\overline{A}, \overline{B}, \overline{C}, \overline{D}$. The input to the Johnson shift counter is a clock oscillator (C_p). Figure 4–17 shows a Johnson shift counter with its input and output waveforms.

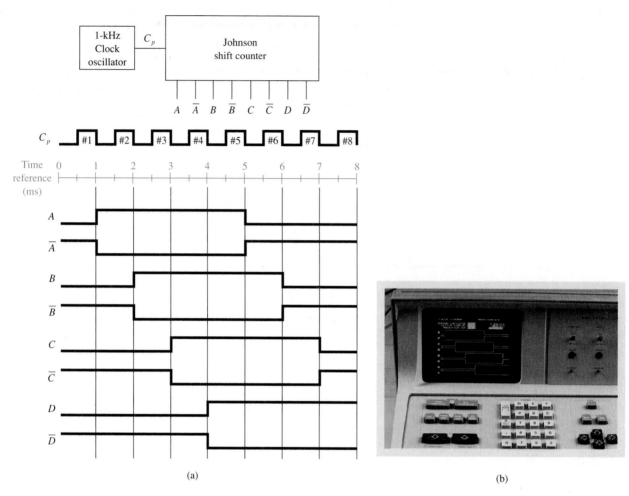

(a)

(b)

Figure 4–17 Johnson shift counter waveform generation: (a) waveform sketch; (b) logic analyzer display.

The clock oscillator produces the C_p waveform, which is input to the Johnson shift counter. The shift counter uses C_p and internal circuitry to generate the eight repetitive output waveforms shown.

Now, if one of those waveforms is exactly what you want, you are all set. But let's say we need a waveform that is HIGH for 3 ms, from 2 until 5 on the millisecond time reference scale. Looking at Figure 4–17, we can see that this waveform is not available.

Using some logic gates, however, will enable us to get any waveform that we desire. In this case, if we feed the A and B waveforms into an AND gate, we will get our HIGH level from 2 to 5, as shown in Figure 4–18.

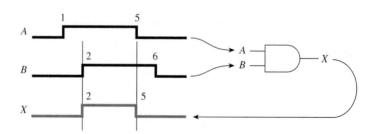

Figure 4–18 Generating a 3-ms HIGH pulse using an AND gate and a Johnson shift counter.

Working through the following examples will help you to understand logic gate operation and waveform generation.

Team Discussion

Could we obtain a LOW pulse from 4 to 5 instead of a HIGH by using the complemented signals of *A* and *D*?

■ **EXAMPLE 4–6** ■

Which Johnson counter outputs will you connect to an AND gate to get a 1-ms HIGH-level output from 4 to 5 ms?

Solution: Referring to Figure 4–17, we see that the two waveforms that are *both* HIGH from 4 to 5 ms are *A* and *D*; therefore, the circuit of Figure 4–19 will give us the required output.

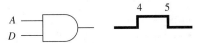

Figure 4–19   Solution to Example 4–6.

■ **EXAMPLE 4–7** ■

Which Johnson counter outputs must be connected to a three-input AND gate to enable just the C_p 4 pulse to be output?

Solution: Referring to Figure 4–17, we see that the *C* and $\overline{D}$ waveforms are both HIGH only during the C_p 4 *period*. To get just the C_p 4 *pulse*, you must provide C_p as the third input. Now, when you look at all three input waveforms, you see that they are all HIGH only during the C_p 4 *pulse* (see Figure 4–20).

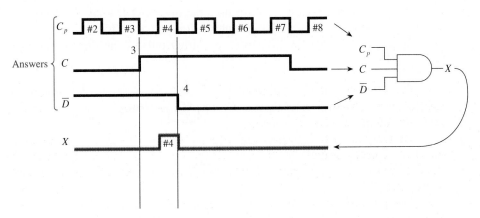

Figure 4–20   Solution to Example 4–7.

■ **EXAMPLE 4–8** ■

Sketch the output waveform that will result from inputting A, $\overline{B}$, and $\overline{C}$ into the three-input OR gate shown in Figure 4–21.

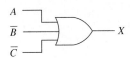

Figure 4–21

Solution: The output of an OR gate is always HIGH unless *all* inputs are LOW. Therefore, the output is always HIGH except between 5 and 6, as shown in Figure 4–22.

Figure 4–22   Solution to Example 4–8.

EXAMPLE 4–9

Sketch the output waveform that will result from inputting C_p, $\overline{B}$, and C into the NAND gate shown in Figure 4–23.

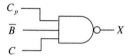

Figure 4–23

Solution: From reviewing the truth table of a NAND gate, we determine that the output is always HIGH unless *all* inputs are HIGH. Therefore, the output will always be HIGH except during pulse 7, as shown in Figure 4–24.

Figure 4–24   Solution to Example 4–9.

EXAMPLE 4–10

Team Discussion

Which of the three inputs could we ground and still get the same answer?

Sketch the output waveforms that will result from inputting A, B, and D into the NOR gate shown in Figure 4–25.

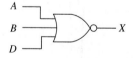

Figure 4–25

Solution: Reviewing the truth table for a NOR gate, we determine that the output is always LOW except when *all* inputs are LOW. Therefore, the output will always be LOW except from 0 to 1, as shown in Figure 4–26.

Figure 4–26   Solution to Example 4–10.

E X A M P L E   4 – 1 1

Sketch the output waveforms for the gates shown in Figure 4–27. The inputs are connected to the Johnson shift counter of Figure 4–17.

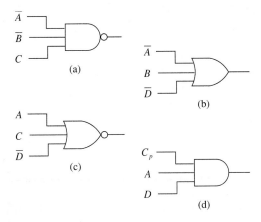

Figure 4–27

Solution: The output waveforms are shown in Figure 4–28.

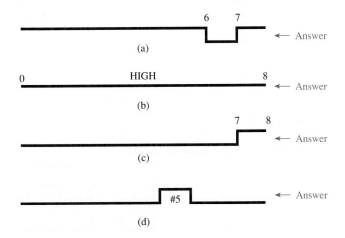

Figure 4–28   Solution to Example 4–11.

Determine which shift counter waveforms from Figure 4–17 will produce the output waveforms shown in Figure 4–29.

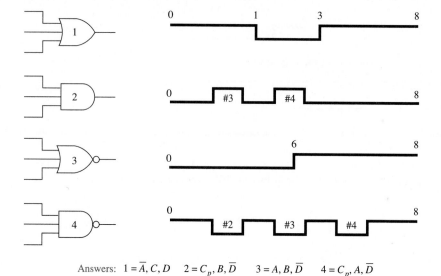

Answers: $1 = \overline{A}, C, D$ $2 = C_p, B, \overline{D}$ $3 = A, B, \overline{D}$ $4 = C_p, A, \overline{D}$

Figure 4–29   Solution to Example 4–12.

By using combinations of gates, we can obtain more specialized waveforms. Sketch the output waveforms for the circuit shown in Figure 4–30.

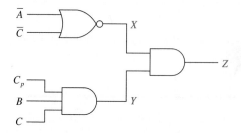

Figure 4–30

Solution: The output waveforms are shown in Figure 4–31.

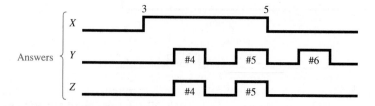

Figure 4–31   Solution to Example 4–13.

CHAPTER 4 I INVERTING LOGIC GATES

Sketch the output waveforms for the circuit shown in Figure 4–32.

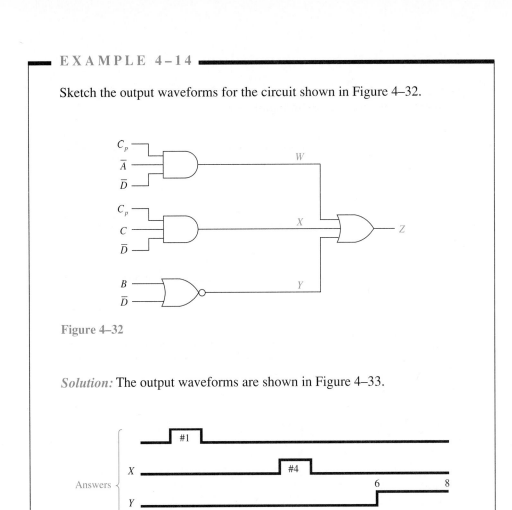

Figure 4–32

Solution: The output waveforms are shown in Figure 4–33.

Figure 4–33 Solution to Example 4–14.

4–5 Using Integrated-Circuit Logic Gates

All the logic gates are available in various configurations in the TTL and CMOS families. To list just a few: The 7404 TTL and the 4049 CMOS are **hex** (six) inverter ICs, the 7400 TTL and the 4011 CMOS are **quad** (four) two-input NAND ICs, and the 7402 TTL and the 4001 CMOS are quad two-input NOR ICs. Other popular NAND and NORs are available in three-, four-, and eight-input configurations. Consult a TTL or CMOS data manual for the availability and pin configuration of these ICs. The pin configurations for the hex inverter, the quad NOR, and the quad NAND are given in Figures 4–34 and 4–35. (High-speed CMOS 74HC04, 74HC00, and 74HC02 have the same pin configuration as the TTL ICs.)

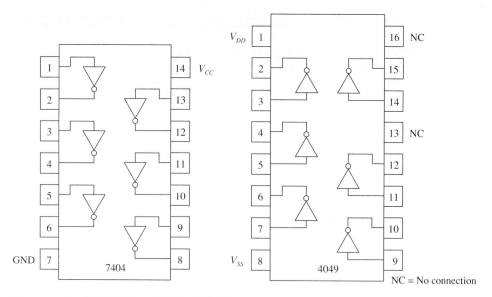

Figure 4–34 7404 TTL and 4049 CMOS inverter pin configurations.

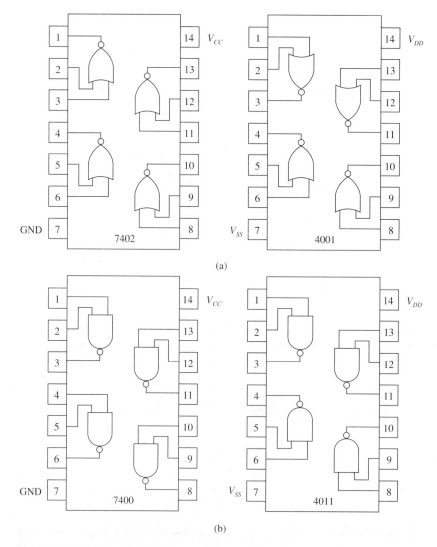

(a)

(b)

Figure 4–35 (a) 7402 TTL NOR and 4001 CMOS NOR pin configurations; (b) 7400 TTL NAND and 4011 CMOS NAND pin configurations.

EXAMPLE 4-15

Draw the external connections to a 4011 CMOS IC to form the circuit shown in Figure 4-36.

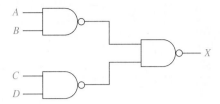

Figure 4-36

Solution: Referring to Figure 4-37, notice that V_{DD} is connected to the +5-V supply and V_{SS}, to ground. According to the CMOS data manual, V_{DD} can be any positive voltage from +3 to +15 V with respect to V_{SS} (usually ground).

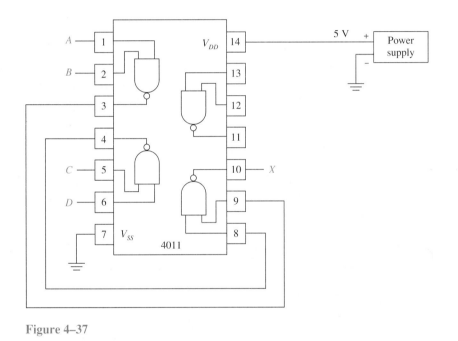

Figure 4-37

4-6 Summary of the Basic Logic Gates and IEEE/IEC Standard Logic Symbols

By now you should have a thorough understanding of the basic logic gates: inverter, AND, OR, NAND, and NOR. In Chapter 5 we will combine several gates to form complex logic functions. Figure 4-38(a) shows the author designing a logic system using a computer-aided design (CAD) system to combine the functions of the NAND, AND, and OR gates shown in Figure 4-38(b).

Because the basic logic gates are the building blocks for larger-scale integrated circuits and digital systems, it is very important that the operation of these gates be second nature to you.

(a)

(b)

Figure 4–38   Computer-aided design (CAD): (a) the author designing with a CAD system; (b) final combinational circuit.

A summary of the basic logic gates is given in Figure 4–39. You should memorize these logic symbols, Boolean equations, and truth tables. Also, a table of the most common integrated-circuit gates in the TTL and CMOS families is given in Table 4–4. You will need to refer to a TTL or CMOS data book for the pin layout and specifications.

Table 4–4   Common IC Gates in the TTL and CMOS Families

Gate Name	Number of Inputs per Gate	Number of Gates per Chip	Standard TTL	Standard CMOS	High-Speed CMOS
			Part Number		
Inverter	1	6	7404	4069	74HC04
AND	2	4	7408	4081	74HC08
	3	3	7411	4073	74HC11
	4	2	7421	4082	—
OR	2	4	7432	4071	74HC32
	3	3	—	4075	74HC4075
	4	2	—	4072	—
NAND	2	4	7400	4011	74HC00
	3	3	7410	4013	74HC10
	4	2	7420	4012	74HC20
	8	1	7430	4068	—
	12	1	74134	—	—
	13	1	74133	—	—
NOR	2	4	7402	4001	74HC02
	3	3	7427	4025	74HC27
	4	2	7425	4002	74HC4002
	5	2	74260	—	—
	8	1	—	4078	—

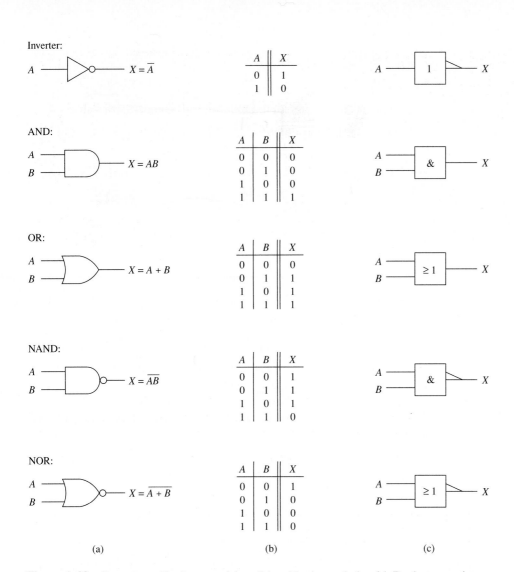

Inverter:

$$A \longrightarrow \!\!\!\!\!\!\!\!\triangleright\!\!o\!\!\longrightarrow X = \overline{A}$$

A	X
0	1
1	0

$$A \longrightarrow \boxed{1} \!\!\!\rangle\!\!\longrightarrow X$$

AND:

$$\begin{array}{c} A \longrightarrow \\ B \longrightarrow \end{array} \!\!\!\!\!\!\!\!\!\!\!\! \longrightarrow X = AB$$

A	B	X
0	0	0
0	1	0
1	0	0
1	1	1

$$\begin{array}{c} A \longrightarrow \\ B \longrightarrow \end{array} \boxed{\&} \longrightarrow X$$

OR:

$$\begin{array}{c} A \longrightarrow \\ B \longrightarrow \end{array} \!\!\!\!\!\!\!\!\!\!\!\! \longrightarrow X = A + B$$

A	B	X
0	0	0
0	1	1
1	0	1
1	1	1

$$\begin{array}{c} A \longrightarrow \\ B \longrightarrow \end{array} \boxed{\geq 1} \longrightarrow X$$

NAND:

$$\begin{array}{c} A \longrightarrow \\ B \longrightarrow \end{array} \!\!\!\!\!\!\!\!\!\!\!\! o\!\!\longrightarrow X = \overline{AB}$$

A	B	X
0	0	1
0	1	1
1	0	1
1	1	0

$$\begin{array}{c} A \longrightarrow \\ B \longrightarrow \end{array} \boxed{\&} \!\!\!\rangle\!\!\longrightarrow X$$

NOR:

$$\begin{array}{c} A \longrightarrow \\ B \longrightarrow \end{array} \!\!\!\!\!\!\!\!\!\!\!\! o\!\!\longrightarrow X = \overline{A + B}$$

A	B	X
0	0	1
0	1	0
1	0	0
1	1	0

$$\begin{array}{c} A \longrightarrow \\ B \longrightarrow \end{array} \boxed{\geq 1} \!\!\!\rangle\!\!\longrightarrow X$$

(a)   (b)   (c)

Figure 4–39   Summary of logic gates: (a) traditional logic symbols with Boolean equation; (b) truth tables; (c) IEEE/IEC standard logic symbols.

Also, in Figure 4–39(c) we introduce the IEEE/IEC standard logic symbols. This new standard for logic symbols was developed in 1984. It uses a method of determining the complete logical operation of a device just by interpreting the notations on the symbol for the device. This includes the basic gates as well as the more complex digital logic functions. Unfortunately, this new standard has not achieved widespread use, but you will see it used in some new designs. Most digital IC data books will show both the traditional and the new standard logic symbols, although most circuit schematics still use the traditional logic symbols. For this reason, the summary in Figure 4–39 shows both logic symbols, but throughout the remainder of this text we will use the traditional logic symbols. (A complete description of the IEEE/IEC standard for logic symbols is provided in Appendix C.)

4–7. What is the function of the Johnson shift counter in this chapter?

4–8. What are the part numbers of a TTL inverter IC and a CMOS NOR IC?

4–9. What type of logic gate is contained within the 7410 IC? the 74HC27 IC?

Summary

In this chapter we have learned that

1. An inverter provides an output that is the complement of its input.

2. A NAND gate outputs a LOW when all of its inputs are HIGH.

3. A NOR gate outputs a HIGH when all of its inputs are LOW.

4. Specialized waveforms can be created by using a repetitive waveform generator and the basic gates.

5. Manufacturers' data manuals are used by the technician to find the pin configuration and operating characteristics for the ICs used in modern circuitry.

Glossary

Complement: A change to the opposite digital state. A 1 becomes a 0, a 0 becomes a 1.

Hex: When dealing with integrated circuits, a term specifying that there are *six* gates on a single IC package.

Inversion Bar: A line over variables in a Boolean equation signifying that the digital state of the variables is to be complemented. For example, the output of a two-input NAND gate is written $X = \overline{AB}$.

Johnson Shift Counter: A digital circuit that produces several repetitive digital waveforms useful for specialized waveform generation.

NOT: When reading a Boolean equation, the word used to signify an inversion bar. For example, the equation $X = \overline{AB}$ is read "X equals NOT AB."

Quad: When dealing with integrated circuits, the term specifying that there are *four* gates on a single IC package.

Repetitive Waveform: A waveform that repeats itself after each cycle.

Waveform Generator: A circuit used to produce specialized digital waveforms.

Problems

4–1. For Figure P4–1, write the Boolean equation at X. If $A = 1$, what is X?

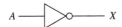

Figure P4–1

4–2. For Figure P4–2, write the Boolean equation at X and Z. If $A = 0$, what is X? What is Z?

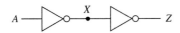

Figure P4–2

4–3. Using Figure P4–2, sketch the output waveform at X and Z if the timing waveform shown in Figure P4–3 is input at A.

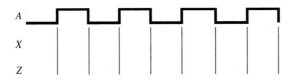

Figure P4–3

Section 4–2

4–4. For Figure P4–4, write the Boolean equation at X and Y.

Figure P4–4

4–5. Build a truth table for each gate in Figure P4–4.

4–6. Using Figure P4–4, sketch the output waveforms for X and Y, given the input waveforms shown in Figure P4–6.

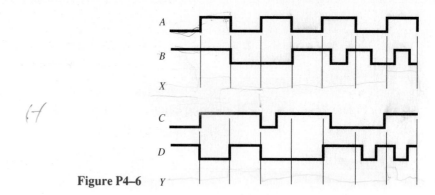

Figure P4–6

Section 4–3

4–7. Using Figure P4–7, sketch the waveforms at X and Y with the switches in the down (0) position. Repeat with the switches in the up (1) position.

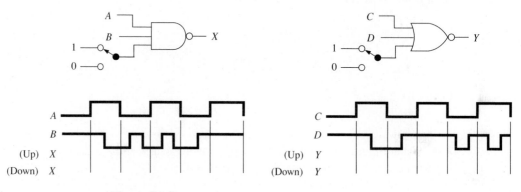

Figure P4–7

4–8. In words, what effect does the switch have on each circuit in Figure P4–7?

4–9. For Figure P4–9, write the Boolean equation at X and Y.

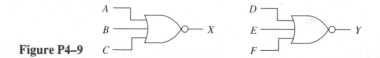

Figure P4–9

4–10. Make a truth table for the first NOR gate in Figure P4–9.

4–11. Referring to Figure P4–9, sketch the output at X and Y, given the input waveforms in Figure P4–11.

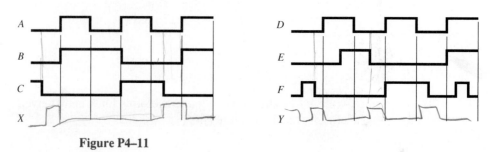

Figure P4–11

4–12. The Johnson shift counter outputs shown in Figure 4–17 are connected to the inputs of the logic gates shown in Figure P4–12. Sketch and label the output waveform at *U, V, W, X, Y,* and *Z*.

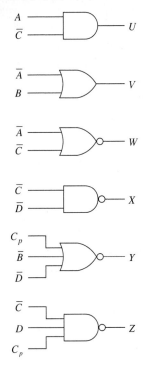

Figure P4–12

4–13. Repeat Problem 4–12 for the gates shown in Figure P4–13.

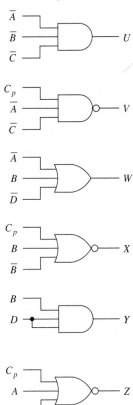

Figure P4–13

4–14. Using the Johnson shift counter outputs from Figure 4–17, label the inputs to the logic gates shown in Figure P4–14 so that they will produce the indicated output.

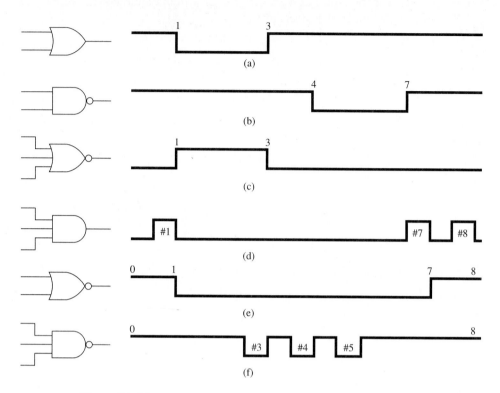

Figure P4–14

C

4–15. Determine which lines from the Johnson shift counter are required at the inputs of the circuits shown in Figure P4–15 to produce the waveforms at U, V, W, and X.

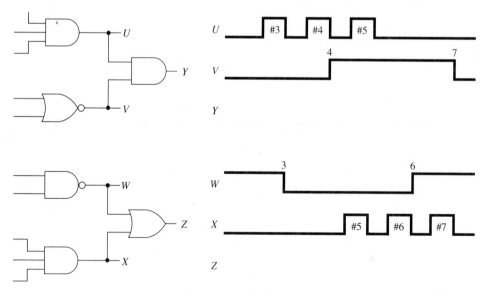

Figure P4–15

C   **4–16.** The waveforms at U, V, W, and X are given in Figure P4–15. Sketch the waveforms at Y and Z.

Section 4–5

4–17. Make the external connections to a 7404 inverter IC and a 7402 NOR IC to implement the function $X = \overline{\overline{A} + B}$.

T   **4–18.** When troubleshooting a NOR gate like the 7402, with the logic pulser applied to one input, should the other input be held HIGH or LOW? Why?

T   **4–19.** When troubleshooting a NAND gate like the 7400, with the logic pulser applied to one input, should the other input be held HIGH or LOW? Why?

T   **4–20.** The following data table was built by putting a logic probe on every pin of the hex inverter shown in Figure P4–20. Are there any problems with the chip? If so, which gate(s) are bad?

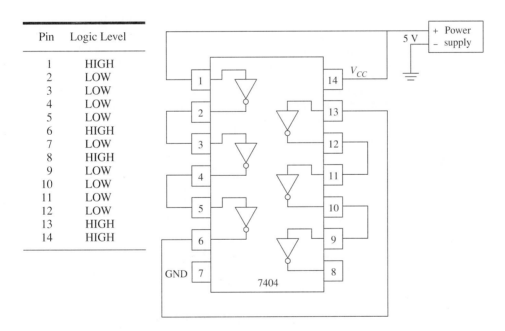

Pin	Logic Level
1	HIGH
2	LOW
3	LOW
4	LOW
5	LOW
6	HIGH
7	LOW
8	HIGH
9	LOW
10	LOW
11	LOW
12	LOW
13	HIGH
14	HIGH

Figure P4–20

C T   **4–21.** The logic probe in Figure P4–21 is always OFF (0) whether the switch is in the up or down position. Is the problem with the inverter or the NOR, or is there no problem?

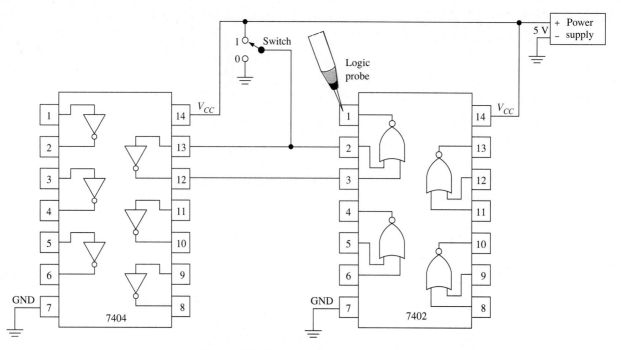

Figure P4–21

C T

4–22. Another circuit constructed the same way as Figure P4–21 causes the logic probe to come on when the switch is in the down (0) position. Further testing with the probe shows that pins 2 and 3 of the NOR IC are both LOW. Is anything wrong? If so, where is the fault?

T

4–23. Your company has purchased several of the 7430 eight-input NANDs shown in Figure P4–23. List the steps that you would follow to determine if they are all good ICs.

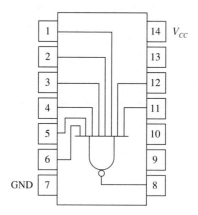

Figure P4–23

T

4–24. The data table above was built by putting a logic probe on every pin of the 7427 NOR IC shown in Figure P4–24 while it was connected in a digital circuit. Which gates, if any, are bad, and why?

CHAPTER 4 | INVERTING LOGIC GATES

Pin	Logic Level
1	LOW
2	LOW
3	LOW
4	LOW
5	LOW
6	HIGH
7	LOW
8	Flashing
9	HIGH
10	LOW
11	Flashing
12	HIGH
13	HIGH
14	HIGH

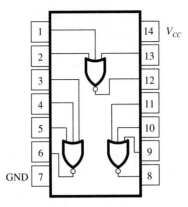

Figure P4–24

Schematic Interpretation Problems

See Appendix G for the schematic diagrams.

S **4–25.** On the 4096/4196 schematic there are several gates labeled U1. Why are they all labeled the same?

S **4–26.** Describe a method that you could use to check the operation of the inverter labeled U4:A of the Watchdog Timer. Assume that you have a dual-trace oscilloscope available for troubleshooting.

S **4–27.** Locate the line labeled RAM_SL at location D8 of the HC11D0 schematic. To get a HIGH level on that line, what level must the inputs to U8 be?

C S **4–28.** Locate the output pins labeled E and R/$\overline{W}$ on U1 of the HC11D0 schematic. During certain operations, line E goes HIGH and line R/$\overline{W}$ is then used to signify a READ operation if it is HIGH or a WRITE operation if it is LOW. For a READ operation, which line goes LOW: WE_B or OE_B?

Answers to Review Questions

4–1. An inverter is used to complement or invert a digital signal.

4–2. A NAND gate is an AND gate with an inverter on its output.

4–3. HIGH

4–4. $X = \overline{A + B + C}$

4–5. HIGH

4–6. LOW

4–7. It is used as a repetitive waveform generator.

4–8. 7404; 4001

4–9. Triple, three-input NAND gates; triple, three-input NOR gates

5 Boolean Algebra and Reduction Techniques

OUTLINE

5–1 Combinational Logic

5–2 Boolean Algebra Laws and Rules

5–3 Simplification of Combinational Logic Circuits Using Boolean Algebra

5–4 De Morgan's Theorem

5–5 The Universal Capability of NAND and NOR Gates

5–6 AND–OR–INVERT Gates for Implementing Sum-of-Products Expressions

5–7 Karnaugh Mapping

5–8 System Design Applications

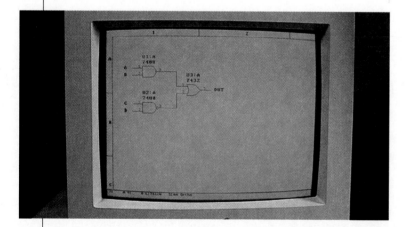

Objectives

Upon completion of this chapter, you should be able to

- Write Boolean equations for combinational logic applications.

- Utilize Boolean algebra laws and rules for simplifying combinational logic circuits.

- Apply De Morgan's theorem to complex Boolean equations to arrive at simplified equivalent equations.

- Design single-gate logic circuits by utilizing the universal capability of NAND and NOR gates.

- Troubleshoot combinational logic circuits.

- Implement sum-of-products expressions utilizing AND–OR–INVERT gates.

- Utilize the Karnaugh mapping procedure to systematically reduce complex Boolean equations to their simplest form.

- Describe the steps involved in solving a complete system design application.

Introduction

Generally, you will find that the simple gate functions AND, OR, NAND, NOR, and INVERT are not enough by themselves to implement the complex requirements of digital systems. The basic gates will be used as the building blocks for the more complex logic that is implemented by using combinations of gates called *combinational logic*.

5–1 Combinational Logic

Combinational logic employs the use of two or more of the basic logic gates to form a more useful, complex function. For example, let's design the logic for an automobile warning buzzer using combinational logic. The criterion for the activation of the warning buzzer is as follows: The buzzer activates if the headlights are on *and* the driver's door is opened, *or* if the key is in the ignition *and* the door is opened.

The logic function for the automobile warning buzzer is illustrated symbolically in Figure 5–1. The figure illustrates a *combination* of logic functions that can be written as a Boolean equation in the form

$$B = K \text{ and } D \quad \text{or} \quad H \text{ and } D$$

which is also written as

$$B = KD + HD$$

This equation can be stated as "*B* is HIGH if *K and D* are HIGH *or* if *H and D* are HIGH."

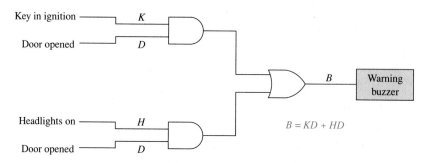

Figure 5–1 Combinational logic requirements for an automobile warning buzzer.

When you think about the operation of the warning buzzer, you may realize that it is activated whenever the door is opened *and* either the key is in the ignition *or* the headlights are on. If you can realize that, you have just performed your first **Boolean reduction** using Boolean algebra. (The systematic reduction of logic circuits is performed using Boolean algebra, named after the nineteenth-century mathematician George Boole.)

The new Boolean equation becomes $B = D$ and $(K$ or $H)$, also written as $B = D(K + H)$. (Notice the use of parentheses. Without them, the equation would imply that the buzzer activates if the door is opened with the key in the ignition or any time the headlights are on, which is invalid. $B \neq DK + H$.) The new equation represents the same logic operation but is a simplified implementation because it requires only two logic gates, as shown in Figure 5–2.

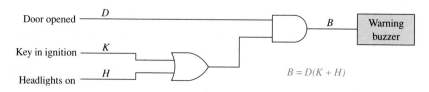

Figure 5–2 Reduced logic circuit for the automobile buzzer.

EXAMPLE 5–1

Write the Boolean logic equation and draw the logic circuit that represent the following function: A bank burglar alarm (*A*) is to activate if it is after banking hours (*H*) *and* the front door (*F*) is opened *or* if it is after banking hours (*H*) and the vault door is opened (*V*).

Solution: A = HF + HV. The logic circuit is shown in Figure 5–3.

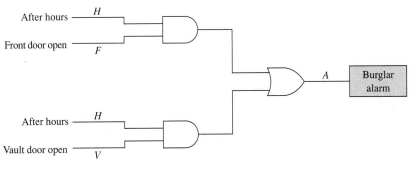

Figure 5–3 Solution to Example 5–1.

EXAMPLE 5–2

Using common reasoning, reduce the logic function described in Example 5–1 to a simpler form.

Solution: The alarm is activated if it is after banking hours *and* if either the front door is opened *or* the vault door is opened (see Figure 5–4). The simplified equation is written as

$$A = H(F + V) \qquad \text{(Notice the use of parentheses.)}$$

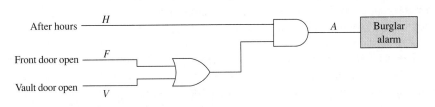

Figure 5–4 Solution to Example 5–2.

EXAMPLE 5–3

Draw the logic circuit that could be used to implement the following Boolean equation:

$$X = AB + C(M + N)$$

Solution: The logic circuit is shown in Figure 5–5.

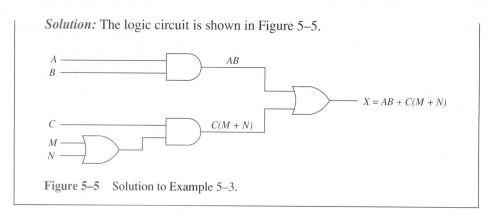

Figure 5–5 Solution to Example 5–3.

E X A M P L E 5 – 4

Write the Boolean equation for the logic circuit shown in Figure 5–6.

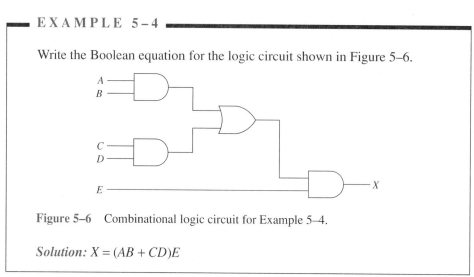

Figure 5–6 Combinational logic circuit for Example 5–4.

Solution: $X = (AB + CD)E$

5–2   Boolean Algebra Laws and Rules

Boolean algebra uses many of the same laws as those of ordinary algebra. The OR function ($X = A + B$) is *Boolean addition*, and the AND function ($X = AB$) is *Boolean multiplication*. The following three laws are the same for Boolean algebra as they are for ordinary algebra:

1. *Commutative law of addition: $A + B = B + A$, and multiplication: $AB = BA$.* These laws mean that the order of ORing or ANDing does not matter.

Helpful Hint

The distributive law shown for four variables is sometimes called the *FOIL* method (first, outside, inside, last).

$$O$$
$$\downarrow F \downarrow$$
$$(A + B)(C + D) =$$
$$\uparrow I \uparrow$$
$$L$$
$$AC + AD + BC + BD$$

2. *Associative law of addition: $A + (B + C) = (A + B) + C$, and multiplication: $A(BC) = (AB)C$.* These laws mean that the grouping of several variables ORed or ANDed together does not matter.

3. *Distributive law: $A(B + C) = AB + AC$, and $(A + B)(C + D) = AC + AD + BC + BD$.* These laws show methods for expanding an equation containing ORs and ANDs.

These three laws hold true for any number of variables. For example, the commutative law can be applied to $X = A + BC + D$ to form the equivalent equation $X = BC + A + D$.

You may wonder when you will need to use one of the laws. Later in this chapter you will see that by using these laws to rearrange Boolean equations you will be able to change some combinational logic circuits to simpler **equivalent circuits** using fewer gates. You can gain a better understanding of the application of these laws by studying Figures 5–7 to 5–12.

Figure 5–7   Using the commutative law of addition to rearrange an OR gate.

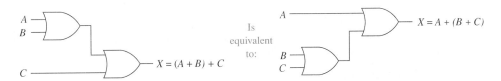

Figure 5–8   Using the commutative law of multiplication to rearrange an AND gate.

Figure 5–9   Using the associative law of addition to rearrange the grouping of OR gates.

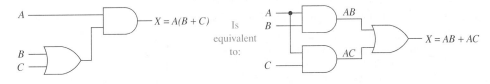

Figure 5–10   Using the associative law of multiplication to rearrange the grouping of AND gates.

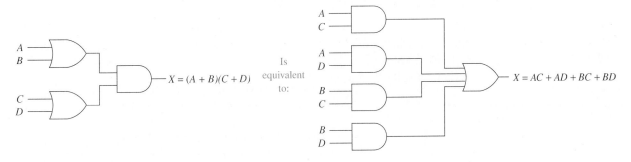

Figure 5–11   Using the distributive law to form an equivalent circuit.

Figure 5–12   Using the distributive law to form an equivalent circuit.

Besides the three basic laws, there are several rules concerning Boolean algebra. The rules of Boolean algebra allow us to combine or eliminate certain variables in the equation to form simpler equivalent circuits.

The following example illustrates the use of the first Boolean rule, which states that anything ANDed with a 0 will always output a 0.

EXAMPLE 5–5

A bank burglar alarm (B) will activate if it is after banking hours (A) and someone opens the front door (D). The logic level of the variable A is 1 after banking hours and 0 during banking hours. Also, the logic level of the variable D is 1 if the door sensing switch is opened and 0 if the door sensing switch is closed. The Boolean equation is therefore $B = AD$. The logic circuit to implement this function is shown in Figure 5–13(a).

Figure 5–13   (a) Logic circuit for a simple burglar alarm; (b) disabling the burglar alarm by making $D = 0$.

Later, a burglar comes along and puts tape on the door sensing switch, holding it closed so that it always puts out a 0 logic level. Now the Boolean equation *(B = AD)* becomes $B = A \lozenge 0$ because the door sensing switch is always 0. The alarm will never sound in this condition because one input to the AND gate is always 0. The burglar must have studied the Boolean rules and realized that anything ANDed with a 0 will output a 0, as shown in Figure 5–13(b).

Example 5–5 helped illustrate the reasoning for Boolean Rule 1. The other nine rules can be derived using common sense and knowing basic gate operation.

Helpful Hint

You should *make sense* out of these 10 rules—not simply memorize them.

Rule 1: Anything ANDed with a 0 is equal to 0 ($A \cdot 0 = 0$).

Rule 2: Anything ANDed with a 1 is equal to itself ($A \cdot 1 = A$). From Figure 5–14 we can see that, with one input tied to a 1, if the A input is 0, the X output is 0; if A is 1, X is 1; therefore, X is equal to whatever the logic level of A is ($X = A$).

$$X = A \cdot 1 = A$$

A	1	X
0	1	0
1	1	1

X equals A

Figure 5–14   Logic circuit and truth table illustrating Rule 2.

Rule 3: Anything ORed with a 0 is equal to itself ($A + 0 = A$). In Figure 5–15, because one input is always 0, if $A = 1$, $X = 1$, and if $A = 0$, $X = 0$; therefore, X is equal to whatever the logic level of A is ($X = A$).

$$X = A + 0 = A$$

A	0	X
0	0	0
1	0	1

X equals A

Figure 5–15   Logic circuit and truth table illustrating Rule 3.

CHAPTER 5 | BOOLEAN ALGEBRA AND REDUCTION TECHNIQUES

Rule 4: Anything ORed with a 1 is equal to 1 ($A + 1 = 1$). In Figure 5–16, because one input to the OR gate is always 1, the output is always 1, no matter what A is ($X = 1$).

A	1		X
0	1		1
1	1		1

$X = A + 1 = 1$ } X equals 1

Figure 5–16   Logic circuit and truth table illustrating Rule 4.

Rule 5: Anything ANDed with itself is equal to itself ($A \cdot A = A$). In Figure 5–17, since both inputs to the AND gate are A, if $A = 1$, 1 and 1 equals 1, and if $A = 0$, 0 and 0 equals 0. Therefore, X is equal to whatever the logic level of A is ($X = A$).

A	A		X
0	0		0
1	1		1

$X = A \cdot A = A$ } X equals A

Figure 5–17   Logic circuit and truth table illustrating Rule 5.

Rule 6: Anything ORed with itself is equal to itself ($A + A = A$). In Figure 5–18, because both inputs to the OR gate are A, if $A = 1$, 1 or 1 equals 1, and if $A = 0$, 0 or 0 equals 0. Therefore, X is equal to whatever the logic level of A is ($X = A$).

A	A		X
0	0		0
1	1		1

$X = A + A = A$ } X equals A

Figure 5–18   Logic circuit and truth table illustrating Rule 6.

Rule 7: Anything ANDed with its own complement equals 0. In Figure 5–19, because the inputs are complements of each other, one of them is always 0. With a zero at the input, the output is always 0 ($X = 0$).

A	$\overline{A}$		X
0	1		0
1	0		0

$X = A \cdot \overline{A} = 0$ } X equals 0

Figure 5–19   Logic circuit and truth table illustrating Rule 7.

Rule 8: Anything ORed with its own complement equals 1. In Figure 5–20, since the inputs are complements of each other, one of them is always 1. With a 1 at the input, the output is always 1 ($X = 1$).

A	$\overline{A}$		X
0	1		1
1	0		1

$X = A + \overline{A} = 1$ } X equals 1

Figure 5–20   Logic circuit and truth table illustrating Rule 8.

Rule 9: A variable that is complemented twice will return to its original logic level. As shown in Figure 5–21, when a variable is complemented once, it changes to the opposite logic level. When it is complemented a second time, it changes back to its original logic level ($\overline{\overline{A}} = A$).

$$X = \overline{\overline{A}} = A$$

A	$\overline{A}$	$\overline{\overline{A}}$	X
0	1	0	0
1	0	1	1

X equals A

Figure 5–21 Logic circuit and truth table illustrating Rule 9.

Rule 10: $A + \overline{A}B = A + B$ and $\overline{A} + AB = \overline{A} + B$. This rule differs from the others because it involves two variables. It is useful because, when an equation is in this form, one or more variables in the second term can be eliminated. The validity of these two equations is proven in Table 5–1. In each case, equivalence is demonstrated by showing that the truth table derived from the expression on the left side of the equation matches that on the right side.

Table 5–1 Using Truth Tables to Prove the Equations in Rule 10

A	B	$A + \overline{A}B$	$A + B$	A	B	$\overline{A} + AB$	$\overline{A} + B$
0	0	0	0	0	0	1	1
0	1	1	1	0	1	1	1
1	0	1	1	1	0	0	0
1	1	1	1	1	1	1	1
		↑	↑			↑	↑
		Equivalent				Equivalent	

Table 5–2 summarizes the laws and rules that relate to Boolean algebra. By using them, we can reduce complicated combinational logic circuits to their simplest form, as we will see in the next sections. The letters used in Table 5–2 are variables and were chosen arbitrarily. For example, $C + \overline{C}D = C + D$ is also a valid use of Rule 10(a).

Table 5–2 Boolean Laws and Rules for the Reduction of Combinational Logic Circuits

Laws
1. $A + B = B + A$
   $AB = BA$
2. $A + (B + C) = (A + B) + C$
   $A(BC) = (AB)C$
3. $A(B + C) = AB + AC$
   $(A + B)(C + D) = AC + AD + BC + BD$

Rules
1. $A \cdot 0 = 0$
2. $A \cdot 1 = A$
3. $A + 0 = A$
4. $A + 1 = 1$
5. $A \cdot A = A$
6. $A + A = A$
7. $A \cdot \overline{A} = 0$
8. $A + \overline{A} = 1$
9. $\overline{\overline{A}} = A$
10. (a) $A + \overline{A}B = A + B$
    (b) $\overline{A} + AB = \overline{A} + B$

5–1. How many gates are required to implement the following Boolean equations?

(a) $X = (A + B)C$

(b) $Y = AC + BC$

(c) $Z = (ABC + CD)E$

5–2. Which Boolean law is used to transform each of the following equations?

(a) $B + (D + E) = (B + D) + E$

(b) $CAB = BCA$

(c) $(B + C)(A + D) = BA + BD + CA + CD$

5–3. The output of an AND gate with one of its inputs connected to 1 will always output a level equal to the level at the other input. True or false?

5–4. The output of an OR gate with one of its inputs connected to 1 will always output a level equal to the level at the other input. True or false?

5–5. If one input to an OR gate is connected to 0, the output will always be 0 regardless of the level on the other input. True or false?

5–6. Use one of the forms of Rule 10 to transform each of the following equations:

(a) $\overline{B} + AB = ?$

(b) $B + \overline{B}C = ?$

5–3   Simplification of Combinational Logic Circuits Using Boolean Algebra

Often in the design and development of digital systems, a designer will start with simple logic rate requirements but add more and more complex gating, making the final design a complex combination of several gates, some having the same inputs. At that point the designer must step back and review the combinational logic circuit that has been developed and see if there are ways of reducing the number of gates without changing the function of the circuit. If an equivalent circuit can be formed with fewer gates, the cost of the circuit is reduced and its reliability is improved. This process is called the *reduction* or *simplification of combinational logic circuits* and is performed by using the laws and rules of Boolean algebra presented in the preceding section.

The following examples illustrate the use of Boolean algebra and present some techniques for the simplification of logic circuits.

EXAMPLE 5–6

The logic circuit shown in Figure 5–22 is used to turn on a warning buzzer at X based on the input conditions at A, B, and C. A simplified equivalent circuit that will perform the same function can be formed by using Boolean algebra. Write the equation of the circuit in Figure 5–22, simplify the equation, and draw the logic circuit of the simplified equation.

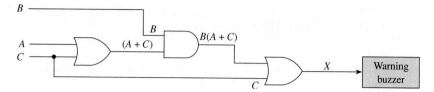

Figure 5–22 Logic circuit for Example 5–6.

Solution: The Boolean equation for X is

$$X = B(A + C) + C$$

To simplify, first apply Law 3 $[B(A + C) = BA + BC]$:

$$X = BA + BC + C$$

Next, factor a C from terms 2 and 3:

$$X = BA + C(B + 1)$$

Apply Rule 4 $(B + 1 = 1)$:

$$X = BA + C \cdot 1$$

Apply Rule 2 $(C \cdot 1 = C)$:

$$X = BA + C$$

Apply Law 1 $(BA = AB)$:

$$X = AB + C \leftarrow \text{simplified equation}$$

The logic circuit of the simplified equation is shown in Figure 5–23.

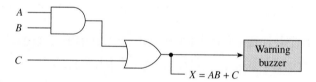

Figure 5–23 Simplified logic circuit for Example 5–6.

EXAMPLE 5–7

Repeat Example 5–6 for the logic circuit shown in Figure 5–24.

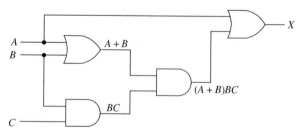

Figure 5–24 Logic circuit for Example 5–7.

Solution: The Boolean equation for X is

$$X = (A + B)BC + A$$

To simplify, first apply Law 3 $[(A + B)BC = ABC + BBC]$:

$$X = ABC + BBC + A$$

Apply Rule 5 $(B \cdot B = B)$:

$$X = ABC + BC + A$$

Factor a BC from terms 1 and 2:

$$X = BC(A + 1) + A$$

Apply Rule 4 $(A + 1 = 1)$:

$$X = BC \cdot 1 + A$$

Apply Rule 2 $(BC \cdot 1 = BC)$:

$$X = BC + A \quad \leftarrow \text{simplified equation}$$

The logic circuit for the simplified equation is shown in Figure 5–25.

Figure 5–25   Simplified logic circuit for Example 5–7.

EXAMPLE 5–8

Repeat Example 5–6 for the logic circuit shown in Figure 5–26.

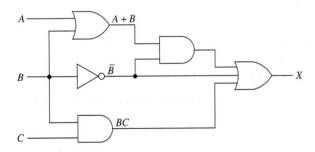

Figure 5–26   Logic circuit for Example 5–8.

Solution: The Boolean equation for X is

$$X = (A + B)\overline{B} + \overline{B} + BC$$

To simplify, first apply Law 3 $[(A + B)\overline{B} = A\overline{B} + B\overline{B}]$:

$$X = A\overline{B} + B\overline{B} + \overline{B} + BC$$

Apply Rule 7 $(B\overline{B} = 0)$:

$$X = A\overline{B} + 0 + \overline{B} + BC$$

Apply Rule 3 $(A\overline{B} + 0 = A\overline{B})$:

$$X = A\overline{B} + \overline{B} + BC$$

Factor a $\overline{B}$ from terms 1 and 2:

$$X = \overline{B}(A + 1) + BC$$

Apply Rule 4 ($A + 1 = 1$):

$$X = \overline{B} \cdot 1 + BC$$

Apply Rule 2 ($\overline{B} \cdot 1 = \overline{B}$):

$$X = \overline{B} + BC$$

Apply Rule 10(b) ($\overline{B} + BC = \overline{B} + C$):

$$X = \overline{B} + C \leftarrow \text{simplified equation}$$

The logic circuit of the simplified equation is shown in Figure 5–27.

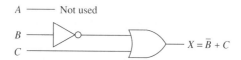

Figure 5–27   Simplified logic circuit for Example 5–8.

EXAMPLE 5–9

Repeat Example 5–6 for the logic circuit shown in Figure 5–28.

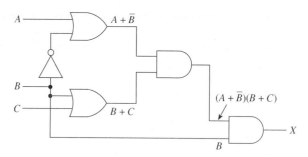

Figure 5–28   Logic circuit for Example 5–9.

Solution: The Boolean equation for X is

$$X = [(A + \overline{B})(B + C)]B$$

To simplify, first apply Law 3:

$$X = (AB + AC + \overline{B}B + \overline{B}C)B$$

The $\overline{B}B$ term can be eliminated using Rule 7 and then Rule 3:

$$X = (AB + AC + \overline{B}C)B$$

Apply Law 3 again:

$$X = ABB + ACB + \overline{B}CB$$

Apply Law 1:

$$X = ABB + ABC + \overline{B}BC$$

Apply Rules 5 and 7:

$$X = AB + ABC + 0 \cdot C$$

Apply Rule 1:

$$X = AB + ABC$$

Factor an AB from both terms:

$$X = AB(1 + C)$$

Apply Rule 4 and then Rule 2:

$$X = AB \quad \leftarrow \text{simplified equation}$$

The logic circuit of the simplified equation is shown in Figure 5–29.

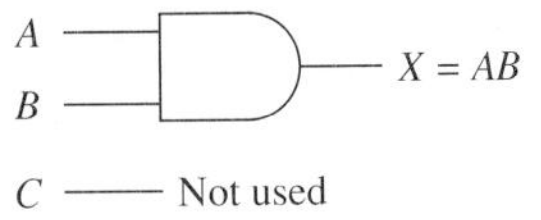

$$A$$
$$B$$
$$X = AB$$
$$C \quad \text{—— Not used}$$

Figure 5–29 Simplified logic circuit for Example 5–9.

5–4　De Morgan's Theorem

You may have noticed that we did not use NANDs or NORs in any of the logic circuits in Section 5–3. To simplify circuits containing NANDs and NORs, we need to use a theorem developed by the mathematician Augustus De Morgan. This theorem allows us to convert an expression having an inversion bar over two or more variables into an expression having inversion bars over single variables only. This allows us to use the rules presented in the preceding section for the simplification of the equation.

In the form of an equation, **De Morgan's theorem** is stated as follows:

$$\overline{A \cdot B} = \overline{A} + \overline{B}$$

$$\overline{A + B} = \overline{A} \cdot \overline{B}$$

Also, for three or more variables,

$$\overline{A \cdot B \cdot C} = \overline{A} + \overline{B} + \overline{C}$$

$$\overline{A + B + C} = \overline{A} \cdot \overline{B} \cdot \overline{C}$$

Basically, to use the theorem you break the bar over the variables and either change the AND to an OR or change the OR to an AND.

To prove to ourselves that this works, let's apply the theorem to a NAND gate and then compare the truth table of the equivalent circuit to that of the original NAND gate. As you can see in Figure 5–30, to use De Morgan's theorem on a NAND gate, first break the bar over the $A \cdot B$; then change the AND symbol to an OR. The new equation becomes $X = \overline{A} + \overline{B}$. Notice that **inversion bubbles** are used on the OR gate instead of inverters. By observing the truth tables of the two equations, we can see that the result in the X column is the same for both, which proves that they provide an equivalent output result.

Also, by looking at the two circuits we can say that *an AND gate with its output inverted is equivalent to an OR gate with its inputs inverted.* Therefore, the OR gate with inverted inputs is sometimes used as an alternative symbol for a NAND gate.

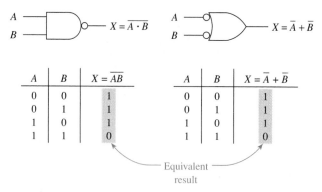

Figure 5–30   De Morgan's theorem applied to NAND gate produces 2 identical truth tables.

By applying De Morgan's theorem to a NOR gate, we will also produce two identical truth tables, as shown in Figure 5–31(a). Therefore, we can also think of an OR gate with its output inverted as being equivalent to an AND gate with its inputs inverted. The inverted input AND gate symbol is also sometimes used as an alternative to the NOR gate symbol.

When you write the equation for an AND gate with its inputs inverted, be careful to keep the inversion bar over each individual variable (not both) because $\overline{A \cdot B}$ is not equal to $\overline{A} \cdot \overline{B}$. (Prove that to yourself by building a truth table for both. Also, $\overline{A + B}$ is not equal to $\overline{A} + \overline{B}$.)

The question always arises: Why would a designer ever use an *inverted-input OR gate* symbol instead of a NAND? Or why use an *inverted-input AND gate* symbol instead of a NOR? In complex logic diagrams you will see both the inverted-input and the inverted-output symbols being used. The designer will use whichever symbol makes more sense for the particular application.

For example, referring to Figure 5–30, let's say you need a HIGH output level whenever either A or B is LOW. It makes sense to think of that function as an OR gate with inverted A and B inputs, but you could save two inverters by just using a NAND gate.

Also, referring to Figure 5–31(a), let's say you need a HIGH output whenever both A and B are LOW. You would probably use the *inverted-input AND gate* for your logic diagram because it makes sense logically, but you would use a NOR gate to actually implement the circuit because you could eliminate the inverters.

The alternative methods of drawing NANDs and NORs are also useful for the simplification of logic circuits. Take, for example, the circuit of Figure 5–31(b). By changing the NOR gate to an *inverted-input AND gate,* the inversion bubbles cancel and the equation becomes simply $X = ABCD$. Figure 5–31(c) summarizes the alternative representations for the inverter, NAND, and NOR gates.

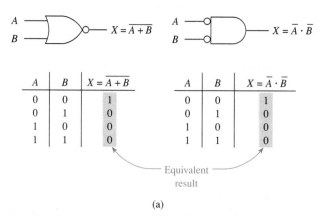

(a)

Figure 5–31   (a) De Morgan's theorem applied to NOR gate produces 2 identical truth tables;

Original circuit   Inversion   Final circuit
                    bubbles
                    cancel

NOR
equivalent

$X \equiv$ $X \equiv$ $X = ABCD$

(b)

Inverter $\equiv$

NAND $\equiv$

NOR $\equiv$

(c)

Figure 5–31 *(Continued)* (b) using the alternative NOR symbol eases circuit simplification; (c) summary of alternative gate symbols.

The following examples illustrate the application of De Morgan's theorem for the simplification of logic circuits.

E X A M P L E 5 – 1 0

Write the Boolean equation for the circuit shown in Figure 5–32. Use De Morgan's theorem and then Boolean algebra rules to simplify the equation. Draw the simplified circuit.

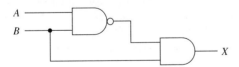

Figure 5–32

Solution: The Boolean equation at X is

$$X = \overline{AB} \cdot B$$

Applying De Morgan's theorem produces

$$X = (\overline{A} + \overline{B}) \cdot B$$

(Notice the use of parentheses to maintain proper grouping.) Using Boolean algebra rules produces

$$X = \overline{A}B + \overline{B}B$$
$$= \overline{A}B + 0$$
$$= \overline{A}B \quad \leftarrow \text{simplified equation}$$

The simplified circuit is shown in Figure 5–33.

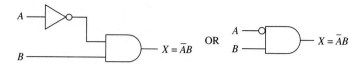

Figure 5–33   Simplified logic circuit for Example 5–10.

EXAMPLE 5–11

Repeat Example 5–10 for the circuit shown in Figure 5–34.

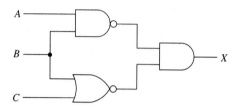

Figure 5–34

Solution: The Boolean equation at X is

$$X = \overline{AB} \cdot \overline{B + C}$$

Applying De Morgan's theorem produces

$$X = (\overline{A} + \overline{B}) \cdot \overline{B}\,\overline{C}$$

(Notice the use of parentheses to maintain proper grouping.) Using Boolean algebra rules produces

$$X = \overline{A}\,\overline{B}\,\overline{C} + \overline{B}\,\overline{B}\,\overline{C}$$
$$= \overline{A}\,\overline{B}\,\overline{C} + \overline{B}\,\overline{C}$$
$$= \overline{B}\,\overline{C}(\overline{A} + 1)$$
$$= \overline{B}\,\overline{C} \quad \leftarrow \text{simplified equation}$$

The simplified circuit is shown in Figure 5–35.

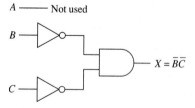

Figure 5–35   Simplified logic circuit for Example 5–11.

Also remember from Figure 5–31(a) that an AND gate with inverted inputs is equivalent to a NOR gate. Therefore, an equivalent solution to Example 5–11 would be a NOR gate with B and C as inputs, as shown in Figure 5–36.

CHAPTER 5 | BOOLEAN ALGEBRA AND REDUCTION TECHNIQUES

$$X = \overline{B} \cdot \overline{C} \qquad \text{Is equivalent to:} \qquad X = \overline{B + C} = \overline{B} \cdot \overline{C}$$

Figure 5–36   Equivalent solution to Example 5–11.

EXAMPLE 5–12

Repeat Example 5–10 for the circuit shown in Figure 5–37.

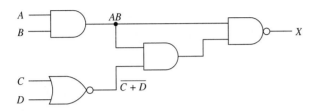

Figure 5–37

Solution:

$$X = \overline{(AB \cdot \overline{C + D}) \cdot AB}$$
$$= \overline{AB \cdot \overline{C + D}} + \overline{AB}$$
$$= \overline{AB} + \overline{\overline{C + D}} + \overline{AB}$$
$$= \overline{A} + \overline{B} + C + D + \overline{A} + \overline{B}$$
$$= \overline{A} + \overline{B} + C + D \leftarrow \text{simplified equation}$$

The simplified circuit is shown in Figure 5–38.

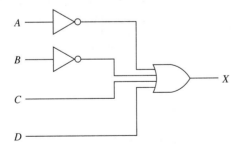

Figure 5–38   Simplified logic circuit for Example 5–12.

EXAMPLE 5–13

Use De Morgan's theorem and Boolean algebra on the circuit shown in Figure 5–39 to develop an equivalent circuit that has inversion bars covering only single variables.

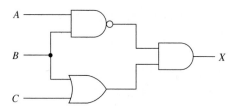

Figure 5–39

Solution: The Boolean equation at X is

$$X = \overline{AB} \cdot (B + C)$$

Applying De Morgan's theorem produces

$$X = (\overline{A} + \overline{B}) \cdot (B + C)$$

(Notice the use of parentheses to maintain proper grouping.) Using Boolean algebra rules produces

$$X = \overline{A}B + \overline{A}C + \overline{B}B + \overline{B}C$$
$$= \overline{A}B + \overline{A}C + \overline{B}C \quad \leftarrow \text{final equation (sum-of-products form)}$$

The equivalent circuit is shown in Figure 5–40.

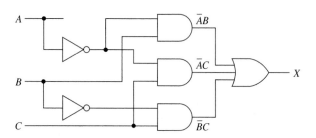

Figure 5–40 Logic circuit equivalent for Example 5–13.

Idea for Discussion

The final circuit in this example is actually *more* complicated than the original. As you will see later, it is in the form for implementation using AND-OR-INVERT gates and programmable logic devices. Besides, it is much easier to fill in a truth table from a sum of products. Build a truth table from the original equation and then from the final SOP to prove the point.

Notice that the final equation actually produces a circuit that is more complicated than the original. In fact, if a technician were to build a circuit, he or she would choose the original because it is simpler and has fewer gates. However, the final equation is in a form called the **sum-of-products (SOP) form.** This form of the equation was achieved by using Boolean algebra and is very useful for building truth tables and Karnaugh maps, which are covered in Section 5–7.

E X A M P L E   5 – 1 4

Using De Morgan's theorem and Boolean algebra, prove that the two circuits shown in Figure 5–41 are equivalent.

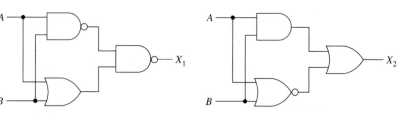

Figure 5–41

Solution: They can be proved to be equivalent if their simplified equations match.

$$X_1 = \overline{\overline{AB} \cdot (A + B)} \qquad X_2 = AB + \overline{A} + B$$
$$= \overline{\overline{AB}} + \overline{A + B} \qquad\qquad = AB + \overline{A}\,\overline{B}$$
$$= AB + \overline{A}\,\overline{B} \quad\longleftarrow \text{Equivalent}$$

EXAMPLE 5–15*

Draw the logic circuit for the following equation, simplify the equation, and construct a truth table for the simplified equation

$$X = \overline{A \cdot \overline{B}} + \overline{A \cdot (\overline{A} + C)}$$

Solution: To draw the circuit, we have to reverse our thinking from the previous examples. When we study the equation, we see that we need two NANDs feeding into an OR gate, as shown in Figure 5–42(a). Then we have to provide the inputs to the NAND gates, as shown in Figure 5–42(b).

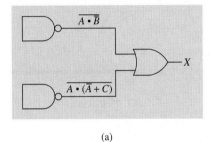

(a)

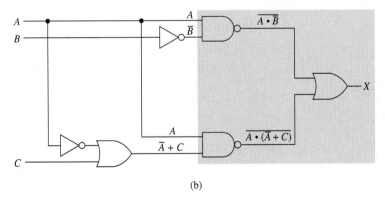

(b)

Figure 5–42 (a) Partial solution to Example 5–15; (b) logic circuit of the equation for Example 5–15.

Next, we use De Morgan's theorem and Boolean algebra to simplify the equation:

$$X = \overline{A \cdot \overline{B}} + \overline{A \cdot (\overline{A} + C)}$$
$$= (\overline{A} + \overline{\overline{B}}) + (\overline{A} + \overline{\overline{A} + C})$$
$$= \overline{A} + B + \overline{A} + \overline{\overline{A}} \cdot \overline{C}$$
$$= \overline{A} + \overline{A} + A\overline{C} + B$$
$$= \overline{A} + A\overline{C} + B$$

---

*This example is also solved using Programmable Array Logic (PAL) in Appendix E.

Apply Rule 10:

$$X = \overline{A} + \overline{C} + B \quad \leftarrow \text{simplified equation}$$

Now, to construct a truth table (Table 5–3), we need three input columns (A, B, C) and eight entries ($2^3 = 8$), and we fill in a 1 for X when $A = 0$, $C = 0$, or $B = 1$.

Table 5–3 Truth Table for Example 5–15

A	B	C	$X = \overline{A} + \overline{C} + B$
0	0	0	1
0	0	1	1
0	1	0	1
0	1	1	1
1	0	0	1
1	0	1	0
1	1	0	1
1	1	1	1

EXAMPLE 5–16

Repeat Example 5–15 for the following equation:

$$X = \overline{\overline{A}\overline{B} \cdot (A + C)} + \overline{A}B \cdot \overline{\overline{A} + \overline{B} + \overline{C}}$$

Solution: The required logic circuit is shown in Figure 5–43. The Boolean equation simplification is

$$\begin{aligned}
X &= \overline{\overline{A}\overline{B} \cdot (A + C)} + \overline{A}B \cdot \overline{\overline{A} + \overline{B} + \overline{C}} \\
&= \overline{A}\overline{B} + \overline{A + C} + \overline{A}B \cdot (\overline{\overline{A}} \cdot \overline{\overline{B}} \cdot \overline{\overline{C}}) \\
&= (\overline{A} + \overline{\overline{B}}) + \overline{A} \cdot \overline{C} + \overline{A}\,\overline{A}BBC \\
&= \overline{A} + B + \overline{A}\,\overline{C} + \overline{A}BC \\
&= \overline{A}(1 + \overline{C}) + B + \overline{A}BC \\
&= \overline{A} + B + \overline{A}BC \\
&= \overline{A} + B(1 + \overline{A}C) \\
&= \overline{A} + B \quad \leftarrow \text{simplified equation}
\end{aligned}$$

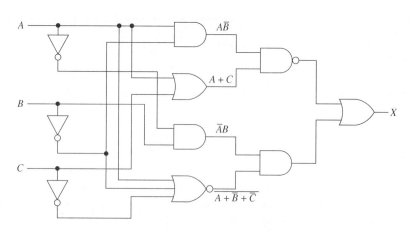

Figure 5–43 Logic circuit for the equation of Example 5–16.

	Table 5–4	Truth Table for Example 5–16	

A	B	C	$X = \overline{A} + B$
0	0	0	1
0	0	1	1
0	1	0	1
0	1	1	1
1	0	0	0
1	0	1	0
1	1	0	1
1	1	1	1

Three columns are used in the truth table (Table 5–4), because the original equation contained three variables (A, B, C). C is considered a **don't care,** however, because it does not appear in the final equation and it does not matter whether it is 1 or 0.

From the simplified equation ($X = \overline{A} + B$), we can determine that $X = 1$ when A is 0 or when B is 1, and we fill in the truth table accordingly.

EXAMPLE 5–17

Complete the truth table and timing diagram for the following simplified Boolean equation:

$$X = AB + B\overline{C} + \overline{A}\,\overline{B}C$$

Solution: The required truth table and timing diagram are shown in Figure 5–44. To fill in the truth table for X, we first put a 1 for X when $A = 1$, $B = 1$. Then $X = 1$ for $B = 1$, $C = 0$; then $X = 1$ for $A = 0$, $B = 0$, $C = 1$. All other entries for X are 0.

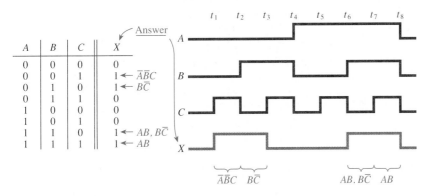

Figure 5–44 Truth table and timing diagram depicting the logic levels at X for all combinations of inputs.

The timing diagram performs the same function as the truth table, except it is a more graphic illustration of the HIGH and LOW logic levels of X as the A, B, and C inputs change over time. The logic levels at X are filled in the same way as they were for the truth table.

EXAMPLE 5–18

Repeat Example 5–17 for the following simplified equation:

$$X = A\overline{B}\,\overline{C} + \overline{A}\,\overline{B}C + ABC$$

Solution: The required truth table and timing diagram are shown in Figure 5–45.

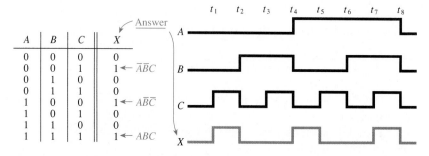

A	B	C	X	
0	0	0	0	
0	0	1	1	← $\overline{A}\overline{B}C$
0	1	0	0	
0	1	1	0	
1	0	0	1	← $A\overline{B}\overline{C}$
1	0	1	0	
1	1	0	0	
1	1	1	1	← ABC

Figure 5–45   Truth table and timing diagram depicting the logic levels at *X* for all combinations of inputs.

Bubble Pushing

Another trick that can be used, based on De Morgan's theorem, is called *bubble pushing* and is illustrated in Figure 5–46. As you can see, to form the equivalent logic circuit, you must

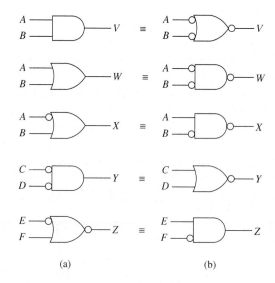

(a)   (b)

Figure 5–46   (a) Original logic circuits; (b) equivalent logic circuits.

1. Change the logic gate (AND to OR or OR to AND).

2. Add bubbles to the inputs and outputs where there were none, and remove the original bubbles.

Prove to yourself that this method works by comparing the truth table of each original circuit to its equivalent.

Notice in Figure 5–46 that we have equivalent logic circuits for the AND and OR gates (*V* and *W*). It is worth pointing out here that you will be seeing these two equivalents often when studying data memory ICs and microprocessor circuitry (Chapters 16 and 17). Figure 5–47 shows part of the gating circuitry that is often used to access microprocessor memory. Microprocessor control signals are usually **active-LOW,** meaning that they issue a LOW when they want to perform their specified task. Also, for the microprocessor to activate the block labeled Memory, the line labeled $\overline{\text{MA}}$ (memory access) must be made LOW. (The overbars on the variables signify that they are active-LOW.)

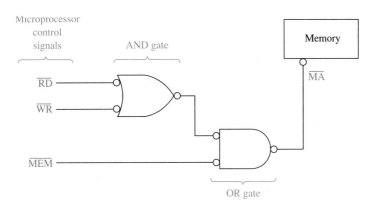

Figure 5–47   Typical gating circuitry used for microprocessor memory access.

The gating shown in Figure 5–47 will provide the LOW at $\overline{\text{MA}}$ if $\overline{\text{MEM}}$ is LOW *and* either $\overline{\text{WR}}$ is LOW *or* $\overline{\text{RD}}$ is LOW. The control signals from the microprocessor meet these conditions whenever the microprocessor is reading ($\overline{\text{RD}}$) or writing ($\overline{\text{WR}}$) from memory ($\overline{\text{MEM}}$). For example, if the microprocessor is to read from memory, it will make the $\overline{\text{RD}}$ line go LOW to signify that it wants to read, and it will make the $\overline{\text{MEM}}$ line go LOW to signify that it wants to read its information from memory. With these two lines LOW, $\overline{\text{MA}}$ is LOW, which activates the block labeled Memory. (When working with circuitry like this, it is better not to think of the bubbles as inverters; instead, think of that line as a part of the circuit that requires a LOW to "do its thing" or satisfy that input.)

The OR gate with three bubbles outputs a LOW if either input is LOW. This symbol makes the logic easy to understand, but to actually implement the circuit, its equivalent (the 7408 AND gate) would be used. Also, the AND gate with three bubbles would actually be an OR gate (the 7432).

Review Questions

5–7. Why is De Morgan's theorem important in the simplification of Boolean equations?

5–8. Using De Morgan's theorem, you can prove that a NOR gate is equivalent to an _____ (OR, AND) gate with inverted inputs.

5–9. Using the bubble-pushing technique, an AND gate with one of its inputs inverted is equivalent to a _____ (NAND, NOR) gate with its other input inverted.

5–10. Using bubble pushing to convert an inverted-input OR gate will yield a(n) _____ (AND, NAND) gate.

5–5 The Universal Capability of NAND and NOR Gates

NAND and NOR gates are sometimes referred to as **universal gates,** because by utilizing a combination of NANDs all the other logic gates (inverter, AND, OR, NOR) can be formed. Also, by utilizing a combination of NORs, all the other logic gates (inverter, AND, OR, NAND) can be formed.

This principle is useful, because often you may have extra NANDs available but actually need some other logic function. For example, let's say that you designed a circuit that required a NAND, an AND, and an inverter. You would probably purchase a 7400 quad NAND TTL IC. This chip has four NANDs in a single package. One of the NANDs will be used directly in your circuit. The AND requirement could actually be fulfilled by connecting the third and fourth NANDs on the chip to form an AND. The inverter can be formed from the second NAND on the chip. How do we convert a NAND into an inverter and two NANDs into an AND? Let's see.

An inverter can be formed from a NAND simply by connecting both NAND inputs, as shown in Figure 5–48. Both inputs to the NAND are therefore connected to A. The equation at X is $X = \overline{A \cdot A} = \overline{A}$, which is the inverter function.

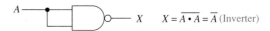

Figure 5–48 Forming an inverter from a NAND.

The next task is to form an AND from two NANDs. Do you have any ideas? What is the difference between a NAND and an AND? If we invert the output of a NAND, it will act like an AND, as shown in Figure 5–49.

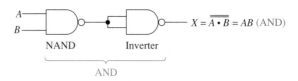

Figure 5–49 Forming an AND from two NANDs.

Now back to the original problem; we wanted to form a circuit requiring a NAND, an AND, and an inverter using a single 7400 quad NAND TTL IC. Let's make the external connections to the 7400 IC to form the circuit of Figure 5–50, which contains a NAND, an AND, and an inverter.

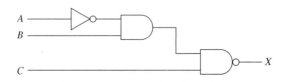

Figure 5–50 Logic circuit to be implemented using only NANDs.

First, let's redraw the logic circuit using only NANDs. Now, using the configuration shown in Figure 5–51, we can make the actual connections to a single 7400 IC, as shown in Figure 5–52, which reduces the chip count from three ICs down to one.

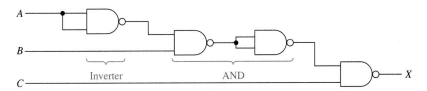

Figure 5–51   Equivalent logic circuit using only NANDs.

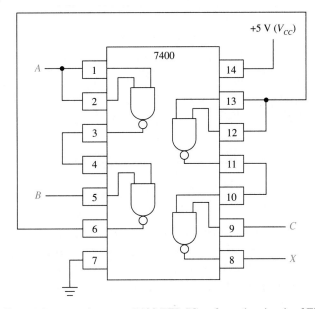

Figure 5–52   External connections to a 7400 TTL IC to form the circuit of Figure 5–51.

Besides forming inverters and ANDs from NANDs, we can form ORs and NORs from NANDs. Remember from De Morgan's theorem that an AND with an inverted output (NAND) is equivalent to an OR with inverted inputs. Therefore, if we invert the inputs to a NAND, we should find that it is equivalent to an OR, as shown in Figure 5–53.

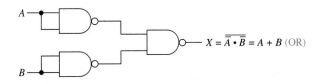

$$X = \overline{\overline{A} \cdot \overline{B}} = A + B \text{ (OR)}$$

Figure 5–53   Forming an OR from three NANDs.

Now, to form a NOR from NANDs, all we need to do is invert the output of Figure 5–53, as shown in Figure 5–54.

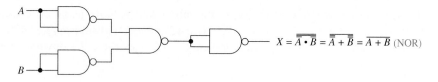

$$X = \overline{\overline{A \cdot B}} = \overline{\overline{\overline{A}} + \overline{\overline{B}}} = \overline{A + B} \text{ (NOR)}$$

Figure 5–54   Forming a NOR from four NANDs.

The procedure for converting NOR gates into an inverter, OR, AND, or NAND is similar to the conversions just discussed for NAND gates. For example, to form an inverter from a NOR gate, just connect the inputs as shown in Figure 5–55.

Helpful Hint

It is instructive for you to make a chart on your own showing how to convert NANDs into any of the other four logic gates. Repeat for NORs.

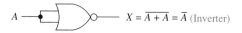

$$X = \overline{A + A} = \overline{A} \text{ (Inverter)}$$

Figure 5–55   Forming an inverter from a NOR gate.

Take some time now to try to convert NORs to an OR, NORs to an AND, and NORs to a NAND. Prove to yourself that your solution is correct by using De Morgan's theorem and Boolean algebra.

EXAMPLE 5–19

Make the external connections to a 4001 CMOS NOR IC to implement the function $X = \overline{A} + B$.

Solution: We will need an inverter and an OR gate to provide the function for X. An inverter can be made from a NOR by connecting the inputs, and an OR can be made by inverting the output of a NOR, as shown in Figure 5–56.

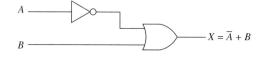

$$X = \overline{A} + B$$

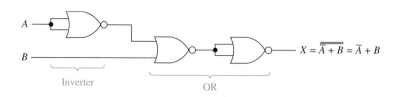

$$X = \overline{\overline{A} + B} = \overline{A} + B$$

Inverter   OR

Figure 5–56   Implementing the function $X = \overline{A} + B$ using only NOR gates.

The pin configuration for the 4001 CMOS quad NOR can be found in a CMOS data book. Figure 5–57 shows the pin configuration and external connections to implement $X = \overline{A} + B$.

Common Misconception

When sketching an inverter constructed from a NOR or a NAND gate, students often mistakenly show only a single input into the gate instead of two inputs tied together.

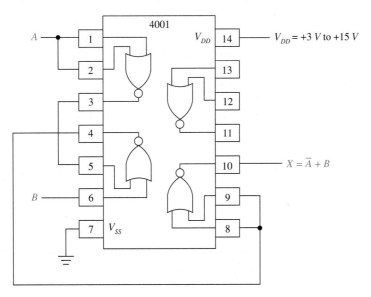

Figure 5–57 External connections to a 4001 CMOS IC to implement the circuit of Figure 5–56.

EXAMPLE 5–20

Troubleshooting

You have connected the circuit of Figure 5–57 and want to test it. Because the Boolean equation is $X = \overline{A} + B$, you first try $A = 0$, $B = 1$ and expect to get a 1 output at X, *but you don't*. V_{DD} is set to + 5 V and V_{SS} is connected to ground. Using a logic probe, you record the results shown in Table 5–5 at each pin. Determine the trouble with the circuit.

Table 5–5 Logic Probe Operation[a]	
Probe on Pin	Indicator Lamp
1	Off
2	Off
3	On
4	Off
5	On
6	On
7	Off
8	Dim
9	Off
10	Off
11	On
12	Dim
13	Dim
14	On

[a]Lamp off, 0; lamp on, 1; lamp dim, float.

Solution: Because $A = 0$, pins 1 and 2 should both be 0, which they are. Pin 3 is a 1, because 0–0 into a NOR will produce a 1 output. Pin 6 is 1, because it is connected to the 1 at B. Pin 5 matches pin 3, as it is supposed to. Pin 4 sends a 0 to pins 8 and 9, but pin 8 is floating (not 0 or 1). That's it! The connection to pin 8 must be broken.

To be sure that the circuit operates properly, the problem at pin 8 should be corrected and all four combinations of inputs at A and B should be tested.

EXAMPLE 5–21

(a) Write the simplified equation that will produce the output waveform at X, given the inputs at A, B, and C shown in Figure 5–58.
(b) Draw the logic circuit for this equation.
(c) Redraw the logic circuit using only NAND gates.

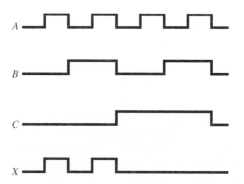

Figure 5–58

Solution: **(a)** The first HIGH pulse at X is produced for $A = 1$, $B = 0$, $C = 0$ ($A\overline{B}\,\overline{C}$). The second HIGH pulse at X happens when $A = 1$, $B = 1$, $C = 0$ ($AB\overline{C}$). Therefore, X is 1 for $A\overline{B}\,\overline{C}$ or $AB\overline{C}$.

$$X = A\overline{B}\,\overline{C} + AB\overline{C}$$

Simplifying yields

$$X = A\overline{C}(\overline{B} + B)$$
$$= A\overline{C}(1)$$
$$= A\overline{C} \quad \leftarrow \text{simplified equation}$$

(b) The logic circuit is shown in Figure 5–59(a).
(c) Redrawing the same circuit using only NANDs produces the circuit shown in Figure 5–59(b).

CHAPTER 5 | BOOLEAN ALGEBRA AND REDUCTION TECHNIQUES

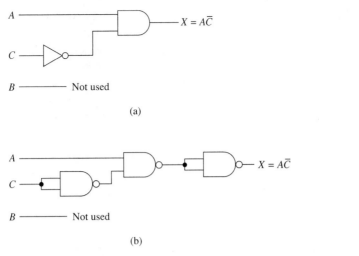

Figure 5–59 (a) Logic circuit that yields the waveform at X; (b) circuit of part (a) redrawn using only NANDs.

Review Questions

5–11. Why are NAND gates and NOR gates sometimes referred to as *universal* gates?

5–12. Why would a designer want to form an AND gate from two NAND gates?

5–13. How many inverters could be formed using a 7400 quad NAND IC?

5–6 AND–OR–INVERT Gates for Implementing Sum-of-Products Expressions

Most Boolean reductions result in an equation in one of two forms:

1. Product-of-sums (POS) form

2. Sum-of-products (SOP) form

The POS expression usually takes the form of two or more ORed variables within parentheses ANDed with two or more other variables within parentheses. Examples of POS expressions are

$$X = (A + \overline{B}) \cdot (B + C)$$

$$X = (B + \overline{C} + \overline{D}) \cdot (BC + \overline{E})$$

$$X = (A + \overline{C}) \cdot (\overline{B} + E) \cdot (C + B)$$

The SOP expression usually takes the form of two or more variables ANDed together ORed with two or more other variables ANDed together. Examples of SOP expressions are

$$X = A\overline{B} + AC + \overline{A}BC$$

$$X = AC\overline{D} + \overline{C}D + B$$

$$X = B\overline{C}\,\overline{D} + A\overline{B}DE + CD$$

The SOP expression is used most often because it lends itself nicely to the development of truth tables and timing diagrams. SOP circuits can also be constructed easily using a special combinational logic gate called the **AND–OR–INVERT gate.**

For example, let's work with the equation

$$X = \overline{A\overline{B} + \overline{C}D}$$

Using De Morgan's theorem yields

$$X = \overline{A\overline{B}} \cdot \overline{\overline{C}D}$$

Using De Morgan's theorem again puts it into a POS format:

$$X = (\overline{A} + B) \cdot (C + \overline{D}) \quad \leftarrow \text{POS}$$

Using the distributive law produces an equation in the SOP format:

$$X = \overline{A}C + \overline{A}\,\overline{D} + BC + B\overline{D} \quad \leftarrow \text{SOP}$$

Now, let's fill in a truth table for X (Table 5–6). Using the SOP expression, we put a 1 at X for $A = 0$, $C = 1$; and for $A = 0$, $D = 0$; and for $B = 1$, $C = 1$; and for $B = 1$, $D = 0$. That wasn't hard, was it?

However, if we were to use the POS expression, it would be more difficult to visualize. We would put a 1 at X for $A = 0$ or $B = 1$ whenever $C = 1$ or $D = 0$. Confusing? Yes, it is much more difficult to deal intuitively with POS expressions.

Table 5–6 Truth Table Completed Using the SOP Expression

A	B	C	D	X
0	0	0	0	1
0	0	0	1	0
0	0	1	0	1
0	0	1	1	1
0	1	0	0	1
0	1	0	1	0
0	1	1	0	1
0	1	1	1	1
1	0	0	0	0
1	0	0	1	0
1	0	1	0	0
1	0	1	1	0
1	1	0	0	1
1	1	0	1	0
1	1	1	0	1
1	1	1	1	1

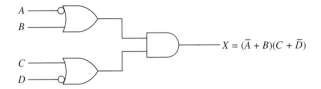

Figure 5–60   Logic circuit for the POS expression.

Drawing the logic circuit for the POS expression involves using OR gates feeding into an AND gate, as shown in Figure 5–60. Drawing the logic circuit for the SOP expression involves using AND gates feeding into an OR gate, as shown in Figure 5–61. The logic circuit for the SOP expression used more gates for this particular example, but the SOP form *is* easier to deal with and, besides, there is an IC gate specifically made to simplify the implementation of SOP circuits.

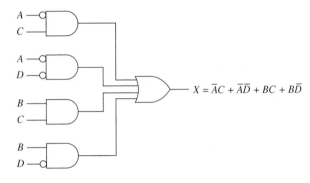

Figure 5–61   Logic circuit for the SOP expression.

That gate is the *AND–OR–INVERT* (AOI). AOIs are available in several different configurations within the TTL or CMOS families. Skim through your TTL and CMOS data books to identify some of the available AOIs. One AOI that is particularly well suited for implementing the logic of Figure 5–61 is the 74LS54 TTL IC. The pin configuration and logic symbol for the 74LS54 are shown in Figure 5–62.

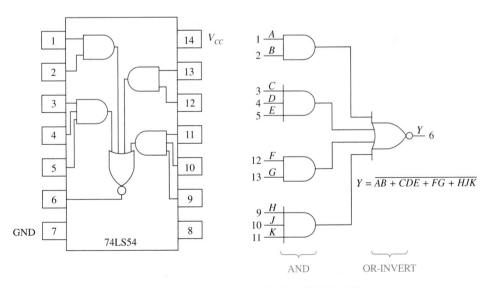

Figure 5–62   Pin configuration and logic symbol for the 74LS54 AOI gate.

Notice that the output at *Y* is inverted, so we have to place an inverter after *Y*. Also, two of the AND gates have *three* inputs instead of just the two-input gates that we need, so we just connect the unused third input to a 1. Figure 5–63 shows the required connections to the AOI to implement the SOP logic circuit of Figure 5–61. Omitting the inverter from Figure 5–63 would provide an active-LOW output function, which may be acceptable, depending on the operation required. (The new equation would be $\overline{X} = \overline{A}C + \overline{A}\,\overline{D} + BC + B\overline{D}$.)

Common Misconception

Students often forget the inverter, which makes the output active-LOW. The equations so far have been active-HIGH, but in later chapters you will see why active-LOW is so common.

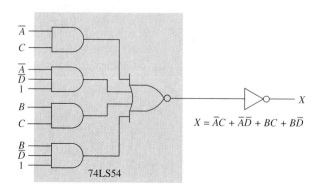

Figure 5–63 Using an AOI IC to implement an SOP equation.

EXAMPLE 5–22

Simplify the circuit shown in Figure 5–64 down to its SOP form; then draw the logic circuit of the simplified form using a 74LS54 AOI gate.

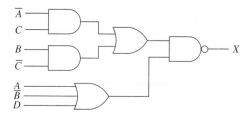

Figure 5–64 Original circuit for Example 5–22.

Solution:

$$X = \overline{(\overline{A}C + B\overline{C}) \cdot (A + \overline{B} + D)}$$
$$= \overline{\overline{A}C + B\overline{C}} + \overline{A + \overline{B} + D}$$
$$= \overline{\overline{A}C} \cdot \overline{B\overline{C}} + \overline{A}B\overline{D}$$
$$= (A + \overline{C})(\overline{B} + C) + \overline{A}B\overline{D}$$
$$= A\overline{B} + AC + \overline{B}\,\overline{C} + \overline{C}C + \overline{A}B\overline{D}$$
$$= A\overline{B} + AC + \overline{B}\,\overline{C} + \overline{A}B\overline{D} \quad \leftarrow \text{SOP}$$

The simplified circuit is shown in Figure 5–65.

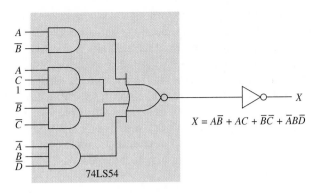

$$X = A\overline{B} + AC + \overline{B}\,\overline{C} + \overline{A}B\overline{D}$$

74LS54

Figure 5–65 Using an AOI IC to implement the simplified SOP equation for Example 5–22.

Team Discussion

What other options are available instead of inputting a 1 to the second AND gate?

Programmable Logic Devices (PLDs)

It is worth mentioning here that the latest technology for implementing large AOI functions is the programmable logic device (PLD) IC. It is covered in detail in Chapter 16 and Appendix E. PLDs provide the designer with a means to specify the exact contents of the gating inside an AOI function. For example, instead of being limited to the four AND terms feeding into a single NOR, as in the 74LS54, PLDs provide much more flexibility. Some PLDs provide for up to 16 inputs into each of 48 AND gates, with all the AND gates capable of being routed to any of eight different OR gate outputs. The exact configuration of the AND–OR combinations is programmable by the designer to meet his or her specific needs. Thus a company needs to stock only a few different PLD parts, and the specific logic function is defined and programmed by the user right at his or her own workstation.

Review Questions

5–14. Which form of Boolean equation is better suited for completing truth tables and timing diagrams, SOP or POS?

5–15. AOI ICs are used to implement _____ (SOP, POS) expressions.

5–16. The equation $X = AB + BCD + DE$ has only three product terms. If a 74LS54 AOI IC is used to implement the equation, what must be done with the three inputs to the unused fourth AND gate?

5–7   Karnaugh Mapping

We learned in previous sections that by using Boolean algebra and De Morgan's theorem we can minimize the number of gates that are required to implement a particular logic function. This is very important for the reduction of circuit cost, physical size, and gate failures. You may have found that some of the steps in the Boolean reduction process require ingenuity on your part and a lot of practice.

Karnaugh mapping, named for its originator, is another method of simplifying logic circuits. It still requires that you reduce the equation to an SOP form, but from there you follow a *systematic approach,* which will always produce the simplest configuration possible for the logic circuit.

A **Karnaugh map** (K-map) is similar to a truth table in that it graphically shows the output level of a Boolean equation for each of the possible input variable combinations. Each output level is placed in a separate **cell** of the K-map. K-maps can be used to simplify equations having two, three, four, five, or six different input variables. Solving five- and six-variable K-maps is extremely cumbersome; they can be more practically solved using advanced computer techniques. In this book we will solve two-, three-, and four-variable K-maps.

Determining the number of cells in a K-map is the same as finding the number of combinations or entries in a truth table. A two-variable map requires $2^2 = 4$ cells. A three-variable map requires $2^3 = 8$ cells. A four-variable map requires $2^4 = 16$ cells. The three different K-maps are shown in Figure 5–66.

Common Misconception

Students sometimes design their own layouts for K-maps by moving the overbars. This move can produce invalid results if it causes more than one variable to change as you move from cell to cell.

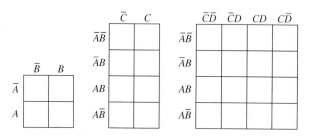

Figure 5–66   Two-, three-, and four-variable Karnaugh maps.

Each cell within the K-map corresponds to a particular combination of the input variables. For example, in the two-variable K-map, the upper left cell corresponds to $\overline{A}\,\overline{B}$, the lower left cell is $A\overline{B}$, the upper right cell is $\overline{A}B$, and the lower right cell is AB.

Also notice that when moving from one cell to an **adjacent cell,** only one variable changes. For example, look at the three-variable K-map. The upper left cell is $\overline{A}\,\overline{B}\,\overline{C}$; the adjacent cell just below it is $\overline{A}B\overline{C}$. In this case the $\overline{A}\,\overline{C}$ remained the same and only the $\overline{B}$ changed, to B. The same holds true for each adjacent cell.

To use the K-map reduction procedure, you must perform the following steps:

1. Transform the Boolean equation to be reduced into an SOP expression.

2. Fill in the appropriate cells of the K-map.

3. Encircle adjacent cells in groups of two, four, or eight. (The more adjacent cells encircled, the simpler the final equation is.)

4. Find each term of the final SOP equation by determining which variables remain constant within each circle.

Now, let's consider the equation

$$X = \overline{A}(\overline{B}C + \overline{B}\,\overline{C}) + \overline{A}B\overline{C}$$

First, transform the equation to an SOP expression:

$$X = \overline{A}\,\overline{B}C + \overline{A}\,\overline{B}\,\overline{C} + \overline{A}B\overline{C}$$

The terms of that SOP expression can be put into a truth table and then transferred to a K-map, as shown in Figure 5–67. Working with the K-map, we now encircle adjacent 1's in groups of two, four, or eight. We end up with two circles of two cells each, as shown in Figure 5–68. The first circle surrounds the two 1's at the top of the K-map, and the second circle surrounds the two 1's in the left column of the K-map.

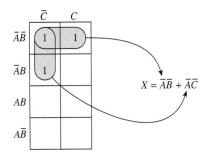

A	B	C	X
0	0	0	1
0	0	1	1
0	1	0	1
0	1	1	0
1	0	0	0
1	0	1	0
1	1	0	0
1	1	1	0

Figure 5–67 Truth table and Karnaugh map of $X = \overline{A}\,\overline{B}\,\overline{C} + \overline{A}\,\overline{B}C + \overline{A}B\overline{C}$.

Figure 5–68 Encircling adjacent cells in a Karnaugh map.

$$X = \overline{A}\,\overline{B} + \overline{A}\,\overline{C}$$

Once the circles have been drawn encompassing all the 1's in the map, the final simplified equation is obtained by determining *which variables remain the same within each circle.* Well, the first circle (across the top) encompasses $\overline{A}\,\overline{B}\,\overline{C}$ and $\overline{A}\,\overline{B}C$. The variables that remain the same within the circle are $\overline{A}\,\overline{B}$. Therefore, $\overline{A}\,\overline{B}$ becomes one of the terms in the final SOP equation. The second circle (left column) encompasses $\overline{A}\,\overline{B}\,\overline{C}$ and $\overline{A}B\overline{C}$. The variables that remain the same within that circle are $\overline{A}\,\overline{C}$. Therefore, the second term in the final equation is $\overline{A}\,\overline{C}$.

Since the final equation is always written in the SOP format, the answer is $X = \overline{A}\,\overline{B} + \overline{A}\,\overline{C}$. Actually, the original equation was simple enough that we could have reduced it using standard Boolean algebra. Let's do it just to check our answer:

$$X = \overline{A}\,\overline{B}C + \overline{A}\,\overline{B}\,\overline{C} + \overline{A}B\overline{C}$$
$$= \overline{A}\,\overline{B}(C + \overline{C}) + \overline{A}B\overline{C}$$
$$= \overline{A}\,\overline{B} + \overline{A}B\overline{C}$$
$$= \overline{A}(\overline{B} + B\overline{C})$$
$$= \overline{A}(\overline{B} + \overline{C})$$
$$= \overline{A}\,\overline{B} + \overline{A}\,\overline{C} \quad \checkmark$$

There are several other points to watch out for when applying the Karnaugh mapping technique. The following examples will be used to illustrate several important points in filling in the map, determining adjacencies, and obtaining the final equation. Work through these examples carefully so that you do not miss any special techniques.

EXAMPLE 5–23

Simplify the following SOP equation using the Karnaugh mapping technique:

$$X = \overline{A}B + \overline{A}\,\overline{B}\,\overline{C} + AB\overline{C} + A\overline{B}\,\overline{C}$$

Solution:

1. Construct an eight-cell K-map (Figure 5–69) and fill in a 1 in each cell that corresponds to a term in the original equation. (Notice that $\overline{A}B$ has no C variable in it. Therefore, $\overline{A}B$ is satisfied whether C is HIGH or LOW, so $\overline{A}B$ will fill in two cells: $\overline{A}BC + \overline{A}B\overline{C}$.)

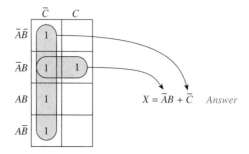

Figure 5–69 Karnaugh map and final equation for Example 5–23.

2. Encircle adjacent cells in the largest group of two or four or eight.

3. Identify the variables that remain the same within each circle and write the final simplified SOP equation by ORing them together.

EXAMPLE 5–24

Simplify the following equation using the Karnaugh mapping procedure:

$$X = \overline{A}B\overline{C}D + A\overline{B}\,\overline{C}D + \overline{A}\,\overline{B}\,\overline{C}D + AB\overline{C}D + ABC\,\overline{D} + ABCD$$

Solution: Since there are four different variables in the equation, we need a 16-cell map ($2^4 = 16$), as shown in Figure 5–70.

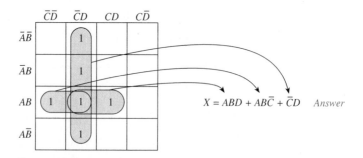

Figure 5–70 Solution to Example 5–24.

EXAMPLE 5-25

Simplify the following equation using the Karnaugh mapping procedure:

$$X = B\overline{C}\overline{D} + \overline{A}B\overline{C}D + AB\overline{C}D + \overline{A}BCD + ABCD$$

Solution: Notice in Figure 5–71 that the $B\overline{C}\overline{D}$ term in the original equation fills in *two* cells: $AB\overline{C}\overline{D} + \overline{A}B\overline{C}\overline{D}$. Also notice in Figure 5–71 that we could have encircled four cells and then two cells, but that would not have given us the simplest final equation. By encircling four cells and then four cells, we are sure to get the simplest final equation. (Always encircle the largest number of cells possible, even if some of the cells have already been encircled in another group.)

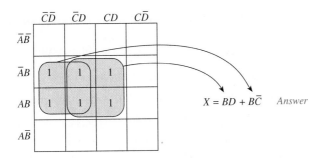

$X = BD + B\overline{C}$ *Answer*

Figure 5–71   Solution to Example 5–25.

Common Misconception

Students often solve a map like this by encircling 4 and 2 instead of 4 and 4. Analyze both results to see why choosing 4 and 4 is better.

EXAMPLE 5-26

Simplify the following equation using the Karnaugh mapping procedure:

$$X = \overline{A}\,\overline{B}\,\overline{C} + A\overline{C}\,\overline{D} + A\overline{B} + ABC\overline{D} + \overline{A}\,\overline{B}C$$

Solution: Notice in Figure 5–72 that a new technique called **wraparound** is introduced. You have to think of the K-map as a continuous cylinder in the horizontal direction, like the label on a soup can. This makes the left row of cells adjacent to the right row of cells. Also, in the vertical direction, a continuous cylinder like a soup can lying on its side makes the top row of cells adjacent to the bottom row of cells. In Figure 5–72, for example, the four top cells are adjacent to the four bottom cells, to combine as eight cells having the variable $\overline{B}$ in common.

Another circle of four is formed by the wraparound adjacencies of the lower left and lower right pairs combining to have $A\overline{D}$ in common. The final equation becomes $X = \overline{B} + A\overline{D}$. Compare that simple equation with the original equation that had five terms in it.

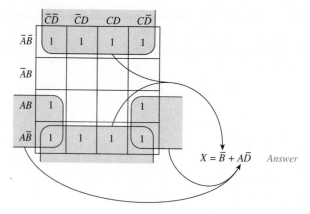

Figure 5–72 Solution to Example 5–26 illustrating the wraparound feature.

E X A M P L E   5 – 2 7

Simplify the following equation using the Karnaugh mapping procedure:

$$X = \overline{B}(CD + \overline{C}) + C\overline{D}(\overline{A + B} + AB)$$

Solution: Before filling in the K-map, an SOP expression must be formed:

$$X = \overline{B}CD + \overline{B}\,\overline{C} + C\overline{D}(\overline{A}\,\overline{B} + AB)$$
$$= \overline{B}CD + \overline{B}\,\overline{C} + \overline{A}\,\overline{B}C\overline{D} + ABC\overline{D}$$

The group of four 1's can be encircled to form $\overline{A}\,\overline{B}$, as shown in Figure 5–73. Another group of four can be encircled using wraparound to form $\overline{B}\,\overline{C}$. That leaves two 1's that are not combined with any others. The unattached 1 in the bottom row can be combined within a group of four, as shown, to form $\overline{B}D$.

The last 1 is not adjacent to any other, so it must be encircled by itself to form $ABC\overline{D}$. The final simplified equation is

$$X = \overline{A}\,\overline{B} + \overline{B}\,\overline{C} + \overline{B}D + ABC\overline{D}$$

![Common Misconception icon]

Common Misconception

Students often neglect to include the single encirclement (four-variable) term in the final equation.

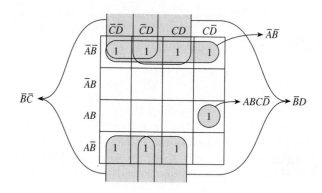

Figure 5–73 Solution to Example 5–27.

E X A M P L E   5 – 2 8

Simplify the following equation using the Karnaugh mapping procedure:

$$X = \overline{A}\,\overline{D} + A\overline{B}\,\overline{D} + \overline{A}\,\overline{C}D + \overline{A}CD$$

Solution: First, the group of eight cells can be encircled, as shown in Figure 5–74. $\overline{A}$ is the only variable present in each cell within the circle, so the circle of eight simply reduces to $\overline{A}$. (Notice that larger circles will reduce to fewer variables in the final equation.)

Team Discussion

What is the final equation of a map that has *all* cells filled in?

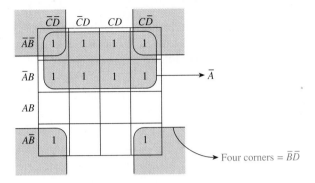

Figure 5–74   Solution to Example 5–28.

Also, all four corners are adjacent to each other because the K-map can be wrapped around in both the vertical *and* horizontal directions. Encircling the four corners results in $\overline{B}\,\overline{D}$. The final equation is

$$X = \overline{A} + \overline{B}\,\overline{D}$$

EXAMPLE 5–29

Simplify the following equation using the Karnaugh mapping procedure:

$$X = \overline{A}\,\overline{B}\,\overline{D} + A\overline{C}\,\overline{D} + \overline{A}B\overline{C} + AB\overline{C}D + A\overline{B}C\overline{D}$$

Solution: Encircling the four corners forms $\overline{B}\,\overline{D}$, as shown in Figure 5–75. The other group of four forms $B\overline{C}$. You may be tempted to encircle the $\overline{C}\,\overline{D}$ group of four as shown by the dotted line, but that would be **redundant** because each of those 1's is already contained within an existing circle. Therefore, the final equation is

$$X = \overline{B}\,\overline{D} + B\overline{C}$$

Team Discussion

So what's wrong with being redundant?

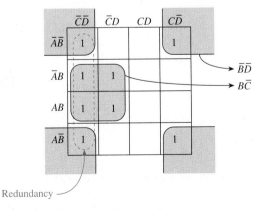

Figure 5–75   Solution to Example 5–29.

5–8 System Design Applications

Let's summarize the entire chapter now by working through two complete design problems. The following examples illustrate practical applications of a K-map to ensure that when we implement the circuit using an AOI we will have the simplest possible solution.

NOTE: The construction of digital circuits with higher complexity than those of these examples will be more practically suited for implementation using PLDs, which are discussed in Chapter 16 and Appendix E.

SYSTEM DESIGN 5–1

Design a circuit that can be built using an AOI and inverters that will output a HIGH (1) whenever the 4-bit hexadecimal input is an odd number from 0 to 9.

Team Discussion

The LSB (variable A) is always HIGH for an odd number. Why can't we just say "odd number = A"?

Table 5–7 Hex Truth Table Used to Determine the Equation for Odd Numbers[a] from 0 to 9

D	C	B	A	DEC	
0	0	0	0	0	
0	0	0	1	1	$\leftarrow A\overline{B}\,\overline{C}\,\overline{D}$
0	0	1	0	2	
0	0	1	1	3	$\leftarrow AB\,\overline{C}\,\overline{D}$
0	1	0	0	4	
0	1	0	1	5	$\leftarrow A\overline{B}C\overline{D}$
0	1	1	0	6	
0	1	1	1	7	$\leftarrow ABC\overline{D}$
1	0	0	0	8	
1	0	0	1	9	$\leftarrow A\overline{B}\,\overline{C}D$

[a]Odd number = $A\overline{B}\,\overline{C}\,\overline{D} + AB\overline{C}\,\overline{D} + A\overline{B}C\overline{D} + ABC\overline{D} + A\overline{B}\,\overline{C}D$.

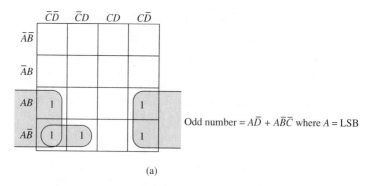

Odd number = $A\overline{D} + A\overline{B}\,\overline{C}$ where A = LSB

(a)

Figure 5–76 (a) Simplified equation derived from a Karnaugh map;

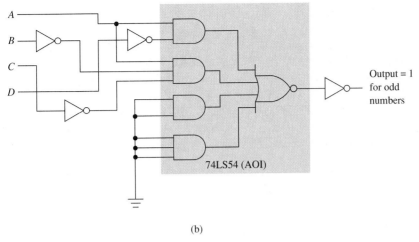

(b)

Figure 5–76   *(Continued)* (b) implementation of the odd-number decoder using an AOI.

Solution: First, build a truth table (Table 5–7) to identify which hex codes from 0 to 9 produce odd numbers. (Use the variable A to represent the 2^0 hex input, B for 2^1, C for 2^2, and D for 2^3.) Next, reduce this equation into its simplest form by using a Karnaugh map, as shown in Figure 5–76(a). Finally, using an AOI with inverters, the circuit can be constructed as shown in Figure 5–76(b).

SYSTEM DESIGN 5–2

A chemical plant needs a microprocessor-driven alarm system to warn of critical conditions in one of its chemical tanks. The tank has four HIGH/LOW (1/0) switches that monitor temperature (T), pressure (P), fluid level (L), and weight (W). Design a system that will notify the microprocessor to activate an alarm when any of the following conditions arise:

1. High fluid level with high temperature and high pressure
2. Low fluid level with high temperature and high weight
3. Low fluid level with low temperature and high pressure
4. Low fluid level with low weight and high temperature

Solution: First, write in Boolean equation form the conditions that will activate the alarm:

$$\text{alarm} = LTP + \overline{L}TW + \overline{L}\,\overline{T}P + \overline{L}\,\overline{W}T$$

Next, factor the equation into its simplest form by using a Karnaugh map, as shown in Figure 5–77(a). Finally, using an AOI with inverters, the circuit can be constructed as shown in Figure 5–77(b).

*Team
Discussion*

By rereading conditions 2 and 4, can you logically explain why the weight is irrelevant and doesn't appear in the final equation?

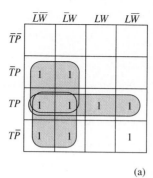

	$\overline{L}\,\overline{W}$	$\overline{L}W$	LW	$L\overline{W}$
$\overline{T}\,\overline{P}$				
$\overline{T}P$	1	1		
TP	1	1	1	1
$T\overline{P}$	1	1		1

$\text{Alarm} = TP + P\overline{L} + T\overline{L}$

(a)

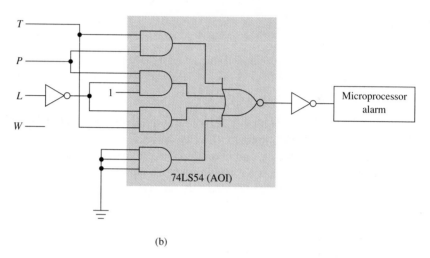

(b)

Figure 5–77 (a) Simplified equation derived from a Karnaugh map; (b) implementation of the chemical tank alarm using an AOI.

Review Questions

5–17. The number of cells in a Karnaugh map is equal to the number of entries in a corresponding truth table. True or false?

5–18. The order in which you label the rows and columns of a Karnaugh map does not matter as long as every combination of variables is used. True or false?

5–19. Adjacent cells in a Karnaugh map are encircled in groups of 2, 4, 6, or 8. True or false?

5–20. Which method of encircling eight adjacent cells in a Karnaugh map produces the simplest equation; two groups of four or one group of eight?

Summary

In this chapter we have learned that

1. Several logic gates can be connected together to form combinational logic.

2. There are several Boolean laws and rules that provide the means to form equivalent circuits.

3. Boolean algebra is used to reduce logic circuits to simpler equivalent circuits that function identically to the original circuit.

4. De Morgan's theorem is required in the reduction process whenever inversion bars cover more than one variable in the original Boolean equation.

5. NAND and NOR gates are sometimes referred to as *universal gates*, because they can be used to form any of the other gates.

6. AND–OR–INVERT (AOI) gates are often used to implement sum-of-products (SOP) equations.

7. Karnaugh mapping provides a systematic method of reducing logic circuits.

Glossary

Active-LOW: An output of a logic circuit that is LOW when activated or an input that needs to be LOW to be activated.

Adjacent Cell: Cells within a Karnaugh map that border each other on one side or the top or bottom of the cell.

AND–OR–INVERT Gate: (AOI) An integrated circuit containing combinational logic consisting of several AND gates feeding into an OR gate and then an inverter. It is used to implement logic equations that are in the SOP format.

Boolean Reduction: An algebraic technique that follows specific rules to convert a Boolean equation into a simpler form.

Cell: Each box within a Karnaugh map. Each cell corresponds to a particular combination of input variable logic levels.

Combinational Logic: Logic circuits formed by combining several of the basic logic gates to form a more complex function.

De Morgan's Theorem: A Boolean law used for equation reduction that allows the user to convert an equation having an inversion bar over several variables into an equivalent equation having inversion bars over single variables only.

Don't Care: A variable appearing in a truth table or timing waveform that will have no effect on the final output regardless of the logic level of the variable. Therefore, don't-care variables can be ignored.

Equivalent Circuit: A simplified version of a logic circuit that can be used to perform the exact logic function of the original complex circuit.

Inversion Bubbles: An alternative to drawing the triangular inversion symbol. The bubble (or circle) can appear at the input or output of a logic gate.

Karnaugh Map: A two-dimensional table of Boolean output levels used as a tool to perform a systematic reduction of complex logic circuits into simplified equivalent circuits.

Product-of-Sums (POS) Form: A Boolean equation in the form of a group of ORed variables ANDed with another group of ORed variables [for example, $X = (A + \overline{B} + C)(B + D)(\overline{A} + \overline{C})$].

Redundancy: Once all filled-in cells in a Karnaugh map are contained within a circle, the final simplified equation can be written. Drawing another circle around a different group of cells is redundant.

Sum-of-Products (SOP) Form: A Boolean equation in the form of a group of ANDed variables ORed with another group of ANDed variables (for example, $X = ABC + \overline{B}DE + \overline{A}\,\overline{D}$).

Universal Gates: The NOR and NAND logic gates are sometimes called universal gates because any of the other logic gates can be formed from them.

Wraparound: The left and right cells and the top and bottom cells of a Karnaugh map are actually adjacent to each other by means of the wraparound feature.

Problems

Section 5–1

5–1. Write the Boolean equation for each of the logic circuits shown in Figure P5–1.

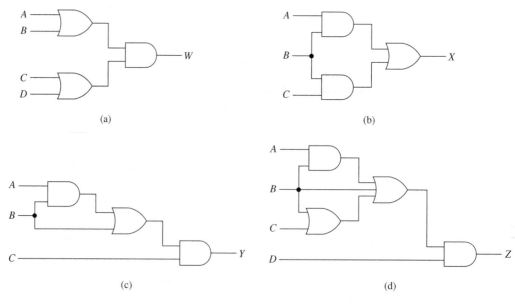

(a)

(b)

(c)

(d)

Figure P5–1

CHAPTER 5 | BOOLEAN ALGEBRA AND REDUCTION TECHNIQUES

5–2. Draw the logic circuit that would be used to implement the following Boolean equations. (*Hint:* Applying Law 3 to some equations will make them easier to use.)

(a) $M = (AB) + (C + D)$

(b) $N = (A + B + C)D$

(c) $P = (AC + BC)(A + C)$

(d) $Q = (A + B)BCD$

(e) $R = BC + D + AD$

(f) $S = B(A + C) + AC + D$

5–3. Construct a truth table for each of the equations given in Problem 5–2.

5–4. Write the Boolean equation and then complete the timing diagram at W, X, Y, and Z for the logic circuits shown in Figure P5–4.

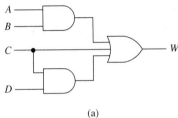

(a)

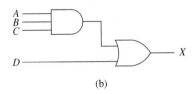

(b)

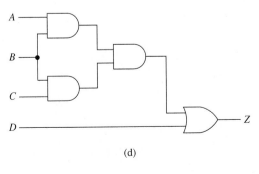

(c)

(d)

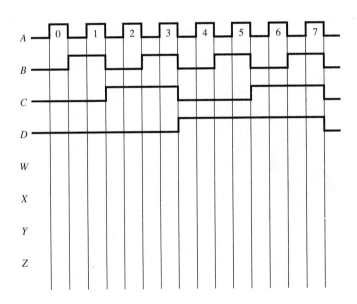

Figure P5–4

5–5. State the Boolean law that makes each of the equivalent circuits shown in Figure P5–5 valid.

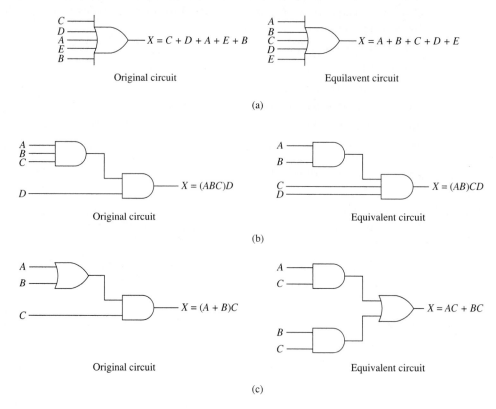

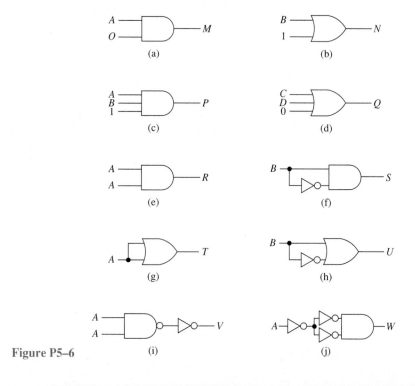

Figure P5–5

5–6. Using the ten Boolean rules presented in Table 5–2, determine the outputs of the logic circuits shown in Figure P5–6.

Figure P5–6

CHAPTER 5 ∣ BOOLEAN ALGEBRA AND REDUCTION TECHNIQUES

5–7. Write the Boolean equation for the circuits of Figure P5–7. Simplify the equations and draw the simplified logic circuit.

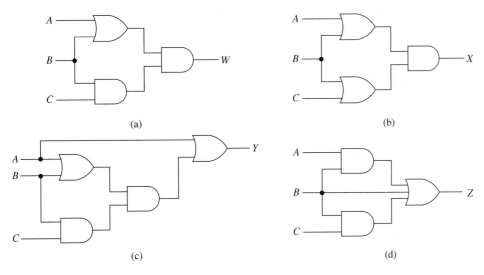

Figure P5–7

5–8. Repeat Problem 5–7 for the circuits shown in Figure P5–8.

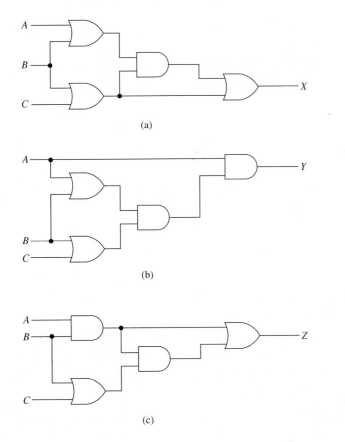

Figure P5–8

5–9. Draw the logic circuit for the following equations. Simplify the equations and draw the simplified logic circuit.

(a) $V = AC + ACD + CD$

(b) $W = (BCD + C)CD$

(c) $X = (B + D)(A + C) + ABD$

(d) $Y = AB + BC + ABC$

(e) $Z = ABC + CD + CDE$

5–10. Construct a truth table for each of the simplified equations of Problem 5–9.

5–11. The pin layouts for a 74HCT08 CMOS AND gate and a 74HCT32 CMOS OR gate are given in Figure P5–11. Make the external connections to the chips to implement the following logic equation. (Simplify the logic equation first.)

$$X = (A + B)(D + C) + ABD$$

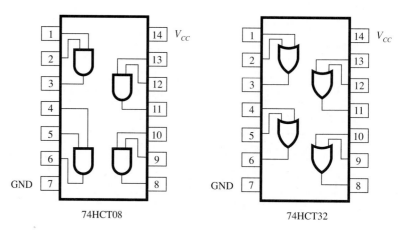

Figure P5–11

5–12. Repeat Problem 5–11 for the following equation:

$$Y = AB(C + BD) + BD$$

Section 5–4

5–13. Write a sentence describing how De Morgan's theorem is applied in the simplification of a logic equation.

5–14. (a) De Morgan's theorem can be used to prove that an OR gate with inverted inputs is equivalent to what type of gate?

(b) An AND gate with inverted inputs is equivalent to what type of gate?

5–15. Which two circuits in Figure P5–15 produce equivalent output equations?

CHAPTER 5 | BOOLEAN ALGEBRA AND REDUCTION TECHNIQUES

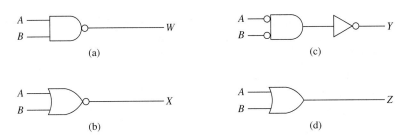

Figure P5–15

5–16. Use De Morgan's theorem to prove that a NOR gate with inverted inputs is equivalent to an AND gate.

5–17. Draw the logic circuit for the following equations. Apply De Morgan's theorem and Boolean algebra rules to reduce them to equations having inversion bars over single variables only. Draw the simplified circuit.

(a) $W = \overline{AB} + \overline{A + C}$

(b) $X = A\overline{B} + C + \overline{BC}$

(c) $Y = \overline{(AB) + C} + B\overline{C}$

(d) $Z = AB + \overline{(\overline{A} + C)}$

5–18. Write the Boolean equation for the circuits of Figure P5–18. Use De Morgan's theorem and Boolean algebra rules to simplify the equation. Draw the simplified circuit.

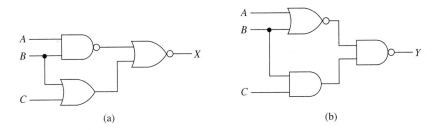

Figure P5–18

C   **5–19.** Repeat Problem 5–17 for the following equations.

(a) $W = \overline{\overline{AB} + CD} + \overline{ACD}$

(b) $X = \overline{\overline{A} + B} \cdot BC + \overline{BC}$

(c) $Y = \overline{AB\overline{C} + D} + \overline{A\overline{B} + B\overline{C}}$

(d) $Z = (C + D)\overline{A\overline{C}D}(\overline{A}C + \overline{D})$

C   **5–20.** Repeat Problem 5–18 for the circuits of Figure P5–20.

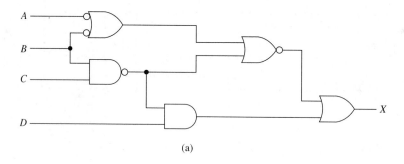

(a)

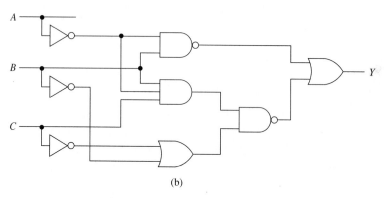

(b)

Figure P5–20

D*   **5–21.** Design a logic circuit that will output a 1 (HIGH) only if A and B are both 1 while either C or D is 1.

D   **5–22.** Design a logic circuit that will output a 0 only if A or B is 0.

D   **5–23.** Design a logic circuit that will output a LOW only if A is HIGH or B is HIGH while C is LOW or D is LOW.

C D   **5–24.** Design a logic circuit that will output a HIGH if only one of the inputs A, B, or C is LOW.

C D   **5–25.** Design a circuit that outputs a 1 when the binary value of $ABCD$ (D = LSB) is >11.

C D   **5–26.** Design a circuit that outputs a LOW when the binary value of $ABCD$ (D = LSB) is >7 and <10.

5–27. Complete a truth table for the following simplified Boolean equations.

(a) $W = A\overline{B}\,\overline{C} + \overline{B}C + \overline{A}B$

(b) $X = \overline{A}\,\overline{B} + A\overline{B}C + B\overline{C}$

(c) $Y = \overline{C}D + \overline{A}\,\overline{B}\,\overline{C}\,\overline{D} + BCD + \overline{A}C\overline{D}$

(d) $Z = \overline{A}BC\overline{D} + \overline{A}C + C\overline{D} + \overline{B}\,\overline{C}$

5–28. Complete the timing diagram in Figure P5–28 for the following simplified Boolean equations.

(a) $X = \overline{A}\,\overline{B}\,\overline{C} + ABC + A\overline{C}$

(b) $Y = \overline{B} + \overline{A}B\overline{C} + AC$

(c) $Z = B\overline{C} + A\overline{B} + \overline{A}BC$

---

*The letter **D** designates a circuit **D**esign problem.

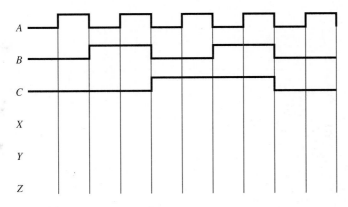

Figure P5–28

5–29. Use the bubble-pushing technique to convert the gates in Figure P5–29.

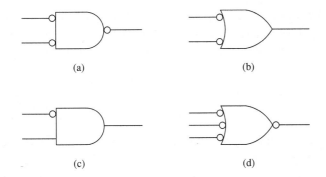

(a)

(b)

(c)

(d)

Figure P5–29

C D   **5–30.** Some computer systems have two disk drives, commonly called drive A and drive B, for storing and retrieving data. Assume that your computer has four control signals provided by its internal microprocessor to enable data to be read and written to either drive. Design a gating scheme similar to that provided in Figure 5–47 to supply an active-LOW drive select signal to drive A $(\overline{DS_a})$ or to drive B $(\overline{DS_b})$ whenever they are read or written to. The four control signals are also active-LOW and are labeled $\overline{RD}$ (Read), $\overline{WR}$ (Write), $\overline{DA}$ (drive A), and $\overline{DB}$ (drive B).

Section 5–5

5–31. Draw the connections required to convert

(a) A NAND gate into an inverter

(b) A NOR gate into an inverter

5–32. Draw the connections required to construct

(a) An OR gate from two NOR gates

(b) An AND gate from two NAND gates

(c) An AND gate from several NOR gates

(d) A NOR gate from several NAND gates

5–33. Redraw the logic circuits of Figure P5–33 to their equivalents *using only* NOR gates.

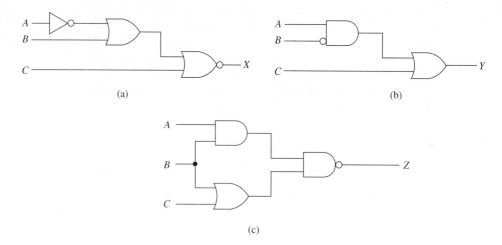

(a)

(b)

(c)

Figure P5–33

C

5–34. Convert the circuits of Figure P5–34 to their equivalents *using only* NAND gates. Next, make the external connections to a 7400 quad NAND to implement the new circuit. (Each new equivalent circuit is limited to *four* NAND gates.)

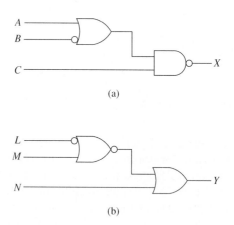

(a)

(b)

Figure P5–34

Section 5–6

5–35. Identify each of the following Boolean equations as a product-of-sums (POS) expression, a sum-of-products (SOP) expression, or both.

(a) $U = A\overline{B}C + BC + \overline{A}C$

(b) $V = (A + C)(\overline{B} + \overline{C})$

(c) $W = A\overline{C}(\overline{B} + C)$

(d) $X = AB + \overline{C} + BD$

(e) $Y = (A\overline{B} + D)(A + \overline{C}D)$

(f) $Z = (A + \overline{B})(BC + A) + \overline{A}B + CD$

5–36. Simplify the circuit of Figure P5–36 down to its SOP form; then draw the logic circuit of the simplified form implemented using a 74LS54 AOI gate.

CHAPTER 5 | BOOLEAN ALGEBRA AND REDUCTION TECHNIQUES

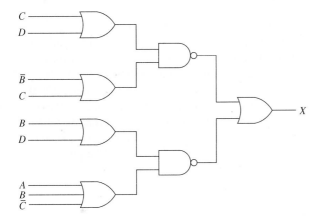

Figure P5–36

Section 5–7

5–37. Using a Karnaugh map, reduce the following equations to a minimum sum-of-products form.

(a) $X = AB\overline{C} + \overline{A}B + \overline{A}\,\overline{B}$

(b) $Y = BC + \overline{A}\,\overline{B}C + B\overline{C}$

(c) $Z = ABC + A\overline{B}\,\overline{C} + \overline{A}\,\overline{B}C + AB\overline{C}$

5–38. Using a Karnaugh map, reduce the following equations to a minimum sum-of-products form.

(a) $W = \overline{B}(C\overline{D} + \overline{A}D) + \overline{B}\,\overline{C}(A + \overline{A}\,\overline{D})$

(b) $X = \overline{A}\,\overline{B}\,\overline{D} + B(\overline{C}\,\overline{D} + ACD) + AB\overline{D}$

(c) $Y = A(C\overline{D} + \overline{C}\,\overline{D}) + A\overline{B}D + \overline{A}\,\overline{B}C\overline{D}$

(d) $Z = \overline{B}\,\overline{C}D + B\overline{C}D + \overline{C}\,\overline{D} + C\overline{D}(B + \overline{A}\,\overline{B})$

C **5–39.** Use a Karnaugh map to simplify the circuits in Figure P5–39.

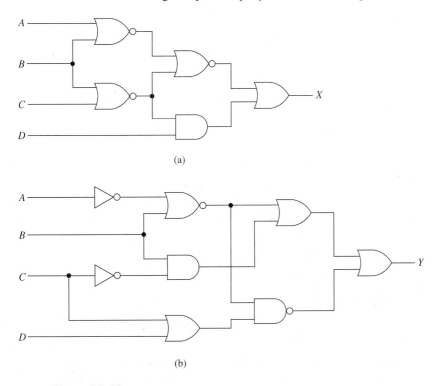

(a)

(b)

Figure P5–39

C D **5–40.** Seven-segment displays are commonly used in calculators to display each decimal digit. Each segment of a digit is controlled separately, and when all seven of the segments are on, the number 8 is displayed. The upper right segment of the display comes on when displaying the numbers 0, 1, 2, 3, 4, 7, 8, and 9. (The numerical designation for each of the digits 0 to 9 is shown in Figure P5–40.) Design a circuit that outputs a HIGH (1) whenever a 4-bit BCD code translates to a number that uses the upper right segment. Use variable A to represent the 2^3 BCD input. Implement your design with an AOI and inverters.

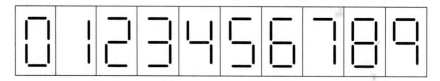

Figure P5–40

C D **5–41.** Repeat Problem 5–40 for the lower left segment of a seven-segment display (0, 2, 6, 8).

T **5–42.** The logic circuit of Figure P5–42(a) is implemented by making connections to the 7400 as shown in Figure P5–42(b). The circuit is not working properly. The problem is in the IC connections or in the IC itself. The data table in Figure P5–42(c) is completed by using a logic probe at each pin. Identify the problem.

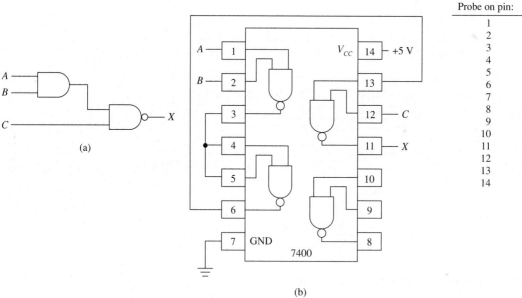

Test conditions
$A = 1$
$B = 1$
$C = 1$
X should equal 0

Probe on pin:	Indicator lamp
1	On
2	On
3	Off
4	Off
5	Off
6	Off
7	Off
8	On
9	Dim
10	Dim
11	On
12	On
13	Off
14	On

(c)

(a)

(b)

Figure P5–42

CHAPTER 5 | BOOLEAN ALGEBRA AND REDUCTION TECHNIQUES

T   **5–43.** Repeat Problem 5–42 for the circuit shown in Figure P5–43.

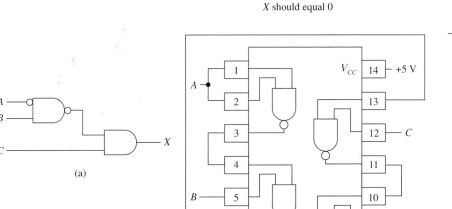

Test conditions
$A = 0$
$B = 1$
$C = 1$
X should equal 0

Probe on pin:	Indicator lamp
1	Off
2	Off
3	On
4	On
5	On
6	Off
7	Off
8	On
9	Off
10	On
11	On
12	On
13	Off
14	On

(c)

(a)

(b)

Figure P5–43

Schematic Interpretation Problems

See Appendix G for the schematic diagrams.

S   **5–44.** Find U8 in the HC11D0 schematic. Pins 11 and 12 are unused so they are connected to V_{CC}. What if they were connected to ground instead?

S   **5–45.** Find U1:A in the Watchdog Timer schematic. This device is called a flip-flop and is explained in Chapter 10. It has two inputs, D and CLK, and two outputs, Q_A and $\overline{Q}_A$. Write the Boolean equation at the output (pin 3) of U2:A.

S   **5–46.** Write the Boolean equation at the output (pin 3) of U12:A in the Watchdog Timer schematic. (*Hint:* Use the information given in Problem 5–45.)

C S   **5–47.** Locate the U14 gates in the 4096/4196 schematic.

(a) Write the Boolean equation of the output at pin 6 of U14.

(b) What kind of gate does it turn into if you use the bubble-pushing technique?

(c) This is a 74HC08. What kind of logic gate is that?

(d) Complete the following sentence: Pin 3 of U14:A goes LOW if _____ OR if _____.

C S   **5–48.** U10 of the 4096/4196 schematic is a RAM memory IC. Its operation is discussed in Chapter 16. To enable the chip to work, the Chip Enable input at pin 20 must be made LOW. Write a sentence describing the logic operation that makes that line go LOW. (*Hint:* Pin 20 of U10 goes LOW if _____.)

Answers to Review Questions

5–1.   (a) 2 (b) 3 (c) 4

5–2.   (a) Associative law of addition (b) commutative law of multiplication (c) distributive law

5–3.   True

5–4.   False

5–5.   False

5–6.   (a) $A + \overline{B}$ (b) $B + C$

5–7.   Because it enables you to convert an expression having an inversion bar over more than one variable into an expression with inversion bars over single variables only

5–8.   AND

5–9.   NOR

5–10.   NAND

5–11.   Because by utilizing a combination of these gates, all other gates can be formed

5–12.   Because in designing a circuit you may have extra NAND gates available and can avoid using extra ICs

5–13.   4

5–14.   SOP

5–15.   SOP

5–16.   They must be connected to ground.

5–17.   True

5–18.   False

5–19.   False

5–20.   One group of 8

6 Exclusive-OR and Exclusive-NOR Gates

OUTLINE

6–1   The Exclusive-OR Gate
6–2   The Exclusive-NOR Gate
6–3   Parity Generator/Checker
6–4   System Design Applications

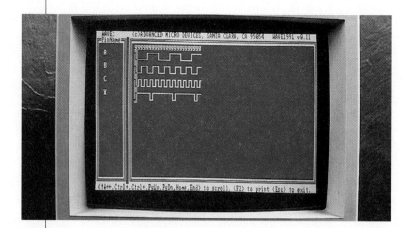

Objectives

Upon completion of this chapter, you should be able to

- Describe the operation and use of exclusive-OR and exclusive-NOR gates.

- Construct truth tables and draw timing diagrams for exclusive-OR and ex-clusive-NOR gates.

- Simplify combinational logic circuits containing exclusive-OR and exclu-sive-NOR gates.

- Design odd- and even-parity generator and checker systems.

- Explain the operation of a binary comparator and a controlled inverter.

Introduction

We have seen in the previous chapters that by using various combinations of the basic gates we can form almost any logic function that we need. Often a particu-lar combination of logic gates provides a function that is especially useful for a wide variety of tasks. The AOI discussed in Chapter 5 is one such circuit. In this chapter we learn about and design systems using two new combinational logic gates: the exclusive-OR and the exclusive-NOR.

6–1   The Exclusive-OR Gate

Remember, the OR gate provides a HIGH output if one input or the other input is HIGH *or if both inputs are HIGH*. The **exclusive-OR,** on the other hand, provides a HIGH output if one input or the other input is HIGH, *but not both*. This point is made more clear by comparing the truth tables for an OR gate versus an exclusive-OR gate, as shown in Table 6–1.

Table 6–1 Truth Tables for an OR Gate Versus an Exclusive-OR Gate					
A	*B*	*X*	*A*	*B*	*X*
0	0	0	0	0	0
0	1	1	0	1	1
1	0	1	1	0	1
1	1	1	1	1	0
(OR)			(Exclusive-OR)		

The Boolean equation for the Ex-OR function is written $X = \overline{A}B + A\overline{B}$ and can be constructed using the combinational logic shown in Figure 6–1. By experimenting and using Boolean reduction, we can find several other combinations of the basic gates that provide the Ex-OR function. For example, the combination of AND, OR, and NAND gates shown in Figure 6–2 will reduce to the "one or the other but not both" (Ex-OR) function.

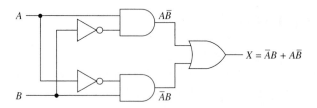

Figure 6–1   Logic circuit for providing the exclusive-OR function.

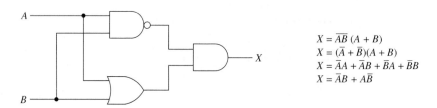

$$X = \overline{AB} \, (A + B)$$
$$X = (\overline{A} + \overline{B})(A + B)$$
$$X = \overline{A}A + \overline{A}B + \overline{B}A + \overline{B}B$$
$$X = \overline{A}B + A\overline{B}$$

Figure 6–2   Exclusive-OR built with an AND–OR–NAND combination.

The exclusive-OR gate is common enough to deserve its own logic symbol and equation, shown in Figure 6–3. (Note the shorthand method of writing the Boolean equation is to use a plus sign with a circle around it.)

$$X = A \oplus B = \overline{A}B + A\overline{B}$$

Figure 6–3   Logic symbol and equation for the exclusive-OR.

CHAPTER 6 | EXCLUSIVE-OR AND EXCLUSIVE-NOR GATES

6–2   The Exclusive-NOR Gate

The **exclusive-NOR** is the complement of the exclusive-OR. A comparison of the truth tables in Table 6–2 illustrates this point.

Table 6–2 Truth Tables of the Exclusive-NOR Versus the Exclusive-OR					
$X \quad AB + \overline{A}\,\overline{B}$			$X = \overline{A}B + A\overline{B}$		
A	B	X	A	B	X
0	0	1	0	0	0
0	1	0	0	1	1
1	0	0	1	0	1
1	1	1	1	1	0
Exclusive-NOR			Exclusive-OR		

The truth table for the Ex-NOR shows a HIGH output for both inputs LOW or both inputs HIGH. The Ex-NOR is sometimes called the *equality gate* because both inputs must be equal to get a HIGH output. The basic logic circuit and symbol for the Ex-NOR are shown in Figure 6–4.

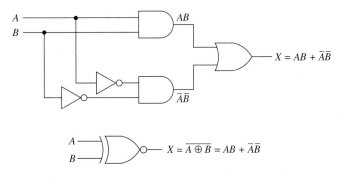

Figure 6–4   Exclusive-NOR logic circuit and logic symbol.

Summary

The exclusive-OR and exclusive-NOR gates are two-input logic gates that provide a very important, commonly used function that we will see in upcoming examples. Basically, the gates operate as follows:

The exclusive-OR gate provides a HIGH output for one or the other inputs HIGH, but not both ($X = \overline{A}B + A\overline{B}$).

The exclusive-NOR gate provides a HIGH output for both inputs HIGH or both inputs LOW ($X = AB + \overline{A}\,\overline{B}$).

Also, the Ex-OR and Ex-NOR gates are available in both TTL and CMOS integrated-circuit packages. For example, the 7486 is a TTL quad Ex-OR and the 4077 is a CMOS quad Ex-NOR.

EXAMPLE 6-1

Determine for each circuit shown in Figure 6–5 if its output provides the Ex-OR function, the Ex-NOR function, or neither.

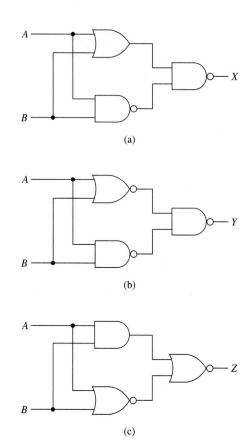

(a)

(b)

(c)

Figure 6–5

Solution:

(a) $X = \overline{(A + B)\overline{AB}}$

$\quad = \overline{A + B} + \overline{\overline{AB}}$

$\quad = \overline{A}\,\overline{B} + AB \quad \leftarrow$ Ex-NOR

(b) $Y = \overline{\overline{\overline{A + B}}\ \overline{AB}}$

$\quad = \overline{\overline{A + B}} + \overline{\overline{AB}}$

$\quad = A + B + AB$

$\quad = A + B(1 + A)$

$\quad = A + B \quad \leftarrow$ neither (OR function)

(c) $Z = \overline{AB + \overline{A + B}}$

$\quad = \overline{AB}\ \overline{\overline{A + B}}$

$\quad = (\overline{A} + \overline{B})(A + B)$

$\quad = \overline{A}B + \overline{A}A + \overline{B}A + \overline{B}B$

$\quad = \overline{A}B + A\overline{B} \quad \leftarrow$ Ex-OR

EXAMPLE 6–2

Write the Boolean equation for the circuit shown in Figure 6–6 and simplify.

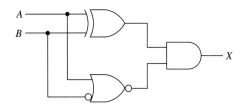

Figure 6–6

Solution:

$$X = (\overline{A}B + A\overline{B})\overline{\overline{A} + \overline{B}}$$
$$= (\overline{A}B + A\overline{B})\overline{\overline{A}}\,\overline{\overline{B}}$$
$$= \overline{A}B\overline{A}B + A\overline{B}\,\overline{A}B$$
$$= \overline{A}B$$

EXAMPLE 6–3

Write the Boolean equation for the circuit shown in Figure 6–7 and simplify.

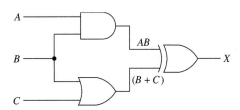

Figure 6–7

Solution:

$$X = \overline{AB}(B + C) + AB(\overline{B + C})$$
$$= (\overline{A} + \overline{B})(B + C) + AB\overline{B}\,\overline{C}$$
$$= \overline{A}B + \overline{A}C + \overline{B}B + \overline{B}C$$
$$= \overline{A}B + \overline{A}C + \overline{B}C$$

Hint:
$$X = \overline{IN_1}\,IN_2 + IN_1\overline{IN_2}$$

Review Questions

6–1. The exclusive-OR gate is the complement (or inverse) of the OR gate. True or false?

6–2. The exclusive-OR gate is the complement of the exclusive-NOR gate. True or false?

6–3. Write the Boolean equation for an exclusive-NOR gate.

6–3   Parity Generator/Checker

Now let's look at some digital systems that use the Ex-OR and Ex-NOR gates. We start by studying the parity generator.

In the **transmission** of binary information from one digital device to another, it is possible for external **electrical noise** or other disturbances to cause an error in the digital signal. For example, if a 4-bit digital system is transmitting a BCD 5 (0101), electrical noise present on the line during the transmission of the LSB may change a 1 to a 0. If so, the receiving device on the other end of the transmission line would receive a BCD 4 (0100), which is wrong. If a parity system is used, this error would be recognized, and the receiving device would signal an error condition or ask the transmitting device to retransmit.

Parity systems are defined as either *odd parity* or *even parity.* The parity system adds an extra bit to the digital information being transmitted. A 4-bit system will require a fifth bit, an 8-bit system will require a ninth bit, and so on.

In a 4-bit system such as BCD or hexadecimal, the fifth bit is the parity bit and will be a 1 or 0, depending on what the other 4 bits are. In an *odd-parity* system, the parity bit that is added must make the *sum of all 5 bits odd.* In an *even-parity* system, the parity bit makes the *sum of all 5 bits even.*

The parity generator is the circuit that creates the parity bit. On the receiving end, a parity checker determines if the 5-bit result is of the right parity. The type of system (odd or even) must be agreed on beforehand so that the parity checker knows what to look for (this is called *protocol*). Also, the parity bit can be placed next to the MSB *or* LSB as long as the device on the receiving end knows which bit is parity and which bits are data.

Let's look at the example of transmitting the BCD number 5 (0101) in an odd-parity system.

As shown in Figure 6–8, the transmitting device puts a 0101 on the BCD lines. The parity generator puts a 1 on the parity-bit line, making the sum of the bits odd $(0 + 1 + 0 + 1 + 1 = 3)$. The parity checker at the receiving end checks to see that the 5 bits are odd and, if so, assumes that the BCD information is valid.

Helpful Hint

Typically, the *error indicator* is actually a signal that initiates a retransmission of the original signal or produces an error message on a computer display.

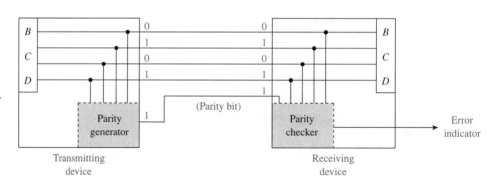

Figure 6–8   Odd-parity generator/checker system.

If, however, the data in the LSB were changed due to electrical noise somewhere in the transmission cable, the parity checker would detect that an even-parity number was received and would signal an error condition on the **error indicator** output.

CHAPTER 6 | EXCLUSIVE-OR AND EXCLUSIVE-NOR GATES

This scheme detects only errors that occur to 1 bit. If 2 bits were changed, the parity checker would think everything is okay. However, the likelihood of 2 bits being affected is highly unusual. An error occurring to even 1 bit is unusual.

EXAMPLE 6–4

Add a parity bit next to the LSB of the following hexadecimal codes to form even parity: 0111, 1101, 1010, 1111, 1000, 0000.

Solution:

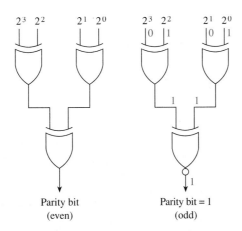

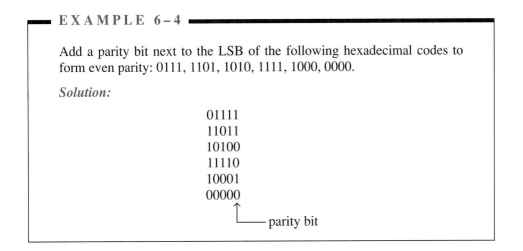

Figure 6–9 Even- and odd-parity generators.

The parity generator and checker can be constructed from exclusive-OR gates. Figure 6–9 shows the connections to form a 4-bit even- and a 4-bit odd-parity generator. The odd-parity generator has the BCD number 5 (0101) at its inputs. If you follow the logic through with these bits, you will see that the parity bit will be a 1, just as we want. Try some different 4-bit numbers at the inputs to both the even- and odd-parity generators to prove to yourself that they work properly. Computer systems generally transmit 8 or 16 bits of parallel data at a time. An 8-bit even-parity generator can be constructed by adding more gates, as shown in Figure 6–10.

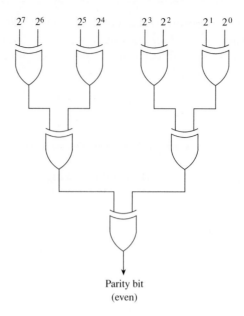

Parity bit
(even)

Figure 6–10   Eight-bit even-parity generator.

A parity checker is constructed in the same way as the parity generator except that in a 4-bit system there must be five inputs (including the parity bit), and the output is used as the error indicator (1 = error condition). Figure 6–11 shows a 5-bit even-parity checker. The BCD 6 with even parity is input. Follow the logic through the diagram to prove to yourself that the output will be 0, meaning "no error."

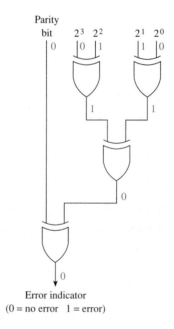

Error indicator
(0 = no error   1 = error)

Figure 6–11   Five-bit even-parity checker.

Integrated-Circuit Parity Generator/Checker

You may have guessed by now that parity generator and checker circuits are available in single integrated-circuit packages. One popular 9-bit parity generator/checker is the

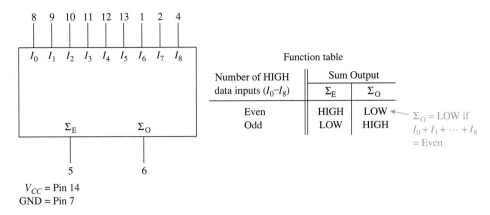

Function table		
Number of HIGH data inputs (I_0–I_8)	Sum Output	
	Σ_E	Σ_O
Even	HIGH	LOW
Odd	LOW	HIGH

Σ_O = LOW if
$I_0 + I_1 + \cdots + I_8$
= Even

V_{CC} = Pin 14
GND = Pin 7

Figure 6–12 Logic symbol and function table for the 74280 9-bit parity generator/checker.

74280 TTL IC (or 74HC280 CMOS IC). The logic symbol and **function table** for the 74280 are given in Figure 6–12.

The 74280 has nine inputs. If used as a parity checker, the first eight inputs would be the data input and the ninth would be the parity-bit input. If your system is looking for even parity, the sum of the nine inputs should be even, which will produce a HIGH at the Σ_E output and a LOW at the Σ_O output.

6–4   System Design Applications

SYSTEM DESIGN 6–1

Parity Error-Detection System

Using 74280s, design a complete parity generator/checking system. It is to be used in an 8-bit, even-parity computer configuration.

Solution: *Parity generator:* Because the 74280 has nine inputs, we have to connect the unused ninth input (I_8) to ground (0) so that it will not affect our result. The 8-bit input data are connected to I_0 through I_7.

Now, the generator sums bits I_0 through I_7 and puts out a LOW on Σ_O and a HIGH on Σ_E if the sum is even. Therefore, the parity bit generated should be taken from the Σ_O output, because we want the sum of all 9 bits sent to the receiving device to be even.

Parity checker: The checker will receive all 9 bits and check if their sum is even. If their sum *is* even, the Σ_E line goes HIGH. We will use the Σ_O output because it will be LOW for "no error" and HIGH for "error." The complete circuit design is shown in Figure 6–13.

Common Misconception

Students often have a hard time understanding why we use the sum-*odd* (Σ_O) output in an even system. The key to understanding that reasoning is found in the function table for the 74280 in Figure 6–12.

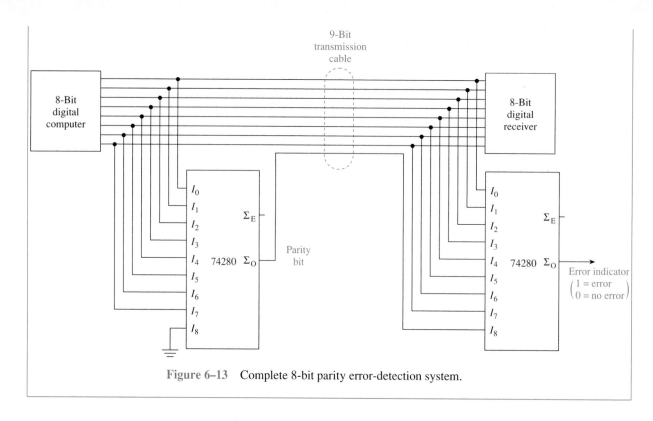

Figure 6–13   Complete 8-bit parity error-detection system.

SYSTEM DESIGN 6–2
Parallel Binary Comparator

Design a system—called a parallel binary **comparator**—that compares the 4-bit **binary string** A to the 4-bit binary string B. If the strings are exactly equal, provide a HIGH-level output to drive a warning buzzer.

Solution: Using four exclusive-NOR gates, we can compare string A to string B, bit by bit. Remember, if both inputs to an exclusive-NOR are the same (0–0 or 1–1), it outputs a 1. If all four Ex-NOR gates are outputting a 1, the 4 bits of string A must match the 4 bits of string B. The complete circuit design is shown in Figure 6–14.

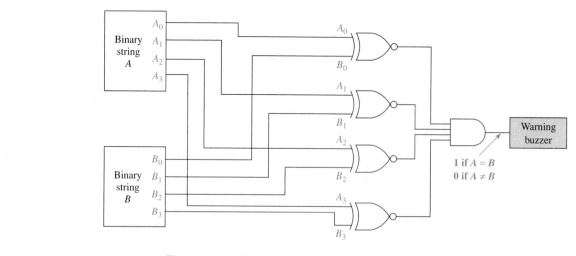

Figure 6–14   Binary comparator system.

Controlled Inverter

Often in binary arithmetic circuits we need to have a device that complements an entire binary string when told to do so by some control signal. Design an 8-bit **controlled inverter** (complementing) circuit. The circuit will receive a control signal that, if HIGH, causes the circuit to complement the 8-bit string and, if LOW, does not.

Solution: The circuit shown in Figure 6–15 can be used to provide the complementing function. If the control signal (C) is HIGH, each of the input data bits is complemented at the output. If the control signal is LOW, the data bits pass through to the output uncomplemented. Two 7486 quad exclusive-OR ICs could be used to implement this design.

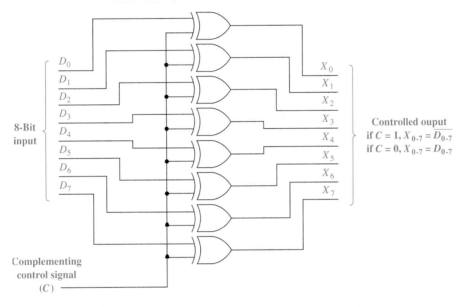

Figure 6–15   Controlled inverter (complementing) circuit.

Review Questions

6–4. An *odd* parity generator produces a 1 if the sum of its inputs is odd. True or false?

6–5. In an 8-bit parallel transmission system, if one or two of the bits are changed due to electrical noise, the parity checker will detect the error. True or false?

6–6. Which output of the 74280 parity generator is used as the parity bit in an *odd* system?

6–7. If all nine inputs to a 74280 are HIGH, the output at Σ_E will be _____ (HIGH, LOW)?

Summary

In this chapter we have learned that

1. The exclusive-OR gate outputs a HIGH if one or the other inputs, but not both, is HIGH.

2. The exclusive-NOR gate outputs a HIGH if both inputs are HIGH or if both inputs are LOW.

3. A parity bit is commonly used for error detection during the transmission of digital signals.

4. Exclusive-OR and NOR gates are used in applications such as parity checking, binary comparison, and controlled complementing circuits.

Glossary

Binary String: Two or more binary bits used collectively to form a meaningful binary representation.

Comparator: A device or system that identifies an equality between two quantities.

Controlled Inverter: A digital circuit capable of complementing a binary string of bits based on an external control signal.

Electrical Noise: Unwanted electrical irregularities that can cause a change in a digital logic level.

Error Indicator: A visual display or digital signal that is used to signify that an error has occurred within a digital system.

Exclusive-NOR: A gate that produces a HIGH output for both inputs HIGH or both inputs LOW.

Exclusive-OR: A gate that produces a HIGH output for one or the other input HIGH, but not both.

Function Table: A chart that illustrates the input/output operating characteristics of an integrated circuit.

Parity: An error-detection scheme used to detect a change in the value of a bit.

Transmission: The transfer of digital signals from one location to another.

Problems

Sections 6–1 and 6–2

6–1. Describe in words the operation of an exclusive-OR gate; an exclusive-NOR gate.

6–2. Describe in words the difference between

(a) An exclusive-OR and an OR gate

(b) An exclusive-NOR and an AND gate

6–3. Complete the timing diagram in Figure P6–3 for the exclusive-OR and the exclusive-NOR.

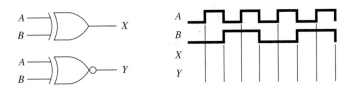

Figure P6–3

6–4. Write the Boolean equations for the circuits in Figure P6–4(a) through (d). Simplify the equations and determine if they function as an Ex-OR, Ex-NOR, or neither.

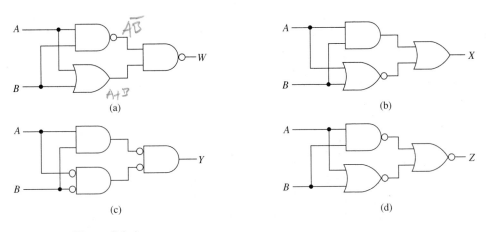

Figure P6–4

D   **6–5.** Design an exclusive-OR gate constructed from all NOR gates.

D   **6–6.** Design an exclusive-NOR gate constructed from all NAND gates.

6–7. Write the Boolean equations for the circuits of Figure P6–7. Reduce the equations to their simplest form.

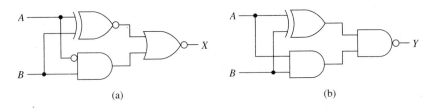

Figure P6–7

C   **6–8.** Repeat Problem 6–7 for the circuits of Figure P6–8.

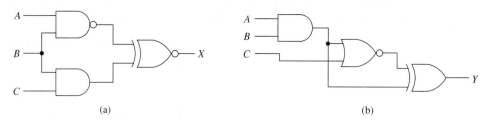

(a)   (b)

Figure P6–8

Section 6–3

6–9. Convert the following hexadecimal numbers to their 8-bit binary code. Add a parity bit next to the LSB to form odd parity.

A7   4C   79   F3   00   FF

6–10. The pin configuration of the 74HC86 CMOS quad exclusive-OR IC is given in Figure P6–10. Make the external connections to the IC to form a 4-bit even-parity generator.

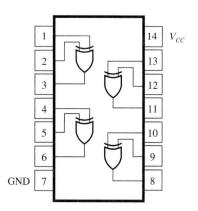

Figure P6–10

6–11. Repeat Problem 6–10 for a 5-bit even-parity checker. Use the pin configuration shown in Figure P6–11.

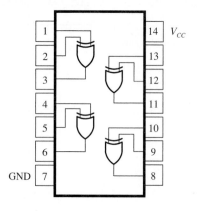

Figure P6–11

CHAPTER 6 | EXCLUSIVE-OR AND EXCLUSIVE-NOR GATES

6–12. Figure P6–12 shows another design used to form a 4-bit parity generator. Determine if the circuit will function as an odd- or even-parity generator.

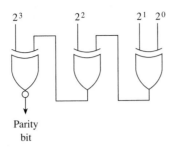

2^3 2^2 2^1 2^0

Parity
bit

Figure P6–12

C D   **6–13.** Referring to Figure 6–13, design and sketch a 4-bit odd-parity error-detection system. Use two 74280 ICs and a five-line transmission cable between the sending and receiving devices.

C D   **6–14.** Design a binary comparator system similar to Figure 6–14 using exclusive-ORs instead of exclusive-NORs.

C   **6–15.** If the exclusive ORs in Figure 6–15 are replaced by exclusive NORs, will the circuit still function as a controlled inverter? If so, should C be HIGH or LOW to complement?

Schematic Interpretation Problems

See Appendix G for the schematic diagrams.

C D S   **6–16.** Find Port 1 (P1.7–P1.0) of U8 in the 4096/4196 schematic. On a separate piece of paper, draw an 8-bit controlled inverter for that output port. The inverting function is to be controlled by the P3.5 output (pin 15).

C D S   **6–17.** Find Port 2 (P2.7–P2.0) of U8 in the 4096/4196 schematic. This port outputs the high-order address bits for the system (A8–A15). (Microcontroller addresses are discussed further in Chapter 16.) On a separate piece of paper, draw a binary comparator that compares the four bits A8–A11 to the four bits A12–A15. The HIGH output for an equal comparison is to be input to P3.4 (pin 14) of U8.

Answers to Review Questions

6–1.   False

6–2.   True

6–3. $X = AB + \overline{A}\,\overline{B}$

6–4.   False

6–5.   False

6–6. Σ_E

6–7.   LOW

7 Arithmetic Operations and Circuits

O U T L I N E

7–1 Binary Arithmetic
7–2 Two's-Complement Representation
7–3 Two's-Complement Arithmetic
7–4 Hexadecimal Arithmetic
7–5 BCD Arithmetic
7–6 Arithmetic Circuits
7–7 Four-Bit Full-Adder ICs
7–8 System Design Applications
7–9 Arithmetic/Logic Units

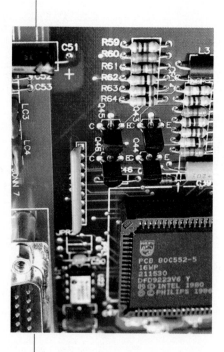

Objectives

Upon completion of this chapter, you should be able to

- Perform the four binary arithmetic functions: addition, subtraction, multiplication, and division.

- Convert positive and negative numbers to signed two's-complement notation.

- Perform two's-complement, hexadecimal, and BCD arithmetic.

- Explain the design and operation of a half-adder and a full-adder circuit.

- Utilize full-adder ICs to implement arithmetic circuits.

- Explain the operation of a two's-complement adder/subtractor circuit and a BCD adder circuit.

- Explain the function of an arithmetic/logic unit (ALU).

Introduction

An important function of digital systems and computers is the execution of arithmetic operations. In this chapter we will see that there is no magic in taking the sum of two numbers electronically. Instead, there is a basic set of logic-circuit building blocks, and the arithmetic operations follow a step-by-step procedure to arrive at the correct answer. All the "electronic arithmetic" will be performed using digital input and output levels with basic combinational logic circuits or medium-scale-integration (MSI) chips.

7–1 Binary Arithmetic

Before studying the actual digital electronic requirements for arithmetic circuits, let's look at the procedures for performing the four basic arithmetic functions: addition, subtraction, multiplication, and division.

Addition

The procedure for adding numbers in binary is similar to adding in decimal except that the binary sum is made up of only 1's and 0's. When the binary **sum** exceeds 1, you must carry a 1 to the next-more-significant column, as in regular decimal addition.

The four possible combinations of adding two binary numbers can be stated as follows:

$$0 + 0 = 0 \quad \text{carry } 0$$
$$0 + 1 = 1 \quad \text{carry } 0$$
$$1 + 0 = 1 \quad \text{carry } 0$$
$$1 + 1 = 0 \quad \text{carry } 1$$

The general form of binary addition in the least significant column can be written

$$A_0 + B_0 = \Sigma_0 + C_{out}$$

The sum output is given by the *summation* symbol (Σ), called sigma, and the **carry-out** is given by C_{out}. The truth table in Table 7–1 shows the four possible conditions when adding two binary digits.

Table 7–1 Truth Table for Addition of Two Binary Digits in the Least Significant Column

A_0	B_0	Σ_0	C_{out}
0	0	0	0
0	1	1	0
1	0	1	0
1	1	0	1

If a carry-out *is* produced, it must be added to the next-more-significant column as a **carry-in** (C_{in}). Figure 7–1 shows this operation and truth table. In the truth table, the C_{in} term comes from the value of C_{out} from the previous addition. Now, with three possible inputs there are eight combinations of outputs ($2^3 = 8$). Review the truth table to be sure that you understand how each sum and carry were determined.

Now let's perform some binary additions. We represent all binary numbers in groups of 8 or 16, because that is the standard used for arithmetic in most digital computers today.

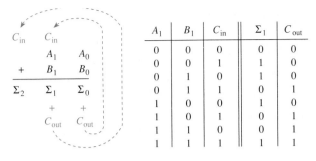

A_1	B_1	C_{in}	Σ_1	C_{out}
0	0	0	0	0
0	0	1	1	0
0	1	0	1	0
0	1	1	0	1
1	0	0	1	0
1	0	1	0	1
1	1	0	0	1
1	1	1	1	1

Figure 7–1   Addition in the more significant columns requires including C_{in} with $A_1 + B_1$.

EXAMPLE 7–1

Perform the following decimal additions. Convert the original decimal numbers to binary and add them. Compare answers. **(a)** 5 + 2; **(b)** 8 + 3; **(c)** 18 + 2; **(d)** 147 + 75; **(e)** 31 + 7.

Solution:

		Decimal	**Binary**
(a)		5	0000 0101
		+2	+0000 0010
		7	0000 0111 = 7_{10} √
(b)		8	0000 1000
		+ 3	+0000 0011
		11	0000 1011 = 11_{10} √
(c)		18	0001 0010
		+ 2	+0000 0010
		20	0001 0100 = 20_{10} √
(d)		147	1001 0011
		+ 75	+0100 1011
		222	1101 1110 = 222_{10} √
(e)		31	0001 1111
		+ 7	+0000 0111
		38	0010 0110 = 38_{10} √

Subtraction

The four possible combinations of subtracting two binary numbers can be stated as follows:

$$0 - 0 = 0 \quad \text{borrow } 0$$

$$0 - 1 = 1 \quad \text{borrow } 1$$

$$1 - 0 = 1 \quad \text{borrow } 0$$

$$1 - 1 = 0 \quad \text{borrow } 0$$

The general form of binary subtraction in the least significant column can be written

$$A_0 - B_0 = R_0 + B_{out}$$

The difference, or **remainder,** from the subtraction is R_0, and if a **borrow** is required, B_{out} is 1. The truth table in Table 7–2 shows the four possible conditions when subtracting two binary digits.

Table 7–2	Truth Table for Subtraction of Two Binary Digits in the Least Significant Column		
A_0	B_0	R_0	B_{out}
0	0	0	0
0	1	1	1
1	0	1	0
1	1	0	0

Borrow required because $A_0 < B_0$

If a borrow *is* required, the A_0 must borrow from A_1 in the next-more-significant column. When A_0 borrows from its left, A_0 increases by 2 (just as in decimal subtraction, where the number increases by 10). For example, let's subtract $2 - 1$ ($10_2 - 01_2$).

Borrow 1 from A_1.

$$
\begin{array}{cc}
A_1 & A_0 \\
-B_1 & B_0 \\
\hline
R_1 & R_0
\end{array}
\qquad
\begin{array}{cc}
\overset{0}{1} & \overset{2}{\cancel{0}} \\
-0 & 1 \\
\hline
0 & 1
\end{array}
$$

Because A_0 was 0, it borrowed 1 from A_1. A_1 becomes a 0 and A_0 becomes 2 (2_{10} or 10_2). Now the subtraction can take place: in the LS column $2 - 1 = 1$, and in the MS column $0 - 0 = 0$.

As you can see, the second column and all more significant columns first have to determine if A was borrowed from before subtracting $A - B$. Therefore, they have three input conditions, for a total of eight different possible combinations, as illustrated in Figure 7–2.

$$
\begin{array}{cc}
B_{in} & B_{in} \\
A_1 & A_0 \\
-\quad B_1 & B_0 \\
\hline
R_1 & R_0 \\
+ & + \\
B_{out} & B_{out}
\end{array}
$$

A_1	B_1	B_{in}	R_1	B_{out}
0	0	0	0	0
0	0	1	1	1
0	1	0	1	1
0	1	1	0	1
1	0	0	1	0
1	0	1	0	0
1	1	0	0	0
1	1	1	1	1

Borrow (B_{out}) required because B_{in} needs to borrow from A_1, which is zero.

Figure 7–2 Subtraction in the more significant columns.

The outputs in the truth table in Figure 7–2 are a little more complicated to figure out. To help you along, let's look at the subtraction 4 minus 1 ($0100_2 - 0001_2$):

$$
\begin{array}{cc}
4_{10} & A_3A_2A_1A_0 \\
-1_{10} & -B_3B_2B_1B_0 \\
\hline
3_{10} & R_3R_2R_1R_0
\end{array}
\qquad
\begin{array}{cccc}
& \overset{0}{} & \overset{2}{1}\!\rightarrow\!\overset{\scriptstyle 1}{2} & \\
0 & \cancel{1} & \cancel{0} & \cancel{0} \\
-0 & 0 & 0 & 1 \\
\hline
0 & 0 & 1 & 1 = 3_{10}\ \checkmark
\end{array}
$$

To subtract $0100 - 0001$, A_0 must borrow from A_1; but A_1 is 0. Therefore, A_1 must first borrow from A_2, making A_2 a 0. Now A_1 is a 2. A_0 borrows from A_1, making A_1 a 1 and A_0 a 2. Now we can subtract to get 0011 (3_{10}). Actually, the process is very similar to the process you learned many years ago for regular decimal subtraction. Work through each entry in the truth table (Figure 7–2) to determine how it was derived.

Fortunately, as we will see in Section 7–2, digital computers use a much easier method for subtracting binary numbers, called two's complement. We do, however, need to know the standard method for subtracting binary numbers. Work through the following example to better familiarize yourself with the binary subtraction procedure.

EXAMPLE 7–2

Perform the following decimal subtractions. Convert the original decimal numbers to binary and subtract them. Compare answers. (a) $27 - 10$; (b) $9 - 4$; (c) $172 - 42$; (d) $154 - 54$; (e) $192 - 3$.

Solution:

		Decimal	**Binary**
(a)		27	0001 1011
		-10	$-0000\ 1010$
		17	$0001\ 0001 = 17_{10}$ ✓
(b)		9	0000 1001
		-4	$-0000\ 0100$
		5	$0000\ 0101 = 5_{10}$ ✓
(c)		172	1010 1100
		$-\ 42$	$-0010\ 1010$
		130	$1000\ 0010 = 130_{10}$ ✓
(d)		154	1001 1010
		$-\ 54$	$-0011\ 0110$
		100	$0110\ 0100 = 100_{10}$ ✓
(e)		192	1100 0000
		$-\ 3$	$-0000\ 0011$
		189	$1011\ 1101 = 189_{10}$ ✓

Multiplication

Binary multiplication is like decimal multiplication except you deal only with 1's and 0's. Figure 7–3 illustrates the procedure for multiplying 13×11.

Decimal   *Binary*

          13   0000 1101 (multiplicand)
        $\times 11$ $\times\ 0000\ \ 1011$ (multiplier)
          13   0000 1101
        $\underline{13}$ $\underline{00001\ \ 101}$
         143   000000 00
                       $\underline{0000110\ \ 1}$
                       0001000 1111 (product)

8-bit answer = 1000 1111 = 143_{10} ✓

Figure 7–3 Binary multiplication procedure.

The procedure for the multiplication in Figure 7–3 is as follows:

1. Multiply the 2^0 bit of the multiplier times the multiplicand.

2. Multiply the 2^1 bit of the multiplier times the multiplicand. Shift the result one position to the left before writing it down.

3. Repeat step 2 for the 2^2 bit of the multiplier. Because the 2^2 bit is a 0, the result is 0.

4. Repeat step 2 for the 2^3 bit of the multiplier.

5. Repeating step 2 for the four leading 0's in the multiplier will have no effect on the answer, so don't bother.

6. Take the sum of the four partial products to get the final **product** of 143_{10}. (Written as an 8-bit number, the product is $1000\ 1111_2$.)

EXAMPLE 7–3

Perform the following decimal multiplications. Convert the original decimal numbers to binary and multiply them. Compare answers. **(a)** 5×3; **(b)** 45×3; **(c)** 15×15; **(d)** 23×9.

Solution:

	Decimal	**Binary**
(a)	5	0000 0101
	× 3	×0000 0011
	15	0000 0101
		+00000 101
		00000 1111 = 0000 1111 = 15_{10} ✓
(b)	45	0010 1101
	× 3	×0000 0011
	135	0010 1101
		+00101 101
		01000 0111 = 1000 0111 = 135_{10} ✓
(c)	15	0000 1111
	×15	×0000 1111
	75	0000 1111
	+15	00001 111
	225	000011 11
		+0000111 1
		0001110 0001 = 1110 0001 = 225_{10} ✓
(d)	23	0001 0111
	× 9	×0000 1001
	207	0001 0111
		00000 000
		000000 00
		0001011 1
		0001100 1111 = 1100 1111 = 207_{10} ✓

Common Misconception

Most errors in binary multiplication occur when students are careless in the vertical alignment of the addition columns.

Team Discussion

Develop a method to determine the value to carry when adding columns with several 1's in them, such as those encountered when multiplying 15 × 15.

Division

Binary division uses the same procedure as decimal division. Example 7–4 illustrates this procedure.

It is beneficial to review the procedure for base 10 long division that you learned in grade school.

EXAMPLE 7–4

Perform the following decimal divisions. Convert the original decimal numbers to binary and divide them. Compare answers. **(a)** $9 \div 3$; **(b)** $35 \div 5$; **(c)** $135 \div 15$; **(d)** $221 \div 17$.

Solution:

Decimal   **Binary**

(a)
$$\begin{array}{r} 3 \\ 3\overline{)\,9} \\ -9 \\ \hline 0 \end{array}$$

$$\begin{array}{r} 11 = 3_{10} \;\checkmark \\ 0000\ 0011\overline{)0000\ 1001} \\ -\underline{11} \\ 11 \\ -\underline{11} \\ 0 \end{array}$$

(b)
$$\begin{array}{r} 7 \\ 5\overline{)\,35} \\ -35 \\ \hline 0 \end{array}$$

$$\begin{array}{r} 111 = 7_{10} \;\checkmark \\ 0000\ 0101\overline{)0010\ 0011} \\ -\underline{1\ 01} \\ 111 \\ -\underline{101} \\ 101 \\ -\underline{101} \\ 0 \end{array}$$

(c)
$$\begin{array}{r} 9 \\ 15\overline{)\,135} \\ -135 \\ \hline 0 \end{array}$$

$$\begin{array}{r} 1001 = 9_{10} \;\checkmark \\ 0000\ 1111\overline{)1000\ 0111} \\ -\underline{111\ 1} \\ 1111 \\ -\underline{1111} \\ 0 \end{array}$$

(d)
$$\begin{array}{r} 13 \\ 17\overline{)\,221} \\ -17 \\ \hline 51 \\ 51 \\ \hline 0 \end{array}$$

$$\begin{array}{r} 1101 = 13_{10} \;\checkmark \\ 0001\ 0001\overline{)1101\ 1101} \\ -\underline{1000\ 1} \\ 101\ 01 \\ -\underline{100\ 01} \\ 1\ 0001 \\ -\underline{1\ 0001} \\ 0 \end{array}$$

Review Questions

7–1. Binary addition in the least significant column deals with how many inputs and how many outputs?

7–2. In binary subtraction, the borrow-out of the least significant column becomes the borrow-in of the next-more-significant column. True or false?

7–3. Binary multiplication and division are performed by a series of additions and subtractions. True or false?

7–2   Two's-Complement Representation

The most widely used method of representing binary numbers and performing arithmetic in computer systems is by using the **two's-complement method.** With this method, both positive and negative numbers can be represented using the same format, and binary subtraction is greatly simplified.

All along we have seen representing binary numbers in groups of eight for a reason. Most computer systems are based on 8-bit or 16-bit numbers. In an 8-bit system, the total number of different combinations of bits is 256 (2^8); in a 16-bit system the number is 65,536 (2^{16}).

To be able to represent both positive *and* negative numbers, the two's-complement format uses the most significant bit (MSB) of the 8- or 16-bit number to signify whether the number is positive or negative. The MSB is therefore called the **sign bit** and is defined as 0 for positive numbers and 1 for negative numbers. *Signed two's-complement* numbers are shown in Figure 7–4.

$$D_7 D_6 D_5 D_4 D_3 D_2 D_1 D_0$$

Sign bit

(a)

$$D_{15} D_{14} D_{13} D_{12} D_{11} D_{10} D_9 D_8 D_7 D_6 D_5 D_4 D_3 D_2 D_1 D_0$$

Sign bit

(b)

Figure 7–4   Two's-complement numbers: (a) 8-bit number; (b) 16-bit number.

The *range of positive numbers* in an 8-bit system is 0000 0000 to 0111 1111 (0 to 127). The *range of negative numbers* is 1111 1111 to 1000 0000 (−1 to −128). In general, the maximum positive number is equal to $2^{N-1} - 1$, and the maximum negative number is $-(2^{N-1})$, where N is the number of bits in the number, including the sign bit (for example, for an 8-bit positive number, $2^{8-1} - 1 = 127$).

A table of two's-complement numbers can be developed by starting with some positive number and continuously subtracting 1. Table 7–3 shows the signed two's-complement numbers from +7 to −8.

Table 7–3	Signed Two's-Complement Numbers +7 Through −8
Decimal	Two's Complement
+7	0000 0111
+6	0000 0110
+5	0000 0101
+4	0000 0100
+3	0000 0011
+2	0000 0010
+1	0000 0001
0	0000 0000
−1	1111 1111
−2	1111 1110
−3	1111 1101
−4	1111 1100
−5	1111 1011
−6	1111 1010
−7	1111 1001
−8	1111 1000

Converting a decimal number to two's complement, and vice versa, is simple and can be done easily using logic gates, as we will see later in this chapter. For now, let's deal with 8-bit numbers; however, the procedure for 16-bit numbers is exactly the same.

Steps for Decimal-to-Two's-Complement Conversion

1. If the decimal number is positive, the two's-complement number is the true binary equivalent of the decimal number (for example, +18 = 0001 0010).

2. If the decimal number is negative, the two's-complement number is found by
   (a) Complementing each bit of the true binary equivalent of the decimal number (this is called the **one's complement**).
   (b) Adding 1 to the one's-complement number to get the magnitude bits. (The sign bit will always end up being 1.)

Team Discussion

Try to represent the number 160_{10} in two's-complement for an 8-bit system. Why doesn't it work?

Steps for Two's-Complement-to-Decimal Conversion

1. If the two's-complement number is positive (sign bit = 0), do a regular binary-to-decimal conversion.

2. If the two's-complement number is negative (sign bit = 1), the decimal sign will be − and the decimal number is found by
   (a) Complementing the entire two's-complement number, bit by bit.
   (b) Adding 1 to arrive at the true binary equivalent.
   (c) Doing a regular binary-to-decimal conversion to get the decimal numeric value.

The following examples illustrate the conversion process.

Common Misconception

As soon as some students see the phrase "convert to two's complement," they go ahead with the procedure for negative numbers whether the original number is positive or negative.

EXAMPLE 7–5

Convert $+35_{10}$ to two's complement.

Solution:

$$
\begin{aligned}
\text{True binary} &= 0010\ \ 0011 \\
\text{Two's complement} &= 0010\ \ 0011 \quad \textit{Answer}
\end{aligned}
$$

EXAMPLE 7–6

Convert -35_{10} to two's complement.

Solution:

$$
\begin{aligned}
\text{True binary} &= 0010\ \ 0011 \\
\text{One's complement} &= 1101\ \ 1100 \\
\text{Add 1} &= \qquad\quad +1 \\
\text{Two's complement} &= 1101\ \ 1101 \quad \textit{Answer}
\end{aligned}
$$

EXAMPLE 7–7

Convert 1101 1101 two's complement back to decimal.

Solution: The sign bit is 1, so the decimal result will be negative.

$$
\begin{aligned}
\text{Two's complement} &= 1101\ \ 1101 \\
\text{Complement} &= 0010\ 0010 \\
\text{Add 1} &= \qquad\quad +1 \\
\text{True binary} &= 0010\ 0011 \\
\text{Decimal equivalent} &= -35 \quad \textit{Answer}
\end{aligned}
$$

EXAMPLE 7–8

Convert -98_{10} to two's complement.

Solution:

$$
\begin{aligned}
\text{True binary} &= 0110\ \ 0010 \\
\text{One's complement} &= 1001\ \ 1101 \\
\text{Add 1} &= \qquad\quad +1 \\
\text{Two's complement} &= 1001\ \ 1110 \quad \textit{Answer}
\end{aligned}
$$

EXAMPLE 7–9

Convert 1011 0010 two's complement to decimal.

Solution: The sign bit is 1, so the decimal result will be negative.

$$
\begin{aligned}
\text{Two's complement} &= 1011\ \ 0010 \\
\text{Complement} &= 0100\ \ 1101 \\
\text{Add 1} &= \qquad\quad +1 \\
\text{True binary} &= 0100\ \ 1110 \\
\text{Decimal equivalent} &= -78 \quad \textit{Answer}
\end{aligned}
$$

Review Questions

7–4. Which bit in an 8-bit two's-complement number is used as the sign bit?

7–5. Are the following two's-complement numbers positive or negative?

(a) 1010 0011

(b) 0010 1101

(c) 1000 0000

7–3   Two's-Complement Arithmetic

All four of the basic arithmetic functions involving positive *and* negative numbers can be dealt with very simply using two's-complement arithmetic. Subtraction is done by *adding* the two two's-complement numbers. Thus the same digital circuitry can be used for additions *and* subtractions, and there is no need always to subtract the smaller number from the larger number. We must be careful, however, not to exceed the *maximum range* of the two's-complement number: +127 to −128 for 8-bit systems and +32,767 to −32,768 for 16-bit systems ($+2^{N-1} - 1$ to -2^{N-1}).

When *adding* numbers in the two's-complement form, simply perform a regular binary addition to get the result. When *subtracting* numbers in the two's-complement form, convert the number being subtracted to a *negative* two's-complement number and perform a regular binary addition [for example, $5 - 3 = 5 + (-3)$]. The result will be a two's-complement number, and if the result is negative, the sign bit will be 1.

Work through the following examples to familiarize yourself with the addition and subtraction procedure.

E X A M P L E   7 – 1 0

Add $19 + 27$ using 8-bit two's-complement arithmetic.

Solution:

$$
\begin{array}{r}
19 = 0001\ 0011 \\
27 = \underline{0001\ 1011} \\
\text{Sum} = \overline{0010\ 1110} = 46_{10}
\end{array}
$$

E X A M P L E   7 – 1 1

Perform the following subtractions using 8-bit two's-complement arithmetic: **(a)** $18 - 7$; **(b)** $21 - 13$; **(c)** $118 - 54$; **(d)** $59 - 96$.

Solution:

(a) $18 - 7$ is the same as $18 + (-7)$, so just add 18 to negative 7.

$$
\begin{array}{r}
+18 = 0001\ 0010 \\
-7 = \underline{1111\ 1001} \\
\text{Sum} = \overline{0000\ 1011} = 11_{10}
\end{array}
$$

Note: The carry-out of the MSB is ignored. (It will always occur for positive sums.) The 8-bit answer is 0000 1011.

(b) $+21 = 0001\ 0101$
$-13 = \underline{1111\ 0011}$
$\text{Sum} = 0000\ 1000 = 8_{10}$

(c) $+118 = 0111\ 0110$
$-54 = \underline{1100\ 1010}$
$\text{Sum} = 0100\ 0000 = 64_{10}$

(d) $+59 = 0011\ 1011$
$-96 = \underline{1010\ 0000}$
$\text{Sum} = 1101\ 1011 = -37_{10}$

Review Questions

7–6. Which of the following decimal numbers cannot be converted to 8-bit two's-complement notation?

(a) 89

(b) 135

(c) −107

(d) −144

7–7. The procedure for subtracting numbers in two's-complement notation is exactly the same as for adding numbers. True or false?

7–8. When subtracting a smaller number from a larger number in two's complement, there will always be a carry-out of the MSB, which will be ignored. True or false?

7–4 Hexadecimal Arithmetic

Hexadecimal representation, as discussed in Chapter 1, is a method of representing groups of 4 bits as a single digit. Hexadecimal notation has been widely adopted by manufacturers of computers and microprocessors because it simplifies the documentation and use of their equipment. Eight- and 16-bit computer system data, program instructions, and addresses use hexadecimal to make them easier to interpret and work with than their binary equivalents.

Hexadecimal Addition

Remember, hexadecimal is a base 16 numbering system, meaning that it has 16 different digits (as shown in Table 7–4). Adding $3 + 6$ in hex equals 9, and $5 + 7$ equals C. But adding $9 + 8$ in hex equals a sum greater than F, which will create a carry. The sum of $9 + 8$ is 17_{10}, which is 1 larger than 16, making the answer 11_{16}.

Table 7–4 Hexadecimal Digits with Their Equivalent Binary and Decimal Values

Hexadecimal	Binary	Decimal
0	0000	0
1	0001	1
2	0010	2
3	0011	3
4	0100	4
5	0101	5
6	0110	6
7	0111	7
8	1000	8
9	1001	9
A	1010	10
B	1011	11
C	1100	12
D	1101	13
E	1110	14
F	1111	15

The procedure for adding hex digits is as follows:

1. Add the two hex digits by working with their decimal equivalents.

2. If the decimal sum is less than 16, write down the hex equivalent.

3. If the decimal sum is 16 or more, subtract 16, write down the hex result in that column, and carry 1 to the next-more-significant column.

Work through the following examples to familiarize yourself with this procedure.

E X A M P L E   7 – 1 2

Add 9 + C in hex.

Solution: C is equivalent to decimal 12.

$$12 + 9 = 21$$

Because 21 is greater than 16: (1) subtract $21 - 16 = 5$, and (2) carry 1 to the next-more-significant column. Therefore,

$$9 + C = 15_{16}$$

E X A M P L E   7 – 1 3

Add 4F + 2D in hex.

Solution:

$$\begin{array}{r} 4\ F \\ +2\ D \\ \hline 7\ C \end{array}$$

Explanation: $F + D = 15 + 13 = 28$, which is 12 with a carry ($28 - 16 = 12$). 12 is written down as C. $4 + 2 +$ carry $= 7$.

Add A7C5 + 2DA8 in hex.

Solution:

$$
\begin{array}{r}
\text{A 7 C 5} \\
+\text{2 D A 8} \\
\hline
\text{D 5 6 D}
\end{array}
$$

Explanation: 5 + 8 = 13, which is D. C + A = 22, which is 6 with a carry. 7 + D + carry = 21, which is 5 with a carry. A + 2 + carry = 13, which is D.

Alternative Method: An alternative method of hexadecimal addition, which you might find more straightforward, is to convert the hex numbers to binary and perform a regular binary addition. The binary sum is then converted back to hex. For example:

$$
\begin{array}{r}
\text{4 F} \\
+\text{2 D} \\
\hline
\end{array}
\Rightarrow
\begin{array}{r}
0100\ 1111_2 \\
+0010\ 1101_2 \\
\hline
0111\ 1100_2 = 7\,C_{16} \quad \checkmark
\end{array}
$$

Hexadecimal Subtraction

Subtraction of hexadecimal numbers is similar to decimal subtraction except that, when you borrow 1 from the left, the borrower increases in value by 16. Consider the hexadecimal subtraction 24 − 0C.

$$
\begin{array}{r}
24 \\
-0C \\
\hline
18
\end{array}
$$

Explanation: We cannot subtract C from 4, so the 4 borrows 1 from the 2. This changes the 2 to a 1, and the 4 increases in value to 20 (4 + 16 = 20). Now, 20 − C = 20 − 12 = 8, and 1 − 0 = 1. Therefore,

$$24 - 0C = 18$$

The next two examples illustrate hexadecimal subtraction.

Subtract D7 − A8 in hex.

Solution:

$$
\begin{array}{r}
\text{D 7} \\
-\text{A 8} \\
\hline
\text{2 F}
\end{array}
$$

Explanation: 7 borrows from the D, which increases its value to 23 (7 + 16 = 23). 23 − 8 = 15, which is an F. D becomes a C, and C − A = 12 − 10 = 2.

EXAMPLE 7–16

Subtract A05C − 24CA in hex.

Solution:

$$
\begin{array}{r}
\text{A } 0 \text{ } 5 \text{ } \text{C} \\
-2 \text{ } 4 \text{ } \text{C } \text{A} \\
\hline
7 \text{ } \text{B } 9 \text{ } 2
\end{array}
$$

Explanation: C − A = 12 − 10 = 2. 5 borrows from the 0, which borrows from the A (5 + 16 = 21); 21 − C = 21 − 12 = 9. The 0 borrowed from the A, but it was also borrowed from, so it is now a 15; 15 − 4 = 11, which is a B. The A was borrowed from, so it is now a 9; 9 − 2 = 7.

Review Questions

7–9. Why is hexadecimal arithmetic commonly used when working with 8-, 16-, and 32-bit computer systems?

7–10. When adding two hex digits, if the sum is greater than _____ (9, 15, 16), the result will be a two-digit answer.

7–11. When subtracting hex digits, if the least significant digit borrows from its left, its value increases by _____ (10, 16).

7–5   BCD Arithmetic

If human beings had 16 fingers and toes, we probably would have adopted hexadecimal as our primary numbering system instead of decimal, and dealing with microprocessor-generated numbers would have been so much easier. (Just think how much better we could play a piano, too!) But, unfortunately, we normally deal in base 10 decimal numbers. Digital electronics naturally works in binary, and we have to group four binary digits together to get enough combinations to represent the 10 different decimal digits. This 4-bit code is called *binary-coded decimal* (BCD).

So what we have is a 4-bit code that is used to represent the decimal digits that we need when reading a display on calculators or computer output. The problem arises when we try to add or subtract these BCD numbers. For example, digital circuitry would naturally like to add the BCD numbers 1000 + 0011 to get 1011, but 1011 is an invalid BCD result. (In Chapter 1 we described the range of valid BCD numbers as 0000 to 1001.) Therefore, when adding BCD numbers, we have to build extra circuitry to check the result to be certain that each group of 4 bits is a valid BCD number.

BCD Addition

Addition is the most important operation because subtraction, multiplication, and division can all be done by a series of additions or two's-complement additions.

The procedure for BCD addition is as follows:

1. Add the BCD numbers as regular true binary numbers.

2. If the sum is 9 (1001) or less, it is a valid BCD answer; leave it as is.

3. If the sum is greater than 9 or if there is a carry-out of the MSB, it is an invalid BCD number; do step 4.

4. If it is invalid, add 6 (0110) to the result to make it valid. Any carry-out of the MSB is added to the next-more-significant BCD number.

5. Repeat steps 1 to 4 for each group of BCD bits.

Use this procedure for the following example.

EXAMPLE 7–17

Convert the following decimal numbers to BCD and add them. Convert the result back to decimal to check your answer. **(a)** $8 + 7$; **(b)** $9 + 9$; **(c)** $52 + 63$; **(d)** $78 + 69$.

Solution:

(a)

$$
\begin{array}{rl}
8 = & 1000 \\
+7 = & \underline{0111} \\
\text{Sum} = & 1111 \quad \text{(invalid)} \\
\text{Add } 6 = & \underline{0110} \\
1 \quad & 0101 = 0001\ 0101_{\text{BCD}} = 15_{10} \quad \checkmark
\end{array}
$$

(b)

$$
\begin{array}{rl}
9 = & 1001 \\
+9 = & \underline{1001} \\
\text{Sum} = & 1\_0010 \quad \text{(invalid because of carry)} \\
& \text{cy} \\
\text{Add } 6 = & \underline{0110} \\
1 \quad & 1000 = 0001\ 1000_{\text{BCD}} = 18_{10} \quad \checkmark
\end{array}
$$

(c)

$$
\begin{array}{rl}
52 = & 0101\ 0010 \\
+63 = & \underline{0110\ 0011} \\
\text{Sum} = & 1011\_0101 \\
\text{Add } 6 = & \underline{0110} \qquad \text{invalid} \\
1 \quad & 0001\ 0101 = 0001\ 0001\ 0101 = 115_{10} \quad \checkmark
\end{array}
$$

(d)

$$
\begin{array}{rl}
78 = & 0111\ 1000 \\
+69 = & \underline{0110\ 1001} \\
\text{Sum} = & 1110\_0001 \quad \text{(Both groups of 4} \\
& \text{cy} \qquad \text{BCD bits are invalid.)} \\
\text{Add } 6 = & \underline{0110} \\
& 1110\ 0111 \\
\text{Add } 6 = & \underline{0110} \\
1 \quad & 0100\ 0111 = 0001\ 0100\ 0111 = 147_{10} \quad \checkmark
\end{array}
$$

When one of the numbers being added is negative (such as in subtraction), the procedure is much more difficult, but basically it follows a complement-then-add procedure, which is not covered in this book but is similar to that introduced in Section 7–3.

Now that we understand the more common arithmetic operations that take place within digital equipment, we are ready for the remainder of the chapter, which explains the actual circuitry used to perform these operations.

7–12. When adding two BCD digits, the sum is invalid and needs correction if it is _____ or if _____.

7–13. What procedure is used to correct the result of a BCD addition if the sum is greater than 9?

7–6 Arithmetic Circuits

All the arithmetic operations and procedures covered in the previous sections can be implemented using adders formed from the basic logic gates. For a large number of digits we can use medium-scale-integration (**MSI**) circuits, which actually have several adders within a single integrated package.

Basic Adder Circuit

By reviewing the truth table in Figure 7–5, we can determine the input conditions that produce each combination of sum and carry output bits. Figure 7–5 shows the addition of two 2-bit numbers. This could easily be expanded to cover 4-, 8-, or 16-bit addition. Notice that addition in the least significant–bit column requires analyzing only two inputs (A_0 plus B_0) to determine the output sum (Σ_0) and carry (C_{out}). But any more

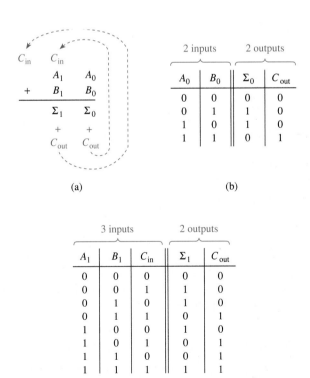

(a)

2 inputs		2 outputs	
A_0	B_0	Σ_0	C_{out}
0	0	0	0
0	1	1	0
1	0	1	0
1	1	0	1

(b)

3 inputs			2 outputs	
A_1	B_1	C_{in}	Σ_1	C_{out}
0	0	0	0	0
0	0	1	1	0
0	1	0	1	0
0	1	1	0	1
1	0	0	1	0
1	0	1	0	1
1	1	0	0	1
1	1	1	1	1

(c)

Figure 7–5 (a) Addition of two 2-bit binary numbers; (b) truth table for the LSB addition; (c) truth table for the more significant column.

significant columns (2^1 column and up) require the inclusion of a third input, which is the carry (C_{in}) from the column to its right. For example, the carry-out (C_{out}) of the 2^0 column becomes the carry-in (C_{in}) to the 2^1 column. Figure 7–5(c) shows the inclusion of a third input for the truth table of the more significant column additions.

Half-Adder

Designing logic circuits to automatically implement the desired outputs for these truth tables is simple. Look at the LSB truth table; for what input conditions is the Σ_0 bit HIGH? The answer is *A or B* HIGH but *not both* (exclusive-OR function). For what input condition is the C_{out} bit HIGH? The answer is *A and B* HIGH (AND function). Therefore, the circuit design to perform addition in the LSB column can be implemented using an exclusive-OR and an AND gate. That circuit is called a **half-adder** and is shown in Figure 7–6. If the exclusive-OR function in Figure 7–6 is implemented using an AND–NOR–NOR configuration, we can tap off the AND gate for the carry, as shown in Figure 7–7. [The AND–NOR–NOR configuration is an Ex-OR, as proved in Figure 6–5(c).]

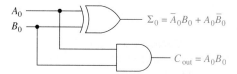

Figure 7–6 Half-adder circuit for addition in the LSB column.

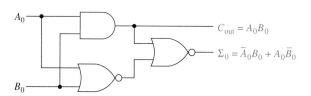

Figure 7–7 Alternative half-adder circuit built from an AND–NOR–NOR configuration.

Full-Adder

As you can see in Figure 7–5, addition in the 2^1 (or higher) column requires three inputs to produce the sum (Σ_1) and carry (C_{out}) outputs. Look at the truth table [Figure 7–5(c)]; for what input conditions is the sum output (Σ_1) HIGH? The answer is that the Σ_1 bit is HIGH whenever the three inputs (A_1, B_1, C_{in}) are *odd*. From Chapter 6 you may remember that an even-parity generator produces a HIGH output whenever the sum of the inputs is odd. Therefore, we can use an even-parity generator to generate our Σ_1 output bit, as shown in Figure 7–8.

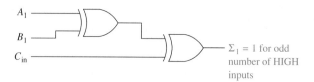

Figure 7–8 The sum (Σ_1) function of the full-adder is generated from an even-parity generator.

How about the carry-out (C_{out}) bit? What input conditions produce a HIGH at C_{out}? The answer is that C_{out} is HIGH whenever any two of the inputs are HIGH. Therefore, we can take care of C_{out} with three ANDs and an OR, as shown in Figure 7–9.

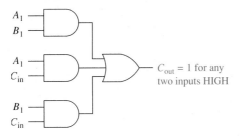

Figure 7–9 Carry-out (C_{out}) function of the full-adder.

The two parts of the full-adder circuit shown in Figures 7–8 and 7–9 can be combined to form the complete *full-adder* circuit shown in Figure 7–10. In the figure the Σ_1 function is produced using the same logic as that in Figure 7–8 (an Ex-OR feeding an Ex-OR). The C_{out} function comes from $A_1 B_1$ or C_{in} ($A_1\overline{B}_1 + \overline{A}_1 B_1$). Prove to yourself that the Boolean equation at C_{out} will produce the necessary result. [*Hint:* Write the equation for C_{out} from the truth table in Figure 7–5(c).] Also, Example 7–18 will help you better understand the operation of the full-adder.

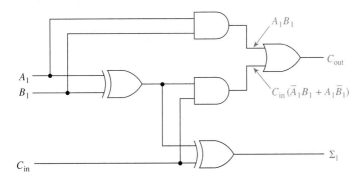

Figure 7–10 Logic diagram of a full-adder.

EXAMPLE 7–18

Apply the following input bits to the full-adder of Figure 7–10 to verify its operation ($A_1 = 0$, $B_1 = 1$, $C_{in} = 1$).

Solution: The full-adder operation is shown in Figure 7–11.

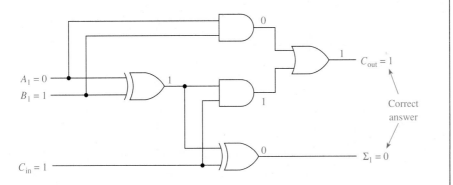

Figure 7–11 Full-adder operation for Example 7–18.

Block Diagrams

Now that we know the construction of half-adder and full-adder circuits, we can simplify their representation by just drawing a box with the input and output lines, as shown in Figure 7–12. When drawing multibit adders, a **block diagram** is used to represent the addition in each column. For example, in the case of a 4-bit adder, the 2^0 column needs only a half-adder, because there will be no carry-in. Each of the more significant columns requires a full-adder, as shown in Figure 7–13.

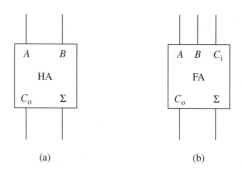

(a)   (b)

Figure 7–12   Block diagrams of (a) half-adder; (b) full-adder.

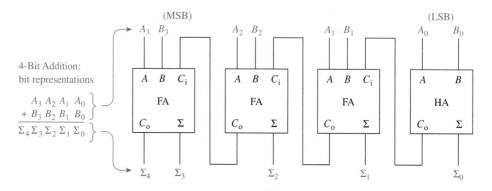

Figure 7–13   Block diagram of a 4-bit binary adder.

Notice in Figure 7–13 that the LSB half-adder has no carry-in. The carry-out (C_{out}) of the LSB becomes the carry-in (C_{in}) to the next full-adder to its left. The carry-out (C_{out}) of the MSB full-adder is actually the highest-order sum output (Σ_4).

Review Questions

7–14. Name the inputs and outputs of a half-adder.

7–15. Why are the input requirements of a full-adder different from those of a half-adder?

7–16. The sum output (Σ) of a full-adder is 1 if the sum of its three inputs is _____ (odd, even).

7–17. What input conditions to a full-adder produce a 1 at the carry-out (C_{out})?

7–7   Four-Bit Full-Adder ICs

Medium-scale-integration (MSI) ICs are available with four full-adders in a single package. Table 7–5 lists the most popular adder ICs. Each adder in the table contains four full-adders, and all are functionally equivalent; however, their pin layouts differ (refer to your data manual for the pin layouts). They each will add two 4-bit **binary words** plus one incoming carry. The binary sum appears on the sum outputs (Σ_1 to Σ_4) and the outgoing carry.

Table 7–5   MSI Adder ICs

Device	Family	Description
7483	TTL	4-Bit binary full-adder, fast carry
74HC283	CMOS	4-Bit binary full-adder, fast carry
4008	CMOS	4-Bit binary full-adder, fast carry

Figure 7–14 shows the functional diagram, the logic diagram, and the logic symbol for the 7483. In the figure the least significant binary inputs (2^0) come into the $A_1 B_1$ terminals, and the most significant (2^3) come into the $A_4 B_4$ terminals. (Be careful;

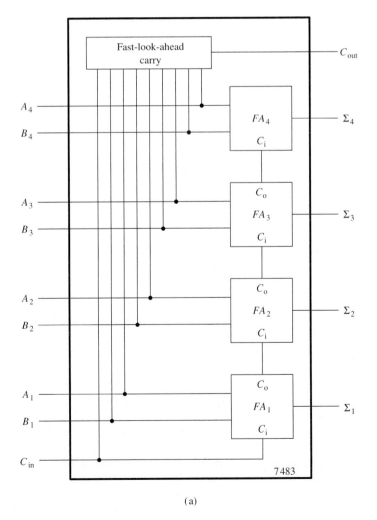

(a)

Figure 7–14   The 7483 4-bit full-adder: (a) functional diagram;

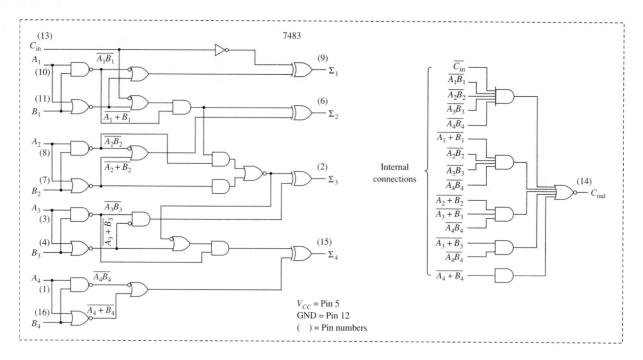

(b)

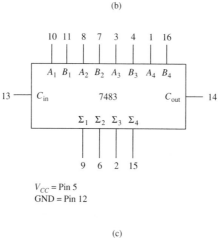

(c)

Figure 7–14 *(Continued)* (b) logic diagram; (c) logic symbol. [(b) Courtesy of Signetics Corporation.]

depending on which manufacturer's data manual you are using, the inputs may be labeled A_1B_1 to A_4B_4 or A_0B_0 to A_3B_3). The carry-out (C_{out}) from each full-adder is *internally connected* to the carry-in of the next full-adder. The carry-out of the last full-adder is brought out to a terminal to be used as the sum$_5$ (Σ_5) output or to be used as a carry-in (C_{in}) to the next full-adder IC if more than 4 bits are to be added (as in Example 7–19).

Something else that we have not seen before is the **fast-look-ahead carry** [see Figure 7–14(a)]. This is very important for speeding up the arithmetic process. For example, if we were adding two 8-bit numbers using two 7483s, the fast-look-ahead carry evaluates the four **low-order** inputs (A_1B_1 to A_4B_4) to determine if they are going to produce a carry-out of the fourth full-adder to be passed on to the next-higher-order adder IC (see Example 7–19). In this way the addition of the **high-order** bits (2^4 to 2^7) can take place concurrently with the low-order (2^0 to 2^3) addition *without having to wait* for the carries to propagate, or **ripple** through FA$_1$ to FA$_2$ to FA$_3$ to FA$_4$ to become available to the high-order addition. A discussion of the connections for the addition of two 8-bit numbers using two 7483s is presented in the following example.

CHAPTER 7 | ARITHMETIC OPERATIONS AND CIRCUITS

EXAMPLE 7–19

Show the external connections to two 4-bit adder ICs to form an 8-bit adder capable of performing the following addition:

$$A_7A_6A_5A_4A_3A_2A_1A_0$$
$$+\ \underline{B_7B_6B_5B_4B_3B_2B_1B_0}$$
$$\Sigma_8\Sigma_7\Sigma_6\Sigma_5\Sigma_4\Sigma_3\Sigma_2\Sigma_1\Sigma_0$$

Solution: We can choose any of the IC adders listed in Table 7–5 for our design. Let's choose the 74HC283, which is the high-speed CMOS version of the 4-bit adder (it has the same logic symbol as the 7483). As you can see in Figure 7–15, the two 8-bit numbers are brought into the A_1B_1-to-A_4B_4 inputs of each chip, and the sum output comes out of the Σ_4-to-Σ_1 outputs of each chip.

Team Discussion

What if you only wanted to add two 6-bit numbers? How could you get at the internal carry to output to E_6?

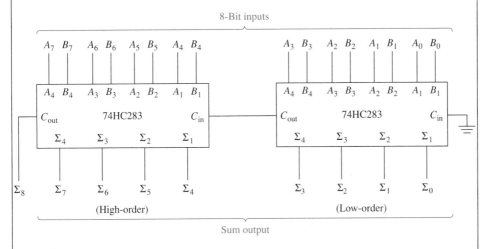

Figure 7–15 Eight-bit binary adder using two 74HC283 ICs.

The C_{in} of the least significant addition ($A_0 + B_0$) is grounded (0) because there is no carry-in (it acts like a half-adder), and if it were left floating, the IC would not know whether to assume a 1 state or 0 state.

The carry-out (C_{out}) from the addition of $A_3 + B_3$ must be connected to the carry-in (C_{in}) of the $A_4 + B_4$ addition, as shown. The fast-look-ahead carry circuit ensures that the carry-out (C_{out}) signal from the low-order addition is provided in the carry-in (C_{in}) of the high-order addition within a very short period of time so that the $A_4 + B_4$ addition can take place without having to wait for all the internal carries to propagate through all four of the low-order additions first. (The actual time requirements for the sum and carry outputs are discussed in Chapter 9 when we look at IC specifications.)

Review Questions

7–18. All the adders in the 7483 4-bit adder are full-adders. What is done with the carry-in (C_{in}) to make the first adder act like a half-adder?

7–19. What is the purpose of the fast-look-ahead carry in the 7483 IC?

7–8   System Design Applications

Each arithmetic operation discussed in Sections 7–1 through 7–5 can be performed by using circuits built from integrated-circuit adders and logic gates. First, we will design a circuit to perform two's-complement arithmetic, and next we will design a BCD adder.

Two's-Complement Adder/Subtractor Circuit

A quick review of Section 7–3 reminds us that positive two's-complement numbers are exactly the same as regular true binary numbers and can be added using regular binary addition. Also, subtraction in two's-complement arithmetic is performed by converting the number to be subtracted to a *negative* number in the two's-complement form and then using regular binary addition. Therefore, once our numbers are in two's-complement form, we can use a binary adder to get the answer whether we are adding *or* subtracting.

For example, to subtract 18 minus 9, we would first convert 9 to a negative two's-complement number by complementing each bit and then adding 1. We would then add $18 + (-9)$:

$$\text{Two's complement of } 18 = 0001\ 0010$$
$$+ \text{ Two's complement of } -9 = \underline{1111\ 0111}$$
$$\text{Sum} = 0000\ 1001 = +9_{10} \quad \textit{Answer}$$

So it looks like all we need for a combination adder/subtractor circuit is an input switch or signal to signify addition or subtraction so that we will know whether to form a positive or a negative two's complement of the second number. Then we will just use a binary adder to get the final result.

To form negative two's complement, we can use the controlled inverter circuit presented in Figure 6–15 and add 1 to its output. Figure 7–16 shows the complete circuit

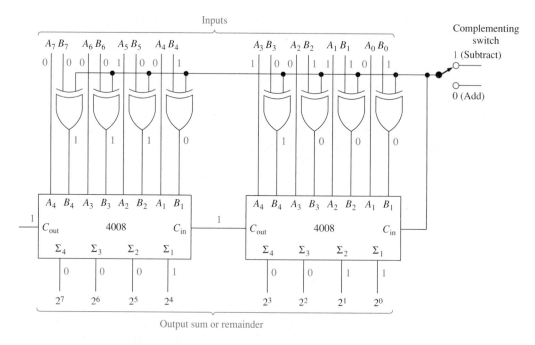

Figure 7–16   Eight-bit two's-complement adder/subtractor illustrating the subtraction 42 − 23 = 19.

used to implement a two's-complement adder/subtractor using two 4008 CMOS adders. The 4008s are CMOS 4-bit binary adders. The 8-bit number on the A inputs (A_7 to A_0) is brought directly into the adders. The other 8-bit binary number comes in on the B_7 to B_0 lines. If the B number is to be subtracted, the complementing switch will be in the up (1) position, causing each bit in the B number to be complemented (one's complement). At the same time, the low-order C_{in} receives a 1, which has the effect of adding a 1 to the already complemented B number, making it a negative two's-complement number.

Now the 4008s perform a regular binary addition. If the complementing switch is up, the number on the B inputs is subtracted from the number on the A inputs. If it is down, the sum is taken. As discussed in Section 7–3, the C_{out} of the MSB is ignored. The result can range from 0111 1111 (+127) to 1000 0000 (−128).

EXAMPLE 7–20

Prove that the subtraction 42 − 23 produces the correct answer at the outputs by labeling the input and output lines on Figure 7–16.

Solution: 42 − 23 should equal 19 (0001 0011). Convert the decimal input numbers to regular binary and label Figure 7–16 (42 = 0010 1010, 23 = 0001 0111). The B input number is complemented, the LSB C_{in} is 1, and the final answer is 0001 0011, which proves that the circuit works for that number.

Try adding and subtracting some other numbers to better familiarize yourself with the operation of the circuit of Figure 7–16.

BCD Adder Circuit

BCD adders can also be formed using the integrated circuit 4-bit binary adders. The problem, as you may remember from Section 7–5, is that when any group-of-four BCD sum exceeds 9, or when there is a carry-out, the number is invalid and must be corrected by adding 6 to the invalid answer to get the correct BCD answer. (The valid range of BCD numbers is 0000 to 1001.)

For example, adding $0111_{BCD} + 0110_{BCD}$ (7 + 6) gives us an invalid result:

$$
\begin{array}{rl}
0111 & \\
+0110 & \\
\hline
1101 & \text{invalid} \\
+0110 & \text{add 6 to correct} \\
\hline
1\ \ 0011 & \\
\end{array}
$$

carry to next BCD digit

The corrected answer is $0001\ 0011_{BCD}$, which equals 13.

Checking for a sum greater than 9, or a carry-out, can be done easily using logic gates. Then, when an invalid sum occurs, it can be corrected by adding 6 (0110) via the connections shown in Figure 7–17. The upper 7483 performs a basic 4-bit addition. If its sum is greater than 9, the Σ_4 (2^3) output *and* either the Σ_3 or Σ_2 (2^2 or 2^1) output will be HIGH. A sum greater than 9 *or* a carry-out will produce a HIGH out of the left OR gate, placing a HIGH–HIGH at the A_3 and A_2 inputs of the correction adder, which has the effect of adding a 6 to the original addition. If there is no carry and the original sum is not greater than 9, the correction adder adds 0000.

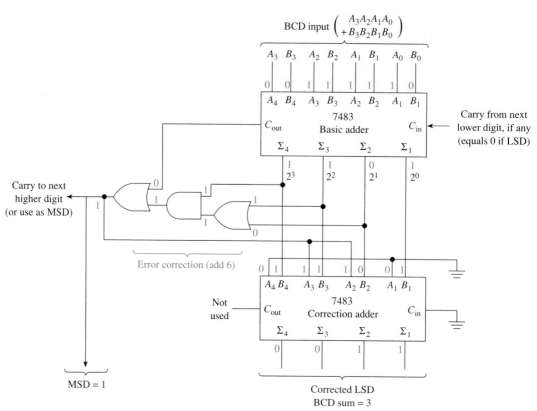

Figure 7–17   BCD adder illustrating the addition $7 + 6 = 13$ ($0111 + 0110 = 0001\ 0011$ BCD).

Common Misconception

Students typically have a hard time seeing where the error correction number 6 (0110) is input to the correction adder at A_4–A_1.

EXAMPLE 7–21

Prove that the BCD addition $0111 + 0110$ ($7 + 6$) produces the correct answer at the outputs by labeling the input and output lines on Figure 7–17.

Solution: The sum out of the basic adder is 13 (1101). Because the 2^3 bit and the 2^2 bit are both HIGH, the error correction OR gate puts out a HIGH, which is added to the next more significant BCD digit and also puts a HIGH–HIGH at A_3, A_2 of the correction adder, which adds 6. The correct answer has a 3 for the least significant digit (LSD) and a 1 in the next more significant digit, for the correct answer of 13.

Familiarize yourself with the operation of Figure 7–17 by testing the addition of several other BCD numbers.

BCD Adder IC: A 4-bit BCD adder is available in a single IC package. The 74HCT583 IC has internal correction circuitry to add two 4-bit numbers and produce a corrected 4-bit answer with carry-out. Refer to a high-speed CMOS data book for an in-depth description of the chip.

7–20. The complementing switch in Figure 7–16 is placed in the 1 position to subtract B from A. Explain how this position converts the binary number on the B inputs into a signed two's-complement number.

7–21. What is the purpose of the AND and OR gates in the BCD adder circuit of Figure 7–17?

7–9 Arithmetic/Logic Units

Arithmetic/logic units (ALUs) are available in large-scale integrated-circuit packages (LSI). Typically, an **ALU** is a multipurpose device capable of providing several different arithmetic and logic operations. The specific operation to be performed is chosen by the user by placing a specific binary code on the mode select inputs. Microprocessors may also have ALUs built in as one of their many operational units. In such cases, the specific operation to be performed is chosen by software instructions.

The ALU that we learn to use in this section is the 74181 (TTL) or 74HC181 (CMOS). The 74181 is a 4-bit ALU that provides 16 arithmetic plus 16 logic operations. Its logic symbol and function table are given in Figure 7–18. The mode control input *(M)* is used to set the mode of operation as either *logic (M = H) or arithmetic (M = L)*. When M is HIGH, all internal carries are disabled, and the device performs *logic operations* on the individual bits (A_0 to A_3, B_0 to B_3), as indicated in the function table.

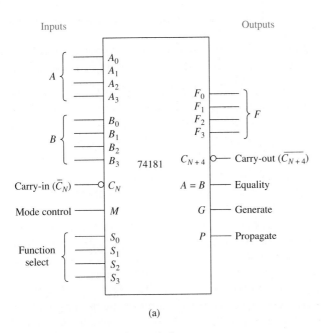

(a)

Figure 7–18 The 74181 ALU: (a) logic symbol.

Selection S_3 S_2 S_1 S_0	$M = H$ Logic functions	$M = L$ Arithmetic operations $\overline{C}_n = H$ (no carry)
L L L L	$F = \overline{A}$	$F = A$
L L L H	$F = \overline{A + B}$	$F = A + B$
L L H L	$F = \overline{A}B$	$F = A + \overline{B}$
L L H H	$F = 0$	$F = $ minus 1 (2's comp.)
L H L L	$F = \overline{AB}$	$F = A$ plus $A\overline{B}$
L H L H	$F = \overline{B}$	$F = (A + B)$ plus $A\overline{B}$
L H H L	$F = A \oplus B$	$F = A$ minus B minus 1
L H H H	$F = A\overline{B}$	$F = A\overline{B}$ minus 1
H L L L	$F = \overline{A} + B$	$F = A$ plus AB
H L L H	$F = \overline{A \oplus B}$	$F = A$ plus B
H L H L	$F = B$	$F = (A + \overline{B})$ plus AB
H L H H	$F = AB$	$F = AB$ minus 1
H H L L	$F = 1$	$F = A$ plus A^*
H H L H	$F = A + \overline{B}$	$F = (A + B)$ plus A
H H H L	$F = A + B$	$F = (A + \overline{B})$ plus A
H H H H	$F = A$	$F = A$ minus 1

*Each bit is shifted to the next-more-significant position.

(b)

Figure 7–18 *(Continued)* (b) function table.

When M is LOW, the internal carries are enabled and the device performs *arithmetic operations* on the two 4-bit binary inputs. Ripple carry output is provided at $\overline{C}_{N+4}$, and fast-look-ahead carry is provided at G and P for high-speed arithmetic operations. The carry-in and carry-out terminals are each active-LOW (as signified by the bubble), which means that a 0 signifies a carry.

Once the **mode control** *(M)* is set, you have 16 choices within either the logic or arithmetic categories. The specific function you want is selected by applying the appropriate binary code to the **function select** inputs (S_3 to S_0).

For example, with $M = H$ and $S_3S_2S_1S_0 = LLLL$, the F outputs will be equal to the complement of A (see the function table). This means that $F_0 = \overline{A}_0$, $F_1 = \overline{A}_1$, $F_2 = \overline{A}_2$, and $F_3 = \overline{A}_3$. Another example is with $M = H$ and $S_3S_2S_1S_0 = HHHL$; the F outputs will be equal to $A + B$ (A or B). This means that $F_0 = A_0 + B_0$, $F_1 = A_1 + B_1$, $F_2 = A_2 + B_2$, and $F_3 = A_3 + B_3$.

From the function table we can see that other logic operations (AND, NAND, NOR, Ex-OR, Ex-NOR, and several others) are available.

The function table in Figure 7–18(b) also shows the result of the 16 different *arithmetic operations* available when $M = L$. Note that the results listed are with carry-in ($\overline{C}_N$) equal to H (no carry). For $\overline{C}_N = L$, just add 1 to all results. All results produced by the device are in two's-complement notation. Also, in the function table, note that the $+$ *sign* means *logical-OR* and the word "PLUS" means *arithmetic-SUM*.

For example, to subtract B from A ($A_3A_2A_1A_0 - B_3B_2B_1B_0$), set $M = L$ and $S_3S_2S_1S_0 = LHHL$. The result at the F outputs will be the two's complement of A minus B minus 1; therefore, to get just A minus B, we need to add for 1. (This can be done automatically by setting $\overline{C}_N = 0$.) Also, as discussed earlier for two's-complement subtraction, a carry-out (borrow) is generated ($\overline{C}_{N+4} = 0$) when the result is positive or zero. Just ignore it.

Read through the function table to see the other 15 arithmetic operations that are available.

EXAMPLE 7-22

Show the external connections to a 74181 to form a 4-bit subtractor. Label the input and output pins with the binary states that occur when subtracting $13 - 7$ ($A = 13$, $B = 7$).

Solution: The 4-bit subtractor is shown in Figure 7–19. The ALU is set in the subtract mode by setting $M = 0$ and $S_3 S_2 S_1 S_0 = 0110$ (*LHHL*). 13 (1101) is input at A and 7 (0111) is input at B.

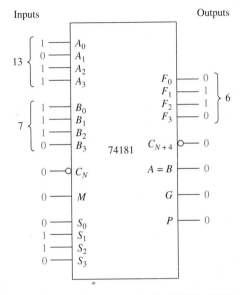

Figure 7–19 Four-bit binary subtractor using the 74181 ALU to subtract $13 - 7$.

By setting $\overline{C_N} = 0$, the output at F_0, F_1, F_2, F_3 will be A minus B instead of A minus B minus 1 as shown in the function table [Figure 7–18(b)]. The result of the subtraction is a positive 6 (0110) with a carry-out ($\overline{C_{N+4}} = 0$). (As before, with two's-complement subtraction, there is a carry-out for any positive or zero answer, which is ignored.)

EXAMPLE 7-23

Place the following values at the inputs of a 74181: A_3–$A_0 = 1001$, B_3–$B_0 = 0011$, S_3–$S_0 = 1101$, and $\overline{C_N} = 1$.

(a) With $M = 1$, determine the output at F (F_3–F_0).

(b) Change M to 0 and determine the output at F (F_3–F_0).

Solution: **(a)** From the chart in Figure 7–18(b), the logic function chosen is $F = A + \overline{B}$ (F equals A ORed with the complement of B).

$$A = 1001$$
$$B = 0011$$
$$\overline{B} = 1100$$
$$F = A + \overline{B} = 1101 \quad \textit{Answer}$$

(b) With $M = 0$, the arithmetic operation is (A ORed with B) with the result added to A.

$$A = 1001$$
$$B = 0011$$
$$A \text{ OR } B = 1011$$
$$A \text{ OR } B \text{ PLUS } A = 0100 \quad \textit{Answer}$$

Review Questions

7–22. What is the purpose of the *mode control* input to the 74181 arithmetic/logic unit?

7–23. If $M = H$ and $S_3, S_2, S_1, S_0 = L, L, H, H$, on the 74181, then F_3, F_2, F_1, F_0 will be set to L, L, L, L. True or false?

7–24. The arithmetic operations of the 74181 include both $F = A + B$ and $F = A$ plus B. How are the two designations different?

Summary

In this chapter we have learned that

1. The binary arithmetic functions of addition, subtraction, multiplication, and division can be performed bit by bit using several of the same rules of regular base 10 arithmetic.

2. The two's-complement representation of binary numbers is commonly used by computer systems for representing positive and negative numbers.

3. Two's-complement arithmetic simplifies the process of subtraction of binary numbers.

4. Hexadecimal addition and subtraction is often required for determining computer memory space and locations.

5. When performing BCD addition a correction must be made for sums greater than 9 or when a carry to the next more significant digit occurs.

6. Binary adders can be built using simple combinational logic circuits.

7. A half-adder is required for addition of the least significant bits.

8. A full-adder is required for addition of the more significant bits.

9. Multibit full-adder ICs are commonly used for binary addition and two's-complement arithmetic.

10. Arithmetic/logic units are multipurpose ICs capable of providing several different arithmetic and logic functions.

Glossary

ALU: Arithmetic/logic unit. A multifunction integrated-circuit device used to perform a variety of user-selectable arithmetic and logic operations.

Binary Word: A group, or string, of binary bits. In a 4-bit system a word is 4 bits long. In an 8-bit system a word is 8 bits long, and so on.

Block Diagram: A simplified functional representation of a circuit or system drawn in a box format.

Borrow: When subtracting numbers, if the number being subtracted from is not large enough, it must "borrow," or take an amount from, the next-more-significant digit.

Carry-In: An amount from a less-significant-digit addition that is applied to the current addition.

Carry-Out: When adding numbers, when the sum is greater than the amount allowed in that position, part of the sum must be applied to the next-more-significant position.

Fast-Look-Ahead Carry: When cascading several full-adders end to end, the carry-out of the last full-adder cannot be determined until each of the previous full-adder additions is completed. The internal carry must ripple or propagate through each of the lower-order adders before reaching the last full-adder. A fast-look-ahead carry system is used to speed up the process in a multibit system by reading all the input bits simultaneously to determine ahead of time if a carry-out of the last full-adder is going to occur.

Full-Adder: An adder circuit having three inputs, used to add two binary digits plus a carry. It produces their sum and carry as outputs.

Function Select: On an ALU the pins used to select the actual arithmetic or logic operation to be performed.

Half-Adder: An adder circuit used in the LS position when adding two binary digits with no carry-in to consider. It produces their sum and carry as outputs.

High Order: In numbering systems, the high-order positions are those representing the larger magnitudes.

Low Order: In numbering systems, the low-order positions are those representing the smaller magnitudes.

Mode Control: On an ALU, the pin used to select either the arithmetic or the logic mode of operation.

MSI: Medium-scale integration. An IC chip containing combinational logic that is packed more densely than a basic logic gate IC (small-scale integration, SSI) but not as dense as a microprocessor IC (large-scale integration, LSI).

One's Complement: A binary number that is a direct (true) complement, bit by bit, of some other number.

Product: The result of the multiplication of numbers.

Remainder: The result of the subtraction of numbers.

Ripple Carry: *See* fast-look-ahead carry.

Sign Bit: The leftmost, or MSB, in a two's-complement number, used to signify the sign of the number (1 = negative, 0 = positive).

Sum: The result of the addition of numbers.

Two's Complement: A binary numbering representation that simplifies arithmetic in digital systems.

Problems

Section 7–1

7–1. Perform the following decimal additions, convert the original decimal numbers to binary, and add them. Compare answers.

(a)	6	(b)	8	(c)	22	(d)	29
	+3		+7		+ 6		+37

(e)	134	(f)	254	(g)	208	(h)	196
	+ 66		+ 36		+127		+156

7–2. Repeat Problem 7–1 for the following subtractions.

(a)	15	(b)	22	(c)	84	(d)	66
	− 4		−11		−36		−31

(e)	126	(f)	113	(g)	109	(h)	111
	− 64		− 88		− 60		−104

7–3. Repeat Problem 7–1 for the following multiplications.

(a)	7	(b)	6	(c)	12	(d)	39
	×3		×7		× 5		× 7

(e)	63	(f)	127	(g)	31	(h)	255
	×125		× 15		×13		×127

7–4. Repeat Problem 7–1 for the following divisions.

(a) $4\overline{)12}$ (b) $3\overline{)15}$ (c) $12\overline{)48}$ (d) $5\overline{)25}$

(e) $5\overline{)125}$ (f) $14\overline{)294}$ (g) $15\overline{)195}$ (h) $12\overline{)228}$

Section 7–2

7–5. Produce a table of 8-bit two's-complement numbers from +15 to −15.

7–6. Convert the following decimal numbers to 8-bit two's-complement notation.

(a) 7   (b) −7   (c) 14   (d) 36   (e) −36

(f) 66   (g) −48   (h) 112   (i) −112   (j) −125

7–7. Convert the following two's-complement numbers to decimal.

(a) 0001 0110   (b) 0000 1111

(c) 0101 1100   (d) 1000 0110

(e) 1110 1110   (f) 1000 0001

(g) 0111 1111   (h) 1111 1111

7–8. What is the maximum positive-to-negative range of a two's-complement number in each?

(a) An 8-bit system   **(b)** A 16-bit system

7–9. Convert the following decimal numbers to two's-complement form and perform the operation indicated.

(a)	5	**(b)**	12	**(c)**	32	**(d)**	32
	+7		– 6		+18		–18

(e)	–28	**(f)**	125	**(g)**	36	**(h)**	–36
	+38		– 66		–48		–48

Section 7–4

7–10. Build a table similar to Table 7–4 for hex digits 0C to 22.

7–11. Add the following hexadecimal numbers.

(a)	A	**(b)**	7	**(c)**	0B	**(d)**	2 3
	+4		+6		+16		+A7

(e)	8A	**(f)**	A7	**(g)**	A0 4 9	**(h)**	0FFF
	+8 2		+BB		+0AFC		+9001

7–12. Subtract the following hexadecimal numbers.

(a)	A	**(b)**	8	**(c)**	1B	**(d)**	A7
	–4		–2		–06		–18

(e)	2A	**(f)**	A 7	**(g)**	4A2D	**(h)**	8BB0
	–07		–1 D		–1A2F		–4AC8

C   **7–13.** Memory locations in personal computers are usually given in hexadecimal. If a computer programmer writes a program that requires 100 memory locations, determine the last memory location that is used if the program starts at location 2C8DH (H = base 16 hexadecimal).

C   **7–14.** A particular model of personal computer indicates in its owner's manual that the following memory locations are used for the storage of operating system subroutines: 07A4BH to 0BD78H inclusive and 02F80H to 03000H inclusive. Determine the total number of memory locations used for that purpose (in hex).

Section 7–5

7–15. Which of the following bit strings cannot be valid BCD numbers?

(a) 0111 1001   **(d)** 0100 1000

(b) 0101 1010   **(e)** 1011 0110

(c) 1110 0010   **(f)** 0100 1001

7–16. Convert the following decimal numbers to BCD and add them. Convert the result back to decimal to check your answer.

(a)	8	**(b)**	12	**(c)**	43	**(d)**	47
	+3		+16		+72		+38

(e)	12	**(f)**	36	**(g)**	99	**(h)**	80
	+89		+22		+11		+23

7–17. Under what circumstances would you use a half-adder instead of a full-adder?

7–18. Reconstruct the half-adder circuit of Figure 7–7 using only NOR gates.

C **7–19.** The circuit in Figure P7–19 is an attempt to build a half-adder. Will the C_{out} and Σ_0 function properly? (*Hint:* Write the Boolean equation at C_{out} and Σ_0.)

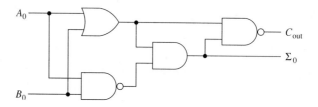

Figure P7–19

C **7–20.** Use a Karnaugh map to prove that the C_{out} function of the full-adder whose truth table is given in Figure 7–5(c) can be implemented using the circuit given in Figure 7–9.

Section 7–7

7–21. Draw the block diagram of a 4-bit full-adder using *four full-adders*.

7–22. In Figure 7–15, the C_{in} to the first adder is grounded; explain why. Also, why isn't the C_{in} to the second adder grounded?

D **7–23.** Design and draw a 6-bit binary adder similar to Figure 7–15 using two 7483 4-bit adders.

7–24. The 7483 has a fast-look-ahead carry. Explain why that is beneficial in some adder designs.

D **7–25.** Design and draw a 16-bit binary adder using four 4008 CMOS 4-bit adders.

Section 7–8

7–26. What changes would have to be made to the adder/subtractor circuit of Figure 7–16 if exclusive-NORs are to be used instead of exclusive-ORs?

T **7–27.** Figure P7–27 is a 4-bit two's-complement adder/subtractor. We are attempting to subtract $9 - 3$ ($1001 - 0011$) but keep getting the wrong answer of 10 (1010). Each test node in the circuit is labeled with the logic state observed using a logic probe. Find the two faults in the circuit.

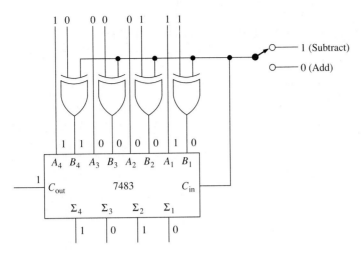

Figure P7–27

Section 7–9

T   **7–28.** Figure P7–28 is supposed to be set up as a one-digit hexadecimal adder. To test it, the values C and 2 (1100 + 0010) are input to the A and B inputs. The answer should be C + 2 = E (1110), but it is not! The figure is labeled with the states observed with a logic probe. Find the problem(s)!

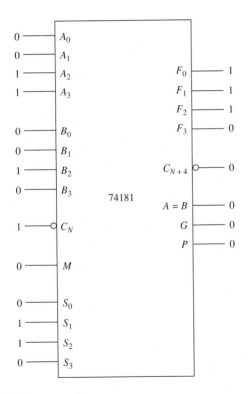

Figure P7–28

C   **7–29.** Re-solve Example 7–23(a) and (b) for S_3–$S_0 = 0100$.

Schematic Interpretation Problems

See Appendix G for the schematic diagrams.

S D   **7–30.** Find U9 and U10 of the Watchdog Timer schematic. These are counter ICs that output their 4-bit binary count to Q_0–Q_3. On a separate piece of paper, draw the connections that you would make to add the output of U9 to the output of U10 using a 74HC283 4-bit adder IC.

S C D   **7–31.** Find Port E (PE0–PE7) of U1 in the HC11D0 schematic. Assume that the low-order bits (PE0–PE3) contain the 4-bit binary number A and the high-order bits (PE4–PE7) contain the 4-bit binary number B. On a separate piece of paper, make the connections necessary to subtract A minus B using a two's-complement adder/subtractor circuit design. The result of the subtraction is to be read by Port A, bits PA4–PA7.

S C D   **7–32.** Repeat Problem 7–31 using a 74181 ALU.

Answers to Review Questions

7–1.   2 inputs, 2 outputs

7–2.   True

7–3.   True

7–4. D_7

7–5.   (a) Negative (b) positive (c) negative

7–6.   b, d

7–7.   True

7–8.   True

7–9.   Because it simplifies the documentation and use of the equipment

7–10. 15

7–11. 16

7–12. Greater than 9; there is a carry out of the MSB.

7–13. Add 6 (0110).

7–14. Inputs: A_0, B_0; outputs: Σ_0, C_{out}

7–15. Because it needs a carry-in from the previous adder

7–16. Odd

7–17. When any two of the inputs are HIGH

7–18. Connect it to zero.

7–19. To speed up the arithmetic process

7–20. It provides a C_{in}, and it puts a 1 on the inputs of the X-OR gates, which inverts B.

7–21. They check for a sum greater than 9 and provide a C_{out}.

7–22. It sets the mode of operation for either logic or arithmetic.

7–23. True

7–24. + means logical OR; *plus* means arithmetic sum.

8 Code Converters, Multiplexers, and Demultiplexers

OUTLINE

8–1 Comparators

8–2 Decoding

8–3 Encoding

8–4 Code Converters

8–5 Multiplexers

8–6 Demultiplexers

8–7 System Design Applications

Objectives

Upon completion of this chapter, you should be able to

- Utilize an integrated-circuit magnitude comparator to perform binary comparisons.

- Describe the function of a decoder and an encoder.

- Design the internal circuitry for encoding and decoding.

- Utilize manufacturers' data sheets to determine the operation of IC decoder and encoder chips.

- Explain the procedure involved in binary, BCD, and Gray code converting.

- Explain the operation of code-converter circuits built from SSI and MSI ICs.

- Describe the function and uses of multiplexers and demultiplexers.

- Design circuits that employ multiplexer and demultiplexer ICs.

Introduction

Information, or data, that is used by digital devices comes in many formats. The mechanisms for the conversion, transfer, and selection of data are handled by combinational logic ICs.

In this chapter we first take a general approach to the understanding of data-handling circuits and then deal with the specific operation and application of practical data-handling MSI chips. The MSI chips covered include comparators, decoders, encoders, code converters, multiplexers, and demultiplexers.

8–1 Comparators

Often in the evaluation of digital information it is important to compare two binary strings (or binary words) to determine if they are exactly equal. This comparison process is performed by a digital **comparator**.

The basic comparator evaluates two binary strings bit by bit and outputs a 1 if they are exactly equal. An exclusive-NOR gate is the easiest way to compare the equality of bits. If both bits are equal (0–0 or 1–1), the Ex-NOR puts out a 1.

To compare more than just 2 bits, we need additional Ex-NORs, and the output of all of them must be 1. For example, to design a comparator to evaluate two 4-bit numbers, we need four Ex-NORs. To determine total equality, connect all four outputs into an AND gate. That way, if all four outputs are 1's, the AND gate puts out a 1. Figure 8–1 shows a comparator circuit built from exclusive-NORs and an AND gate.

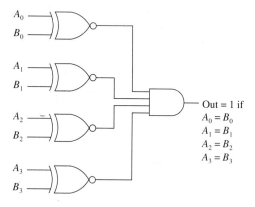

Figure 8–1 Binary comparator for comparing two 4-bit binary strings.

Studying Figure 8–1, you should realize that if A_0–B_0 equals 1–1 or 0–0, the top Ex-NOR will output a 1. The same holds true for the second, third, and fourth Ex-NOR gates. If all of them output a 1, the AND gate outputs a 1, indicating equality.

EXAMPLE 8–1

Referring to Figure 8–1, determine if the following pairs of input binary numbers will output a 1.

(a) $A_3A_2A_1A_0 = 1\ 0\ 1\ 1$
$B_3B_2B_1B_0 = 1\ 0\ 1\ 1$

(b) $A_3A_2A_1A_0 = 0\ 1\ 1\ 0$
$B_3B_2B_1B_0 = 0\ 1\ 1\ 1$

Solution: **(a)** When the *A* and *B* numbers are applied to the inputs, each of the four Ex-NORs will output 1's, so the output of the AND gate will be 1 (equality).

(b) For this case, the first three Ex-NORs will output 1's, but the last Ex-NOR will output a 0 because its inputs are not equal. The AND gate will output a 0 (inequality).

Integrated-circuit *magnitude comparators* are available in both the TTL and CMOS families. A magnitude comparator not only determines if *A* equals *B*, but also if *A* is *greater than B* or *A* is *less than B*.

The 7485 is a TTL 4-bit magnitude comparator. The pin configuration and logic symbol for the 7485 are given in Figure 8–2. The 7485 can be used just like the basic comparator of Figure 8–1 by using the A inputs, B inputs, and the equality output ($A = B$). The 7485 has the additional feature of telling you which number is larger if the equality is not met. The $A > B$ output is 1 if A is larger than B, and the $A < B$ output is 1 if B is larger than A.

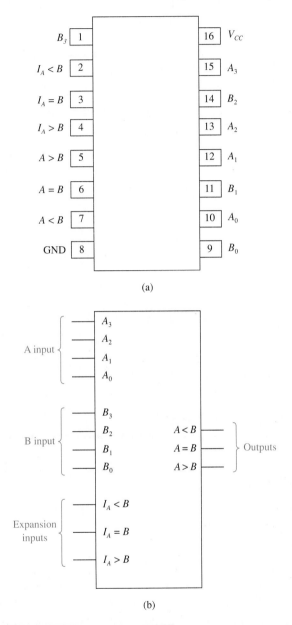

(a)

(b)

Figure 8–2 The 7485 4-bit magnitude comparator: (a) pin configuration; (b) logic symbol.

Common Misconception

Students often think that the I inputs have a priority over the A and B inputs. However, the I inputs are used by the 7485 only if the A inputs are equal to the B inputs. (To illustrate, try $A = 1100\ 0111$ and $B = 1100\ 0011$ in Figure 8–3).

The expansion inputs $I_A < B$, $I_A = B$, and $I_A > B$ are used for expansion to a system capable of comparisons greater than 4 bits. For example, to set up a circuit capable of comparing two *8-bit words*, two 7485s are required. The $A > B$, $A = B$, and $A < B$ outputs of the low-order (least significant) comparator are connected to the expansion inputs of the high-order comparator. That way the comparators act together, comparing two entire 8-bit words and outputting the result from the high-order comparator outputs. For proper operation, the expansion inputs to the low-order comparator should be tied as follows: $I_A > B =$ LOW, $I_A = B =$ HIGH, and $I_A < B =$ LOW. Expansion to greater than 8 bits using multiple 7485s is also possible. Figure 8–3 shows the connections for magnitude comparison of two 8-bit binary strings. If the high-order A inputs are equal to the high-order B inputs, then the expansion inputs are used as a tie breaker.

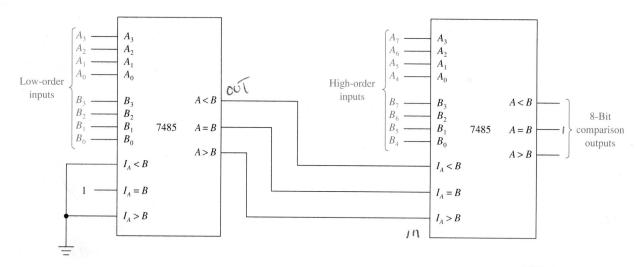

Figure 8–3 Magnitude comparison of two 8-bit binary strings (or binary words).

Review Questions

8–1. More than one output of the 7485 comparator can be simultaneously HIGH. True or false?

8–2. If all inputs to a 7485 comparator are LOW except for the $I_A < B$ input, what will the output be?

8–2 Decoding

Decoding is the process of converting some code (such as binary, BCD, or hex) into a singular active output representing its numeric value. Take, for example, a system that reads a 4-bit BCD code and converts it to its appropriate decimal number by turning on a decimal indicating lamp. Figure 8–4 illustrates such a system. This **decoder** is made up of a combination of logic gates that produces a HIGH at one of the 10 outputs, based on the levels at the four inputs.

In this section we learn how to use decoder ICs by first looking at the combinational logic that makes them work and then by selecting the actual decoder IC and making the appropriate pin connections.

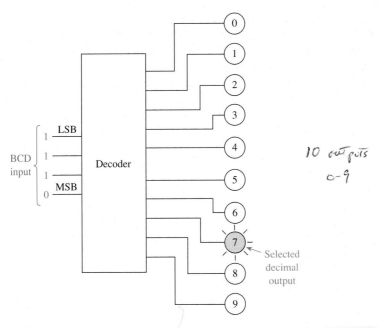

Figure 8-4 A BCD decoder selects the correct decimal-indicating lamp based on the BCD input.

3-Bit Binary-to-Octal Decoding

To design a decoder, it is useful first to make a truth table of all possible input/output combinations. An octal decoder must provide eight outputs, one for each of the eight different combinations of inputs, as shown in Table 8-1.

Table 8-1 Truth Tables for an Octal Decoder

(a) Active-HIGH Outputs

Input			Output							
2^2	2^1	2^0	0	1	2	3	4	5	6	7
0	0	0	1	0	0	0	0	0	0	0
0	0	1	0	1	0	0	0	0	0	0
0	1	0	0	0	1	0	0	0	0	0
0	1	1	0	0	0	1	0	0	0	0
1	0	0	0	0	0	0	1	0	0	0
1	0	1	0	0	0	0	0	1	0	0
1	1	0	0	0	0	0	0	0	1	0
1	1	1	0	0	0	0	0	0	0	1

(b) Active-LOW Outputs

Input			Output							
2^2	2^1	2^0	0	1	2	3	4	5	6	7
0	0	0	0	1	1	1	1	1	1	1
0	0	1	1	0	1	1	1	1	1	1
0	1	0	1	1	0	1	1	1	1	1
0	1	1	1	1	1	0	1	1	1	1
1	0	0	1	1	1	1	0	1	1	1
1	0	1	1	1	1	1	1	0	1	1
1	1	0	1	1	1	1	1	1	0	1
1	1	1	1	1	1	1	1	1	1	0

Before the design is made, we must decide if we want an *active-HIGH-level* output or an *active-LOW-level* output to indicate the value selected. For example, the *active-HIGH* truth table in Table 8–1(a) shows us that, for an input of 011 (3), output 3 is HIGH, and all other outputs are LOW. The *active-LOW* truth table is just the opposite (output 3 is LOW, all other outputs are HIGH).

Therefore, we have to know whether the indicating lamp (or other receiving device) requires a HIGH level to activate or a LOW level. We will learn in Chapter 9 that most devices used in digital electronics are designed to activate from a LOW-level signal, so most decoder designs use *active-LOW* outputs, as shown in Table 8–1(b). The combinational logic requirements to produce a LOW at output 3 for an input of 011 are shown in Figure 8–5.

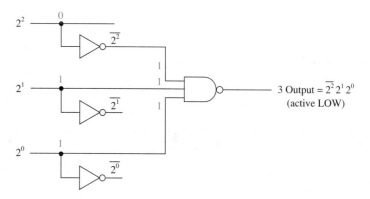

Figure 8–5 Logic requirements to produce a LOW at output 3 for a 011 input.

To design the complete octal decoder, we need a separate NAND gate for each of the eight outputs. The input connections for each of the NAND gates can be determined by referring to Table 8–1(b). For example, the NAND gate 5 inputs are connected to the $2^2 - \overline{2^1} - 2^0$ input lines, NAND gate 6 is connected to the $2^2 - 2^1 - \overline{2^0}$ input lines, and so on. The complete circuit is shown in Figure 8–6. Each NAND gate

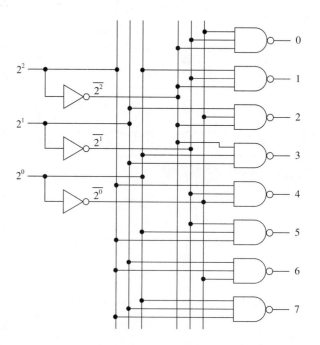

Figure 8–6 Complete circuit for an active-LOW output octal (1-of-8) decoder.

CHAPTER 8 | CODE CONVERTERS, MULTIPLEXERS, AND DEMULTIPLEXERS

in Figure 8–6 is wired so that its output goes LOW when the correct combination of input levels is present at its input. BCD and hexadecimal decoders can be designed in a similar manner.

The octal decoder is sometimes referred to as a *1-of-8 decoder* because, based on the input code, one of the eight outputs will be active. It is also known as a *3-line-to-8-line decoder,* because it has three input lines and eight output lines.

Integrated-circuit decoder chips provide basic decoding as well as several other useful functions. Manufacturers' data books list several decoders and give function tables illustrating the input/output operation and special functions. Rather than designing decoders using combinational logic, it is much more important to be able to use a data book to find the decoder that you need and to determine the proper pin connections and operating procedure to perform a specific decoding task. Table 8–2 lists some of the more popular TTL decoder ICs. (Equivalent CMOS ICs are also available.)

Table 8–2	Decoder ICs
Device Number	Function
74138	1-of-8 octal decoder (3-line-to-8-line)
7442	1-of-10 BCD decoder (4-line-to-10-line)
74154	1-of-16 hex decoder (4-line-to-16-line)
7447	BCD-to-seven-segment decoder (covered in Chapter 12)

Octal Decoder IC

The 74138[*] is an octal decoder capable of decoding the eight possible octal codes into eight separate active-LOW outputs, just like our combinational logic design. It also has three enable inputs for additional flexibility. Figure 8–7 shows information presented in a data book for the 74138.

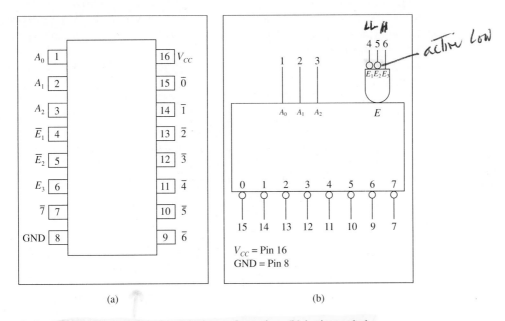

(a)   (b)

Figure 8–7 The 74138 octal decoder: (a) pin configuration; (b) logic symbol;

---

[*]A PAL design of a 74138 is given in Appendix E.

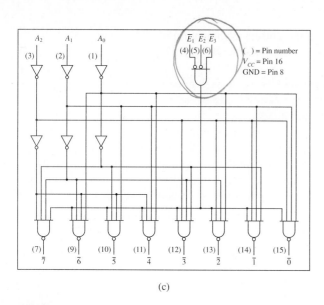

(c)

	Inputs						Outputs							
$\overline{E}_1$	$\overline{E}_2$	E_3	A_0	A_1	A_2	$\overline{0}$	$\overline{1}$	$\overline{2}$	$\overline{3}$	$\overline{4}$	$\overline{5}$	$\overline{6}$	$\overline{7}$
H	X	X	X	X	X	H	H	H	H	H	H	H	H
X	H	X	X	X	X	H	H	H	H	H	H	H	H
X	X	L	X	X	X	H	H	H	H	H	H	H	H
L	L	H	L	L	L	L	H	H	H	H	H	H	H
L	L	H	H	L	L	H	L	H	H	H	H	H	H
L	L	H	L	H	L	H	H	L	H	H	H	H	H
L	L	H	H	H	L	H	H	H	L	H	H	H	H
L	L	H	L	L	H	H	H	H	H	L	H	H	H
L	L	H	H	L	H	H	H	H	H	H	L	H	H
L	L	H	L	H	H	H	H	H	H	H	H	L	H
L	L	H	H	H	H	H	H	H	H	H	H	H	L

Notes
H = HIGH voltage level
L = LOW voltage level
X = Don't care
A_2 = MSB

(d)

(handwritten: active low ↓↓ ; Disabled ; enabled)

Figure 8–7 *(Continued)* (c) logic diagram; (d) function table. (Courtesy of Signetics Corporation.)

Helpful Hint

Decoding the A inputs in Figure 8–7(c) could have been done using just three inverters, similar to Figure 8–6. This is a good time to start thinking about gate loading, which is covered in Chapter 9. (Using six inverters ensures that each A input drives only one gate load.)

Helpful Hint

This is a good time to begin realizing the meaning of overbars in schematics. Don't develop the bad habit of thinking that $\overline{E}_1$ and $\overline{E}_2$ are inverted as they enter the IC. Instead, realize that $\overline{E}_1$ and $\overline{E}_2$ require a LOW to be satisfied. Also, the eight outputs each become active by going LOW.

Just by looking at the logic symbol [Figure 8–7(b)] and function table [Figure 8–7(d)], we can figure out the complete operation of the chip. First, the inversion bubbles on the decoded outputs indicate active-LOW operation. The three inputs $\overline{E}_1$, $\overline{E}_2$, and E_3 are used to *enable* the chip. The function table shows that the chip is disabled (all outputs HIGH) *unless* $\overline{E}_1 = $ LOW *and* $\overline{E}_2 = $ LOW *and* $E_3 = $ HIGH. The enables are useful for go/no-go operation of the chip based on some external control signal.

When the chip is *disabled*, the ×'s in the binary input columns A_0, A_1, and A_2 indicate **don't-care** levels, meaning the outputs will all be HIGH no matter at what level A_0, A_1, and A_2 are. When the chip is *enabled*, the binary inputs A_0, A_1, and A_2 are used to select which output goes LOW. In this case, A_0 is the least significant bit (LSB) input. Be aware that some manufacturers label the inputs A, B, C instead of A_0, A_1, A_2 and assume that A is the LSB.

The logic diagram in Figure 8–7(c) shows the actual internal combinational logic required to perform the decoding. The extra inverters on the inputs are required to prevent excessive loading of the driving source(s). These internal inverters supply the driving current to the eight NAND gates instead of the driving source(s) having to do it. (Gate loading is discussed in Chapter 9.) The three enable inputs ($\overline{E}_1$, $\overline{E}_2$, and E_3) are connected to an AND gate, which can disable all the output NANDs by sending them a LOW input level if $\overline{E}_1$, $\overline{E}_2$, and E_3 are not 001. Example 8–2 shows a waveform analysis of the 74138, and Section 8–7 discusses its use in a microcomputer application.

EXAMPLE 8–2

Sketch the output waveforms of the 74138 in Figure 8–8. Figure 8–9 shows the input waveforms to the 74138.

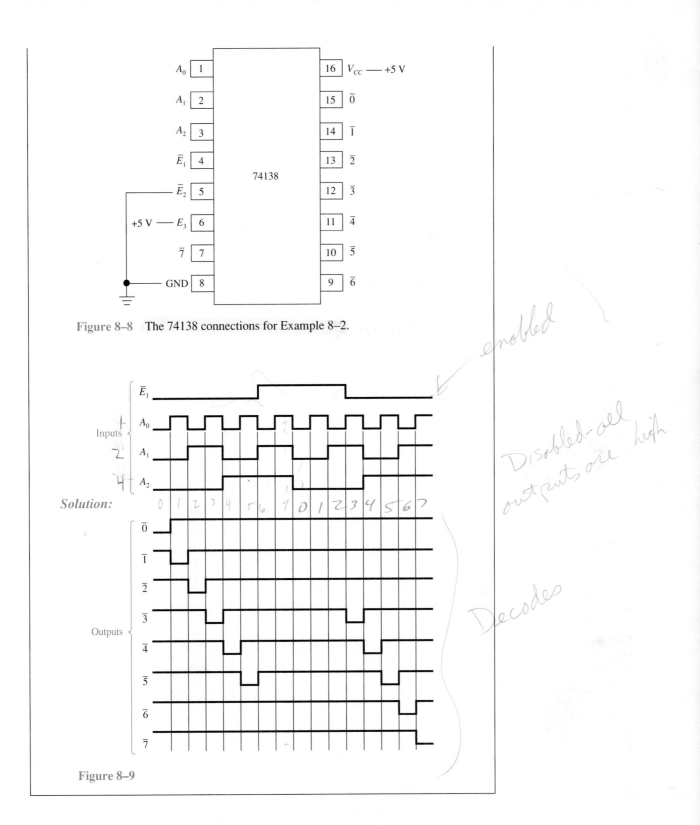

Figure 8–8   The 74138 connections for Example 8–2.

Figure 8–9

BCD Decoder IC

The 7442 is a BCD-to-decimal (1-of-10) decoder. It has four pins for the BCD input bits (0000 to 1001) and 10 active-LOW outputs for the decoded decimal numbers. Figure 8–10 gives the operational information for the 7442 from a manufacturer's data book.

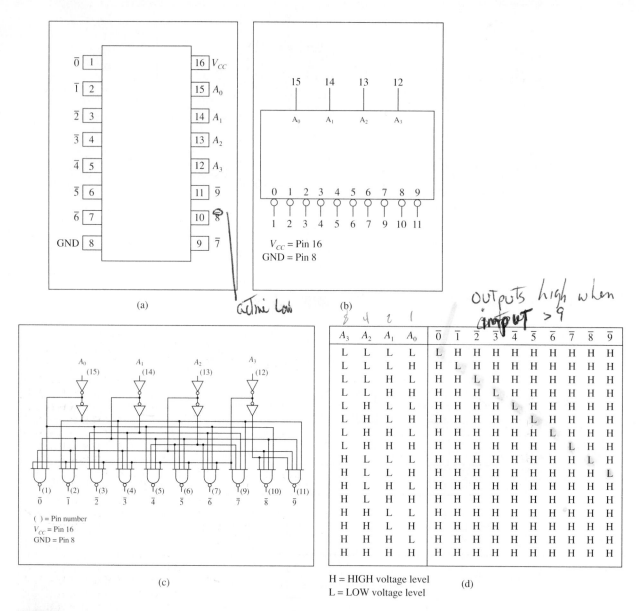

active Low

outputs high when input > 9

(a)

(b)

(c)

(d)

H = HIGH voltage level
L = LOW voltage level

Figure 8–10 The 7442 BCD-to-DEC decoder: (a) pin configuration; (b) logic symbol; (c) logic diagram; (d) function table. (Courtesy of Signetics Corporation.)

Team Discussion

What happens when you enter an invalid BCD string [see Figure 8–10(d)]?

Hexadecimal 1-of-16 Decoder IC

The 74154 is a 1-of-16 decoder. It accepts a 4-bit binary input (0000 to 1111), decodes it, and provides an active-LOW output to one of the 16 output pins. It also has a two-input active-LOW enable gate for disabling the outputs. If either enable input ($\overline{E_0}$ or $\overline{E_1}$) is made HIGH, the outputs are forced HIGH regardless of the A_0 to A_3 inputs. The operational information for the 74154 is given in Figure 8–11.

CHAPTER 8 | CODE CONVERTERS, MULTIPLEXERS, AND DEMULTIPLEXERS

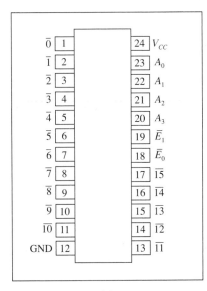

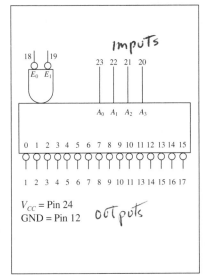

(a)   (b)

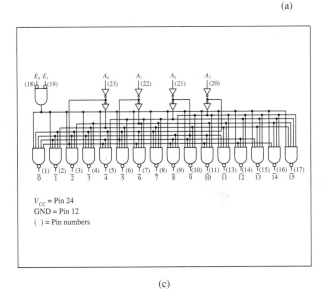

(c)

Inputs						Outputs															
$\bar{E_1}$	$\bar{E_2}$	A_3	A_2	A_1	A_0	$\bar{0}$	$\bar{1}$	$\bar{2}$	$\bar{3}$	$\bar{4}$	$\bar{5}$	$\bar{6}$	$\bar{7}$	$\bar{8}$	$\bar{9}$	$\overline{10}$	$\overline{11}$	$\overline{12}$	$\overline{13}$	$\overline{14}$	$\overline{15}$
L	H	X	X	X	X	H	H	H	H	H	H	H	H	H	H	H	H	H	H	H	H
H	L	X	X	X	X	H	H	H	H	H	H	H	H	H	H	H	H	H	H	H	H
H	H	X	X	X	X	H	H	H	H	H	H	H	H	H	H	H	H	H	H	H	H
L	L	L	L	L	L	L	H	H	H	H	H	H	H	H	H	H	H	H	H	H	H
L	L	L	L	L	H	H	L	H	H	H	H	H	H	H	H	H	H	H	H	H	H
L	L	L	L	H	L	H	H	L	H	H	H	H	H	H	H	H	H	H	H	H	H
L	L	L	L	H	H	H	H	H	L	H	H	H	H	H	H	H	H	H	H	H	H
L	L	L	H	L	L	H	H	H	H	L	H	H	H	H	H	H	H	H	H	H	H
L	L	L	H	L	H	H	H	H	H	H	L	H	H	H	H	H	H	H	H	H	H
L	L	L	H	H	L	H	H	H	H	H	H	L	H	H	H	H	H	H	H	H	H
L	L	L	H	H	H	H	H	H	H	H	H	H	L	H	H	H	H	H	H	H	H
L	L	H	L	L	L	H	H	H	H	H	H	H	H	L	H	H	H	H	H	H	H
L	L	H	L	L	H	H	H	H	H	H	H	H	H	H	L	H	H	H	H	H	H
L	L	H	L	H	L	H	H	H	H	H	H	H	H	H	H	L	H	H	H	H	H
L	L	H	L	H	H	H	H	H	H	H	H	H	H	H	H	H	L	H	H	H	H
L	L	H	H	L	L	H	H	H	H	H	H	H	H	H	H	H	H	L	H	H	H
L	L	H	H	L	H	H	H	H	H	H	H	H	H	H	H	H	H	H	L	H	H
L	L	H	H	H	L	H	H	H	H	H	H	H	H	H	H	H	H	H	H	L	H
L	L	H	H	H	H	H	H	H	H	H	H	H	H	H	H	H	H	H	H	H	L

H = HIGH voltage level
L = LOW voltage level
X = Don't care

(d)

Figure 8–11 The 74154 1-of-16 decoder: (a) pin configuration; (b) logic symbol; (c) logic diagram; (d) function table. (Courtesy of Signetics Corporation.)

Team Discussion

Why are the A inputs listed as don't cares in the first three entries of the function table?

The logic diagram in Figure 8–11(c) shows the actual combinational logic circuit that is used to provide the decoding. The inverted-input AND gate is used in the circuit to disable all output NAND gates if either $\bar{E_0}$ or $\bar{E_1}$ is made HIGH. Follow the logic levels through the circuit for several combinations of inputs to A_0 through A_3 to prove its operation.

Review Questions

8-3. A BCD-to-decimal decoder has how many inputs and how many outputs? _4 inputs_ _10 active Low outputs_

8-4. An octal decoder with active-LOW outputs will output seven LOWs and one HIGH for each combination of inputs. True or false? _F_

8-5. A hexadecimal decoder is sometimes called a 4-line-to-10-line decoder. True or false? _F_

8-6. Only one of the three *enable* inputs must be satisfied to enable the 74138 decoder IC. True or false? _F_

8-7. The 7442 BCD decoder has active-_____ (LOW, HIGH) inputs and active-_____ (LOW, HIGH) outputs.

8-3   Encoding

Encoding is the opposite process from decoding. Encoding is used to generate a coded output (such as BCD or binary) from a singular active numeric input line. For example, Figure 8-12 shows a typical block diagram for a decimal-to-BCD **encoder** and an octal-to-binary encoder.

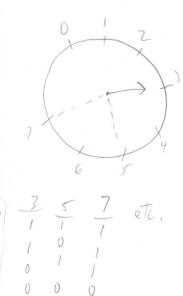

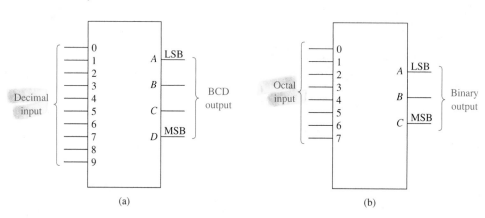

Figure 8-12   Typical block diagrams for encoders: (a) decimal-to-BCD encoder; (b) octal-to-binary encoder.

The design of encoders using combinational logic can be done by reviewing the truth table (see Table 8-3) for the operation to determine the relationship each output has with the inputs. For example, by studying Table 8-3 for a decimal-to-BCD encoder, we can see that the A output (2^0) is HIGH for all odd decimal input numbers (1, 3, 5, 7, and 9). The B output (2^1) is HIGH for decimal inputs 2, 3, 6, and 7. The C output (2^2) is HIGH for decimal inputs 4, 5, 6, and 7, and the D output (2^3) is HIGH for decimal inputs 8 and 9.

Now, from what we have just observed, it seems that we can design a decimal-to-BCD encoder with just four OR gates; the A output OR gate goes HIGH for any odd decimal input, the B output goes HIGH for 2 *or* 3 *or* 6 *or* 7, and so on, for the C output and D output. The complete design of a basic decimal-to-BCD encoder is given in Figure 8-13. The design for an octal-to-binary encoder uses the same procedure, but, of course, these encoders are available in integrated-circuit form: the 74147 decimal to BCD and the 74148 octal to binary.

Decimal Input	BCD Output			
	D	C	B	A
0	0	0	0	0
1	0	0	0	1
2	0	0	1	0
3	0	0	1	1
4	0	1	0	0
5	0	1	0	1
6	0	1	1	0
7	0	1	1	1
8	1	0	0	0
9	1	0	0	1

Table 8–3 Decimal-to-BCD Encoder Truth Table

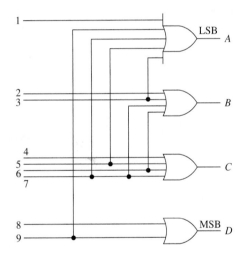

Figure 8–13 Basic decimal-to-BCD encoder.

Helpful Hint

Once you have seen this encoder design, it should be a confidence booster for you to sketch an octal encoder circuit with the book closed.

The 74147 Decimal-to-BCD Encoder

The 74147 operates similarly to our basic design from Figure 8–13 except for two major differences.

1. The inputs *and* outputs are all active-LOW [see the bubbles on the logic symbol, Figure 8–14(a)].

2. The 74147 is a *priority* encoder, which means that if more than one decimal number is input, the highest numeric input has *priority* and will be encoded to the output [see the function table, Figure 8–14(b)]. For example, looking at the second line in the function table, if $\overline{I_9}$ is LOW (decimal 9), all other inputs are don't care (could be HIGH *or* LOW), and the BCD output is 0110 (active-LOW BCD-9).

Team Discussion

Discuss how you would use the function table to determine when the 74147 would encode a decimal zero output.

					Input						Output		
$\overline{I_1}$	$\overline{I_2}$	$\overline{I_3}$	$\overline{I_4}$	$\overline{I_5}$	$\overline{I_6}$	$\overline{I_7}$	$\overline{I_8}$	$\overline{I_9}$		$\overline{A_3}$	$\overline{A_2}$	$\overline{A_1}$	$\overline{A_0}$
H	H	H	H	H	H	H	H	H		H	H	H	H
X	X	X	X	X	X	X	X	L		L	H	H	L
X	X	X	X	X	X	X	L	H		L	H	H	H
X	X	X	X	X	X	L	H	H		H	L	L	L
X	X	X	X	X	L	H	H	H		H	L	L	H
X	X	X	X	L	H	H	H	H		H	L	H	L
X	X	X	L	H	H	H	H	H		H	L	H	H
X	X	L	H	H	H	H	H	H		H	H	L	L
X	L	H	H	H	H	H	H	H		H	H	L	H
L	H	H	H	H	H	H	H	H		H	H	H	L

H = HIGH voltage level
L = LOW voltage level
X = Don't care

(b)

Logic symbol:

11 12 13 1 2 3 4 5 10
I_1 I_2 I_3 I_4 I_5 I_6 I_7 I_8 I_9

A_3 A_2 A_1 A_0
14 6 7 9

V_{CC} = Pin 16
GND = Pin 8

(a)

Figure 8–14 The 74147 decimal-to-BCD (10-line-to-4-line) encoder: (a) logic symbol; (b) function table.

EXAMPLE 8–3

For simplicity, the 74147 IC shown in Figure 8–15 is set up for encoding just three of its inputs (7, 8, and 9). Using the function table from Figure 8–14(b), sketch the outputs at $\overline{A_0}$, $\overline{A_1}$, $\overline{A_2}$, and $\overline{A_3}$ as the $\overline{I_7}$, $\overline{I_8}$, and $\overline{I_9}$ inputs are switching as shown in Figure 8–16.

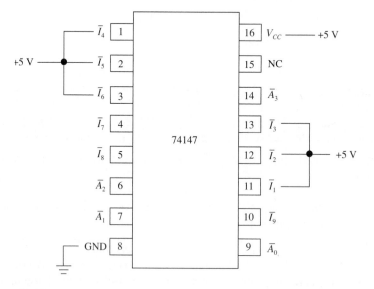

Figure 8–15

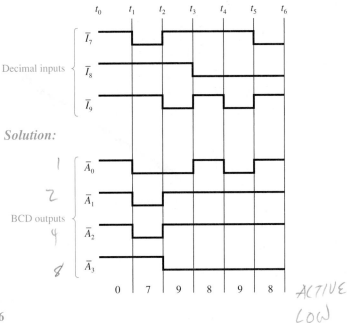

Figure 8–16

Explanation: The $\overline{I_1}$ to $\overline{I_6}$ inputs are all tied HIGH and have no effect on the output.

t_0–t_1: Dec inputs are all HIGH; BCD outputs represent a 0.

t_1–t_2: $\overline{I_7}$ is LOW; BCD outputs represent a 7.

t_2–t_3: $\overline{I_9}$ is LOW; BCD outputs represent a 9.

t_3–t_4: $\overline{I_8}$ is LOW; BCD outputs represent an 8.

t_4–t_5: $\overline{I_8}$ *and* $\overline{I_9}$ are LOW; $\overline{I_9}$ has priority; BCD outputs represent a 9.

t_5–t_6: $\overline{I_7}$ *and* $\overline{I_8}$ are LOW; $\overline{I_8}$ has priority; BCD outputs represent an 8.

The 74148 Octal-to-Binary Encoder

The 74148 encoder accepts data from eight active-LOW inputs and provides a binary representation on three active-LOW outputs. It is also a **priority** encoder, so when two or more inputs are active simultaneously, the input with the highest priority is represented on the output, with input line $\overline{I_7}$ having the highest priority. The logic symbol and function table in Figure 8–17 give us some other information as well.

The 74148 can be expanded to any number of inputs by using several 74148s and their $\overline{EI}$, $\overline{EO}$, and $\overline{GS}$ pins. These special pins are defined as follows:

$\overline{EI}$ Active-LOW enable input: a HIGH on this input forces all outputs ($\overline{A_0}$ to $\overline{A_2}$, $\overline{EO}$, $\overline{GS}$) to their inactive (HIGH) state.

$\overline{EO}$ Active-LOW enable output: this output pin goes LOW when all inputs ($\overline{I_0}$ to $\overline{I_7}$) are inactive (HIGH) and $\overline{EI}$ is LOW.

$\overline{GS}$ Active-LOW group signal output: this output pin goes LOW whenever any of the inputs ($\overline{I_0}$ to $\overline{I_7}$) are active (LOW) and $\overline{EI}$ is LOW.

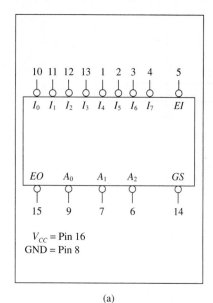

Inputs									Outputs				
$\overline{EI}$	$\overline{I_0}$	$\overline{I_1}$	$\overline{I_2}$	$\overline{I_3}$	$\overline{I_4}$	$\overline{I_5}$	$\overline{I_6}$	$\overline{I_7}$	$\overline{GS}$	$\overline{A_0}$	$\overline{A_1}$	$\overline{A_2}$	$\overline{EO}$
H	X	X	X	X	X	X	X	X	H	H	H	H	H
L	H	H	H	H	H	H	H	H	H	H	H	H	L
L	X	X	X	X	X	X	X	L	L	L	L	L	H
L	X	X	X	X	X	X	L	H	L	H	L	L	H
L	X	X	X	X	X	L	H	H	L	L	H	L	H
L	X	X	X	X	L	H	H	H	L	H	H	L	H
L	X	X	X	L	H	H	H	H	L	L	L	H	H
L	X	X	L	H	H	H	H	H	L	H	L	H	H
L	X	L	H	H	H	H	H	H	L	L	H	H	H
L	L	H	H	H	H	H	H	H	L	H	H	H	H

H = HIGH voltage level
L = LOW voltage level
X = Don't care

(b)

(a)

Figure 8–17 The 74148 octal-to-binary (8-line-to-3-line) encoder: (a) logic symbol; (b) functional table. (Courtesy of Signetics Corporation.)

The following example illustrates the use of these pins.

EXAMPLE 8–4

Sketch the output waveforms for the 74148 connected as shown in Figure 8–18. The input waveforms to $\overline{I_6}$, $\overline{I_7}$, and $\overline{EI}$ are given in Figure 8–19. (Inputs $\overline{I_0}$ to $\overline{I_5}$ are tied HIGH for simplicity.)

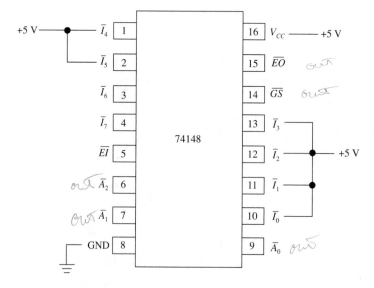

Figure 8–18 The 74148 connections for Example 8–4.

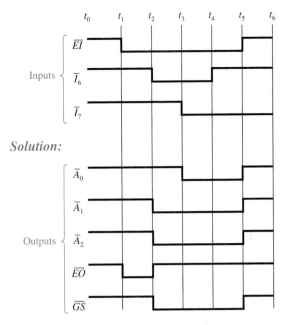

Figure 8–19

t_0–t_1: All outputs are forced HIGH by the HIGH on $\overline{EI}$.

t_1–t_2: $\overline{EI}$ is LOW to enable the inputs, but $\overline{I_0}$ to $\overline{I_7}$ are all HIGH (inactive), so $\overline{EO}$ goes LOW.

t_2–t_3: $\overline{GS}$ goes LOW because one of the inputs ($\overline{I_6}$) is active; the active-LOW binary output is equal to 6.

t_3–t_4: $\overline{I_7}$ and $\overline{I_6}$ are LOW; $\overline{I_7}$ has priority; output = 7.

t_4–t_5: $\overline{I_7}$ is LOW; output = 7.

t_5–t_6: All outputs are forced HIGH by the HIGH on $\overline{EI}$.

Review Questions

8–8. How does an encoder differ from a decoder? *giving us its opposite states*

8–9. If more than one input to a *priority* encoder is active, which input will be encoded? *highest #*

8–10. (a) If all inputs to a 74147 encoder are HIGH, what will the A_3–A_0 outputs be? *all be high*

    (b) Repeat part (a) for all inputs being LOW. *encoding highest one (its a nine)*

8–11. What are the five outputs of the 74148? Are they active-LOW or active-HIGH? *$\overline{GS}$ A_0 A_1 A_2 $\overline{EO}$ active low*

8–4   Code Converters

Often it is important to convert a coded number into another form that is more usable by a computer or digital system. The prime example of this is with binary-coded decimal (BCD). We have seen that BCD is very important for visual display communication between a computer and human beings. But BCD is very difficult to deal with arithmetically. Algorithms, or procedures, have been developed for the conversion of BCD to binary by computer programs (**software**) so that the computer will be able to perform all arithmetic operations in binary.

Another way to convert BCD to binary, the **hardware** approach, is with MSI integrated circuits. Additional circuitry is involved, but it is much faster to convert using hardware rather than software. We look at both methods for the conversion of BCD to binary.

BCD-to-Binary Conversion

If you were going to convert BCD to binary using software program statements, you would first have to develop a procedure, or algorithm, for the conversion. Take, for example, the number 26 in BCD.

$$2 \qquad 6$$
$$0010 \quad 0110$$

If you simply apply regular binary weighting to each bit, you would come up with 38 ($2^1 + 2^2 + 2^5 = 38$). You must realize that the second group of BCD positions has a new progression of powers of 2 but with a **weighting factor** of 10, as shown in Figure 8–20. Now, if we go back and apply the proper weighting factors to 26 in BCD, we should get the correct binary equivalent.

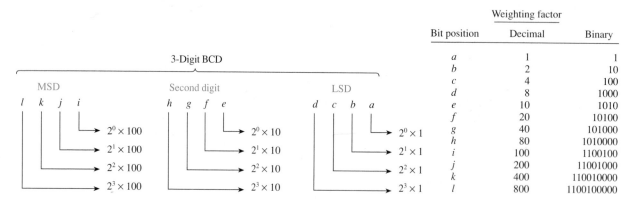

Bit position	Weighting factor	
	Decimal	Binary
a	1	1
b	2	10
c	4	100
d	8	1000
e	10	1010
f	20	10100
g	40	101000
h	80	1010000
i	100	1100100
j	200	11001000
k	400	110010000
l	800	1100100000

Figure 8–20 Weighting factors for BCD bit positions.

EXAMPLE 8–5

Using the weighting factors given in Figure 8–20, convert the BCD equivalent of 26_{10} to binary.

Solution:

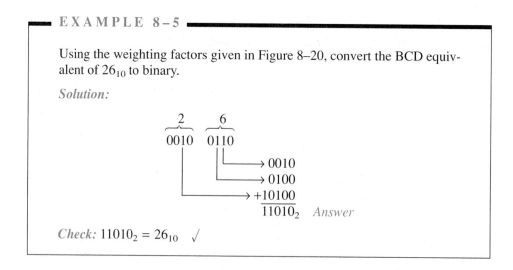

$$\frac{\begin{array}{cc} 2 & 6 \\ 0010 & 0110 \end{array}}{}$$

0010
0100
+10100
11010₂   *Answer*

Check: $11010_2 = 26_{10}$ ✓

EXAMPLE 8–6

Convert the BCD equivalent of 348 to binary.

Solution:

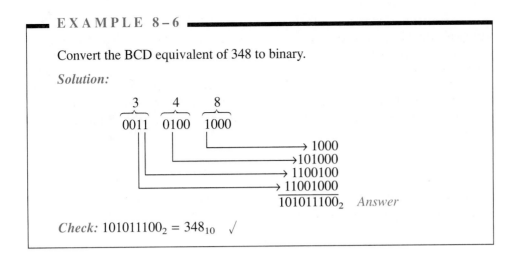

$$\overline{101011100}_2 \quad \textit{Answer}$$

Check: $101011100_2 = 348_{10}$ ✓

Conversion of BCD to Binary Using the 74184

Examples 8–5 and 8–6 illustrate one procedure of conversion that can be used as an algorithm for a computer program (software). The hardware approach using the 74184 IC is another way to accomplish BCD-to-binary conversion.

The logic symbol in Figure 8–21 shows eight active-HIGH binary outputs. Y_1 to Y_5 are outputs for regular BCD-to-binary conversion. Y_6 to Y_8 are used for a special BCD code called nine's complement and ten's complement.

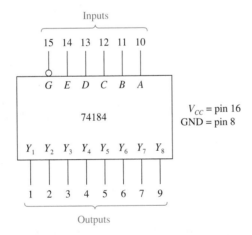

Figure 8–21   Logic symbol for the 74184 BCD-to-binary converter.

The active-HIGH BCD bits are input on A through E. The $\overline{G}$ is an active-LOW enable input. When $\overline{G}$ is HIGH, all outputs are forced HIGH.

Figure 8–22 shows the connections to form a 6-bit BCD converter. Because the LSB of the BCD input is always equal to the LSB of the binary output, the connection is made straight from input to output. The other BCD bits are connected to the A to E inputs. They have the weighting of $A = 2$, $B = 4$, $C = 8$, $D = 10$, and $E = 20$. Because only 2 bits are available for the MSD BCD input, the largest BCD digit in that position will be 3 (11). More useful setups, providing for the input of two or three complete BCD digits, are shown in Figure 8–23(a) and (b).

A companion chip, the 74185, is used to work the opposite way, binary to BCD. Figure 8–23(c) and (d) shows the 74185 used to perform binary-to-BCD conversions.

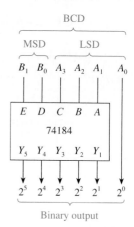

Figure 8–22 Six-bit BCD-to-binary converter.

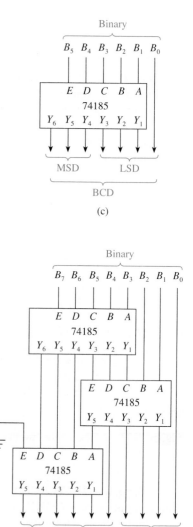

Figure 8–23 BCD-to-binary conversions using the 74184 and binary-to-BCD conversions using the 74185: (a) BCD-to-binary converter for two BCD decades; (b) BCD-to-binary converter for three BCD decades; (c) 6-bit binary-to-BCD converter; (d) 8-bit binary-to-BCD converter. (Courtesy of Texas Instruments, Inc.)

CHAPTER 8 | CODE CONVERTERS, MULTIPLEXERS, AND DEMULTIPLEXERS

Show how the BCD code of the number 65 is converted by the circuit of Figure 8–23(a) by placing 1's and 0's at the inputs and outputs.

Solution: The BCD-to-binary conversion is shown in Figure 8–24. The upper 74184 is used to convert the least significant 6 bits, which are 100101. Using the proper binary weighting, 100101 becomes 011001.

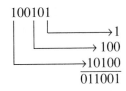

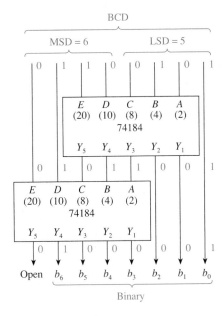

Figure 8–24 Solution to Example 8–7.

The lower 74184 is used to convert the most significant 2 bits plus 4 bits from the upper 74184, which are 010110. Using proper binary weighting, 010110 becomes 010000.

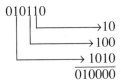

The final binary result is 01000001, which checks out to be equal to 65_{10}.

Helpful Hint

A complete design of a BCD-to-seven-segment decoder using a PLD is given in Section 16–6. Because you are so accustomed to calculator displays, this is a good time to start thinking about the necessity of seven-segment decoders.

BCD-to-Seven-Segment Converters

Calculators and other devices with numeric displays use another form of code conversion involving BCD-to-seven-segment conversion. The term *seven segment* comes from the fact that these displays utilize seven different illuminating segments to make up each of the 10 possible numeric digits. A **code converter** must be employed to convert the 4-bit BCD into a 7-bit code to drive each digit. A commonly used BCD-to-seven-segment converter is the 7447 IC. An in-depth discussion of displays and display converter/drivers will be given in Section 12–6 after you have a better understanding of the circuitry used to create the numeric data to be displayed.

Gray Code

The **Gray code** is another useful code used in digital systems. It is used primarily for indicating the angular position of a shaft on rotating machinery, such as automated lathes and drill presses. This code is like binary in that it can have as many bits as necessary, and the more bits, the more possible combinations of output codes (number of combinations = 2^N). A 4-bit Gray code, for example, has $2^4 = 16$ different representations, giving a resolution of 1 out of 16 possible angular positions at 22.5° each $(360/16 = 22.5)$.

The difference between the Gray code and the regular binary code is illustrated in Table 8–4. Notice in the table that the Gray code varies by only 1 bit from one entry

Table 8–4	Four-Bit Gray Code	
Decimal	Binary	Gray
0	0000	0000
1	0001	0001
2	0010	0011
3	0011	0010
4	0100	0110
5	0101	0111
6	0110	0101
7	0111	0100
8	1000	1100
9	1001	1101
10	1010	1111
11	1011	1110
12	1100	1010
13	1101	1011
14	1110	1001
15	1111	1000

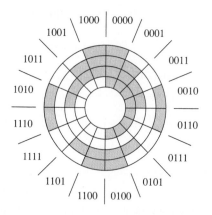

Figure 8–25 Gray code wheel.

CHAPTER 8 | CODE CONVERTERS, MULTIPLEXERS, AND DEMULTIPLEXERS

to the next and from the last entry (15) back to the beginning (0). Now, if each Gray code represents a different position on a rotating wheel, as the wheel turns, the code read from one position to the next will vary by only 1 bit (see Figure 8–25).

If the same wheel were labeled in *binary,* as the wheel turned from 7 to 8, the code would change from 0111 to 1000. If the digital machine happened to be reading the shaft position just as the code was changing, it might see 0111 or 1000, but since all 4 bits are changing (0 to 1 or 1 to 0), the code that it reads may be anything from 0000 to 1111. Therefore, the potential for an error using the regular binary system is great.

With the Gray code wheel, on the other hand, when the position changes from 7 to 8, the code changes from 0100 to 1100. The MSB is the only bit that changes, so if a read is taken right on the border between the two numbers, either a 0100 is read or a 1100 is read (no problem).

Gray Code Conversions

The determination of the Gray code equivalents and the conversions between Gray code and binary code are done very simply with exclusive-OR gates, as shown in Figures 8–26 and 8–27.

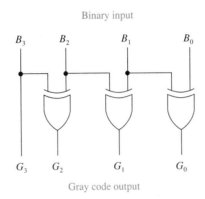

Figure 8–26 Binary-to–Gray code converter.

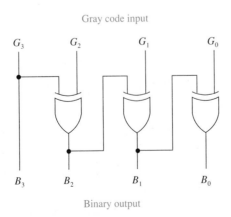

Figure 8–27 Gray code–to-binary converter.

EXAMPLE 8-8

Test the operation of the binary-to–Gray code converter of Figure 8–26 by labeling the inputs and outputs with the conversion of binary 0110 to Gray code.

Solution: The operation of the converter is shown in Figure 8–28.

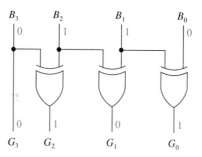

Figure 8–28

EXAMPLE 8-9

Repeat Example 8–8 for the Gray code–to-binary converter of Figure 8–27 by converting a Gray code 0011 to binary.

Solution: The operation of the converter is shown in Figure 8–29.

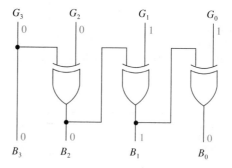

Figure 8–29

Review Questions

8–12. What is the binary weighting factor of the MSB of a two-digit (8-bit) BCD number?

8–13. How many 74184 ICs are required to convert a three-digit BCD number to binary?

8–14. Why is Gray code used for indicating the shaft position of rotating machinery rather than regular binary code?

8–5   Multiplexers

A **multiplexer** is a device capable of funneling several data lines into a single line for transmission to another point. The multiplexer has two or more digital input signals connected to its input. Control signals are also input to tell which data-input line to select for transmission (data selection). Figure 8–30 illustrates the function of a multiplexer.

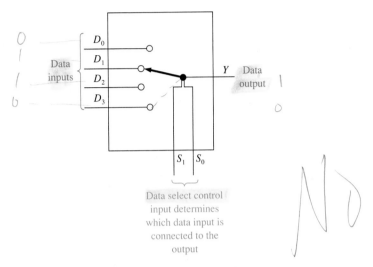

Figure 8–30   Functional diagram of a four-line multiplexer.

The multiplexer is also known as a *data selector*. Figure 8–30 shows that the *data select control inputs* (S_1, S_0) are responsible for determining which data input (D_0 to D_3) is selected to be transmitted to the data-output line (Y). The S_1, S_0 inputs will be a binary code that corresponds to the data-input line that you want to select. If $S_1 = 0$, $S_0 = 0$, then D_0 is selected; if $S_1 = 0$, $S_0 = 1$, then D_1 is selected; and so on. Table 8–5 lists the codes for input data selection.

Table 8–5   Data Select Input Codes for Figure 8–30

Data Select Control Inputs		Data Input Selected
S_1	S_0	
0	0	D_0
0	1	D_1
1	0	D_2
1	1	D_3

A sample four-line multiplexer built from SSI logic gates is shown in Figure 8–31. The control inputs (S_1, S_0) take care of enabling the correct AND gate to pass just one of the data inputs through to the output. In Figure 8–31, 1's and 0's were placed on the diagram to show the levels that occur when selecting data input D_1. Notice that AND gate 1 is enabled, passing D_1 to the output, whereas all other AND gates are disabled.

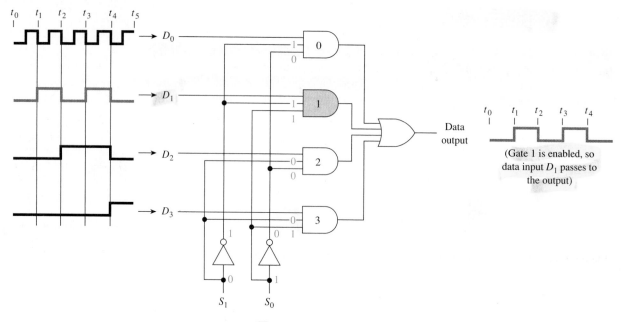

Figure 8–31 Logic diagram for a four-line multiplexer.

Two-, 4-, 8-, and 16-input multiplexers are readily available in MSI packages. Table 8–6 lists some popular TTL and CMOS multiplexers. (H-CMOS, high-speed CMOS, is compared with the other logic families in detail in Chapter 9.)

Table 8–6 TTL and CMOS Multiplexers

Function	Device	Logic Family
Quad two-input	74157	TTL
	74HC157	H-CMOS
	4019	CMOS
Dual eight-input	74153	TTL
	74HC153	H-CMOS
	4539	CMOS
Eight-input	74151	TTL
	74HC151	H-CMOS
	4512	CMOS
Sixteen-input	74150	TTL

The 74151 Eight-Line Multiplexer

The logic symbol and logic diagram for the 74151 are given in Figure 8–32. Because the 74151 has eight lines to select from (I_0 to I_7), it requires three data select inputs (S_2, S_1, S_0) to determine which input to choose ($2^3 = 8$). True (Y) and complemented ($\overline{Y}$) outputs are provided. The active-LOW enable input ($\overline{E}$) disables all inputs when it is HIGH and forces Y LOW regardless of all other inputs.

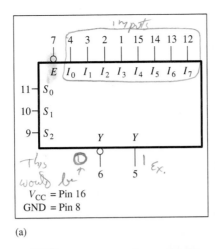

(a)

Figure 8–32 The 74151 eight-line multiplexer: (a) logic symbol;

V_{CC} = Pin 16
GND = Pin 8

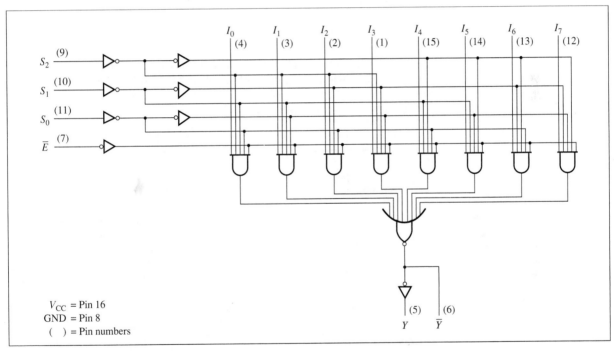

(b)

Figure 8–32 *(Continued)* (b) logic diagram. (Courtesy of Signetics Corporation.)

EXAMPLE 8–10

Sketch the output waveforms at Y for the 74151 shown in Figure 8–33. For this example, the eight input lines (I_0 to I_7) are each connected to a constant level, and the data select lines (S_0 to S_2) and input enable ($\overline{E}$) are given as input waveforms.

SECTION 8–5 I MULTIPLEXERS

245

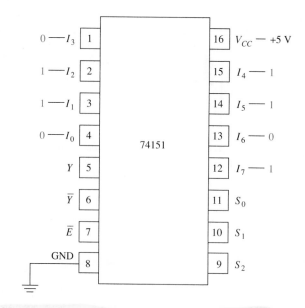

Figure 8–33 The 74151 multiplexer pin connections for Example 8–10.

See Figure 8–34. From t_0 to t_8 the waveforms at S_0, S_1, and S_2 form a binary counter from 000 to 111. Therefore, the output at Y will be selected from I_0, then I_1, then I_2, and so on, up to I_7. From t_8 to t_9 the S_0, S_1, and S_2 inputs are back to 000, so I_0 will be selected for output. From t_9 to t_{11} the $\overline{E}$ enable line goes HIGH, disabling all inputs and forcing Y LOW.

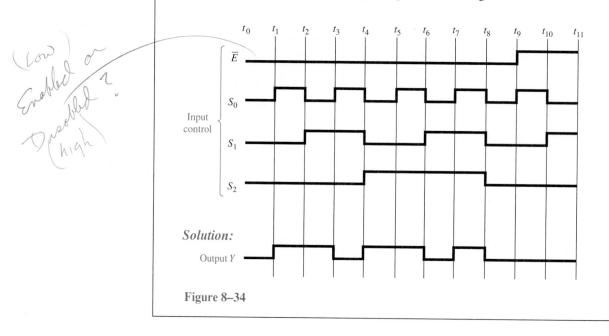

Figure 8–34

EXAMPLE 8–11

Using two 74151s, design a 16-line multiplexer controlled by four data select control inputs.

CHAPTER 8 I CODE CONVERTERS, MULTIPLEXERS, AND DEMULTIPLEXERS

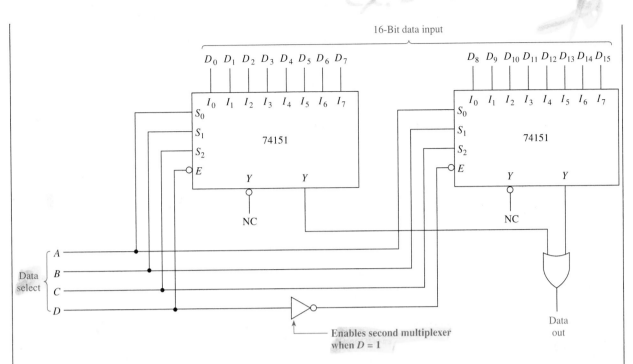

Figure 8–35   Design solution for Example 8–11.

Solution: The multiplexer is shown in Figure 8–35. Because there are 16 data input lines, we must use four data select inputs ($2^4 = 16$). (*A* is the LSB data select line and *D* is the MSB.)

When the data select is in the range from 0000 to 0111, the *D* line is 0, which enables the low-order (left) multiplexer selecting the D_0 to D_7 inputs and disables the high-order (right) multiplexer.

When the data select inputs are in the range from 1000 to 1111, the *D* line is 1, which disables the low-order multiplexer and enables the high-order multiplexer, allowing D_8 to D_{15} to be selected. Since the *Y* output of a disabled multiplexer is 0, an OR gate is used to combine the two outputs, allowing the output from the enabled multiplexer to pass through.

Providing Combination Logic Functions with a Multiplexer

Multiplexers have many other uses besides functioning as data selectors. Another important role of a multiplexer is for implementing combinational logic circuits. One multiplexer can take the place of several SSI logic gates, as shown in the following example.

EXAMPLE 8–12

Use a multiplexer to implement the function

$$X = \overline{A}\,\overline{B}\,\overline{C}D + A\overline{B}\,\overline{C}\,D + AB\overline{C}\,\overline{D} + \overline{A}BC + \overline{A}\,\overline{B}C$$

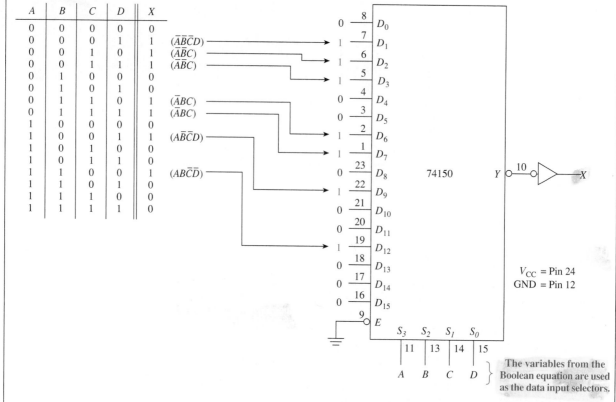

Figure 8–36 Truth table and solution for the implementation of the Boolean equation $X = \overline{A}\,\overline{B}\,\overline{C}\,D + A\overline{B}\,\overline{C}\,D + ABC\,\overline{D} + \overline{A}\,BC + \overline{A}\,\overline{B}\,C$ using a 16-line multiplexer.

Solution: The equation is in the sum-of-products form. Each term in the equation, when fulfilled, will make $X = 1$. For example, when $A = 0$, $B = 0$, $C = 0$, and $D = 1$ ($\overline{A}\,\overline{B}\,\overline{C}D$), X will receive a 1. Also, if $A = 1$, $B = 0$, $C = 0$, and $D = 1$ ($A\overline{B}\,\overline{C}\,D$), X will receive a 1, and so on.

If the $A, B, C,$ and D variables are used as the data input selectors of a 16-line multiplexer (four input variables can have 16 possible combinations) and the appropriate digital levels are placed at the multiplexer data inputs, we can implement the function for X. We will use the 74150 16-line multiplexer. A 1 must be placed at each data input that satisfies any term in the Boolean equation. The truth table and the 74150 connections to implement the function are given in Figure 8–36.

The logic symbol for the 74150 is similar to the 74151 except that it has 16 input data lines, four data select lines, and only the complemented output. To test the operation of the circuit, let's try some entries from the truth table to see that X is valid. For example, if $A = 0$, $B = 0$, $C = 0$, and $D = 1$ ($\overline{A}\,\overline{B}\,\overline{C}D$), the multiplexer will select D_1, which is 1. It gets inverted twice before reaching X, so X receives a 1, which is correct. Work through the rest of them yourself and you will see that the Boolean function is fulfilled.

8–6   Demultiplexers

Demultiplexing is the opposite procedure from multiplexing. We can think of a **demultiplexer** as a *data distributor*. It takes a single input data value and routes it to one of several outputs, as illustrated in Figure 8–37.

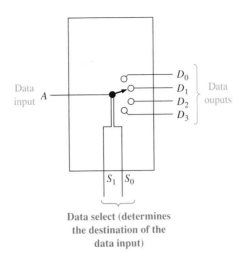

Data input A

D_0
D_1 Data
D_2 ouputs
D_3

S_1 | S_0

Data select (determines
the destination of the
data input)

Figure 8–37   Functional diagram of a 4-line demultiplexer.

Integrated-circuit demultiplexers come in several configurations of inputs/outputs. The two that we discuss in this section are the 74139 dual 4-line demultiplexer and the 74154 16-line demultiplexer.

The logic diagram and logic symbol for the 74139 are given in Figure 8–38. Notice that the 74139 is divided into two equal sections. By looking at the logic diagram, you will see that the schematic is the same as that of a 2-line-to-4-line decoder. Decoders and demultiplexers are the same, except with a decoder you hold the $\overline{E}$ enable line LOW and enter a code at the A_0A_1 inputs. As a demultiplexer, the A_0A_1 inputs are used to select the destination of input data. The input data are brought in via the $\overline{E}$ line. The 74138 3-line-to-8-line decoder that we covered earlier in this chapter can also function as an 8-line demultiplexer.

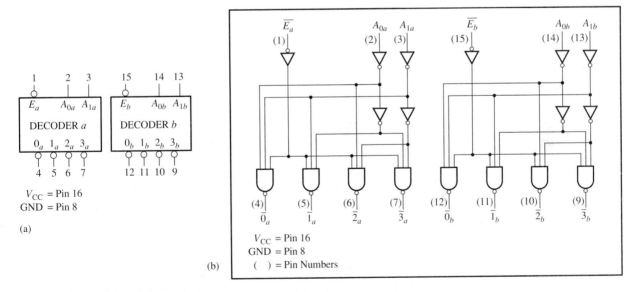

Figure 8–38   The 74139 dual 4-line demultiplexer: (a) logic symbol; (b) logic diagram.
(Courtesy of Signetics Corporation.)

To use the 74139 as a demultiplexer to route some input data signal to, let's say, the $\overline{2a}$ output, the connections shown in Figure 8–39 would be made. In the figure the destination $\overline{2a}$ is selected by making $A_{1a} = 1$ and $A_{0a} = 0$. The input signal is brought into the enable line ($\overline{E_a}$). When $\overline{E_a}$ goes LOW, the selected output line goes LOW; when $\overline{E_a}$ goes HIGH, the selected output line goes HIGH. (All nonselected lines remain HIGH continuously.)

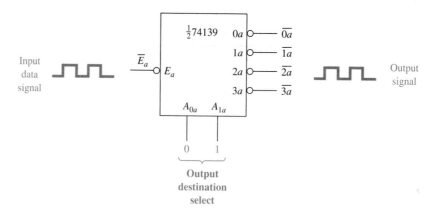

Figure 8–39   Connections to route an input data signal to the $\overline{2}_a$ output of a 74139 demultiplexer.

The 74154 was used earlier in the chapter as a 4-line-to-16-line hexadecimal decoder. It can also be used as a 16-line demultiplexer. Figure 8–40 shows how it can be connected to route an input data signal to the $\overline{5}$ output.

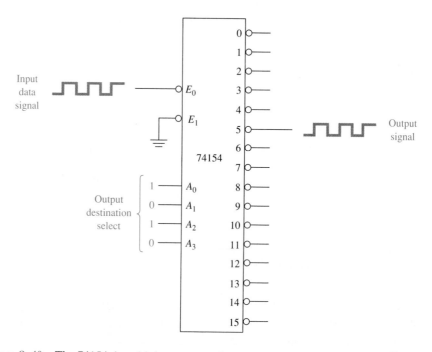

Figure 8–40   The 74154 demultiplexer connections to route an input signal to the $\overline{5}$ output.

Analog Multiplexer/Demultiplexer

Several analog multiplexers/demultiplexers are available in the CMOS family. The 4051, 4052, and 4053 are combination multiplexer *and* demultiplexer CMOS ICs. (High-speed CMOS versions, such as the 74HCT4051, are also available.) They can function in either configuration because their inputs and outputs are **bidirectional,** meaning that the flow can go in either direction. Also, they are *analog,* meaning that they can input and output levels other than just 1 and 0. The input/output levels can be any analog voltage between the positive and negative supply levels.

The functional diagram for the 4051 eight-channel multiplexer/demultiplexer is given in Figure 8–41. The eight square boxes in the functional diagram represent the bidirectional I/O lines. Used as a multiplexer, the analog levels come in on the Y_0 to Y_7

CHAPTER 8 | CODE CONVERTERS, MULTIPLEXERS, AND DEMULTIPLEXERS

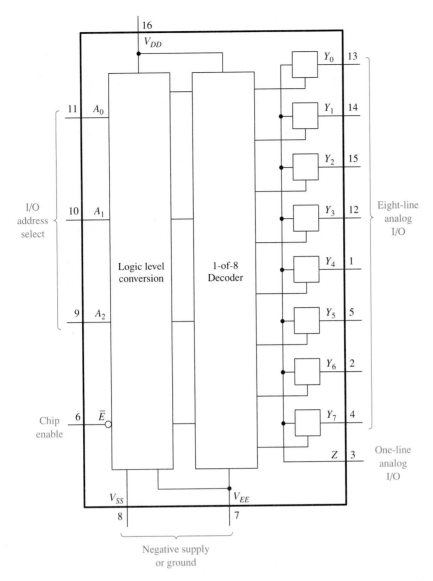

Figure 8–41 The 4051 CMOS analog multiplexer/demultiplexer. (Courtesy of Signetics Corporation.)

lines, and the decoder selects which of these inputs are output to the Z line. As a demultiplexer, the connections are reversed, with the input coming into the Z line and the output going out on one of the Y_0 to Y_7 lines.

Review Questions

8–15. Why is a *multiplexer* sometimes called a data selector?

8–16. Why is a *demultiplexer* sometimes called a data distributor?

8–17. What is the function of the S_0, S_1, and S_2 pins on the 74151 multiplexer?

8–18. What is the function of the A_0, A_1, A_2, and A_3 pins on the 74154 demultiplexer?

8–7 System Design Applications

Helpful Hint

The applications that follow demonstrate how the MSI chips covered in this chapter interface with practical microprocessor-based systems. They are meant to provide a positive experience for you by showing that you can comprehend circuits of higher complexity now that you have mastered some of the building blocks.

Microprocessor Address Decoding

The 74138 and its CMOS version, the 74HCT138, are popular choices for decoding the address lines in **microprocessor** circuits. A typical 8-bit microprocessor such as the Intel 8085A or the Motorola 6809 has 16 address lines (A_0–A_{15}) for designating unique **addresses** for all the peripheral devices and memory connected to it. When a microprocessor-based system has a large amount of memory connected to it, a designer often chooses to set the memory up in groups, called *memory banks*. For example, Figure 8–42 shows a decoding scheme that can be used to select one of eight separate memory banks within a microprocessor-based system. The high-order bits of the address (A_{12}–A_{15}) are output by the microprocessor to designate which memory bank is to be accessed. In this design, A_{15} must be LOW for the decoder IC to be enabled. The three other high-order bits, A_{12}–A_{14}, are then used to select the designated memory bank.

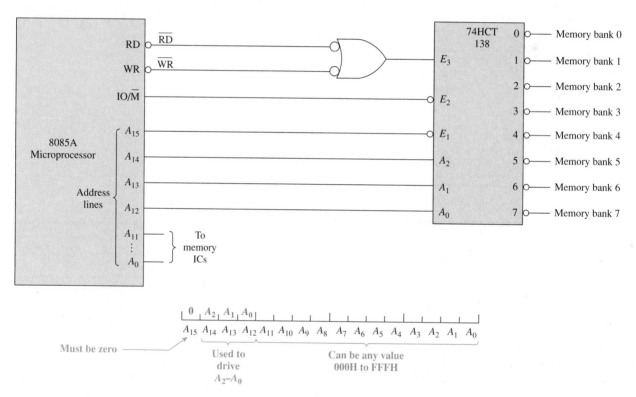

Figure 8–42 Using the 74HCT138 for a memory address decoder in an 8085A microprocessor system.

The 8085A also outputs control signals that are used to enable/disable memory operations. First, if we are performing a memory operation, we must be doing a read ($\overline{RD}$) or a write ($\overline{WR}$). The inverted-input OR gate (NAND) provides the HIGH to the E_3 enable if it receives a LOW $\overline{RD}$ *or* a LOW $\overline{WR}$. The other control signal, IO/$\overline{M}$, is used by the 8085A to distinguish between input/output (IO) to peripheral devices versus memory operations. If IO/$\overline{M}$ is HIGH, an IO operation is to take place, and if it is LOW, a memory operation is to occur. Therefore, one of the memory banks will be selected if IO/$\overline{M}$ is LOW and address line A_{15} is LOW while either $\overline{RD}$ is LOW or $\overline{WR}$ is LOW.

EXAMPLE 8–13

What is the range of addresses that can be specified by the 8085A in Figure 8–42 to access memory within bank 2?

Solution: Referring to the chart in Figure 8–42, address bit A_{13} must be HIGH in order to make A_1 in the decoder HIGH to select bank 2. The other address bits, A_0–A_{11}, are not used by the decoder IC and can be any value. Therefore, any address within the range of 2000H through 2FFFH will select memory bank 2.

Alarm Encoder for a Microcontroller

This design application uses a 74148 encoder to monitor the fluid level of eight chemical tanks. If any level exceeds a predetermined height, a sensor outputs a LOW level to the input of the encoder. The encoder encodes the active input into a 3-bit binary code to be read by a **microcontroller.** This way, the microcontroller needs to use only three input lines to monitor eight separate points. Figure 8–43 shows the circuit connections.

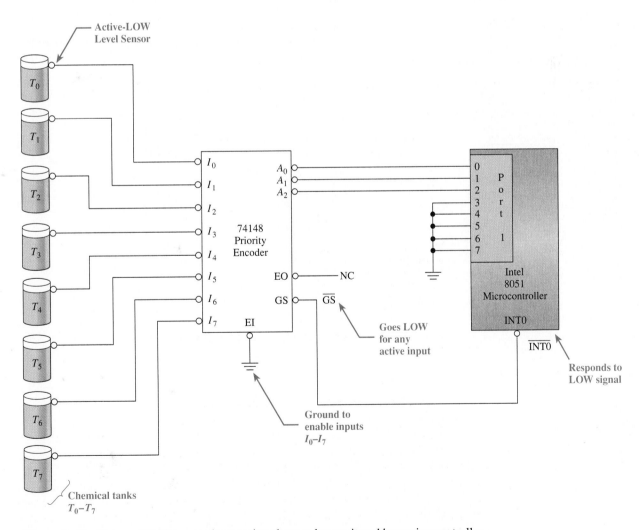

Figure 8–43   Using a 74148 to encode an active alarm to be monitored by a microcontroller

A microcontroller differs from a microprocessor in that it has several input/output ports and memory built into its architecture, making it better suited for monitoring and control applications. The microcontroller used here is the Intel 8051.* We will use one of its 8-bit ports to read the encoded alarm code, and we will use its interrupt input, $\overline{INT0}$, to receive notification that an alarm has occurred. The 8051 will be programmed to be in a HALT mode (or, in some versions, a low-power SLEEP mode) until it receives an interrupt signal to take a specific action. In this case it will perform the desired response to the alarm when it receives a LOW at $\overline{INT0}$. This LOW interrupt signal is provided by $\overline{GS}$, which goes LOW whenever any of the 74148 inputs becomes active.

Serial Data Multiplexing for a Microcontroller

Multiplexing and demultiplexing are very useful for data communication between a computer system and serial data terminals. The advantage is that only one serial receive line and one serial transmit line are required by the computer to communicate with several data terminals. A typical configuration is shown in Figure 8–44.

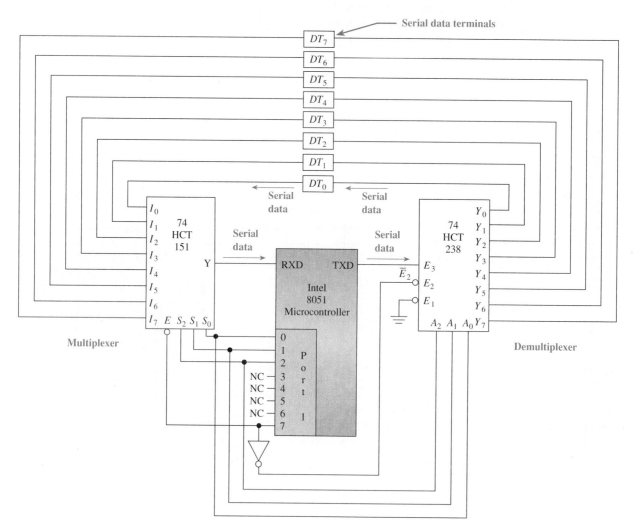

Figure 8–44   Using a multiplexer, demultiplexer, and microcontroller to provide communication capability to several serial data terminals.

---

*For an in-depth study of the 8085A microprocessor and the 8051 microcontroller refer to William Kleitz, *Digital and Micro-processor Fundamentals*, Prentice Hall, Englewood Cliffs, N.J., 1990.

Again the 8051 is used because of its built-in control and communication capability. Its RXD and TXD pins are designed to receive (RXD) and transmit (TXD) serial data at a speed and in a format dictated by a computer program written by the user.

The selected data terminal to be read from is routed through the 74HCT151 multiplexer to the microcontroller's serial input terminal, RXD. First, the computer program writes the appropriate hex code to Port 1 to enable the '151 to route the serial data stream from the selected data terminal through to its Y output. The 8051 then reads the serial data at its RXD input and performs the desired action.

To output to one of the data terminals, the 8051 must first output the appropriate hex code to Port 1 to select the correct data terminal; then it outputs serial data on its TXD pin. The 74HCT238 is used as a demultiplexer (data distributor) in this application. The 74HCT238 is identical to the 74138 decoder/demultiplexer except the '238 has noninverting outputs. The HCT versions are used to match the high speed of the 8051 and keep the power requirements to a minimum (more on speed and power requirements in Chapter 9). The hex code output at Port 1 must provide a LOW to $\overline{E_2}$ and the proper data-routing select code to A_2–A_0. The selected Y output then duplicates the HIGH/LOW levels presented at the E_3 pin.

EXAMPLE 8–14

Determine the correct hex codes that must be output to Port 1 in Figure 8–44 to accomplish the following action: (a) read from data terminal 3 (DT3); (b) write to data terminal 6 (DT6).

Solution: The 0, 1, and 2 outputs of Port 1 are used to control the S_0, S_1, and S_2 input-select pins of the '151 and the A_0, A_1, and A_2 output-select pins of the '238. Output 7 of Port 1 is used to determine which IC is enabled; 0 enables the '151 multiplexer, and 1 enables the '238 demultiplexer. (a) Assuming that the NC (no connection) lines are zeros, the hex code to read from data terminal 3 is 03H (0000 0011). (b) The hex code to write to data terminal 6 is 86H (1000 0110).

Analog Multiplexer Application

The 4051 is very versatile for controlling analog voltages with digital signals. One use is in the design of multitrace oscilloscope displays for displaying as many as eight traces on the same display screen. To do that, each input signal to be displayed must be **superimposed** on (added to) a different voltage level so that each trace will be at a different Y-axis level on the display screen.

The 4051 (and its high-speed version, the 74HCT4051) can be set up to sequentially output eight different voltage levels repeatedly if connected as shown in Figure 8–45. The resistor voltage-divider network in Figure 8–45 is set up to drop 0.5 V across each 100-Ω resistor. This will put 0.5 V at Y_0, 1.0 V at Y_1, and so on. The binary counter outputs a binary progression from 000 up to 111 at its 2^0, 2^1, and 2^2 outputs, which causes each of the Y_0 to Y_7 inputs to be selected for Z out, one at a time, in order. The result is the staircase waveform shown in Figure 8–45, which can superimpose a different voltage level on each of eight separate digital input signals that are brought in via the 74151 8-line *digital* multiplexer (not shown) driven by the same binary counter.

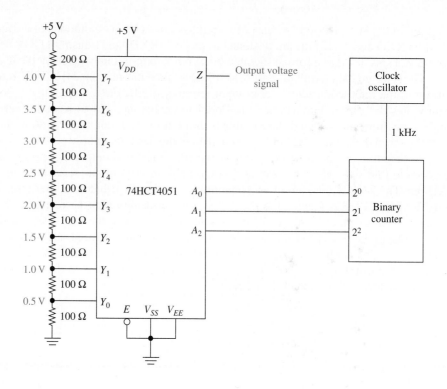

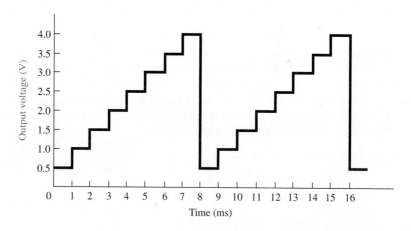

Figure 8–45 The 74HCT4051 analog multiplexer used as a staircase generator.

Multiplexed Display Application

Figure 8–46 shows a common method of using multiplexing to reduce the cost of producing a multidigit display in a digital system or computer. Multiplexing multidigit displays reduces circuit cost and failure rate by *sharing* common ICs, components, and conductors. The seven-segment digit displays, decoders, and drivers are covered in detail in Chapter 12. For now, we need to know that a decoding process must take place to convert the BCD digit information to a recognizable digit display. We will also assume that the "arithmetic circuitry" takes care of loading the four digit registers with the proper data.

The digit **bus** and display bus are each just a common set of conductors *shared by* the digit storage registers and display segments. The four-digit registers are therefore multiplexed into a single-digit bus, and the display bus is demultiplexed into the four-digit displays.

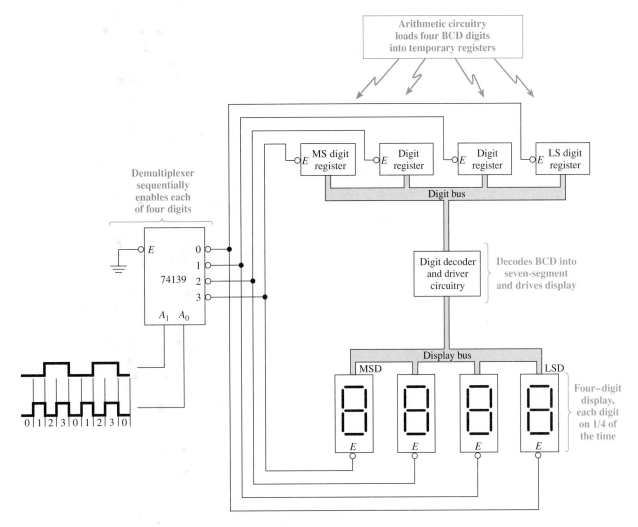

Figure 8-46 Multiplexed four-digit display block diagram.

The 74139 four-line demultiplexer takes care of sequentially accessing each of the four digits. It first outputs a LOW on the $\bar{0}$ line. This enables the LS digit register *and* the LS digit display. The LS BCD information travels down the digit bus to the decoder/driver, which *decodes* the BCD into the special seven-segment code used by the LS digit display and *drives* the LS digit display.

Next, the second digit register and display are enabled, then the third, and then the fourth. This process continues repeatedly, each digit being on one-fourth of the time. The circulation is set up fast enough (1 kHz or more) that it appears that all four digits are on at the same time. The external arithmetic circuitry is free to change the display at any time simply by reloading the temporary digit registers.

Review Questions

8–19. In the address decoding circuit of Figure 8–42, the A_{15} address bit must be HIGH to access memory bank 7. True or false?

8–20. What IC would be used to implement the inverted-input OR gate in Figure 8–42; a 7400, a 7402, a 7408, or a 7432?

8–21. To read from memory bank 0 in Figure 8–42, the microprocessor will output _____, _____, and _____ on the $\overline{RD}$, $\overline{WR}$, and IO/$\overline{M}$ lines.

8–22. In Figure 8–43, how is the microcontroller notified that there is a high fluid level at one of the chemical tanks?

8–23. The circuit of Figure 8–43 would work properly but have inverted outputs at A_0–A_2 if the level sensor outputs were active-HIGH instead of active-LOW. True or false?

8–24. The circuit of Figure 8–44 does not allow for both transmitting and receiving serial data simultaneously. True or false?

8–25. In Figure 8–44, the '151 could be switched with the '238 and still work properly. True or false?

8–26. What is the purpose of the binary counter in Figure 8–45?

8–27. Describe the circuit operation required to display the number 5 in the MSD position of the display in Figure 8–46.

Summary

In this chapter we have learned that

1. Comparators can be used to determine equality or which of two binary strings is larger.

2. Decoders can be used to convert a binary code into a singular active output representing its numeric value.

3. Encoders can be used to generate a coded output from a singular active numeric input line.

4. ICs are available to convert BCD to binary and binary to BCD.

5. The Gray code is useful for indicating the angular position of a shaft on a rotating device, such as a motor.

6. Multiplexers are capable of funneling several data lines into a single line for transmission to another point.

7. Demultiplexers are used to take a single data value or waveform and route it to one of several outputs.

Glossary

Address: A unique binary value that is used to distinguish the location of individual memory bytes or peripheral devices.

Bidirectional: A device capable of functioning in either of two directions, thus being able to reverse its input/output functions.

Bus: A group of conductors that have a common purpose and are shared by several devices or ICs.

Code Converter: A device that converts one type of binary representation to another, such as BCD to binary or binary to Gray code.

Comparator: A device used to compare the magnitude or size of two binary bit strings or words.

Decoder: A device that converts a digital code such as hex or octal into a single output representing its numeric value.

Demultiplexer: A device or circuit capable of routing a single data-input line to one of several data-output lines; sometimes referred to as a data distributor.

Don't Care ($\times$): A variable that is signified in a function table as a don't care, or $\times$, can take on either value, HIGH *or* LOW, without having any effect on the output.

Encoder: A device that converts a weighted numeric input line to an equivalent digital code, such as hex or octal.

Gray Code: A binary coding system used primarily in rotating machinery to indicate a shaft position. Each successive binary string within the code changes by only 1 bit.

Hardware/Software: Sometimes solutions to digital applications can be done using hardware *or* software. The *software* approach uses computer program statements to solve the application, whereas the *hardware* approach uses digital electronic devices and ICs.

Microcontroller: Sometimes referred to as "a computer on a chip," it is especially well suited for data acquisition and control applications. In a single integrated-circuit package, it will contain a microprocessor, memory, I/O ports, and communication capability, among other features.

Microprocessor: A large-scale integration (LSI) integrated circuit that is the fundamental building block of a digital computer. It is controlled by software programs that allow it to do all digital arithmetic, logic, and I/O operations.

Multiplexer: A device or circuit capable of selecting one of several data input lines for output to a single line; sometimes referred to as a data selector.

Priority: When more than one input to a device is active and only one can be acted on, the one with the highest priority will be acted on.

Superimpose: Combining two waveforms together such that the result is the sum of their levels at each point in time.

Weighting Factor: The digit within a numeric string of data is worth more or less depending on which position it is in. A weighting factor is applied to determine its worth.

Problems

Section 8–1

D   **8–1.** Design a binary comparator circuit using exclusive-ORs and a NOR gate that will compare two 8-bit binary strings.

8–2. Label all the lines in your design for Problem 8–1 with the digital levels that will occur when comparing $A = 1101\ 1001$ and $B = 1101\ 1001$.

8–3. Label the digital levels on all the lines in Figure 8–3 that would occur when comparing the two 8-bit strings $A = 1011\ 0101$ and $B = 1100\ 0011$.

Section 8–2

8–4. Write a two-sentence description of the function of a decoder.

8–5. Construct a truth table similar to Table 8–1 for an active-LOW output BCD (1-of-10) decoder.

8–6. What state must the inputs $\overline{E_1}$, $\overline{E_2}$, and E_3 be in in order to *enable* the 74138 decoder?

8–7. What does the $\times$ signify in the function table for the 74138?

8–8. Describe the difference between active-LOW outputs and active-HIGH outputs.

8–9. Sketch the output waveforms ($\overline{0}$ to $\overline{7}$) given the inputs shown in Figure P8–9(b) to the 74138 of Figure P8–9(a).

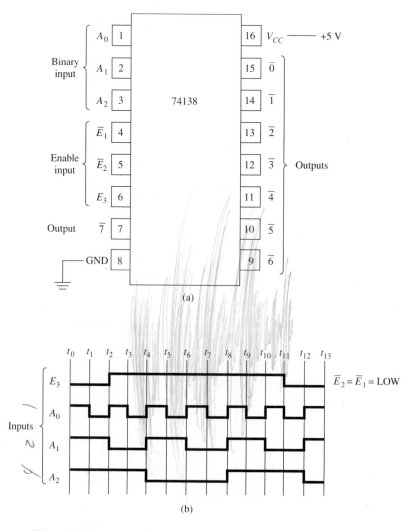

(a)

(b)

Figure P8–9

8–10. Repeat Problem 8–9 for the input waveforms shown in Figure P8–10.

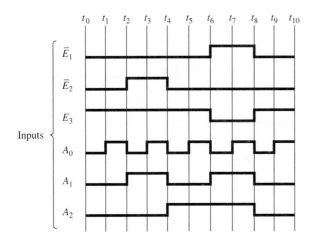

Figure P8–10

8–11. What state do the outputs of a 7442 BCD decoder go to when an invalid BCD number (10 to 15) is input to A_0 to A_3?

D   **8–12.** Design a circuit, based on a 74154 4-line-to-16-line decoder, that will output a HIGH whenever the 4-bit binary input is greater than 12. (When the binary input is less than or equal to 12, it will output a LOW.)

Section 8–3

8–13. With the 74147 priority encoder, if two different decimal numbers are input at the same time, which will be encoded?

8–14. A 74147 is connected with $\overline{I_1} = \overline{I_2} = \overline{I_3} = $ LOW and $\overline{I_4} = \overline{I_5} = \overline{I_6} = \overline{I_7} = \overline{I_8} = \overline{I_9} = $ HIGH. Determine $\overline{A_0}, \overline{A_1}, \overline{A_2}$, and $\overline{A_3}$.

8–15. Sketch the output waveforms $(\overline{A_0}, \overline{A_1}, \overline{A_2}, \overline{EO}, \overline{GS})$ given the inputs shown in Figure P8–15(b) to the 74148 of Figure P8–15(a).

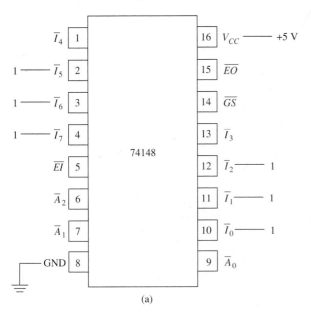

(a)

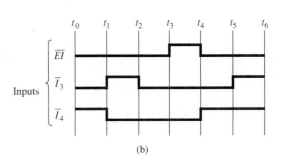

(b)

Figure P8–15

8–16. Repeat Problem 8–15 for the waveforms shown in Figure P8–16.

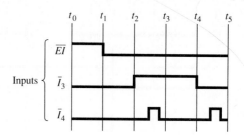

Figure P8–16

8–17. Two 74148s are connected in Figure P8–17 to form an active-LOW input, active-LOW output hexadecimal (16-line-to-4-line) priority encoder. Show the logic levels on each line in Figure P8–17 for encoding an input hexadecimal C (12) to an output binary 1100 (active-LOW 0011).

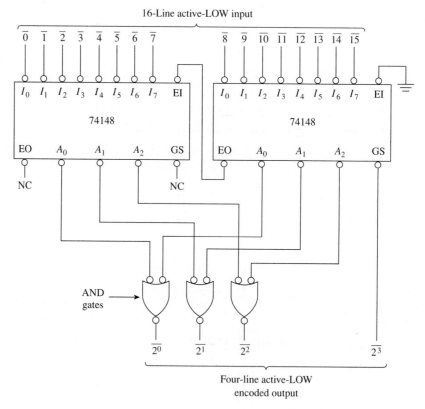

Figure P8–17

8–18. Repeat Problem 8–17 for encoding an input hexadecimal 6 to an output binary six (active-LOW 1001).

CHAPTER 8 I CODE CONVERTERS, MULTIPLEXERS, AND DEMULTIPLEXERS

8–19. Using the weighting factors given in Figure 8–20, convert the following decimal numbers to BCD and then to binary.

(a) 32   (c) 55

(b) 46   (d) 68

C   **8–20.** Figure P8–20 is a two-digit BCD-to-binary converter. Show how the number 49 (0100 1001$_{\text{BCD}}$) is converted to binary by placing 1's and 0's at the inputs and outputs.

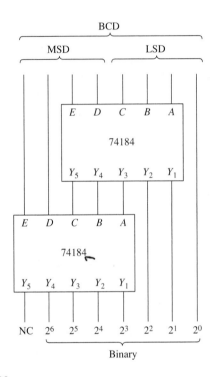

Figure P8–20

C   **8–21.** Repeat Problem 8–20 for the number 73.

8–22. Convert the following Gray codes to binary using the circuit of Figure 8–27.

(a) 1100   (c) 1110

(b) 0101   (d) 0111

8–23. Convert the following binary numbers to Gray code using the circuit of Figure 8–26.

(a) 1010   (c) 0011

(b) 1111   (d) 0001

8–24. The connections shown in Figure P8–24 are made to the 74151 eight-line multiplexer. Determine Y and $\overline{Y}$.

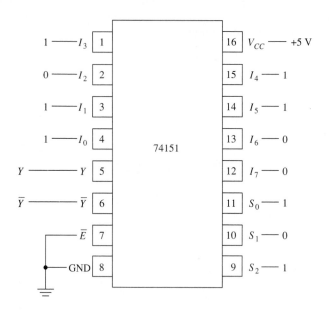

Figure P8–24

C D   **8–25.** Using a technique similar to that presented in Figure 8–35, design a 32-bit multiplexer using four 74151's.

C D   **8–26.** Design a circuit that will output a LOW whenever a month has 31 days. The month number (1 to 12) is input as a 4-bit binary number (January = 0001, and so on). (*Hint:* Use a 74150.)

Section 8–6

C D   **8–27.** Design an 8-bit demultiplexer using one 74139.

C D   **8–28.** Design a 16-bit demultiplexer using two 74138's.

T   **8–29.** There is a malfunction in a digital system that contains several multiplexer and demultiplexer ICs. A reading was taken at each pin with a logic probe, and the results were recorded in Table 8–7. Which IC or ICs are not working correctly?

Section 8–7

8–30. Which memory bank is accessed in Figure 8–42 for each of the following hex addresses output by the microprocessor?

(a) 3000H   (c) 507CH

(b) 6000H   (d) 8001H

Table 8–7 IC Logic States for Troubleshooting Problem 8–29

74150		74151		74139		74154	
Pin	Level	Pin	Level	Pin	Level	Pin	Level
1	0	1	1	1	0	1	1
2	1	2	0	2	1	2	1
3	1	3	0	3	0	3	1
4	0	4	1	4	0	4	1
5	1	5	1	5	1	5	1
6	0	6	0	6	1	6	1
7	1	7	0	7	1	7	1
8	1	8	0	8	0	8	1
9	0	9	0	9	0	9	1
10	0	10	0	10	0	10	1
11	0	11	0	11	1	11	1
12	0	12	0	12	0	12	0
13	1	13	1	13	1	13	1
14	1	14	0	14	0	14	1
15	1	15	0	15	1	15	1
16	0	16	1	16	1	16	1
17	1					17	1
18	0					18	1
19	1					19	0
20	1					20	0
21	1					21	0
22	0					22	1
23	1					23	1
24	1					24	1

8–31. A logic analyzer was used to monitor the 19 microprocessor lines shown in Figure 8–42. Describe the operation that was taking place if the following levels were observed from top to bottom.

(a) 100 0101 0000 0000 0011   (b) 010 0110 0000 1100 0111

8–32. What hex number will be read by port 1 of the 8051 microcontroller in Figure 8–43 if the fluid level is high in the following chemical tanks?

(a) Tank 1   (d) All tanks at a high fluid level

(b) Tank 6   (e) No tanks at a high fluid level

(c) Tanks 2 and 7

8–33. What hex code must be output by port 1 of the 8051 in Figure 8–44 to perform the following operations?

(a) Read from data terminal 5

(b) Read from data terminal 7

(c) Write to data terminal 2

(d) Write to data terminals 1 and 2 at the same time

8–34. Determine the voltage level of the output voltage signal in Figure 8–45 when the binary count reaches 110 (6_{10}).

8–35. What are the output levels at $\overline{0}$, $\overline{1}$, $\overline{2}$, and $\overline{3}$ of the demultiplexer in Figure 8–46 when the circuit is displaying the number 3 in the LSD position?

Schematic Interpretation Problems

See Appendix G for the schematic diagrams.

S   **8–36.** Find the two 4-bit-magnitude comparators, U7 and U8, in the Watchdog Timer schematic. Which IC receives the high-order binary data, U7 or U8? (*Hint:* The bold lines in that schematic represent a *bus,* which is a group of conductors that are shared by several ICs. It simplifies the diagram by showing a single bold line instead of several separate lines. When the individual lines are taken off the bus they are labeled, appropriately, 0–1–2–3 and 4–5–6–7 in this application.)

S   **8–37.** Where is the final output of the comparison made by U7, U8 used in the Watchdog Timer schematic?

S   **8–38.** Find the octal decoder U5 in the HC11D0 schematic. Determine the levels on AS, AD13, AD14, and AD15 required to provide an active-LOW signal on the line labeled MON_SL.

S C   **8–39.** Locate the address decoder section (U3: C, U5, and U9) in the HC11D0 schematic. The octal decoder (U9) is used to determine if the LCD (LCD_SL) or the keyboard (KEY_SL) is to be active.

   (a) Determine the levels on AD3-5, AD11-15, and AS required to select the LCD.

   (b) Repeat for selecting the keyboard.

S C   **8–40.** Find the decoders U28 and U29 on sheet 2 of the 4096/4196 schematic. They are cascaded together to form a 1-of-18 decoder for the lines labeled ICS1–ICS18.

   (a) Determine the levels on pins 2, 5, 6, 9, and 12 of U31 to provide an active-LOW output at ICS5.

   (b) Repeat for ICS18.

S C D   **8–41.** Design a circuit using a 74151 multiplexer that will allow you to input the status of the four temperatures and four pressures in the chemical monitoring system in Figure 1–4 (Chapter 1). Assume that the 68HC11 microcontroller in the HC11D0 schematic will be used to read each of the status bits, one at a time (P_D, then T_D, then P_C, and so on). Also assume that the only input and output pins available for use are at port D, bits PD2, PD3, PD4, and PD5.

Answers to Review Questions

8–1. False

8–2. $A < B = 1$

8–3. 4 inputs, 10 outputs

8–4. False

8–5. False

8–6. False

8–7. HIGH, LOW

8–8. An encoder creates a coded output from a singular active numeric input line. A decoder is the opposite, creating a single output from a numeric input code.

8–9. The highest numeric input

8–10. (a) All HIGH (b) $A_3 = L$, $A_2 = H, A_1 = H, A_0 = L$

8–11. A_0, A_1, A_2: binary outputs; EO: enable output; GS: group signal output. All outputs are active LOW.

8–12. 80

8–13. 6

8–14. Because it varies by only 1 bit when the shaft is rotated from one position to the next

8–15. Because it selects which data input is to be sent to the data output

8–16. Because it takes a single input data line and routes it to one of several outputs

8–17. They select which one of the input lines is sent to the output.

8–18. They select which one of the output lines the input data are sent to.

8–19. False

8–20. 7400

8–21. 0, 1, 0

8–22. $\overline{GS}$ goes LOW.

8–23. False

8–24. True

8–25. False

8–26. It outputs a binary progression, allowing each of the Y_0 to Y_7 inputs to appear at the Z output in a staircase fashion.

8–27. The number 5 is loaded into the MSD register. When the A_0 and A_1 lines of the 74139 reach 1–1, the number 3 line outputs a LOW, enabling the MSD register and the MSD of the display. The number 5 is then transferred to the display by the decoder/driver circuitry.

9 Logic Families and Their Characteristics

OUTLINE

9–1 The TTL Family
9–2 TTL Voltage and Current Ratings
9–3 Other TTL Considerations
9–4 Improved TTL Series
9–5 The CMOS Family
9–6 Emitter-Coupled Logic
9–7 Comparing Logic Families
9–8 Interfacing Logic Families

Objectives

Upon completion of this chapter, you should be able to

- Analyze the internal circuitry of a TTL NAND gate for both the HIGH and LOW output states.

- Determine IC input and output voltage and current ratings from the manufacturer's data manual.

- Explain gate loading, fan-out, noise margin, and time parameters.

- Design wired-output circuits using open-collector TTL gates.

- Discuss the differences and proper use of the various subfamilies within both the TTL and CMOS lines of ICs.

- Describe the reasoning and various techniques for interfacing between the TTL, CMOS, and ECL families of ICs.

Introduction

Integrated-circuit logic gates (small-scale integration, SSI), combinational logic circuits (medium-scale integration, MSI), and microprocessor systems (large-scale integration and very large scale integration, LSI and VLSI) are readily available from several manufacturers through distributors and electronic parts suppliers. Basically, there are three commonly used families of digital IC logic: **TTL** (transistor–transistor logic), **CMOS** (complementary metal oxide semiconductor), and **ECL** (emitter-coupled logic). Within each family, several subfamilies (or series) of logic types are available, with different ratings for speed, power consumption, temperature range, voltage levels, and current levels.

Fortunately, the different manufacturers of digital logic ICs have standardized a numbering scheme so that basic part numbers will be the same regardless of the manufacturer. The prefix of the part number, however, will differ because it is the manufacturer's abbreviation. For example, a typical TTL part number might be S74F08N. The 7408 is the basic number used by all manufacturers for a quad *AND* gate. The F stands for the *FAST* TTL subfamily, and the S prefix is the manufacturer's code for Signetics. National Semiconductor uses the prefix DM, and Texas Instruments uses the prefix SN. The N suffix at the end of the part number is used to specify the package type. N is used for the plastic dual-in-line (DIP), W is used for the ceramic flatpack, and D is used for the surface-mounted SO plastic package. The best sources of information on available package styles and their dimensions are the manufacturers' data manuals. Most data manuals list the 7408 as 5408/7408. The 54XX series is the military version, which has less stringent power supply requirements and an extended temperature range of −55° to +125°C, whereas the 74XX is the commercial version with a temperature range of 0° to +70°C and strict power supply requirements.

For the purposes of this text, reference is usually made to the 74XX commercial version, and the manufacturer's prefix code and package-style suffix code are ignored. The XX is used in this book to fill the space normally occupied by the actual part number. For example, the part number for an inverter in the 74XX series is 7404.

9–1 The TTL Family

The standard 74XX TTL IC family has evolved through several stages since the late 1960s. Along the way, improvements have been made to reduce the internal time delays and power consumption. At the same time, each manufacturer has introduced chips with new functions and applications.

The fundamental operation of a TTL chip can be explained by studying the internal circuitry of the basic two-input 7400 NAND gate shown in Figure 9–1. The diodes D_1 and D_2 are negative clamping diodes used to protect the inputs from any short-term negative input voltages. The input transistor, Q_1, acts like an AND gate and is usually fabricated with a *multiemitter* transistor, which characterizes TTL technology. (To produce two-, three-, four-, and eight-input NAND gates, the manufacturer uses two-, three-, four-, and eight-emitter transistors.) Q_2 provides control and current boosting to the totem-pole output stage.

The reasoning for the totem-pole setup was discussed in Chapter 2. Basically, when the output is HIGH (1), Q_4 is OFF (open) and Q_3 is ON (short). When the output is LOW (0), Q_4 is ON and Q_3 is OFF. Because one or the other transistor is always OFF, the current flow from V_{CC} to ground in that section of the circuit is minimized.

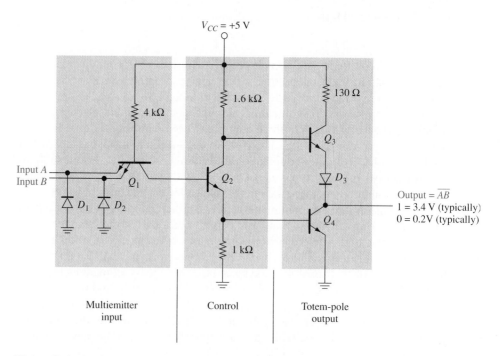

Figure 9–1 Internal circuitry of a 7400 two-input NAND gate.

CHAPTER 9 | LOGIC FAMILIES AND THEIR CHARACTERISTICS

To study the operation of the circuit in more detail, let's first review some basic electronics. An *NPN* transistor is basically two diodes; a *P* to *N* from base to emitter and another *P* to *N* from base to collector, as shown in Figure 9–2. The base-to-emitter diode is forward biased by applying a positive voltage on the base with respect to the emitter. A forward-biased base-to-emitter diode will have 0.7 V across it and will cause the collector-to-emitter junction to become almost a short circuit with approximately 0.3 V across it.

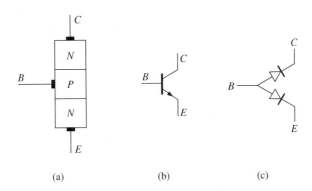

(a) (b) (c)

Figure 9–2 *NPN* transistor: (a) physical layout; (b) symbol; (c) diode equivalent.

Now, referring to Figure 9–3, we see the circuit conditions for the 0 output state and 1 output state. In Figure 9–3(a) ($A = 0$, $B = 0$, output = 1), with $A = 0$ or $B = 0$ or both equal to 0, the base-to-emitter diode of Q_1 is forward biased, saturating (turning on) Q_1 and placing 0.3 V with respect to ground at the base of Q_2. 0.3 V is not enough to turn Q_2 on, so no current flows through Q_2; instead, a small current flows through the 1.6-kΩ resistor to the base of Q_3, turning Q_3 on. (*Note:* The dashed lines in Figure 9–3 indicate the direction of *conventional* current flow. Electron flow is in the opposite direction.) The HIGH-level output voltage is typically 3.4 V, which is the 4.8 V at the base of Q_3 minus the 0.7-V diode drop at the base-to-emitter diode of Q_3 and the 0.7-V drop across D_3. (*Note:* These voltages are approximations used to illustrate circuit operation. Actual voltages will vary depending on the connected output load.)

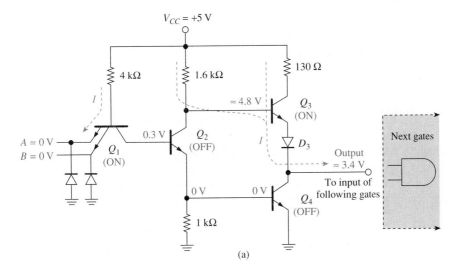

(a)

Figure 9–3 Equivalent circuits for a TTL NAND in the (a) HIGH;

Team Discussion

Discuss why V_{out} drops as gate loads are added to the output of Figure 9–3(a).

Common Misconception

You have probably never seen a transistor on its side, like Q_1. Just realize that it is still turned ON the same way, by applying a positive current flow from base to emitter.

Team Discussion

Think about what happens to the voltage level and current demand on the V_{CC} supply at the switching point if both Q_3 and Q_4 are momentarily ON at the same time.

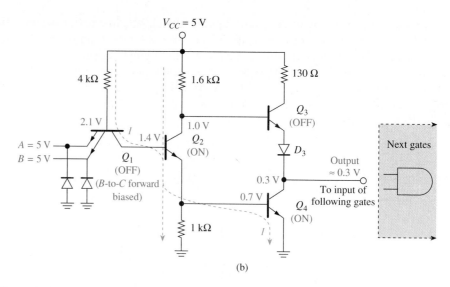

(b)

Figure 9–3 *(Continued.)* (b) LOW output states (I = conventional current flow).

In Figure 9–3(b) ($A = 1$, $B = 1$, output = 0), with $A = 1$ *and* $B = 1$, the base-to-emitter diode of Q_1 is reverse biased, *but* the base-to-collector diode [see Figure 9–2(c)] of Q_1 is forward biased. Current will flow down through the base to collector of Q_1, turning Q_2 on with a positive base voltage and turning Q_4 on with a positive base voltage. The output voltage will be approximately 0.3 V. Q_3 is kept off because there is not enough voltage between the base of Q_3 (1.0 V) to the cathode of D_3 (0.3 V) to overcome the two 0.7-V diode drops required to allow current flow.

Review Questions

9–1. The part number for a basic logic gate varies from manufacturer to manufacturer. True or false?

9–2. The input signal to a TTL NAND gate travels through three stages of internal circuitry: *input, control,* and _____.

9–3. A forward-biased NPN transistor will have approximately _____ volts across its base–emitter junction and _____ volts across its collector–emitter junction.

9–2 TTL Voltage and Current Ratings

Basically, we like to think of TTL circuits as operating at 0- and 5-V levels, but, as you can see in Figure 9–3, that just is not true. As we draw more and more current out of the HIGH-level output, the output voltage drops lower and lower, until finally it will not be recognized as a HIGH level anymore by the other TTL gates that it is feeding.

Input/Output Current and Fan-Out

The **fan-out** of a subfamily is defined as the number of gate inputs of the same subfamily that can be connected to a single output without exceeding the current ratings

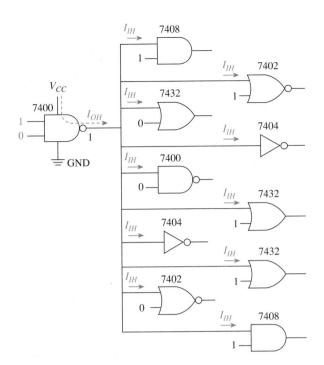

Common Misconception

Students often think that I_{OL} (or I_{OH}) is the *actual* output current, when really it is the *maximum limit not to be exceeded.*

Team Discussion

Describe the difference between the I_{out} of a 7400 that is feeding five inverter inputs versus one that is feeding five inverters connected end to end.

Figure 9–4 Ten gates driven from a single source.

of the gate. (A typical fan-out for most TTL subfamilies is 10.) Figure 9–4 shows an example of fan-out with 10 gates driven from a single gate.

To determine fan-out, you must know how much input current a gate load draws (I_I) and how much output current the driving gate can supply (I_O). In Figure 9–4 the single 7400 is the driving gate, supplying current to 10 other gate loads. The output current capability for the HIGH condition is abbreviated I_{OH} and is called a **source current.** I_{OH} for the 7400 is –400 μA maximum. (The minus sign signifies conventional current *leaving* the gate.)

The input current requirement for the HIGH condition is abbreviated I_{IH} and for the 74XX subfamily is equal to 40 μA maximum. To find the fan-out, divide the source current (–400 μA) by the input requirements for a gate (40 μA). The fan-out is 400 μA/40 μA = 10.

For the LOW condition, the maximum output current for the 74XX subfamily is 16 mA, and the input requirement for each 74XX gate is –1.6 mA maximum, also for a fan-out of 10. The fan-out is usually the same for both the HIGH and LOW conditions for the 74XX gates; if not, we use the lower of the two.

Because a LOW output level is close to 0 V, the current actually flows into the output terminal and sinks down to ground. This is called a **sink current** and is illustrated in Figure 9–5. In the figure, two gates are connected to the output of gate 1. The total current that gate 1 must sink in this case is 2 × 1.6 mA = 3.2 mA. Because the maximum

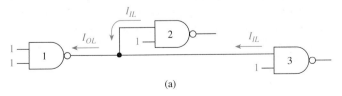

(a)

Figure 9–5 Totem-pole LOW output of a TTL gate sinking the input currents from two gate inputs: (a) logic gate symbols;

Team Discussion

Why does the sink current in Q_4 of gate 1 remain unchanged if *both* inputs of Q_1 gate 3 are connected together, as they are when forming an inverter from a NAND?

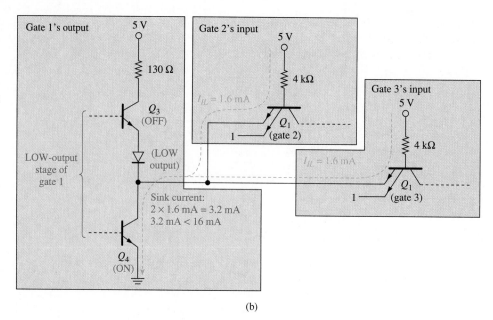

(b)

Figure 9–5 *(Continued)* (b) logic gate internal circuitry.

current a gate can sink in the LOW condition (I_{OL}) is 16 mA, gate 1 is well within its maximum rating of I_{OL}. (Gate 1 could sink the current from as many as *10* gate inputs.)

For the HIGH-output condition, the circuitry is the same, but the current flow is reversed, as shown in Figure 9–6. In the figure you can see that the 40 μA going into

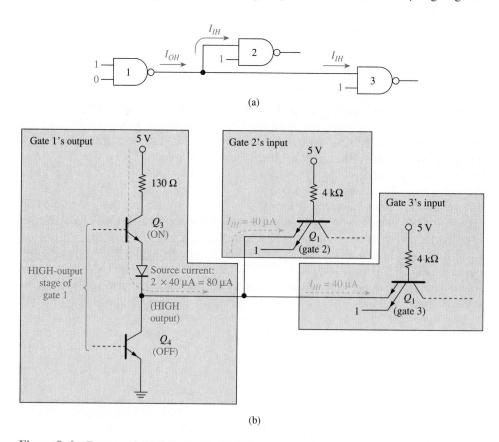

Figure 9–6 Totem-pole HIGH output of a TTL gate sourcing current to two gate inputs: (a) logic gate symbols; (b) logic gate internal circuitry.

each input is actually a small reverse leakage current flowing against the emitter arrow. In this case, the output of gate 1 is sourcing -80 μA to the inputs of gates 2 and 3. -80 μA is well below the maximum allowed HIGH-output current rating of -400 μA.

Summary of Input/Output Current and Fan-Out

1. The maximum current that an input to a *standard* (that is, a 74XX) TTL gate can sink or source is

$$I_{IL}\text{—low-level input current} = -1.6 \text{ mA } (-1600 \text{ } \mu\text{A})$$

$$I_{IH}\text{—high-level input current} = 40 \text{ } \mu\text{A}$$

(The minus sign signifies current *leaving* the gate.)

2. The maximum current that the output of a *standard* TTL gate can sink or source is

$$I_{OL}\text{—low-level output current} = 16 \text{ mA } (16{,}000 \text{ } \mu\text{A})$$

$$I_{OH}\text{—high-level output current} = -400 \text{ } \mu\text{A } (-800 \text{ } \mu\text{A for some})$$

(*Note:* This is *not* the actual amount of current leaving or entering a gate's output, but rather it is the *maximum capability* of the gate to sink or source current. The *actual* output current that flows depends on the number and type of loads connected.)

3. The maximum number of gate inputs that can be connected to a *standard* TTL gate output is 10 (fan-out = 10). Fan-out is determined by taking the smaller result of I_{OL}/I_{IL} or I_{OH}/I_{IH}.

Input/Output Voltages and Noise Margin

We must also concern ourselves with the specifications for the acceptable input and output voltage levels. For the *LOW output condition,* the lower transistor (Q_4) in the totem-pole output stage is saturated (ON) and the upper one (Q_3) is cut off (OFF). V_{out} for the LOW condition (V_{OL}) is the voltage across the saturated Q_4, which has a typical value of 0.2 V and a maximum value of 0.4 V, as specified in the manufacturer's data manual.

For the *HIGH output condition,* the upper transistor (Q_3) is saturated and the lower transistor (Q_4) is cut off. The voltage that reaches the output (V_{OH}) is V_{CC} minus the drop across the 130-Ω resistor, minus the C-E drop, minus the diode drop. Manufacturers' data sheets specify that the HIGH-level output is typically 3.4 V, and they guarantee that the worst-case minimum value will be 2.4 V. This means that the next gate input must interpret any voltage from 2.4 V up to 5.0 V as a HIGH level. Therefore, we must also consider the *input* voltage-level specifications (V_{IH}, V_{IL}).

Manufacturers guarantee that any voltage between a minimum of 2.0 V up to 5.0 V will be interpreted as a HIGH (V_{IH}). Also, any voltage from a maximum of 0.8 V down to 0 V will be interpreted as a LOW (V_{IL}).

These values leave us a little margin for error, what is called the **noise margin.** For example, V_{OL} is guaranteed not to exceed 0.4 V, and V_{IL} can be as high as 0.8 V and still be interpreted as a LOW. Therefore, we have 0.4 V (0.8 V $-$ 0.4 V) of leeway (noise margin), as illustrated in Figure 9–7(a) and (b). Input voltages that fall within the "uncertain region" in Figure 9–7(b) will produce unpredictable results.

Table 9–1 is a summary of input/output voltage levels and noise margins for the standard family of TTL ICs. These numbers are the most common, but be sure that you

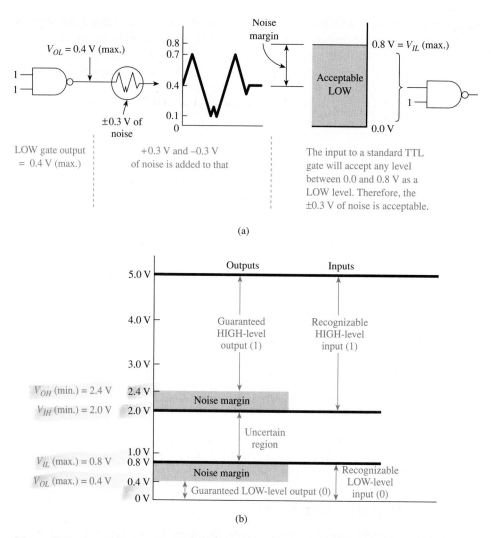

Figure 9–7 (a) Adding noise to a LOW-level output; (b) graphical illustration of the input/output voltage levels for the standard 74XX TTL series.

Table 9–1	Standard 74XX Series Voltage Levels				
Parameter	Minimum	Typical	Maximum		
V_{OL}		0.2 V	0.4 V	}	Noise margin
V_{IL}			0.8 V		= 0.4 V
V_{OH}	2.4 V	3.4 V		}	Noise margin
V_{IH}	2.0 V				= 0.4 V

Noise margin (HIGH) = V_{OH} (min.) − V_{IH} (min.)
Noise margin (LOW) = V_{IL} (max.) − V_{OL} (max.)

can locate these values on a data sheet (The data sheet for a 7400 NAND gate is given in Appendix B.) The prudent designer will always assume worst-case values to ensure that his or her design will always work for any conditions that may arise.

The following examples illustrate the use of the current and voltage ratings for establishing acceptable operating conditions for TTL logic gates.

■ EXAMPLE 9–1 ■

Find the voltages and currents that are asked for in Figure 9–8 if the gates are all standard (74XX) TTLs.

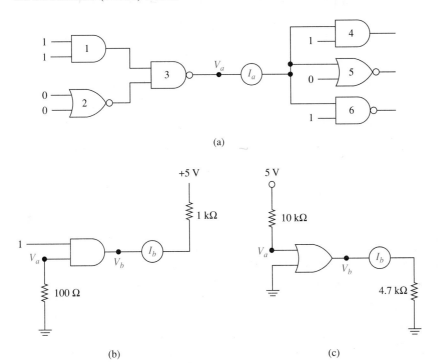

(a)

(b)   (c)

Figure 9–8 Voltage and current ratings.

(a) Find V_a and I_a for Figure 9–8(a).

(b) Find V_a, V_b, and I_b for Figure 9–8(b).

(c) Find V_a, V_b, and I_b for Figure 9–8(c).

Solution: **(a)** The input to gate 3 is a 1–1, so the output will be LOW. Using the *typical* value, $V_a = 0.2$ V. Because gate 3 is LOW, it will be sinking current from the three other gates: 4, 5, and 6. The typical value for each I_{IL} is −1.6 mA; therefore, $I_a = -4.8$ mA (−1.6 mA − 1.6 mA − 1.6 mA).

   (b) The 100-Ω resistor to ground will place a LOW level at that input. I_{IL} typically is −1.6 mA, which flows down through the 100-Ω resistor, making $V_a = 0.16$ V (1.6 mA × 100 Ω). The 0.16 V at V_a will be recognized as a LOW level ($V_{IL} = 0.8$ V max.), so the AND gate will output a LOW level; $V_b = 0.2$ V (typ.). The AND gate will sink current from the 1-kΩ resistor; $I_b = 4.8$ mA [(5 V − 0.2 V)/1 kΩ]. 4.8 mA is well below the maximum allowed current of 16 mA (I_{OL}), so the AND gate will not burn out.

   (c) I_{IH} into the OR gate is 40 μA; therefore, the voltage at $V_a = 4.6$ V [5 V − (10 kΩ × 40 μA)]. The output level of the OR gate will be HIGH (V_{OH}), making $V_b = 3.4$ V and $I_b = 3.4$ V/4.7 kΩ = 723 μA. 723 μA is below the maximum rating of the OR gate ($I_{OH} = -800$ μA max.). Therefore, the OR gate will not burn out.

9–4. Why aren't the HIGH/LOW output levels of a TTL gate exactly 5.0 and 0 V? *There are residue voltage drops.*

9–5. Describe what is meant by fan-out. *# of inputs a single output can drive*

9–6. List the names and abbreviations of the four input and output currents of a digital IC.

9–7. Describe the difference between *sink* and *source* output current.

9–8. Determine if the following input voltages will be interpreted as HIGH, LOW, or undetermined logic levels in a standard TTL IC.

H (a) 3.0 V *L* (c) 1.0 V

H (b) 2.2 V *L* (d) 0.6 V *L*

9–3 Other TTL Considerations

Pulse-Time Parameters: Rise Time, Fall Time, and Propagation Delay

We have been using ideal pulses for the input and output waveforms up until now. Actually, however, the pulse is not perfectly square; it takes time for the digital level to rise from 0 up to 1 and to fall from 1 down to 0.

As shown in Figure 9–9, the **rise time** (t_r) is the length of time it takes for a pulse to rise from its 10% point up to its 90% point. For a 5-V pulse, the 10% point is 0.5 V ($10\% \times 5$ V) and the 90% point is 4.5 V ($90\% \times 5$ V). The **fall time** (t_f) is the length of time it takes to fall from the 90% point to the 10% point.

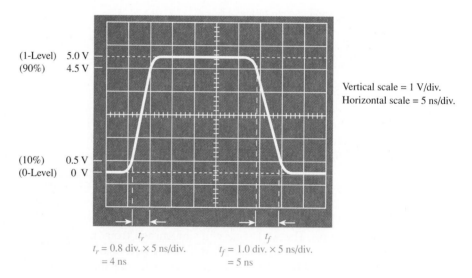

(1-Level)	5.0 V
(90%)	4.5 V
(10%)	0.5 V
(0-Level)	0 V

Vertical scale = 1 V/div.
Horizontal scale = 5 ns/div.

t_r
$t_r = 0.8$ div. × 5 ns/div.
= 4 ns

t_f
$t_f = 1.0$ div. × 5 ns/div.
= 5 ns

Figure 9–9 Oscilloscope display of pulse rise and fall times.

Not only are input and output waveforms sloped on their rising and falling edges, but there is also a delay time for an input wave to propagate through an IC to the output, called the **propagation delay** (t_{PLH} and t_{PHL}). The propagation delay is due to limitations in transistor switching speeds caused by undesirable internal capacitive stored charges.

Figure 9–10 shows that it takes a certain length of time for an input pulse to reach the output of an IC gate. A specific measurement point (1.5 V for the standard TTL series) is used as a reference. The propagation delay time for the *output* to respond in the LOW-to-HIGH direction is labeled t_{PLH}, and in the HIGH-to-LOW direction it is labeled t_{PHL}.

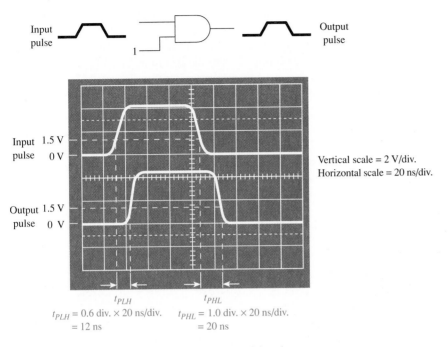

Input pulse   1

Output pulse

Input pulse   1.5 V   0 V

Output pulse   1.5 V   0 V

Vertical scale = 2 V/div.
Horizontal scale = 20 ns/div.

t_{PLH} t_{PHL}

t_{PLH} = 0.6 div. × 20 ns/div. t_{PHL} = 1.0 div. × 20 ns/div.
     = 12 ns   = 20 ns

Figure 9–10   Oscilloscope display of propagation delay times.

EXAMPLE 9–2

The propagation delay times for the 7402 NOR gate shown in Figure 9–11 are listed in a TTL data manual as t_{PLH} = 22 ns and t_{PHL} = 15 ns. Sketch and label the input and output pulses to a 7402.

7402

Figure 9–11

Solution: The input and output pulses are shown in Figure 9–12.

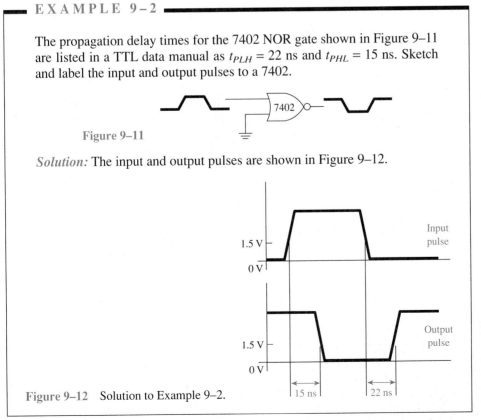

1.5 V

0 V

Input pulse

1.5 V

0 V

Output pulse

15 ns   22 ns

Figure 9–12   Solution to Example 9–2.

Common Misconception

The subscripts *LH* and *HL* pertain to the output, not the input.

Helpful Hint

It is very important that you can find these specs, as well as others, in a data book. (Data sheets for some common ICs are provided in Appendix B.)

Power Dissipation

Another operating characteristic of integrated circuits that has to be considered is the **power dissipation.** The power dissipated (or consumed) by an IC is equal to the total power supplied to the IC power supply terminals (V_{CC} to ground). The current that enters the V_{CC} supply terminal is called I_{CC}. Two values are given for the supply current: I_{CCH} and I_{CCL} for use when the outputs are HIGH or when the outputs are LOW. Since the outputs are usually switching between HIGH and LOW, if we assume a 50% duty cycle (HIGH half of the time, LOW half of the time), then an average I_{CC} can be used and the power dissipation is determined from the formula $P_D = V_{CC} \times I_{CC}$ (av.).

EXAMPLE 9–3

The total supply current for a 7402 NOR IC is given as $I_{CCL} = 14$ mA, $I_{CCH} = 8$ mA. Determine the power dissipation of the IC.

Solution:

$$P_D = V_{CC} \times I_{CC} \text{ (av.)}$$

$$= 5.0 \text{ V} \times \frac{14 \text{ mA} + 8 \text{ mA}}{2} = 55 \text{ mW}$$

Open-Collector Outputs

Instead of using a totem-pole arrangement in the output stage of a TTL gate, another arrangement, called the **open-collector** (OC) **output,** is available. Remember that, with the totem-pole output stage, for a LOW output the lower transistor is ON and the upper transistor is OFF, and vice versa for a HIGH output, whereas with the open-collector output the upper transistor is *removed*, as shown in Figure 9–13. Now the output will be *LOW* when Q_4 is ON and the output will *float* (not HIGH or LOW) when Q_4 is OFF. This means that an open-collector (OC) output can sink current, but it *cannot* source current.

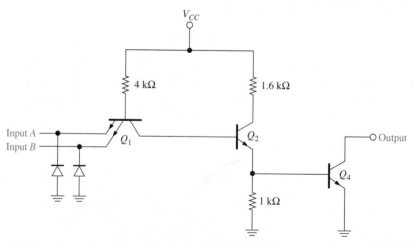

Figure 9–13   TTL NAND with an open-collector output.

To get an OC output to produce a HIGH, an external resistor (called a **pull-up resistor**) must be used, as shown in Figure 9–14. Now when Q_4 is OFF (open) the output is approximately 5 V (HIGH), and when Q_4 is ON (short) the output is approximately 0 V (LOW). The optimum size for a pull-up resistor depends on the size of the output's load and the leakage current through Q_4 (I_{OH}) when it is OFF. Usually, a good size for a pull-up resistor is 10 kΩ; 10 kΩ is not too small to allow excessive current flow when Q_4 is ON and it is not too large to cause an excessive voltage drop across itself when Q_4 is OFF.

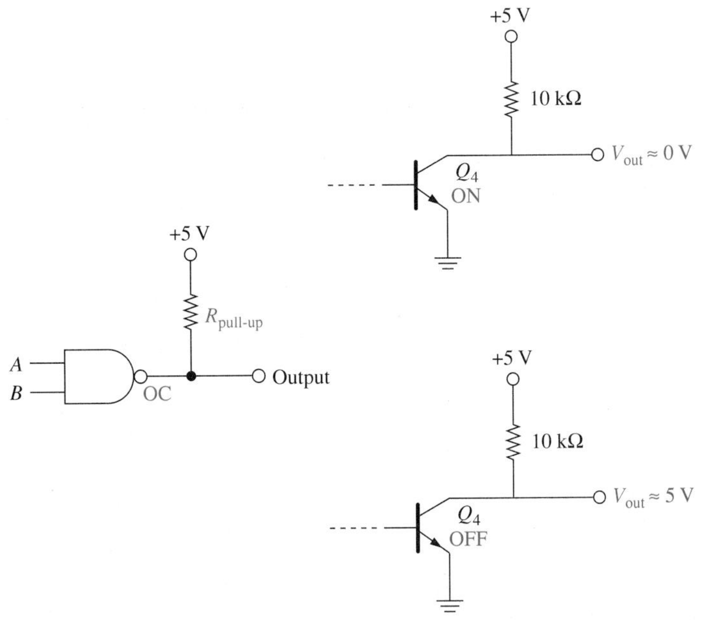

Figure 9–14   Using a pull-up resistor with an open-collector output.

Open-collector **buffer**/driver ICs are available for output loads requiring large sink currents, such as displays, relays, or motors. The term *buffer/driver* signifies the ability to provide high output currents to drive heavy loads. Typical ICs of this type are the 7406 OC inverter buffer/driver and the 7407 OC buffer/driver. They are each capable of sinking up to 40 mA, which is 10 times greater than the 4-mA capability of the standard 7404 inverter.

Wired-Output Operation

The main use of the open-collector gates is when the outputs from two or more gates or other devices have to be tied together. Using the regular totem-pole output gates, if a gate having a HIGH output (5 V) is connected to another gate having a LOW output (0 V), you would have a direct short circuit, causing either or both gates to burn out.

Using open-collector gates, outputs can be connected without worrying about the 5 V–0 V conflict. When connected, they form **wired-AND** logic, as shown in Figure 9–15. The 7405 IC has six OC inverters in a single package. By tying their outputs together, as shown in Figure 9–15(a), we have in effect *ANDed* all the inverters. The outputs of all six inverters must be floating (all inputs must be LOW) to get a HIGH output ($X = 1$ if $A = 0$ AND $B = 0$ AND $C = 0$, and so on). If any of the inverter output transistors (Q_4) turn on, the output will go LOW. The result of this wired-AND connection is the six-input NOR function shown in Figure 9–15(c).

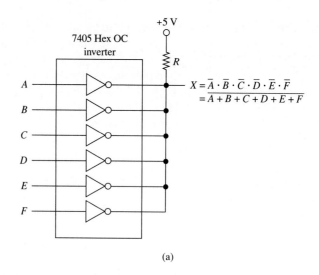

$$X = \overline{A} \cdot \overline{B} \cdot \overline{C} \cdot \overline{D} \cdot \overline{E} \cdot \overline{F}$$
$$= \overline{A + B + C + D + E + F}$$

(a)

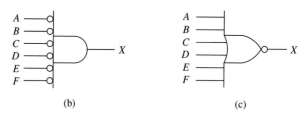

(b)   (c)

Figure 9–15 (a) Wired-AND connections to a hex OC inverter to form a six-input NOR gate; (b) AND gate representation; (c) alternative NOR gate representation.

EXAMPLE 9–4

Write the Boolean equation at the output of Figure 9–16(a).

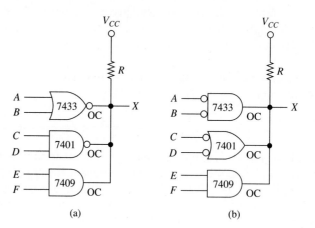

(a)   (b)

Figure 9–16 Wired-ANDing of open-collector gates for Example 9–4: (a) original circuit; (b) alternative gate representations used for clarity.

Solution: The output of all three gates in either circuit must be floating in order to get a HIGH output at X. Using Figure 9–16(b),

$$X = \overline{A}\,\overline{B} \cdot (\overline{C} + \overline{D}) \cdot EF$$

Disposition of Unused Inputs and Unused Gates

Electrically, *open inputs* degrade ac noise immunity as well as the switching speed of a circuit. For example, if two inputs to a three-input NAND gate are being used and the third is allowed to float, unpredictable results will occur if the third input picks up electrical noise from surrounding circuitry.

Unused inputs on AND and NAND gates should be tied HIGH, and on OR and NOR gates they should be tied to ground. An example of this is a three-input AND gate that is using only two of its inputs. Also, the outputs of *unused gates* on an IC should be forced HIGH to reduce the I_{CC} supply current and thus reduce power dissipation. To do this, tie AND and OR inputs HIGH, and tie NAND and NOR inputs LOW. An example of this is a quad NOR IC, where only three of the NOR gates are being used.

Power Supply Decoupling

In digital systems, there are heavy current demands on the main power supply. TTL logic tends to create spikes on the main V_{CC} line, especially at the logic-level transition point (LOW to HIGH or HIGH to LOW). At the logic-level transition there is a period of time that the conduction in the upper and lower **totem-pole output** transistors overlaps. This drastically changes the demand for I_{CC} current, which causes sharp high-frequency spikes to occur on the V_{CC} (power supply) line. These spikes cause false switching of other devices connected to the same power supply line and can also induce magnetic fields that radiate electromagnetic interference (EMI).

Decoupling of IC power supply spikes from the main V_{CC} line can be accomplished by placing a 0.01- to 0.1-μF capacitor directly across the V_{CC}-to-ground pins on each IC in the system. The capacitors tend to hold the V_{CC} level at each IC constant, thus reducing the amount of EMI radiation that is emitted from the system and the likelihood of false switching. Locating these small capacitors close to the IC ensures that the current spike will be kept local to the chip instead of radiating though the entire system back to the power supply.

Review Questions

9–9. The *rise time* is the length of time required for a digital signal to travel from 0 V to its HIGH level. True or false?

9–10. The letters L and H in the abbreviation t_{PLH} refer to the transition in the _____ (input, output) signal.

9–11. Describe the function of a pull-up resistor when it is used with an *open-collector* TTL output.

9–4 Improved TTL Series

Integrated-circuit design engineers have constantly worked to improve the standard TTL series. In fact, very few new designs will incorporate the use of the original 7400 series technology. A simple improvement that was made early on was simply reducing all the internal resistor values of the standard TTL series. This increased the power consumption (or dissipation), which was bad, but it reduced the internal $R \times C$ time constants that cause propagation delays. The result was the 74HXX series, which has almost half the propagation delay time but almost double the power consumption of the standard TTL series. The product of delay time × power (the speed–power product), which is a figure of merit for IC families, remained approximately the same, however.

Another series, the 74LXX, was developed using just the opposite approach. The internal resistors were increased, thus reducing power consumption, but the propagation delay increased, keeping the speed–power product about the same. The 74HXX and 74LXX series have, for the most part, been replaced now by the Schottky TTL and CMOS series of ICs.

Schottky TTL

The major speed limitation of the standard TTL series is due to the capacitive charge in the base region of the transistors. The transistors basically operate at either cutoff or saturation. When the transistor is saturated, charges build up at the base region, and when it is time to switch to cutoff, the stored charges must be dissipated, which takes time, thus causing propagation delay.

Schottky logic overcomes the saturation and stored charges problem by placing a Schottky diode across the base-to-collector junction, as shown in Figure 9–17. With the Schottky diode in place, any excess charge on the base is passed on to the collector, and the transistor is held just below saturation. The Schottky diode has a special metal junction that minimizes its own capacitive charge and increases its switching speed. Using Schottky-clamped transistors and decreased resistor values, the propagation delay is reduced by a factor of 4 and power consumption is only doubled. Therefore, the speed–power product of the 74SXX TTL series is improved to about half that of the 74XX TTL series (the lower, the better).

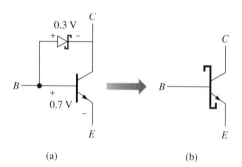

Figure 9–17   Schottky-clamped transistor: (a) Schottky diode reduces stored charges; (b) symbol.

Low-Power Schottky (LS): By using different integration techniques and increasing the values of the internal resistors, the power dissipation of the Schottky TTL is reduced significantly. The speed–power product of the 74LSXX TTL series is about one-third that of the 74SXX series and about one-fifth that of the 74XX series.

Advanced Low-Power Schottky (ALS): Further improvement of the 74LSXX series reduced the propagation delay time from 9 to 4 ns and the power dissipation from 2 to 1 mW per gate. The 74ALSXX and 74LS series are rapidly replacing the standard 74XX and 74SXX series because of the speed and power improvements. As with any new technology, they are slightly more expensive and do not yet provide all the functions available from the standard 74XX series.

Fast (F)

It was long clear to TTL IC design engineers that new processing technology was needed to improve the speed of the LS series. A new process of integration, called *oxide isolation* (also used by the ALS series), has reduced the propagation delay in the 74FXX series to below 3 ns. In this process, transistors are isolated from each other, not by a reverse-biased junction, but by an actual channel of oxide. This dramatically reduces the size of the devices, which in turn reduces their associated capacitances and thus reduces propagation delay.

9–5   The CMOS Family

The CMOS family of integrated circuits differs from TTL by using an entirely different type of transistor as its basic building block. The TTL family uses **bipolar transistors** *(NPN* and *PNP)*. CMOS (complementary metal oxide semiconductor) uses complementary pairs of transistors (*N* type and *P* type) called **MOSFETs** (metal oxide semiconductor field-effect transistors). MOSFETs are also used in other families of MOS ICs, including **PMOS** and **NMOS,** which are most commonly used for large-scale memories and microprocessors in the LSI and VLSI (large-scale and very large scale integration) category. One advantage that MOSFETs have over bipolar transistors is that the input to a MOSFET is electrically isolated from the rest of the MOSFET [see Figure 9–18(b)], giving it a high input impedance, which reduces the input current and power dissipation.

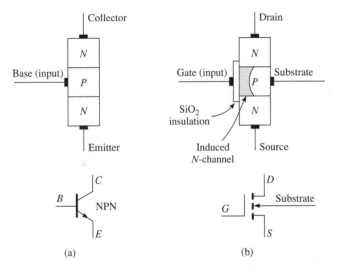

Figure 9–18   Simplified diagrams of bipolar and field-effect transistors: (a) *NPN* bipolar transistor used in TTL ICs; (b) *N*-channel MOSFET used in CMOS ICs.

The *N*-channel MOSFET is similar to the *NPN* bipolar transistor in that it is two back-to-back *N–P* junctions, and current will not flow down through it until a positive voltage is applied to the base (or gate in the case of the MOSFET). The silicon dioxide (SiO$_2$) layer between the gate material and the *P* **substrate** (structure) of the MOSFET prevents any gate current from flowing, which provides a high input impedance and low power consumption.

The MOSFET shown in Figure 9–18(b) is a normally OFF device, because there are no negative carriers in the *P* material for current flow to occur. However, conventional current will flow down from drain to source if a positive voltage is applied to the gate with respect to the substrate. This voltage induces an electric field across the SiO$_2$ layer, which repels enough of the positive charges in the *P* material to form a channel of negative charges on the left side of the *P* material. This allows electrons to flow from source to drain (conventional current flows from drain to source). The channel that is formed is called an *N* channel because it contains negative carriers.

P-channel MOSFETs are just the opposite, constructed from *P–N–P* materials. The channel is formed by placing a *negative* voltage at the gate with respect to the substrate.

There are three major MOS technology families: PMOS (made up of *P*-channel MOSFETs), NMOS (made up of *N*-channel MOSFETs), and CMOS (made up of complementary *P*-channel and *N*-channel MOSFETs). MOS technology provides a higher packing density than bipolar TTL and therefore allows IC manufacturers to provide thousands of logic functions on a single IC chip (VLSI circuitry). For example, it is common to find computer systems with MOS memory ICs containing over 1 million memory cells per chip and MOS microcontroller ICs containing the combined logic of more than 10 LSI ICs. Between NMOS and PMOS, NMOS was historically more widely used because of its higher speed and packing density.

Fabricating both *P*- and *N*-channel transistors in the same package previously made CMOS slightly more expensive and limited its use in VLSI circuits. However, recent advances in fabrication techniques and mass production have reduced its price. Today CMOS competes very favorably with the best bipolar TTL circuitry in the SSI and MSI arena and is by far the most popular technology in LSI and VLSI memory and microprocessor ICs.

Using an *N*-channel MOSFET with its complement, the *P*-channel MOSFET, a simple complementary MOS (CMOS) inverter can be formed as shown in Figure 9–19.

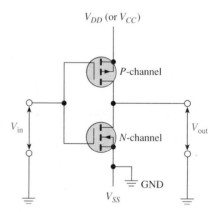

Figure 9–19   CMOS inverter formed from complementary *N*-channel/*P*-channel transistors.

Table 9–2 Basic MOSFET Switching Characteristics		
Gate Level[a]	*N*-Channel	*P*-Channel
1	ON	OFF
0	OFF	ON

[a] $1 \equiv V_{DD}$ (or V_{CC}); $0 \equiv V_{SS}$ (Gnd).

We can think of MOSFETs as ON/OFF switches, just as we did for bipolar transistors. Table 9–2 summarizes the ON/OFF operation of *N*- and *P*-channel MOSFETs. We can use Table 9–2 to prove that the circuit of Figure 9–19 operates as an inverter. With $V_{in} = 1$, the *N*-channel transistor is ON and the *P*-channel transistor is OFF, so $V_{out} = 0$. With $V_{in} = 0$, the *N*-channel transistor is OFF and the *P*-channel is ON, so

$V_{out} = 1$. Therefore, $V_{out} = \overline{V_{in}}$. Notice that this complementary action is very similar to the TTL totem-pole output stage, but much simpler to understand.

The other basic logic gates can also be formed using complementary MOSFET transistors. The operation of CMOS NAND and NOR gates can easily be understood by studying the schematics and data tables presented in Figure 9–20(a) and (b).

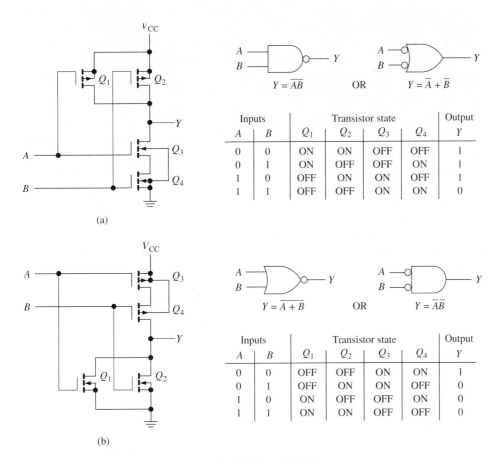

Figure 9–20 CMOS gate schematics: (a) NAND; (b) NOR.

Handling MOS Devices

The silicon dioxide layer that isolates the gate from the substrate is so thin that it is very susceptible to burn-through from electrostatic charges. You must be very careful and use the following guidelines when handling MOS devices:

1. Store the integrated circuits in a conductive foam or leave them in their original container.

2. Work on a conductive surface (for example, a metal tabletop) that is properly grounded.

3. Ground all test equipment and soldering irons.

4. Wear a wrist strap to connect your wrist to ground with a length of wire and a 1-MΩ series resistor (see Figure 9–21).

5. Do not connect signals to the inputs while the device power supply is off.

6. Connect all unused inputs to V_{DD} or ground.

Figure 9–21   Wearing a commercially available wrist strap dissipates static charges from the technician's body to a ground connection while he or she is handling CMOS ICs.

7. Don't wear electrostatic-prone clothing such as wool, silk, or synthetic fibers.

8. Don't remove or insert an IC with the power on.

CMOS Availability

The CMOS family of integrated circuits provides almost all the same functions that are available in the TTL family, plus CMOS has available several special-purpose functions not provided by TTL. Like TTL, the CMOS family has evolved into several different subfamilies, or series, each having better performance specifications than the previous one.

4000 Series: The 4000 series (or the improved 4000B) is the original CMOS line. It became popular because it offered very low power consumption and could be used in battery-powered devices. It is much slower than any of the TTL series and has a low level of electrostatic discharge protection. The power supply voltage to the IC can range anywhere from +3 to +15 V, with the minimum 1-level input equal to $\frac{2}{3}V_{CC}$ and the maximum 0-level input equal to $\frac{1}{3}V_{CC}$.

40H00 Series: This series was designed to be faster than the 4000 series. It did overcome some of the speed limitations, but it is still much slower than LSTTL.

74C00 Series: This series was developed to be pin compatible with the TTL family, making interchangeability easier. It uses the same numbering scheme as TTL except that it begins with 74C. It has a low-power advantage over the TTL family, but it is still much slower.

74HC00 and 74HCT00 Series: The 74HC00 (high-speed CMOS) and 74HCT00 (high-speed CMOS, TTL compatible) offer a vast improvement over the original 74C00 series. The HC/HCT series are as speedy as the LSTTL series and still consume less power, depending on the operating frequency. They are pin compatible (the HCT is also input/output voltage-level compatible) with the TTL family, yet they offer greater noise immunity and greater voltage and temperature operating ranges. Further improvements to the HC/HCT series have led to the Advanced CMOS Logic (ACL) and Advanced CMOS Technology (ACT) series, which have even better operating characteristics.

74-BiCMOS Series: Several IC manufacturers have developed technology that combines the best features of bipolar transistors and CMOS transistors, forming **BiCMOS** logic. The high-speed characteristics of bipolar $P–N$ junctions are integrated with the low-power characteristics of CMOS to form an extremely low power, high speed family of digital logic. Each manufacturer uses different suffixes to identify their BiC-MOS line. For example, Texas Instruments uses 74BCTXXX, Harris uses 74FCTXXX, and Signetics (Phillips) uses 74ABTXXX.

The product line is especially well suited for and is mostly limited to microprocessor bus interface logic. This logic is mainly available in octal (8-bit) configurations used to interface 8-, 16-, and 32-bit microprocessors with high-speed peripheral devices such as memories and displays. An example is the 74ABT244 octal buffer from Signetics. Its logic is equivalent to the 74244 of other families, but it has several advanced characteristics. It has TTL-compatible input and output voltages, gate input currents less than 0.01 μA, and output sink and source current capability of 64 and −32 mA, respectively. It is extremely fast, having a typical propagation delay of 2.9 ns.

One of the most desirable features of these bus-interface ICs is the fact that, when their outputs are inactive ($\overline{OE} = 1$) or HIGH, the current draw from the power supply (I_{CCZ} or I_{CCH}) is only 0.5 μA. Since interface logic spends a great deal of its time in an inactive (idle) state, this can translate into a power dissipation as low as 2.5 μW! The actual power dissipation depends on how often the IC is inactive and on the HIGH/LOW duty cycle of its outputs when it is active.

74-Low Voltage Series: A new series of logic using a nominal supply voltage of 3.3 V has been developed to meet the extremely low power design requirements of battery-powered and hand-held devices. These ICs are being designed into the circuits of notebook computers, mobile radios, hand-held video games, telecom equipment, and high-performance workstation computers. Some of the more common Low-Voltage families are identified by the following suffixes:

LV—Low-Voltage HCMOS

LVC—Low-Voltage CMOS

LVT—Low-Voltage Technology

ALVC—Advanced Low-Voltage CMOS

HLL—High-speed Low-power Low-voltage

The power consumption of CMOS logic ICs decreases approximately with the square of power supply voltage. The propagation delay increases slightly at this reduced voltage but the speed is restored, and even increased, by using finer geometry and sub-micron CMOS technology that is tailored for low-power and low-voltage applications.

The supply voltage of LV logic can range from 1.2 to 3.6 V, which makes it well suited for battery-powered applications. When operated between 3.0 and 3.6 V, it can be interfaced directly with TTL levels. The switching speed of LV logic is extremely fast, ranging from about 9 ns for the LV series, down to 2.1 ns for the ALVC. Like

BiCMOS logic, the power dissipation of LV logic is negligible in the idle state or at low frequencies. At higher frequencies the power dissipation is down to half as much as BiCMOS, depending on the power supply voltage used on the LV logic. Another key benefit of LV logic is its high output drive capability. The highest capability is provided by the LVT series, which can sink up to 64 mA and source up to 32 mA.

9–6   Emitter-Coupled Logic

Another family designed for extremely high speed applications is emitter-coupled logic (ECL). ECL comes in two series, ECL 10K and ECL 100K. ECL is extremely fast, with propagation delay times as low as 0.8 ns. This speed makes it well suited for large mainframe computer systems that require a high number of operations per second, but that are not as concerned about an increase in power dissipation.

The high speed of ECL is achieved by never letting the transistors saturate; in fact, the whole basis for HIGH and LOW levels is determined by which transistor in a **differential amplifier** is conducting more.

Figure 9–22 shows a simplified diagram of the differential amplifier used in ECL circuits. The HIGH and LOW logic-level voltages (−0.8 and −1.7 V, respectively) are somewhat unusual and cause problems when interfacing to TTL and CMOS logic.

An ECL IC uses a supply voltage of −5.2 V at V_{EE} and 0 V at V_{CC}. The reference voltage on the base of Q_3 is set up by internal circuitry and determines the threshold between HIGH and LOW logic levels. In Figure 9–22(a), the base of Q_3 is at a more positive potential with respect to the emitter than Q_1 and Q_2 are. This causes Q_3 to conduct, placing a LOW at V_{out}.

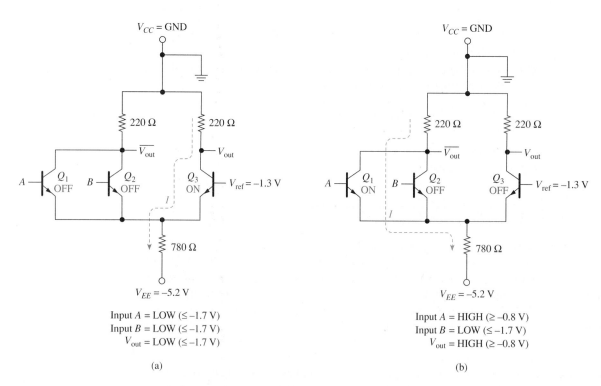

Figure 9–22   Differential amplifier input stage to an ECL OR/NOR gate: (a) LOW output; (b) HIGH output.

If *either* input A or B is raised to -0.8 V (*HIGH*), the base of Q_1 or Q_2 will be at a higher potential than the base of Q_3, and Q_3 will stop conducting, making V_{out} HIGH. Figure 9–22(b) shows what happens when -0.8 V is placed on the A input.

In any case, the transistors never become saturated, so capacitive charges are not built up on the base of the transistors to limit their switching speed. Figure 9–23 shows the logic symbol and truth table for the OR/NOR ECL gate.

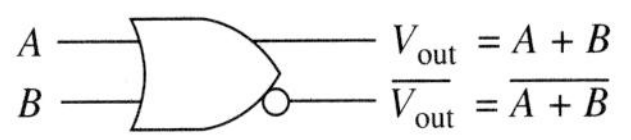

Inputs		Outputs	
A	B	V_{out}	$\overline{V_{out}}$
0	0	0	1
0	1	1	0
1	0	1	0
1	1	1	0

$V_{out} = A + B$
$\overline{V_{out}} = \overline{A + B}$

Figure 9–23   ECL OR/NOR symbol and truth table.

Developing New Digital Logic Technologies

The quest for logic devices that can operate at even higher frequencies and can be packed more densely in an IC package is a continuing process. Designers have high hopes for other new technologies, such as integrated injection logic (I^2L), silicon-on-sapphire (SOS), gallium arsenide (GaAs), and Josephen junction circuits. Eventually, propagation delays will be measured in picoseconds and circuit densities will enable the supercomputer of today to become the desktop computer of tomorrow.

Review Questions

9–12. What effect did the Schottky-clamped transistor have on the operation of the standard TTL IC?

9–13. The earlier 4000 series of CMOS ICs provided what advantage over earlier TTL ICs? What was their disadvantage?

9–14. The BiCMOS family of ICs is fabricated using both CMOS transistors and _____ transistors.

9–15. The high speed of ECL ICs is achieved by fully saturating the ON transistor. True or false?

9–7   Comparing Logic Families

Throughout the years, system designers have been given a wide variety of digital logic to choose from. The main parameters to consider include speed, power dissipation, availability, types of functions, noise immunity, operating frequency, output-drive capability, and interfacing. First and foremost, however, are the basic speed and power concerns. Table 9–3 on page 292 shows the propagation delay, power dissipation, and speed–power product for the most popular families.

The speed–power product is a type of figure of merit but does not necessarily tell the ranking within a specific application. For example, to say that the speed–power product of 15 pW-s for the 74HC family is better than 32 pW-s for the 100K ECL family totally ignores the fact that ECL is a better choice for ultrahigh-speed applications.

Table 9–3 Typical Single-Gate Performance Specifications

Family	Propagation Delay (ns)	Power Dissipation (mW)	Speed–Power Product pW-s (picowatt-seconds)
74	10	10	100
74S	3	20	60
74LS	9	2	18
74ALS	4	1	4
74F	2.7	4	11
4000B (CMOS)	105	1 at 1 MHz	105
74HC (CMOS)	10	1.5 at 1 MHz	15
74BCT (BiCMOS)	2.9	0.0003 to 7.5	0.00087 to 22
100K (ECL)	0.8	40	32

Courtesy of Signetics Corporation

Another way to view the speed–power relationships is with the graph shown in Figure 9–24. From the graph you can see the wide spectrum of choices available. 4000B CMOS and 100K ECL are at opposite ends of the spectrum of speed versus power, while 74ALS and 74F seem to offer the best of both worlds.

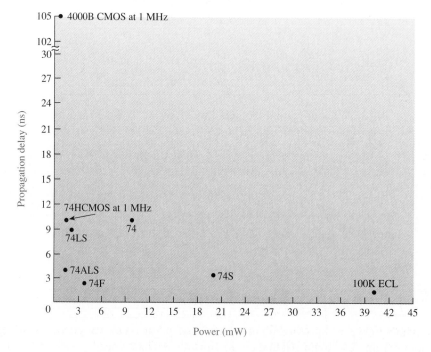

Figure 9–24 Graph of propagation delay versus power. (Courtesy of Signetics Corporation.)

The operating frequency for CMOS devices is critical for determining power dissipation. At very low frequencies, CMOS devices dissipate almost no power at all, but at higher switching frequencies, charging and discharging the gate capacitances draws a heavy current from the power supply (I_{CC}) and thus increases the power dissipation ($P_D = V_{CC} \times I_{CC}$), as shown in Figure 9–25. We see that at high frequencies the power dissipations of 74HC CMOS and 74LS TTL are comparable. At today's microprocessor clock rates, 74HC CMOS ICs actually dissipate more power

CHAPTER 9 I LOGIC FAMILIES AND THEIR CHARACTERISTICS

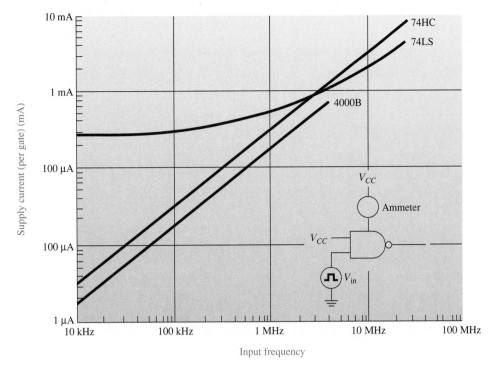

Figure 9–25 Power supply current versus frequency. (Courtesy of Signetics Corporation.)

than 74LS or 74ALS. However, in typical systems, only a fraction of the gates are connected to switch as fast as the clock rate, so significant power savings can be realized by using the 74HC CMOS series.

9–8 Interfacing Logic Families

Often, the need arises to interface (connect) between the various TTL and CMOS families. You have to make sure that a HIGH out of a TTL gate looks like a HIGH to the input of a CMOS gate, and vice versa. The same holds true for the LOW logic levels. You also have to make sure that the driving gate can sink or source enough current to meet the input current requirements of the gate being driven.

TTL to CMOS

Let's start by looking at the problems that might arise when interfacing a standard 7400 series TTL to a 4000B series CMOS. Figure 9–26 shows the input and output voltage specifications for both, assuming that the 4000B is powered by a 5-V supply.

When the TTL gate is used to drive the CMOS gate, there is no problem for the LOW-level output, because the TTL guarantees a maximum LOW-level output of 0.4 V and the CMOS will accept any voltage up to 1.67 V ($\frac{1}{3}V_{CC}$) as a LOW-level input.

But for the HIGH level, the TTL may output as little as 2.4 V as a HIGH. The CMOS expects at least 3.33 V as a HIGH-level input. Therefore, 2.4 V is unacceptable, because it falls within the uncertain region. However, resistor can be connected between the CMOS input to V_{CC}, as shown in Figure 9–27, to solve the HIGH-level input problem.

In Figure 9–27, with V_{out1} *LOW*, the 7404 will sink current from the 10-kΩ resistor and the I_{IL} from the 4069B, making V_{out2} HIGH. With V_{out1} *HIGH*, the 10-kΩ resistor will pull the voltage at V_{in2} up to 5.0 V, causing V_{out2} to go LOW.

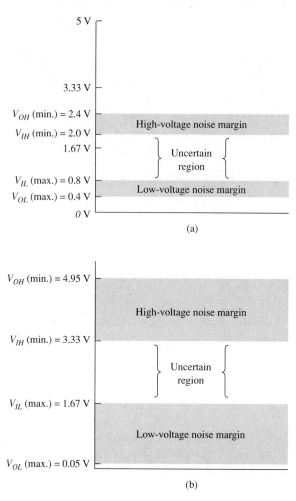

(a)

(b)

Figure 9–26 Input and output voltage specifications: (a) 7400 series TTL; (b) 4000B series CMOS (5-V supply).

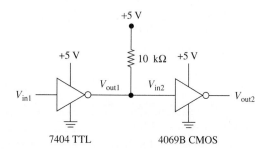

Figure 9–27 Using a pull-up resistor to interface TTL to CMOS.

The 10-kΩ resistor is called a *pull-up resistor* and is used to raise the output of the TTL gate closer to 5 V when it is in a HIGH output state. With V_{out1} HIGH, the voltage at V_{in2} will be almost 5 V because current into the 4069B is so LOW ($\approx 1 \ \mu A$) that the voltage drop across the 10 kΩ is insignificant, leaving almost 5.0 V at V_{in2} ($V_{in2} = 5 \ V - 1\mu A \times 10 \ k\Omega = 4.99 \ V$).

The other thing to look at when interfacing is the current levels of all gates that are involved. In this case, the 7404 can sink (I_{OL}) 16 mA, which is easy enough for the

I_{IL} of the 4069B (1 μA) plus the current from the 10-kΩ resistor (5 V/10 kΩ = 0.5 mA). I_{OH} of the 7404 (−400 μA) is no problem either, because with the pull-up resistor the 7404 will not have to source current.

CMOS to TTL

When driving TTL from CMOS, the voltage levels are no problem because the CMOS will output about 4.95 V for a HIGH and 0.05 V for a LOW, which is easily interpreted by the TTL gate.

But the current levels can be a real concern because 4000B CMOS has severe output-current limitations. (The 74C and 74HC series have much better output-current capabilities, however.) Figure 9–28 shows the input/output currents that flow when interfacing CMOS to TTL.

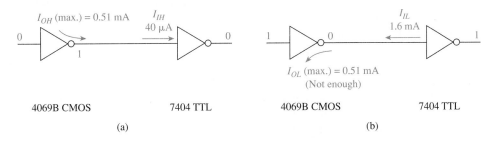

Figure 9–28 Current levels when interfacing CMOS to TTL: (a) CMOS I_{OH}; (b) CMOS I_{OL}.

For the HIGH output condition [Figure 9–28(a)], the 4069B CMOS can source a maximum current of 0.51 mA, which is enough to supply the HIGH-level input current (I_{IH}) to one 7404 inverter. But for the LOW output condition, the 4069B can also sink only 0.51 mA, which is not enough for the 7404 LOW-level input current (I_{IL}).

Most of the 4000B series has the same problem of low-output drive current capability. To alleviate the problem, two special gates, the 4050 buffer and the 4049 inverting buffer, are specifically designed to provide high output current to solve many interfacing problems. They have drive capabilities of I_{OL} = 4.0 mA and I_{OH} = −0.9 mA, which is enough to drive two 74XXTTL loads, as shown in Figure 9–29.

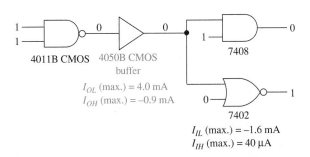

Figure 9–29 Using the 4050B CMOS buffer to supply sink and source current to two standard TTL loads.

If the CMOS buffer were used to drive another TTL series, let's say, the 74LS series, we would have to refer to a TTL data book to determine how many loads could be connected without exceeding the output current limits. (The 4050B can actually drive

Table 9–4 Worst-Case Values for Interfacing Considerations[a]

Parameter	4000B CMOS	74HCMOS	74HCTMOS	74TTL	74LSTTL	74ALSTTL
V_{IH} (min.) (V)	3.33	3.5	2.0	2.0	2.0	2.0
V_{IL} (max.) (V)	1.67	1.0	0.8	0.8	0.8	0.8
V_{OH} (min.) (V)	4.95	4.9	4.9	2.4	2.7	2.7
V_{OL} (max.) (V)	0.05	0.1	0.1	0.4	0.4	0.4
I_{IH} (max.) (μA)	1	1	1	40	20	20
I_{IL} (max.) (μA)	−1	−1	−1	−1600	−400	−100
I_{OH} (max.) (mA)	−0.51	−4	−4	−0.4	−0.4	−0.4
I_{OL} (max.) (mA)	0.51	4	4	16	8	4

[a]All values are for $V_{supply} = 5.0$ V.

Team Discussion

Discuss several attributes that make the HCT family an excellent choice for interfacing situations.

10 74LS loads.) Table 9–4 summarizes the input/output voltage and current specifications of some popular TTL and CMOS series, which easily enables us to determine interface parameters and family characteristics.

By reviewing Table 9–4, we can see that the 74HCMOS has relatively low input-current requirements compared to the bipolar TTL series. Its HIGH output can source 4 mA, which is 10 times the capability of the TTL series. Also, the noise margin for the 74HCMOS is much wider than any of the TTL series (1.4 V HIGH, 0.9 V LOW).

Because of the low input-current requirements, any of the TTL series can drive several of the 74HCMOS loads. An interfacing problem occurs in the voltage level, however. The 74HCMOS logic expects 3.5 V at a minimum for a HIGH-level input. The worst case (which we must always assume could happen) for the HIGH output level of a 74LSTTL is 2.7 V, so we will need to use a pull-up resistor at the 74LSTTL output to ensure an adequate HIGH level for the 74HCMOS input as shown in Figure 9–30.

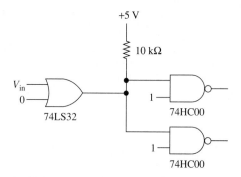

Figure 9–30 Interfacing 74LSTTL to 74HCMOS.

The combinations of interfacing situations are extensive (74HCMOS to 74ALSTTL, 74TTL to 74LSTTL, and so on). In each case, reference to a data book must be made to check the worst-case voltage and current parameters, as we will see in upcoming examples. In general, a pull-up resistor is required when interfacing TTL to CMOS to bring the HIGH-level TTL output up to a suitable level for the CMOS input. (The exception to the rule is when using *74HCTMOS,* which is designed for TTL voltage levels.) The disadvantages of using a pull-up resistor are that it takes up valuable room on a printed-circuit board and it dissipates power in the form of heat.

Different series within the TTL family and the TTL-compatible 74HCTMOS series can be interfaced directly. The main concern then is determining how many gate loads can be connected to a single output.

CHAPTER 9 | LOGIC FAMILIES AND THEIR CHARACTERISTICS

Level Shifting

Another problem arises when you interface families that have different supply voltages. For example, the 4000B series can use anywhere from +3 to +15 V for a supply, and the ECL series uses −5.2 V for a supply.

This problem is solved by using **level-shifter** ICs. The 4049B and 4050B buffer ICs that were introduced earlier are also used for voltage-level shifting. Figure 9–31 shows the connections for interfacing 15-V CMOS to 5-V TTL.

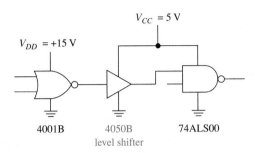

Figure 9–31   Using a level shifter to convert 0-V/15-V logic to 0-V/5-V logic.

The 4050B level-shifting buffer is powered from a 5-V supply and can actually accept 0-V/15-V logic levels at the input, and output the corresponding 0-V/5-V logic levels at the output. For an inverter function, use the 4049B instead of the 4050B.

The reverse conversion, 5-V TTL to 15-V CMOS, is accomplished with the 4504B CMOS level shifter, as shown in Figure 9–32. The 4504B level-shifting buffer requires two power supply inputs: the 5-V V_{CC} supply to enable it to recognize the 0-V/5-V input levels and the 15-V supply to enable it to provide 0-V/15-V output levels.

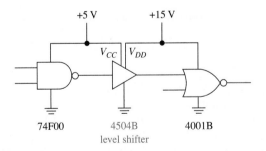

Figure 9–32   Level shifting 0-V/5-V TTL logic to 0-V/15-V CMOS logic.

ECL Interfacing

Interfacing 0-V/5-V logic levels to −5.2 V/0-V ECL circuitry requires another set of level shifters (or translators): the ECL 10125 and the ECL 10124, whose connections are shown in Figure 9–33.

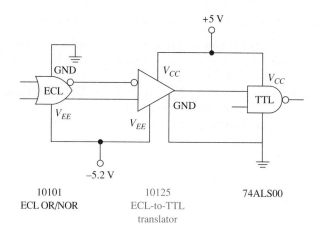

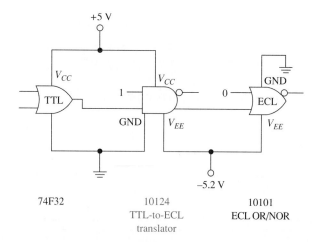

Figure 9–33 Circuit connections for translating between TTL and ECL levels.

EXAMPLE 9–5

Determine from Table 9–4 how many 74LSTTL logic gates can be driven by a single 74TTL logic gate.

Solution: The output voltage levels (V_{OL}, V_{OH}) of the 74TTL series are compatible with the input voltage levels (V_{IL}, V_{IH}) of the 74LSTTL series. The voltage noise margin for the HIGH level is 0.4 V (2.4 − 2.0), and for the LOW level it is 0.4 V (0.8 − 0.4).

The HIGH-level output current (I_{OH}) for the 74TTL series is −400 μA. Each 74LSTTL gate draws 20 μA of input current for the HIGH level (I_{IH}), so one 74TTL gate can drive 20 74LSTTL loads in the HIGH state (400 μA/20 μA = 20).

For the LOW state, the 74TTL I_{OL} is 16 mA and the 74LSTTL I_{IL} is −400 μA, meaning that, for the LOW condition, one 74TTL can drive 40 74LSTTL loads (16 mA/400 μA = 40). Therefore, considering both the LOW and HIGH conditions, a single 74TTL can drive 20 74LSTTL gates.

CHAPTER 9 I LOGIC FAMILIES AND THEIR CHARACTERISTICS

EXAMPLE 9-6

One 74HCT04 inverter is to be used to drive one input to each of the following gates: 7400 (NAND), 7402 (NOR), 74LS08 (AND), and 74ALS32 (OR). Draw the circuit and label input and output worst-case voltages and currents. Will there be total voltage and current compatibility?

Solution: The circuit is shown in Figure 9-34. Figure 9-34(a) shows the worst-case HIGH-level values. If you sum all the input currents, the total that the 74HCT04 must supply is 120 μA (40 μA + 40 μA + 20 μA + 20 μA), which is well below the -4-mA maximum source capability of the 74HCT04. Also, the 4.9-V output voltage of the 74HCT04 *is* compatible with the 2.0-V minimum requirement of the TTL inputs, leaving a noise margin of 2.9 V (4.9 V - 2.0 V).

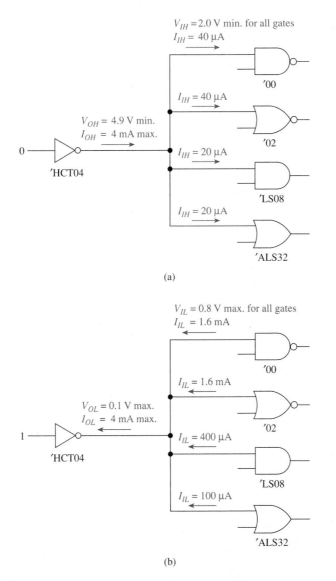

Figure 9-34   Interfacing a 74HCTMOS to several different TTL series: (a) HIGH-level values; (b) LOW-level values.

Figure 9–34(b) shows the worst-case LOW-level values. The sum of all the TTL input currents is 3.7 mA (1.6 mA + 1.6 mA + 400 μA + 100 μA), which is less than the 4-mA maximum sink capability of the 74HCT04. Also, the 0.1-V output of the 74HCT04 *is* compatible with the 0.8-V maximum requirement of the TTL inputs, leaving a noise margin of 0.7 V (0.8 V – 0.1 V).

Review Questions

9–16. Which logic family is faster, the 74ALS or 74HC?

9–17. Which logic family has a lower power dissipation, the 100K ECL or 74LS?

9–18. In the graph of Figure 9–24, the IC families plotted closest to the origin have the best speed–power products. True or false?

9–19. Use the graph in Figure 9–25 to determine which logic family has a lower power dissipation at 100 kHz, the 74HC CMOS or 74LS TTL?

9–20. What is the function of a pull-up resistor when interfacing a TTL IC to a CMOS IC?

9–21. What problem arises when interfacing 4000 series CMOS ICs to standard TTL ICs?

9–22. Which family has more desirable output voltage specifications, the 74HCTMOS or 74ALSTTL? Why?

9–23. What IC specifications are used to determine how many gates of one family can be driven from the output of another family?

Summary

In this chapter we have learned that

1. There are basically three stages of internal circuitry in a TTL (transistor–transistor logic) IC: input, control, and output.

2. The input current (I_{IL} or I_{IH}) to an IC gate is a constant value specified by the IC manufacturer.

3. The output current of an IC gate depends on the size of the load connected to it. Its value cannot exceed the maximum rating of the chip, I_{OL} or I_{OH}.

4. The HIGH- and LOW-level output voltages of the standard TTL family are *not* 5 V and 0 V but typically are 3.4 V and 0.2 V.

5. The propagation delay is the length of time that it takes for the output of a gate to respond to a stimulus at its input.

6. The rise and fall times of a pulse describe how long it takes for the voltage to travel between its 10% and 90% levels.

7. Open-collector outputs are required whenever logic outputs are connected to a common point.

8. Several improved TTL families are available and continue to be introduced each year providing decreased power consumption and decreased propagation delay.

9. The CMOS family uses complementary metal oxide semiconductor transistors instead of the bipolar transistors used in TTL ICs. Traditionally, the CMOS family consumed less power but was slower than TTL. However, recent advances in both technologies have narrowed the differences.

10. The BiCMOS family combines the best characteristics of bipolar technology and CMOS technology to provide logic functions that are optimized for the high-speed, low-power characteristics required in microprocessor systems.

11. Emitter-coupled logic (ECL) provides the highest-speed ICs. Its drawback is its very high power consumption.

12. A figure of merit of IC families is the product of their propagation delay and power consumption, called the speed–power product (the lower, the better).

13. When interfacing logic families, several considerations must be made. The output voltage level of one family must be high and low enough to meet the input requirements of the receiving family. Also, the output current capability of the driving gate must be high enough for the input draw of the receiving gate or gates.

Glossary

BiCMOS: A logic family that is fabricated from a combination of bipolar transistors and complementary MOSFETs. It is an extremely fast low-power family.

Bipolar Transistor: Three-layer *N–P–N* or *P–N–P* junction transistor.

Buffer: A device placed between two other devices that provides isolation and current amplification. The input logic level is equal to the output logic level.

CMOS: Complementary metal oxide semiconductor.

Decoupling: A method of isolating voltage irregularities on the V_{CC} power supply line from an IC V_{CC} input.

Differential Amplifier: An amplifier that basically compares two inputs and provides an output signal based on the *difference* between the two input signals.

ECL: Emitter-coupled logic.

EMI: Electromagnetic interference. Undesirable radiated energy from a digital system caused by magnetic fields induced by high-speed switching.

Fall Time: The time required for a digital pulse to fall from 90% down to 10% of its maximum voltage level.

Fan-Out: The number of logic gate inputs that can be driven from a single gate output of the same subfamily.

Level Shifter: A device that provides an interface between two logic families having different power supply voltages.

MOSFET: Metal oxide semiconductor field-effect transistor.

NMOS: A family of ICs fabricated with N-channel MOSFETs.

Noise Margin: The voltage difference between the guaranteed output voltage level and the required input voltage level of a logic gate.

Open-Collector Output: A special output stage of the TTL family that has the upper transistor of a totem-pole configuration removed.

PMOS: A family of ICs fabricated with P-channel MOSFETs.

Power Dissipation: The electrical power (watts) that is consumed by a device and given off (dissipated) in the form of heat.

Propagation Delay: The time required for a change in logic level to travel from the input to the output of a logic gate.

Pull-Up Resistor: A resistor with one end connected to V_{CC} and the other end connected to a point in a logic circuit that needs to be raised to a voltage level closer to V_{CC}.

Rise Time: The time required for a digital pulse to rise from 10% up to 90% of its maximum voltage level.

Sink Current: Current entering the output or input of a logic gate.

Source Current: Current leaving the output or input of a logic gate.

Substrate: The silicon supporting structure or framework of an integrated circuit.

Totem-Pole Output: The output stage of the TTL family having two opposite-acting transistors, one above the other.

TTL: Transistor–transistor logic.

Wired-AND: The AND function that results from connecting several open-collector outputs together.

Problems

Section 9–1

9–1. What is the purpose of diodes D_1 and D_2 in Figure 9–1?

9–2. In Figure 9–1, when input A is connected to ground (0 V), calculate the approximate value of emitter current in Q_1.

9–3. In Figure 9–1, when the output is HIGH, how do you account for the output voltage being only about 3.4 V instead of 5.0 V?

9–4. In Figure 9–1, describe the state (ON or OFF) of Q_3 and Q_4 for

(a) Both inputs A and B LOW

(b) Both inputs A and B HIGH

Section 9–2

9–5. What does the negative sign in the rating of source current (for example, $I_{OH} = -400\ \mu A$) signify?

9–6. For TTL outputs, which is higher, the source current or the sink current?

C   **9–7.** **(a)** Find V_a and I_a in the circuits of Figure P9–7 using the following specifications:

$$I_{IL} = -1.6 \text{ mA} \qquad I_{IH} = 40 \text{ } \mu A$$
$$V_{IL} = 0.8 \text{ V max} \qquad V_{IH} = 2.0 \text{ V min}$$
$$I_{OL} = 16 \text{ mA} \qquad I_{OH} = -400 \text{ } \mu A$$
$$V_{OL} = 0.2 \text{ V typ.} \qquad V_{OH} = 3.4 \text{ V typ.}$$

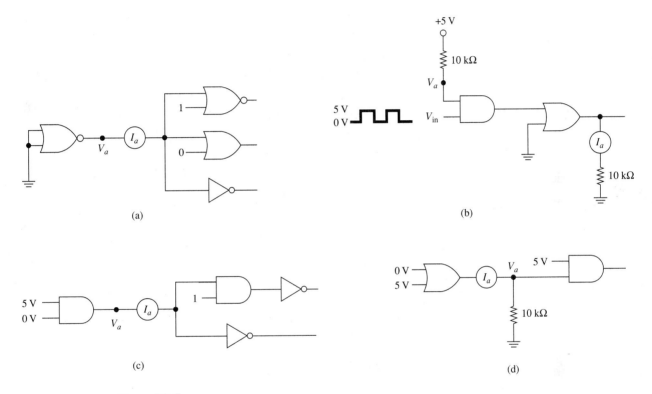

Figure P9–7

(b) Repeat part (a) using input/output specifications that you gather from a TTL data book, assuming that all gates are 74LSXX series.

Section 9–3

9–8. The input and output waveforms to an OR gate are given in Figure P9–8. Determine

(a) The period and frequency of V_{in}

(b) The rise and fall times (t_r, t_f) of V_{in}

(c) The propagation delay times of (t_{PLH}, t_{PHL}) of the OR gate

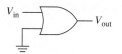

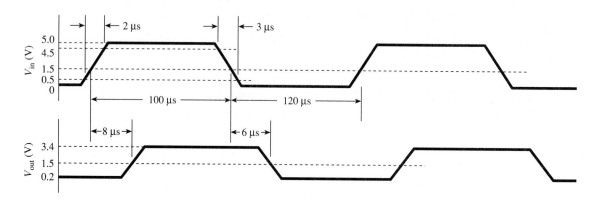

Figure P9–8

9–9. The propagation delay times for a 74LS08 AND gate (Figure P9–9) are $t_{PLH} = 15$ ns, $t_{PHL} = 20$ ns and for a 7402 NOR gate they are $t_{PLH} = 22$ ns, $t_{PHL} = 15$ ns. Sketch V_{out1} and V_{out2} showing the effects of propagation delay. (Assume 0 ns for the rise and fall times.)

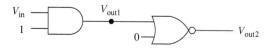

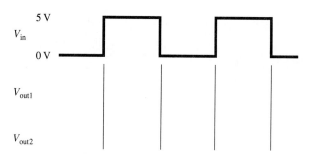

Figure P9–9

9–10. **(a)** Repeat Problem 9–9 for the circuit of Figure P9–10(a).

C

(b) Repeat Problem 9–9 for V_a, V_b, V_c, and V_d in Figure P9–10(b). (Use a TTL data book to determine the propagation delay times.)

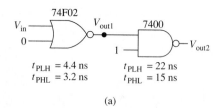

(a)

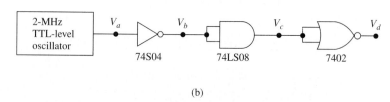

(b)

Figure P9–10

CHAPTER 9 | LOGIC FAMILIES AND THEIR CHARACTERISTICS

9–11. Refer to a TTL data book or the data sheets in Appendix B. Use the total supply current (I_{CCL}, I_{CCH}) to compare the power dissipation of a 7400 versus a 74LS00.

9–12. Refer to a TTL data sheet to compare the typical LOW-level output voltage (V_{OL}) at maximum output current for a 7400 versus a 74LS00.

9–13. **(a)** Refer to a TTL data sheet to determine the noise margins for the HIGH and LOW states of both the 7400 and 74LS00.

   (b) Which has better noise margins, the 7400 or 74LS00?

9–14. **(a)** Refer to a TTL data sheet (or Appendix B) to determine which can sink more current at its output, the commercial 74LS00 or the military 54LS00.

   (b) Which has a wider range of recommended V_{CC} supply voltage, the 7400 or the 5400?

9–15. Why is a pull-up resistor required at the output of an open-collector gate to achieve a HIGH-level output?

9–16. The wired-AND circuits in Figure P9–16 use all open-collector gates. Write the simplified Boolean equations at X and Y.

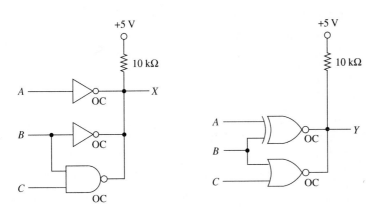

Figure P9–16

Sections 9–4 and 9–5

9–17. Make a general comparison of both the switching speed and power dissipation of the 7400 TTL series versus the 4000B CMOS series.

9–18. Which type of transistor, bipolar or field effect, is used in TTL ICs? CMOS ICs?

9–19. Why is it important to store MOS ICs in antistatic conductive foam?

Sections 9–6 and 9–7

9–20. What is the principal reason that ECL ICs reach such high switching speeds?

9–21. Table 9–3 shows that the speed–power product of the 74ALS family is much better than the 100K ECL family. Why, then, are some large mainframe computers based on ECL technology?

9–22. The graph in Figure 9–24 shows the 4000B CMOS family in the opposite corner from the 100K ECL family. What is the significance of this?

9–23. Referring to Figure 9–25, which logic family dissipates less power at low frequencies, the 74LS or 74HC?

Section 9–8

C

9–24. **(a)** Using the data in Table 9–4, draw a graph of input and output specifications similar to Figure 9–26 for the 74HCMOS and the 74ALSTTL IC series.

(b) From your graphs of the two IC series, compare the HIGH- and LOW-level noise margins.

(c) From your graphs, can you see a problem in *directly* interfacing:

(1) The 74HCMOS to the 74ALSTTL?

(2) The 74ALSTTL to the 74HCMOS?

9–25. Refer to Table 9–4 to determine which of the following interfacing situations (driving gate-to-gate load) will require a pull-up resistor, and why?

(a) 74TTL to 74ALSTTL

(b) 74HCMOS to 74TTL

(c) 74TTL to 74HCMOS

(d) 74LSTTL to 74HCTMOS

(e) 74LSTTL to 4000B CMOS

9–26. Of the interfacing situations given in Problem 9–25, will any of the driving gates have trouble sinking or sourcing current to a single connected gate load?

9–27. From Table 9–4, determine

(a) How many 74LSTTL loads can be driven by a single 74HCTMOS gate?

(b) How many 74HCTMOS loads can be driven by a single 74LSTTL gate?

Schematic Interpretation Problems

See Appendix G for the schematic diagrams.

S

9–28. Assume that the inverter U4:A in the Watchdog Timer schematic has the following propagation delay times: $t_{PHL} = 7.0$ nS, $t_{PLH} = 9.0$ nS. Also assume that WATCHDOG_CLK is a 10-MHz square wave. Sketch the waveforms at WATCHDOG_CLK and the input labeled CLK on U1:B on the same time axis.

S

9–29. Repeat Problem 9–28 with a 7404 used in place of the 74HC04. Assume that the 7404 has the following propagation delay times: $t_{PHL} = 15.0$ nS, $t_{PLH} = 22.0$ nS.

CHAPTER 9 | LOGIC FAMILIES AND THEIR CHARACTERISTICS

S C D   **9–30.** Find U9 in the HC11D0 schematic. LCD_SL and KEY_SL are active-LOW outputs that signify that either the LCD is selected or the keyboard is selected. Add a logic gate to this schematic that outputs a LOW level called I/O_SEL whenever either the LCD *or* the keyboard is selected.

Answers to Review Questions

9–1. False

9–2. Totem-pole output

9–3. 0.7, 0.3

9–4. Because of the voltage drops across the internal transistors of the gate

9–5. It is the number of gates of the same subfamily that can be connected to a single output without exceeding the current rating of the gate.

9–6. Input current HIGH condition (I_{IH}), input current LOW condition (I_{IL}), output current HIGH condition (I_{OH}), output current LOW condition (I_{OL})

9–7. Sink current flows into the gate and goes to ground. Source current flows out of the gate and supplies the other gates.

9–8. (a) HIGH (b) HIGH (c) undetermined (d) LOW

9–9. False

9–10. Output

9–11. It pulls the output of the gate up to 5 V when the output transistor is off (float).

9–12. It reduces the propagation delay to achieve faster speeds.

9–13. Lower power dissipation, slower speeds.

9–14. Bipolar

9–15. False

9–16. 74ALS

9–17. 74LS

9–18. True

9–19. 74HC CMOS

9–20. To raise the output level of the TTL gate so it is recognized as a HIGH by the CMOS gate

9–21. The CMOS gate has severe output current limitations.

9–22. 74HCTMOS, because its voltage is almost perfect at 4.9 V HIGH and 0.1 V LOW.

9–23. Input current requirements (I_{IL}, I_{IH}) and output current capability (I_{OL}, I_{OH})

10

Flip-Flops and Registers

OUTLINE

10–1 *S-R* Flip-Flop
10–2 Gated *S-R* Flip-Flop
10–3 Gated *D* Flip-Flop
10–4 Integrated-Circuit *D* Latch (7475)
10–5 Integrated-Circuit *D* Flip-Flop (7474)
10–6 Master–Slave *J-K* Flip-Flop
10–7 Edge-Triggered *J-K* Flip-Flop
10–8 Integrated-Circuit *J-K* Flip-Flop (7476, 74LS76)
10–9 Using an Octal *D* Flip-Flop in a Microcontroller Application

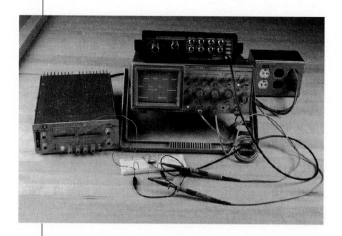

Objectives

Upon completion of this chapter, you should be able to

- Explain the internal circuit operation of S-R and gated S-R flip-flops.

- Compare the operation of D latches and D flip-flops by using timing diagrams.

- Describe the difference between pulse-triggered and edge-triggered flip-flops.

- Explain the theory of operation of master–slave devices.

- Connect integrated-circuit J-K flip-flops as toggle and D flip-flops.

- Use timing diagrams to illustrate the synchronous and asynchronous operation of J-K flip-flops.

Introduction

The logic circuits that we have studied in the previous chapters have consisted mainly of logic gates (AND, OR, NAND, NOR, INVERT) and **combinational logic.** Starting in this chapter, we will deal with data storage circuitry that will **latch** on to (remember) a digital state (1 or 0).

This new type of digital circuitry is called **sequential logic,** because it is controlled by and is used for controlling other circuitry in a specific sequence dictated by a control clock or enable/disable control signals.

The simplest form of data storage is the Set–Reset *(S-R)* flip-flop. These circuits are called **transparent latches** because the outputs respond immediately to changes at the input, and the input state will be remembered, or latched onto. The latch will sometimes have an enable input, which is used to control the latch to accept or ignore the *S-R* input states.

More sophisticated flip-flops use a clock as the control input and are used wherever the input and output signals must occur within a particular sequence.

10–1   S-R Flip-Flop

The *S-R* **flip-flop** is a data storage circuit that can be constructed using the basic gates covered in previous chapters. Using a cross-coupling scheme with two NOR gates, we can form the flip-flop shown in Figure 10–1.

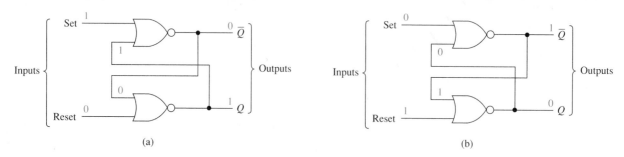

(a)   (b)

Figure 10–1   Cross-NOR *S-R* flip-flop: (a) Set condition; (b) Reset condition.

Common Misconception

You may feel that you can't solve this circuit because one input to each NOR is unknown. Remember that you don't need to know both inputs. If one input is HIGH, the output is LOW.

Let's start our analysis by placing a 1 (HIGH) on the **Set** and a 0 (LOW) on **Reset** [Figure 10–1(a)]. This is defined as the Set condition and should make the Q output 1 and $\overline{Q}$ output 0. A HIGH on the Set will make the output of the upper NOR equal 0 ($\overline{Q} = 0$) and that 0 is fed down to the lower NOR, which together with a LOW on the Reset input will cause the lower NOR's output to equal a 1 ($Q = 1$). [Remember, a NOR gate is always 0 output except when *both* inputs are 0 (Chapter 4).]

Now, when the 1 is removed from the Set input, the flip-flop should remember that it is Set (that is, $Q = 1$, $\overline{Q} = 0$). So with Set = 0, Reset = 0, and $Q = 1$ from previously being Set, let's continue our analysis. The upper NOR has a 0–1 at its inputs, making $\overline{Q} = 0$, while the lower NOR has a 0–0 at its inputs, keeping $Q = 1$. Great—the flip-flop remained Set even after the Set input was returned to 0 (called the *Hold* condition).

Now we should be able to Reset the flip-flop by making $S = 0$, $R = 1$ [Figure 10–1(b)]. With $R = 1$, the lower NOR will output a 0 ($Q = 0$), placing a 0–0 on the upper NOR, making its output 1 ($\overline{Q} = 1$); thus the flip-flop "flipped" to its Reset state.

The only other input condition is when both S and R inputs are HIGH. In this case, both NORs will put out a LOW, making Q *and* $\overline{Q}$ equal 0, which is a condition that is not used. (Why would anyone want to Set *and* Reset at the same time, anyway!) Also, when you return to the Hold condition from $S = 1$, $R = 1$, you will get unpredictable results unless you know which input returned LOW last.

From the previous analysis we can construct the *S-R* flip-flop **function table** shown in Table 10–1, which lists all input and output conditions.

Table 10–1		Function Table for Figure 10–1		
S	R	Q	$\overline{Q}$	Comments
0	0	Q	$\overline{Q}$	*Hold* condition (no change)
1	0	1	0	Flip-flop Set
0	1	0	1	Flip-flop Reset
1	1	0	0	Not used

An *S-R* flip-flop can also be made from cross-NAND gates, as shown in Figure 10–2. Prove to yourself that Figure 10–2 will produce the function table shown in Table 10–2. (Start with $S = 1$, $R = 0$ and remember that a NAND is LOW out only

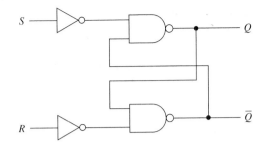

Figure 10–2 Cross-NAND *S-R* flip-flop.

Team Discussion

Will we get the same function table if we eliminate the inverters and switch the *S* and *R* inputs?

Table 10–2 Function Table for Figure 10–2

S	*R*	*Q*	$\overline{Q}$	Comments
0	0	*Q*	$\overline{Q}$	Hold condition
1	0	1	0	Flip-flop Set
0	1	0	1	Flip-flop Reset
1	1	1	1	Not used

when *both* inputs are HIGH.) The symbols used for an *S-R* flip-flop are shown in Figure 10–3. The symbols show that both *true* and **complemented** *Q* outputs are available. The second symbol is technically more accurate, but the first symbol is found most often in manufacturers' data manuals and throughout this book.

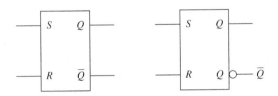

Figure 10–3 Symbols for an *S-R* flip-flop.

Now let's get practical and find an integrated-circuit TTL NOR gate and draw the actual wiring connections to form a cross-NOR like Figure 10–1 so that we may check it in the lab.

The TTL data manual shows a quad NOR gate 7402. Looking at its pin layout in conjunction with Figure 10–1, we can draw the circuit of Figure 10–4. To check out the operation of Figure 10–4 in the lab, apply 5 V to pin 14 and ground pin 7. Set the flip-flop by placing a HIGH (5 V) to the Set input and a LOW (0 V, ground) to the Reset input. A logic probe attached to the *Q* output should register a HIGH. When the *S-R* inputs are returned to the 0–0 state, the *Q* output should remain latched in the 1 state. The Reset function can be checked using the same procedure.

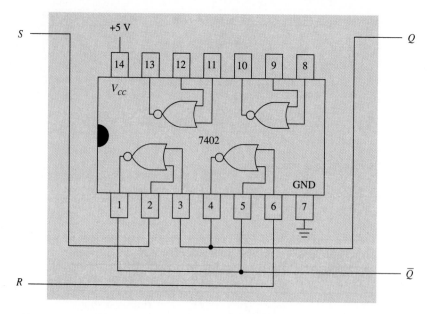

Figure 10–4 *S-R* flip-flop connections using a 7402 TTL IC.

S-R Timing Analysis

By performing a timing analysis on the *S-R* flip-flop, we can see why it is called transparent and also observe the latching phenomenon.

EXAMPLE 10–1

To the *S-R* flip-flop shown in Figure 10–5, we connect the *S* and *R* waveforms given in Figure 10–6. Sketch the *Q* output waveform that will result.

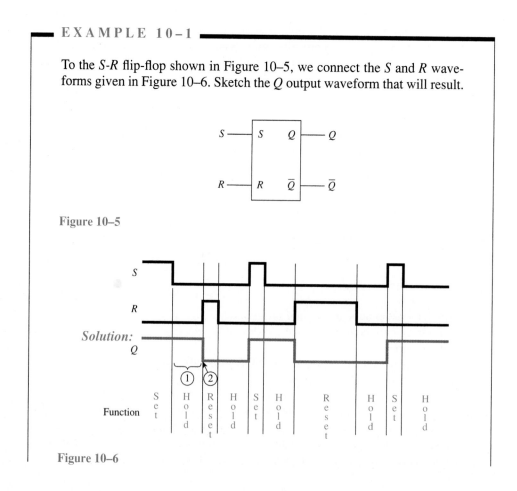

Figure 10–5

Figure 10–6

Notes:

1. The flip-flop is latched in the Set condition even after the HIGH is removed from the *S* input.

2. The flip-flop is considered transparent because the *Q* output responds immediately to input changes.

S-R Flip-Flop Application

Let's say that we need a storage register that will *remember* the value of a binary number ($2^3 2^2 2^1 2^0$) that represents the time of day at the instant a momentary temperature limit switch goes into a HIGH (1) state. Figure 10–7 could be used to implement such a circuit. Because a 4-bit binary number is to be stored, we need four *S-R* flip-flops. We will look at their *Q* outputs with a logic probe to read the stored values.

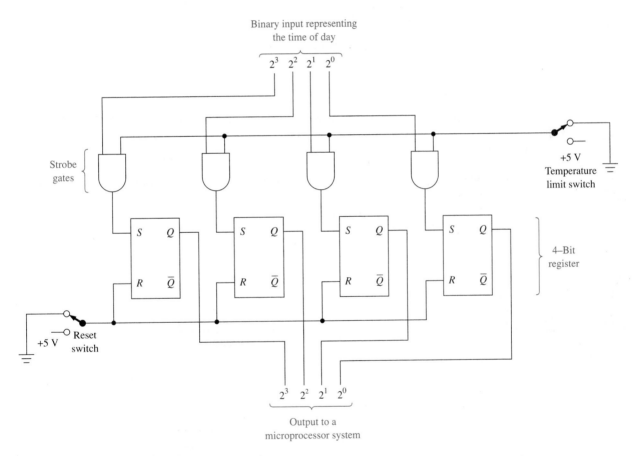

Figure 10–7 *S-R* flip-flop used as a storage register.

Team Discussion

Discuss some of the problems that may occur if the circuit is operated improperly by forgetting to reset or by allowing multiple temperature switch closures.

With the Reset switch in the up position, the *R* inputs will be zero. With the temperature limit switch in the up position, one input to each AND gate is grounded, keeping the *S* inputs at 0 also. To start the operation, first the Reset switch is momentarily pressed down, placing 5 V (1) on all four *R* inputs and thus resetting all flip-flops to 0.

Meanwhile, the binary input number is not allowed to reach the *S* inputs because there is a 0 at the other input of each AND gate. (Gates used in this method are referred to as **strobe gates** because they let information pass only when they are enabled.)

When the temperature limit switch momentarily goes down, 5 V (1) will be placed at each strobe gate, allowing the binary number (1's and 0's) to pass through to the S inputs, thus setting the appropriate flip-flops. (Assume that the switch will go down only once.)

The binary input number representing the time of day that the temperature switch went down will be stored in the 4-bit register and could later be read by a logic probe or automatically by a microprocessor system.

Review Questions

10–1. A flip-flop is different from a basic logic gate because it *remembers* the state of the inputs after they are removed. True or false?

10–2. What levels must be placed on S and R to Set an S-R flip-flop?

10–3. What effect do $S = 0$ and $R = 0$ have on the output level at Q?

10–2   Gated S-R Flip-Flop

Simple gate circuits, combinational logic, and transparent S-R flip-flops are called **asynchronous** (not synchronous) because the output responds immediately to input changes. **Synchronous** circuits operate sequentially, in step, with a control input. To make an S-R flip-flop synchronous, we add a gated input to enable and disable the S and R inputs. Figure 10–8 shows the connections that make the cross-NOR S-R flip-flop into a gated S-R flip-flop.

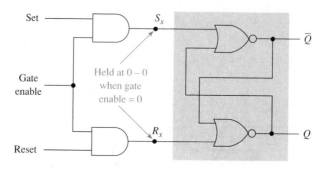

Figure 10–8   Gated S-R flip-flop.

The S_x and R_x lines in Figure 10–8 are the original Set and Reset inputs. With the addition of the AND gates, however, the S_x and R_x lines will be kept LOW–LOW (Hold condition) as long as the Gate Enable is LOW. The flip-flop will operate normally while the Gate Enable is HIGH. The function chart [Figure 10–9(b)] and Example 10–2 illustrate the operation of the gated S-R flip-flop.

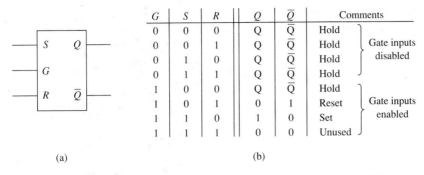

G	S	R	Q	Q̄	Comments	
0	0	0	Q	Q̄	Hold	
0	0	1	Q	Q̄	Hold	Gate inputs
0	1	0	Q	Q̄	Hold	disabled
0	1	1	Q	Q̄	Hold	
1	0	0	Q	Q̄	Hold	
1	0	1	0	1	Reset	Gate inputs
1	1	0	1	0	Set	enabled
1	1	1	0	0	Unused	

(a)   (b)

Figure 10–9 Function table and symbol for the gated *S-R* flip-flop of Figure 10–8.

EXAMPLE 10–2

Feed the *G, S,* and *R* inputs in Figure 10–10 into the gated *S-R* flip-flop, sketch the output wave at *Q*, and list the flip-flop functions.

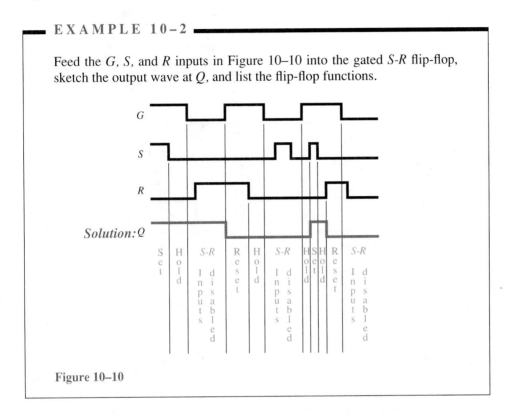

Figure 10–10

EXAMPLE 10–3

Feed the *G, S,* and *R* inputs in Figure 10–11 into the gated *S-R* flip-flop and sketch the output wave at *Q*.

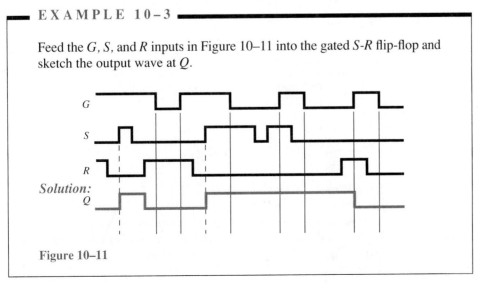

Figure 10–11

10–3   Gated *D* Flip-Flop

Another type of flip-flop is the *D* flip-flop (*Data* flip-flop). It can be formed from the gated *S-R* flip-flop by the addition of an inverter. This enables just a single input *(D)* to both Set *and* Reset the flip-flop.

In Figure 10–12 we see that *S* and *R* will be complements of each other, and *S* is connected to a single line labeled *D* (Data). The operation is such that *Q* will be the same as *D* while *G* is HIGH, and *Q* will remain latched when *G* goes LOW.

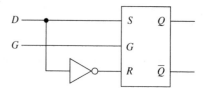

Figure 10–12   Gated *D* flip-flop.

EXAMPLE 10–4

Sketch the output waveform at *Q* for the inputs at *D* and *G* of the gated *D* flip-flop in Figure 10–13.

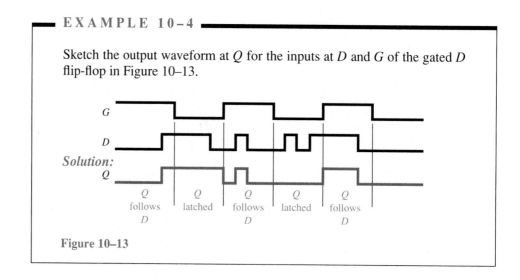

Solution:

Figure 10–13

Review Questions

10–4. Explain why the *S-R* flip-flop is called asynchronous and the *gated S-R* flip-flop is called synchronous.

10–5. Changes in *S* and *R* while a gate is enabled have no effect on the *Q* output of a gated *S-R* flip-flop. True or false?

10–6. What procedure would you use to Reset the *Q* output of a gated *D* flip-flop?

10–4   Integrated-Circuit *D* Latch (7475)

The 7475 is an example of an integrated-circuit *D* latch (also called a *bistable latch*). It contains *four* transparent *D* latches. Its logic symbol and pin configuration are given in Figure 10–14. Latches 0 and 1 share a common enable (E_{0-1}), and latches 2 and 3 share a common enable (E_{2-3}).

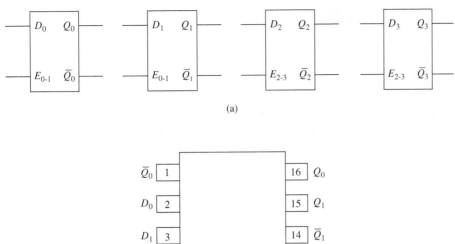

(a)

(b)

Figure 10–14 The 7475 quad bistable D latch: (a) logic symbol; (b) pin configuration.

From the function table (Table 10–3) we can see that the Q output will follow D (transparent) as long as the enable line (E) is HIGH (called active-HIGH enable). When E goes LOW, the Q output will become *latched* to the value that D was just before the HIGH-to-LOW transition of E.

Table 10–3 Function Table for a 7475[a]

Operating Mode	Inputs		Outputs	
	E	D	Q	$\overline{Q}$
Data	H	L	L	H
enabled	H	H	H	L
Data				
latched	L	x	q	$\overline{q}$

[a]q = state of Q before the HIGH-to-LOW edge of E; x = don't care.

Helpful Hint

Some students use the function table as a crutch. This can become a dangerous habit as the functions get more complex. You must get into the habit of understanding the *description* provided by the manufacturer. (In this case, what is meant by *transparent* and *latched*?)

EXAMPLE 10–5

For the inputs at D_0 and E_{0-1} for the 7475 D latch shown in Figure 10–15, sketch the output waveform at Q_0 in Figure 10–16.

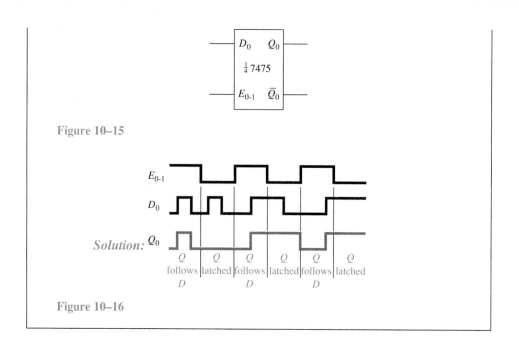

Figure 10–15

Solution:

Q follows D | latched | Q follows D | latched | Q follows D | latched

Figure 10–16

Review Questions

10–7. The 7475 IC contains how many D latches?

10–8. The Q output of the 7475 D latch follows the level on the D input as long as E is _____ (HIGH or LOW).

10–9. Changes to D are ignored by the 7475 while E is LOW. True or false?

10–5 Integrated-Circuit D Flip-Flop (7474)

The 7474 D flip-flop differs from the 7475 D latch in several ways. Most important, the 7474 is an **edge-triggered** device. This means that **transitions** in Q will occur only at the edge of the input **trigger** pulse. The trigger pulse is usually a **clock** or timing signal instead of an enable line. In the case of the 7474, the trigger point is at the **positive edge** of C_p (LOW-to-HIGH transition). The small triangle on the D flip-flop symbol [Figure 10–17(a)] is used to indicate that it is edge triggered.

Edge-triggered devices are made to respond to only the *edge* of the clock signal by converting the positive clock input pulse into a single, narrow spike. Figure 10–18 shows a circuit similar to that inside the 7474 to convert the rising edge of C_p into a positive spike. This is called a *positive edge-detection circuit.*

In Figure 10–18, the original clock, C_p, is input to an inverter whose purpose is to invert the signal and delay it by the propagation delay time of the inverter. This inverted, delayed signal, $\overline{C_{pd}}$, is then fed into the AND gate along with the original clock, C_p. By studying the waveforms you can see that the output waveform, $C_p{}'$, is a very narrow pulse (called a *spike*) that lines up with the positive edge of C_p. This is now used as the trigger signal inside the D flip-flop. Therefore, even though a very wide pulse is entered at C_p of the 7474, the edge-detection circuitry converts it to a spike so that the D flip-flop reacts only to data entered at D at the positive edge of C_p.

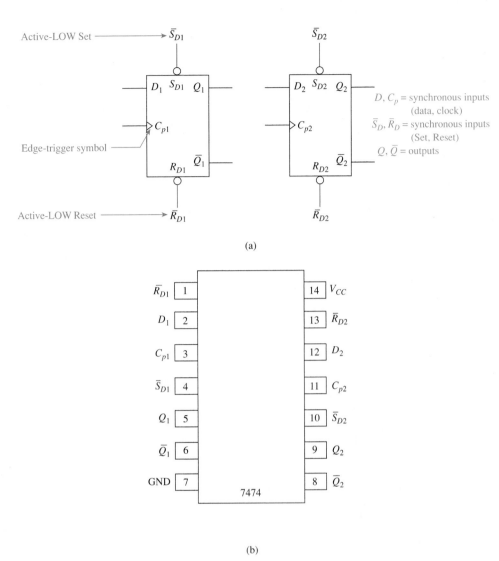

Active-LOW Set ⟶ $\overline{S}_{D1}$

Edge-trigger symbol

Active-LOW Reset ⟶ $\overline{R}_{D1}$

D_1 S_{D1} Q_1

C_{p1}

R_{D1} $\overline{Q}_1$

$\overline{S}_{D2}$

D_2 S_{D2} Q_2

C_{p2}

R_{D2} $\overline{Q}_2$

$\overline{R}_{D2}$

D, C_p = synchronous inputs (data, clock)
$\overline{S}_D, \overline{R}_D$ = synchronous inputs (Set, Reset)
$Q, \overline{Q}$ = outputs

(a)

Common Misconception

Don't mistakenly think that you can leave the asynchronous inputs disconnected if they are not to be used (they must be tied HIGH).

Common Misconception

Students often mistakenly think that applying a LOW to $\overline{S}_D$ will make Q = LOW.

$\overline{R}_{D1}$	1	14	V_{CC}
D_1	2	13	$\overline{R}_{D2}$
C_{p1}	3	12	D_2
$\overline{S}_{D1}$	4	11	C_{p2}
Q_1	5	10	$\overline{S}_{D2}$
$\overline{Q}_1$	6	9	Q_2
GND	7	8	$\overline{Q}_2$

7474

(b)

Figure 10–17   The 7474 dual D flip-flop: (a) logic symbol; (b) pin configuration.

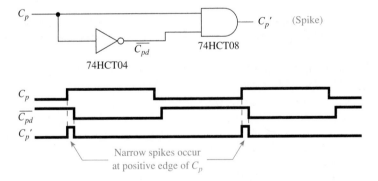

C_p

$\overline{C_{pd}}$

74HCT04

74HCT08

$C_p{}'$ (Spike)

C_p

$\overline{C}_{pd}$

$C_p{}'$

Narrow spikes occur at positive edge of C_p

Figure 10–18   Positive edge-detection circuit and waveforms.

The 7474 has two distinct types of inputs: synchronous and asynchronous. The *synchronous inputs* are the D (Data) and C_p (Clock) inputs. The state at the D input will be transferred to Q at the positive edge of the input trigger (LOW-to-HIGH edge of C_p). The *asynchronous inputs* are $\overline{S}_D$ (Set) and $\overline{R}_D$ (Reset), which operate independently of D and C_p. Being asynchronous means that they are *not* in sync with the clock

SECTION 10–5 I INTEGRATED CIRCUIT D FLIP-FLOP (7474)

pulse, and the Q outputs will respond *immediately* to input changes at $\overline{S_D}$ and $\overline{R_D}$. The little circle at S_D and R_D means that they are **active-LOW** inputs, and because the circles act like inverters, the external pin on the IC is labeled as the complement of the internal label.

This all sounds complicated, but it really is not. Just realize that *a LOW on $\overline{S_D}$ will immediately Set the flip-flop, and a LOW on $\overline{R_D}$ will immediately Reset the flip-flop, regardless of the states at the synchronous (D, C_p) inputs.*

The function table (Table 10–4) and following examples will help illustrate the operation of the 7474 D flip-flop.

Team Discussion

Discuss how the 7474 might be used to remember that a pedestrian had pressed a crosswalk push button. (*Hint:* See Figure 12–37.)

Table 10–4   Function Table for a 7474 D Flip-Flop[a]

Operating Mode	Inputs				Outputs	
	$\overline{S_D}$	$\overline{R_D}$	C_p	D	Q	$\overline{Q}$
Asynchronous Set	L	H	x	x	H	L
Asynchronous Reset	H	L	x	x	L	H
Not used	L	L	x	x	H	H
Synchronous Set	H	H	↑	h	H	L
Synchronous Reset	H	H	↑	l	L	H

[a]↑ = positive edge of clock; H = HIGH; h = HIGH level one setup time prior to positive clock edge; L = LOW; l = LOW level one setup time prior to positive clock edge; x = don't care.

The lowercase h in the D column indicates that, in order to do a synchronous Set, the D must be in a HIGH state at least one setup time prior to the positive edge of the clock. The same rules apply for the lowercase l (Reset).

The **setup time** for this flip-flop is 20 ns, which means that if D is changing while C_p is LOW, that's okay, but D must be held stable (HIGH or LOW) at least 20 ns *before* the LOW-to-HIGH transition of C_p. (We discuss setup time in greater detail in Chapter 11.) Also realize that the only digital level on the D input that is used is the level that is present at the positive edge of C_p.

We have learned a lot of new terms in regard to the 7474 (active-LOW, edge-triggered, asynchronous, and others). These terms are important because they apply to almost all the ICs that are used in the building of sequential circuits.

■ E X A M P L E   1 0 – 6 ■

Sketch the output waveform at Q for the 7474 D flip-flop shown in Figure 10–19(a) whose input waveforms are as given in Figure 10–19(b).

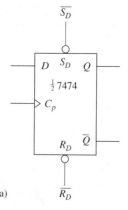

(a)

Figure 10–19

CHAPTER 10 | FLIP-FLOPS AND REGISTERS

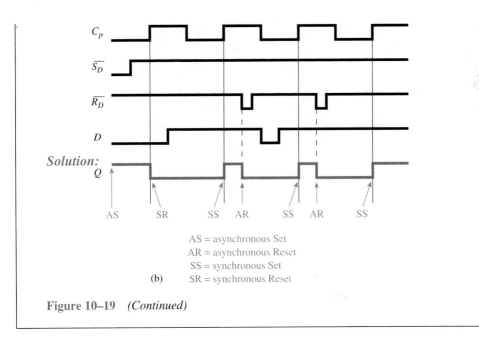

AS   SR   SS   AR   SS   AR   SS

AS = asynchronous Set
AR = asynchronous Reset
SS = synchronous Set
(b)   SR = synchronous Reset

Figure 10–19   *(Continued)*

Helpful Hint

Two ways that you can remember that it takes a LOW to asynchronously Set the flip flop are (1) the overbar on $\overline{S_D}$ and (2) the bubble on $\overline{S_D}$.

EXAMPLE 10–7

Sketch the output waveforms at Q for the 7474 D flip-flops shown in Figure 10–20 whose input waveforms are given in Figure 10–21.

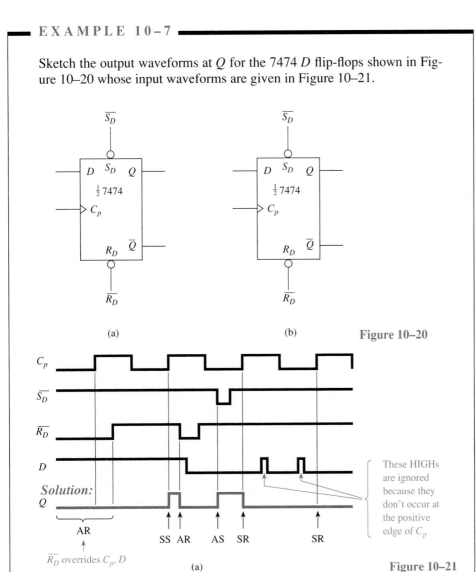

(a)   (b)   **Figure 10–20**

These HIGHs are ignored because they don't occur at the positive edge of C_p

AR   SS AR   AS SR   SR

$\overline{R_D}$ overrides C_p, D
(a)   **Figure 10–21**

SECTION 10–5 | INTEGRATED CIRCUIT *D* FLIP-FLOP (7474)   321

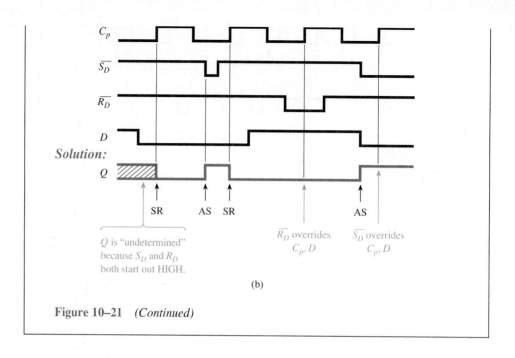

(b)

Figure 10–21   *(Continued)*

Review Questions

10–10. The 7474 is an edge-triggered device. What does this mean?

10–11. Which are the synchronous and which are the asynchronous inputs to the 7474 D flip-flop?

10–12. To perform an asynchronous Set, the $\overline{S_D}$ line must be made HIGH. True or false?

10–13. The purpose of the inverter in the edge-detection circuit of Figure 10–18 is to _____ and _____ the signal from C_p.

10–6   Master–Slave *J-K* Flip-Flop

Another type of flip-flop is the *J-K* flip-flop. It differs from the *S-R* flip-flop in that it has one new mode of operation, called **toggle**. Toggle means that Q and $\overline{Q}$ will switch to their *opposite* states at the active clock edge. (Q will switch from a 1 to a 0 or from a 0 to a 1.) The synchronous inputs to the *J-K* flip-flop are labeled J, K, and C_p. J acts like the S input to an *S-R* flip-flop and K acts like the R input to an *S-R* flip-flop. The toggle mode is achieved by making *both* J and K HIGH before the active clock edge. Table 10–5 shows the four synchronous operating modes of *J-K* flip-flops.

A number of the older flip-flops (74H71, 7472, 7473, 7476, 7478, 74104, 74105) are of the **master–slave** variety. They consist of two latches: a master *S-R* latch (*S-R* flip-flop) that receives data while the input trigger clock is HIGH, and a slave *S-R* latch that receives data from the master and outputs it when the clock goes LOW. Figure 10–22 shows a simplified equivalent circuit and logic symbol for a master–slave *J-K* flip-flop.

Table 10–5 Synchronous Operating Modes of a *J-K* Flip-Flop

Operating Mode	J	K
Hold	0	0
Set	1	0
Reset	0	1
Toggle	1	1

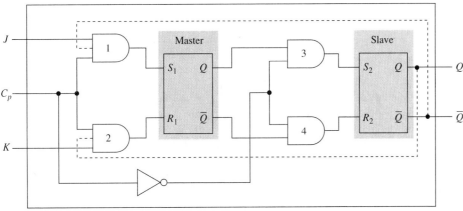

Note: The dashed lines are internal feedback connections that enable the toggle operation.

(a)

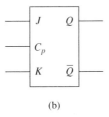

(b)

Figure 10–22 Positive pulse-triggered master–slave *J-K* flip-flop: (a) equivalent circuit; (b) logic symbol.

From Figure 10–22 we can see that the master latch will be loaded with the state of the *J* and *K* inputs, while AND gates 1 and 2 are enabled by a HIGH C_p (that is, the *master* is loaded while C_p is HIGH). For now, let's ignore the feedback connections shown in Figure 10–22 as dashed lines.

When C_p goes LOW, gates 1 and 2 are **disabled,** but gates 3 and 4 are enabled by the HIGH from the inverter, allowing the digital state at the master to pass through to the slave latch inputs.

When C_p goes HIGH again, gates 3 and 4 will be disabled, thus keeping the slave latch at its current **digital state.** Also, with C_p HIGH again, the master will be loaded with the digital states of the *J* and *K* inputs, and the cycle repeats (see Figure 10–23). Master–slave flip-flops are called **pulse-triggered** or **level-triggered** devices because *input data are read during the entire time that the clock pulse is at a HIGH level.*

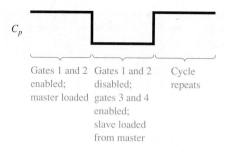

C_p

Gates 1 and 2 Gates 1 and 2 Cycle
enabled; disabled; repeats
master loaded gates 3 and 4
enabled;
slave loaded
from master

Figure 10–23 Enable/disable operation of the C_p line of a master–slave flip-flop.

If you analyze the logic in Figure 10–22, *including* the dashed lines, you will see how the *toggle* operation occurs. With $J = 1$ and $K = 1$, let's assume that $Q = 1$ ($\overline{Q} = 0$). The dashed feedback connection from $Q = 1$ will enable gate 2 (gate 1 is disabled by the 0 on $\overline{Q}$), allowing the master to get reset when C_p goes HIGH. Therefore, Q (of the slave) will toggle to a 0 when C_p returns LOW.

With J and K still 1 and $Q = 0$, the next time C_p is HIGH gate 1 will be enabled because $\overline{Q} = 1$. This will set the master. Then when C_p returns LOW, the Q output of the slave will toggle to a 1. In other words, the feedback connections allow only the *opposite* state to enter the master when $J = 1$ and $K = 1$. Notice that even if J and K are only *momentarily* made HIGH or if they are pulsed HIGH at different times while C_p is HIGH, the master will toggle and pass the toggle on to the slave when C_p goes LOW.

Occasionally, unwanted pulses or short glitches caused by electrostatic **noise** appear on J and K while C_p is HIGH. This phenomenon of interpreting unwanted signals on J and K while C_p is HIGH is called **ones catching** and is eliminated by the newer J-K flip-flops, which use an edge-triggering technique instead of pulse triggering.

EXAMPLE 10–8

To illustrate the master–slave operation, for the master–slave J-K flip-flop shown in Figure 10–24, draw the Q output in Figure 10–25. (Assume that Q is initially 0.)

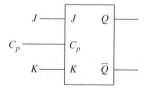

Figure 10–24

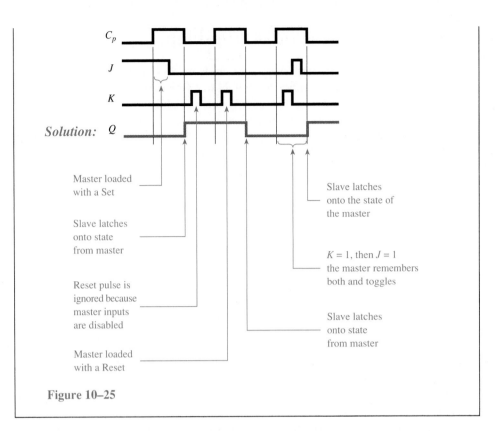

Figure 10–25

EXAMPLE 10–9

For the master–slave *J-K* flip-flop shown in Figure 10–26, sketch the waveform at *Q* in Figure 10–27. (Assume that *Q* is initially 0.)

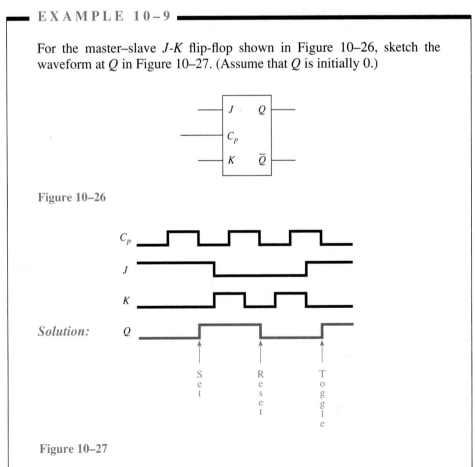

Figure 10–26

Figure 10–27

10–7 Edge-Triggered *J-K* Flip-Flop

With edge triggering, the flip-flop accepts data only on the *J* and *K* inputs that are present at the active clock edge (either the HIGH-to-LOW edge of C_p or the LOW-to-HIGH edge of C_p). This gives the design engineer the ability to accept input data on *J* and *K* at a precise instant in time. Transitions of the level *J* and *K* before or after the active clock trigger edge are ignored. The logic symbols for edge-triggered flip-flops use a small triangle at the clock input to signify that it is an edge-triggered device (see Figure 10–28).

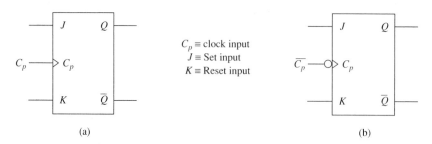

$C_p \equiv$ clock input
$J \equiv$ Set input
$K \equiv$ Reset input

(a)   (b)

Figure 10–28   Symbols for edge-triggered *J-K* flip-flops: (a) positive edge triggered; (b) negative edge triggered.

Transitions of the *Q* output for the positive edge-triggered flip-flop shown in Figure 10–28(a) will occur when the C_p input goes from LOW to HIGH (positive edge). Figure 10–28(b) shows a negative edge-triggered flip-flop. The input clock signal will connect to the IC pin labeled $\overline{C_p}$. The small circle indicates that transitions in the output will occur at the HIGH-to-LOW edge (**negative edge**) of the $\overline{C_p}$ input.

The function table for a negative edge-triggered *J-K* flip flop is shown in Figure 10–29.

	Inputs			Outputs	
Operating mode	$\overline{C_p}$	*J*	*K*	*Q*	$\overline{Q}$
Hold	↓	0	0	No change	
Set	↓	1	0	1	0
Reset	↓	0	1	0	1
Toggle	↓	1	1	Opposite state	

↓ ≡ HIGH- to-LOW

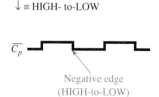

$\overline{C_p}$

Negative edge
(HIGH-to-LOW)

Figure 10–29   Function table for a negative edge-triggered *J-K* flip-flop.

The downward arrow in the $\overline{C_p}$ column indicates that the flip-flop is triggered by the HIGH-to-LOW transition (negative edge) of the clock.

EXAMPLE 10–10

To illustrate edge triggering, let's draw the *Q* output in Figure 10–31 for the negative edge-triggered *J-K* flip-flop shown in Figure 10–30. (Assume that *Q* is initially 0.)

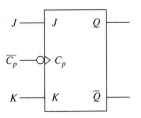

Figure 10–30

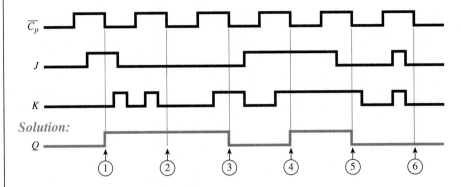

Solution:

① $J = 1, K = 0$ at the negative clock edge; Q is Set

② $J = 0, K = 0$ at the negative clock edge; Q is held
    (transitions in K before the edge are ignored)

③ $J = 0, K = 1$ at the negative clock edge; Q is Reset

④ $J = 1, K = 1$ at the negative clock edge; Q toggles

⑤ $J = 0, K = 1$ at the negative clock edge; Q is Reset

⑥ $J = 0, K = 0$ at the negative clock edge; Q is held

Figure 10–31

Review Questions

10–14. Describe why master–slave flip-flops are called *ones catching*.

10–15. The *Set* input to a *J-K* flip-flop is _____ *(J, K)* and the *Reset* input is _____ *(J, K)*.

10–16. The *edge-triggered J-K* flip-flop looks only at the *J-K* inputs that are present during the active clock edge on C_p. True or false?

10–17. What effect does the *toggle* operation of a *J-K* flip-flop have on the Q output?

10–8 Integrated-Circuit *J-K* Flip-Flop (7476, 74LS76)

Now let's take a look at actual *J-K* flip-flop ICs. The 7476 and 74LS76 are popular *J-K* flip-flops because they are both dual flip-flops (two flip-flops in each IC package) and they have asynchronous inputs ($\overline{R_D}$ and $\overline{S_D}$) as well as synchronous inputs ($\overline{C_p}$, *J*, *K*). The 7476 is a positive pulse-triggered (master–slave) flip-flop, and the 74LS76 is a negative edge-triggered flip-flop, a situation that can trap the unwary technician who attempts to replace the 7476 with the 74LS76!

From Figure 10–32(a) and Table 10–6, we see that the asynchronous inputs $\overline{S_D}$ and $\overline{R_D}$ are *active LOW*. That is, a LOW on $\overline{S_D}$ (Set) will Set the flip-flop ($Q = 1$), and a LOW on $\overline{R_D}$ will Reset the flip-flop ($Q = 0$). Remember, the asynchronous inputs will cause the flip-flop to respond immediately *without* regard to the clock trigger input.

Helpful Hint

The 7476 master-slave can be demonstrated in lab to be a one's catcher. The 74LS76, on the other hand, will ignore *J* and *K* except at the trigger edge. Try it.

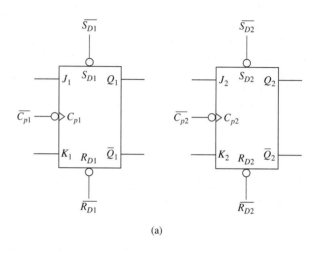

(a)

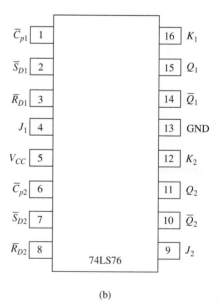

(b)

Figure 10–32 The 74LS76 negative edge-triggered flip-flop: (a) logic symbol; (b) pin configuration.

CHAPTER 10 | FLIP-FLOPS AND REGISTERS

Table 10–6   Function Table for the 74LS76[a]

Operating Mode	Inputs					Outputs	
	$\overline{S_D}$	$\overline{R_D}$	$\overline{C_p}$	J	K	Q	$\overline{Q}$
Asynchronous Set	L	H	x	x	x	H	L
Asynchronous Reset	H	L	x	x	x	L	H
Synchronous Hold	H	H	↓	l	l	q	$\overline{q}$
Synchronous Set	H	H	↓	h	l	H	L
Synchronous Reset	H	H	↓	l	h	L	H
Synchronous Toggle	H	H	↓	h	h	$\overline{q}$	q

[a]H = HIGH-voltage steady state; L = LOW-voltage steady state; h = HIGH voltage one setup time prior to negative clock edge; l = LOW voltage one setup time prior to negative clock edge; x = don't care; q = state of Q prior to negative clock edge; ↓ = HIGH-to-LOW (negative) clock edge.

For synchronous operations using J, K, and $\overline{C_p}$, the asynchronous inputs must be disabled by putting a HIGH level on both $\overline{S_D}$ and $\overline{R_D}$. The J and K inputs are read one setup time prior to the HIGH-to-LOW edge of the clock $(\overline{C_p})$. One setup time for the 74LS76 is 20 ns. This means that the state of J and K 20 ns *before* the negative edge of the clock is used to determine the synchronous operation to be performed. (Of course, the 7476 master–slave will read the state of J and K during the entire positive clock pulse.)

Also notice that in the toggle mode ($J = K = 1$), after a negative clock edge, Q becomes whatever $\overline{Q}$ was before the clock edge, and vice versa (that is, if $Q = 1$ before the negative clock edge, then $Q = 0$ after the negative clock edge).

Now let's work through several timing analysis examples to be sure that we fully understand the operation of *J-K* flip-flops.

EXAMPLE 10–11

Sketch the Q waveform for the 74LS76 negative edge-triggered *J-K* flip-flop shown in Figure 10–33, with the input waveforms given in Figure 10–34.

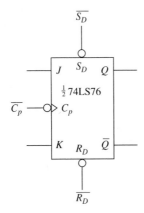

Figure 10–33

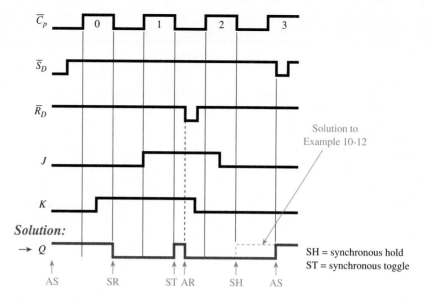

Figure 10–34

Note: Q changes only on the negative edge of $\overline{C_p}$ except when asynchronous operations $(\overline{S_D}, \overline{R_D})$ are taking place.

EXAMPLE 10–12

How would the Q waveform of Example 10–11 be different if we used a 7476 pulse-triggered master–slave flip-flop instead of the 74LS76?

Solution: During positive pulse 2, J is HIGH for a short time. The master latch within the 7476 will remember that and cause the flip-flop to do a synchronous Set ($Q = 1$) when $\overline{C_p}$ returns LOW. (See Figure 10–34.)

EXAMPLE 10–13

Sketch the Q waveform for the 7476 positive pulse-triggered flip-flop shown in Figure 10–35 with input waveforms given in Figure 10–36.

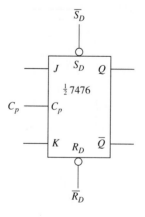

Figure 10–35

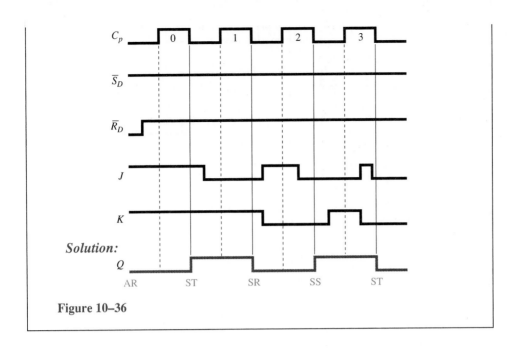

Figure 10–36

EXAMPLE 10–14

The 74109 is a positive edge-triggered $J\text{-}\overline{K}$ flip-flop. The logic symbol (Figure 10–37) and input waveforms (Figure 10–38) are given; sketch Q.

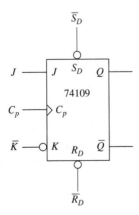

Figure 10–37

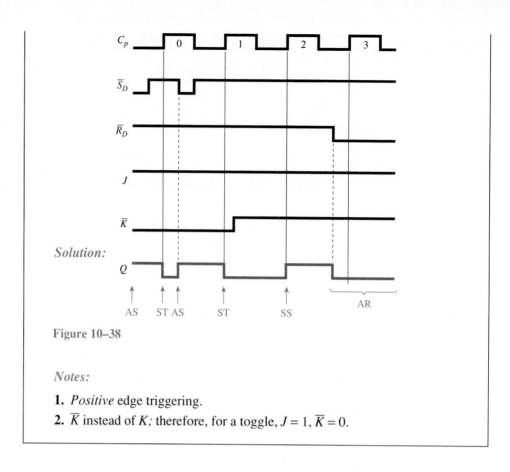

Figure 10–38

Notes:

1. *Positive* edge triggering.

2. $\overline{K}$ instead of K; therefore, for a toggle, $J = 1, \overline{K} = 0$.

The *J-K* flip-flop can be used to form other flip-flops by making the appropriate external connections. For example, to form a *D* flip-flop, add an inverter between the *J* and *K* inputs and bring the data into the *J* input, as shown in Figure 10–39.

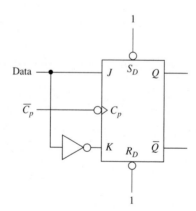

Figure 10–39 *D* flip-flop made from a *J-K* flip-flop.

The flip-flop in Figure 10–39 will operate as a *D* flip-flop because the data are brought in on the *J* terminal and its complement is at the *K*; so if Data = 1, the flip-flop will be Set after the clock edge; if Data = 0, the flip-flop will be Reset after the clock edge. (*Note:* You lose the toggle mode and hold mode using this configuration.)

Also, it is often important for a flip-flop to operate in the toggle mode. This can be done simply by connecting both *J* and *K* to 1. This will cause the flip-flop to change states at each active clock edge, as shown in Figure 10–40. Notice that the frequency of the output waveform at Q will be one-half the frequency of the input waveform at $\overline{C}_p$.

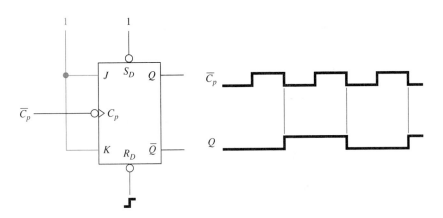

Helpful Hint

It is interesting for you to see an application of the D flip-flop (check out the shift register in Figure 13–1) and the toggle flip-flop (check out the counter in Figure 12–9).

Figure 10–40 *J-K* connected as a toggle flip-flop.

Figure 10–41 shows the test apparatus used to display the input and output of a toggle flip-flop. The oscilloscope display is used to accurately show the timing and frequency relationship between the two waveforms.

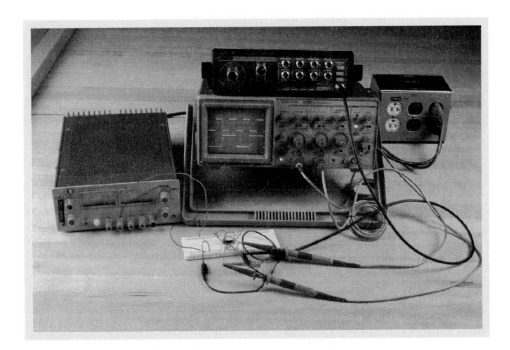

Figure 10–41   Test apparatus used to analyze the input and output of a toggle flip-flop.

As we have seen, there is a variety of flip-flops, each with its own operating characteristics. In Chapters 11 through 13, we learn how to use these ICs to perform sequential operations such as counting, data shifting, and sequencing.

SECTION 10–8 I INTEGRATED-CIRCUIT *J-K* FLIP-FLOP (7476, 74LS76)

First, let's summarize what we have learned about flip-flops by utilizing four common flip-flops in the same circuit, and supplying input signals and sketching the Q outputs of each (Example 10–15).

EXAMPLE 10–15

For each of the flip-flops shown in Figure 10–42, sketch the Q outputs in Figure 10–43.

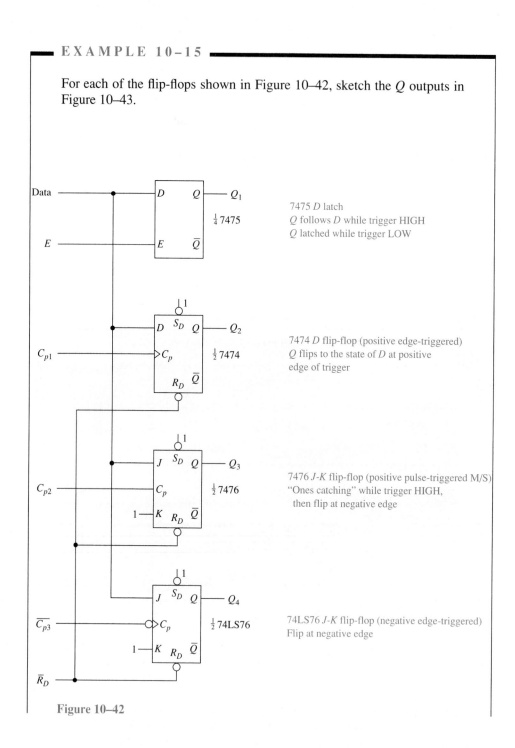

7475 D latch
Q follows D while trigger HIGH
Q latched while trigger LOW

7474 D flip-flop (positive edge-triggered)
Q flips to the state of D at positive edge of trigger

7476 J-K flip-flop (positive pulse-triggered M/S)
"Ones catching" while trigger HIGH,
    then flip at negative edge

74LS76 J-K flip-flop (negative edge-triggered)
Flip at negative edge

Figure 10–42

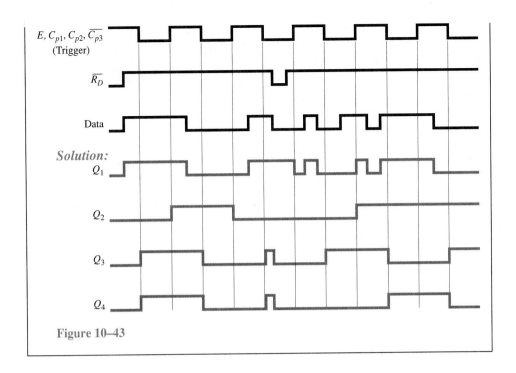

$E, C_{p1}, C_{p2}, \overline{C_{p3}}$
(Trigger)

$\overline{R_D}$

Data

Solution:
Q_1

Q_2

Q_3

Q_4

Figure 10–43

Review Questions

10–18. How do you *asynchronously* Reset the 74LS76 flip-flop?

10–19. The synchronous inputs to the 74LS76 override the asynchronous inputs. True or false?

10–20. To operate a 74LS76 flip-flop synchronously, the $\overline{S_D}$ and $\overline{R_D}$ inputs must be held _____ (HIGH, LOW).

10–21. What is the distinction between uppercase and lowercase letters when used in the function table for the 74LS76 flip-flop?

10–9 Using an Octal *D* Flip-Flop in a Microcontroller Application

Most of the basic latches and flip-flops are also available as **octal** ICs. In this configuration, there are eight latches or flip-flops in a single IC package. If all eight latches or flip-flops are controlled by a common clock, it is called an 8-bit **register.** An example of an 8-bit *D* flip-flop register is the high-speed CMOS 74HCT273 (also available in the TTL LS and S families). The '273 contains eight *D* flip-flops, all controlled by a common edge-triggered clock, C_p (see Figure 10–44). At the positive edge of C_p, the 8 bits of data at D_0 through D_7 are stored in the eight *D* flip-flops and output at Q_0 through Q_7. The '273 also has an active-LOW master reset ($\overline{MR}$), which provides asynchronous Reset capability to all eight flip-flops.

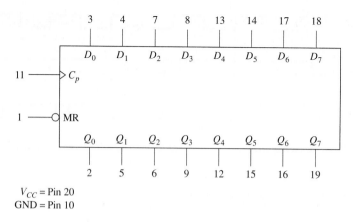

V_{CC} = Pin 20
GND = Pin 10

Figure 10–44   Logic diagram for a 74HCT273 octal D flip-flop.

An application of the '273 octal D flip-flop is shown in Figure 10–45. Here it is used as an *update and hold* register. Every 10 s it receives a clock pulse from the Motorola 68HC11 microcontroller. The data that are on D_0–D_7 at each positive clock edge are stored in the register and output at Q_0–Q_7.

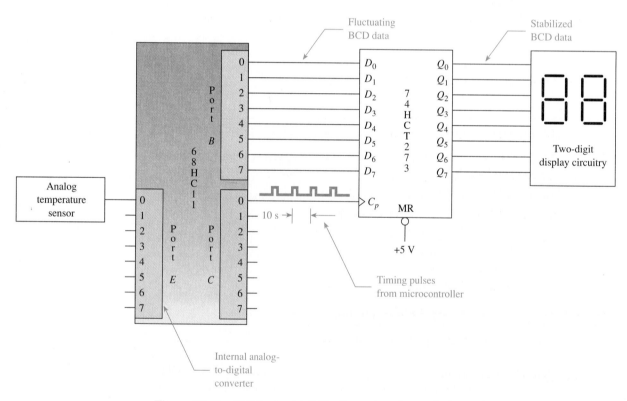

Figure 10–45   Using an octal D flip-flop to interface a display to a microcontroller.

The analog temperature sensor is designed to output a voltage that is proportional to degrees centigrade. (See Section 15–12 for the design of temperature sensors.) The 68HC11 microcontroller has the capability to read analog voltages and convert them into their equivalent digital value. A software program is written for the microcontroller to translate this digital string into a meaningful two-digit BCD output for the display.

The BCD output of the 68HC11 is constantly changing as the temperature fluctuates. One way to stabilize this fluctuating data is to use a **storage register** like the 74HCT273. Because the '273 only accepts the BCD data every 10 s, it will hold the two-digit display constant for that length of time, making it easier to read.

Summary

In this chapter we have learned that

1. The *S-R* flip-flop is a single-bit data storage circuit that can be constructed using basic gates.

2. Adding gate enable circuitry to the *S-R* flip-flop makes it *synchronous*. This means that it will operate only under the control of a clock or enable signal.

3. The *D* flip-flop operates similar to the *S-R,* except it has only a single data input, *D*.

4. The 7475 is an integrated-circuit *D* latch. The output *(Q)* follows *D* while the enable *(E)* is HIGH. When *E* goes LOW, *Q* remains latched.

5. The 7474 is an integrated-circuit *D* flip-flop. It has two *synchronous* inputs, *D* and C_P, and two *asynchronous* inputs, $\overline{S_D}$ and $\overline{R_D}$. *Q* changes to the level of *D* at the *positive edge* of C_p. *Q* responds *immediately* to the asynchronous inputs regardless of the synchronous operations.

6. The *J-K* flip-flop differs from the *S-R* flip-flop because it can also perform a *toggle* operation. Toggling means that *Q* flips to its opposite state.

7. The master–slave *J-K* flip-flop consists of two latches: a *master* that receives data while the clock trigger is HIGH, and a *slave* that receives data from the master and outputs it to *Q* when the clock goes LOW.

8. The 74LS76 is an edge-triggered *J-K* flip-flop IC. It has synchronous and asynchronous inputs. The 7476 is similar, except it is a pulse-triggered master–slave type.

9. The 74HCT273 is an example of an *octal D* flip-flop. It has *eight D* flip-flops in a single IC package, making it ideal for microprocessor applications.

Glossary

Active-LOW: Means that the input to or the output from a terminal must be LOW to be enabled, or "active."

Asynchronous: (Not synchronous.) A condition in which the output of a device will switch states instantaneously as the inputs change without regard to an input clock signal.

Clock: A device used to produce a periodic digital signal that repeatedly switches from LOW to HIGH and back at a predetermined rate.

Combinational Logic: The use of several of the basic gates (AND, OR, NOR, NAND) together to form more complex logic functions.

Complement: Opposite digital state (that is, the complement of 0 is 1, and vice versa).

Digital State: The logic levels within a digital circuit (HIGH level = 1 state and LOW level = 0 state).

Disabled: The condition in which a digital circuit's inputs or outputs are not allowed to accept or transmit digital states.

Edge Triggered: The term given to a digital device that can accept inputs and change outputs only on the positive or negative *edge* of some input control signal or clock.

Enabled: The condition in which a digital circuit's inputs or outputs are allowed to accept or transmit digital states normally.

Flip-Flop: A circuit capable of storing a digital 1 or 0 level based on sequential digital levels input to it.

Function Table: A table that illustrates all the possible combinations of input and output states for a given digital IC or device.

Latch: The ability to *hold* onto a particular digital state. A latch circuit will hold the level of a digital pulse even after the input is removed.

Level Triggered: *See* Pulse triggered.

Master–Slave: A storage device consisting of two sections: the master section, which accepts input data while the clock is HIGH, and the slave section, which receives the data from the master when the clock goes LOW.

Negative Edge: The edge on a clock or trigger pulse that is making the transition from HIGH to LOW.

Noise: Any fluctuations in power supply voltages, switching surges, or electrostatic charges will cause irregularities in the HIGH- and LOW-level voltages of a digital signal. These irregularities or fluctuations in voltage levels are called electrical noise and can cause false readings of digital levels.

Octal: A group of eight. An octal flip-flop IC has eight flip-flops in a single package.

Ones Catching: A feature of the master–slave flip-flop that allows the master section to latch on to any 1 level that is felt at the inputs at any time while the input clock pulse is HIGH and then transfer those levels to the slave when the clock goes LOW.

Positive Edge: The edge on a clock or trigger pulse that is making the transition from LOW to HIGH.

Pulse Triggered: The term given to a digital device that can accept inputs during an entire positive or negative pulse of some input control signal or clock. (Also called level triggered.)

Register: A group of several flip-flops or latches that is used to store a binary string and is controlled by a common clock or enable signal.

Reset: A condition that produces a digital LOW (0) state.

Sequential Logic: Digital circuits that involve the use of a sequence of timing pulses in conjunction with storage devices such as flip-flops and latches and functional ICs such as counters and shift registers.

Set: A condition that produces a digital HIGH (1) state.

Setup Time: The length of time before the active edge of a trigger pulse (control signal) that the inputs of a digital device must be in a stable digital state. [That is, if the setup time of a device is 20 ns, the inputs must be held stable (and will be read) 20 ns before the trigger edge.]

Storage Register: Two or more data storage circuits (such as flip-flops or latches) used in conjunction with each other to hold several bits of information.

Strobe Gates: A control gate used to enable or disable inputs from reaching a particular digital device.

Synchronous: A condition in which the output of a device will operate only in synchronization with (in step with) a specific HIGH or LOW timing pulse or trigger signal.

Toggle: In a flip-flop, a toggle is when Q changes to the level of $\overline{Q}$ and $\overline{Q}$ changes to the level of Q.

Transition: The instant of change in digital state from HIGH to LOW or LOW to HIGH.

Transparent Latch: An asynchronous device whose outputs will hold onto the most recent digital state of the inputs. The outputs immediately follow the state of the inputs without regard to trigger input and remain in that state even after the inputs are removed or disabled.

Trigger: The input control signal to a digital device that is used to specify the instant that the device is to accept inputs or change outputs.

Problems

Section 10–1

10–1. Make the necessary connections to the 7400 quad NAND gate IC in Figure P10–1 to form the cross-NAND S-R flip-flop of Figure 10–2. [Remember that an inverter can be formed from a NAND (Chapter 4).]

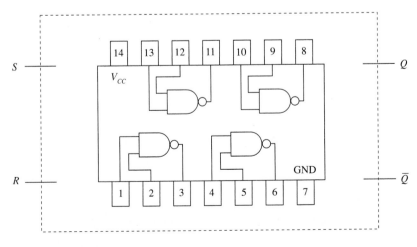

Figure P10–1

10–2. Sketch the Q output waveform for a gated S-R flip-flop (Figure 10–8), given the inputs at S, R, and G shown in Figure P10–2.

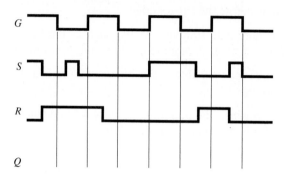

Figure P10–2

10–3. Repeat Problem 10–2 for the input waves shown in Figure P10–3.

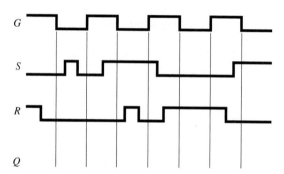

Figure P10–3

10–4. Repeat Problem 10–2 for the input waves shown in Figure P10–4.

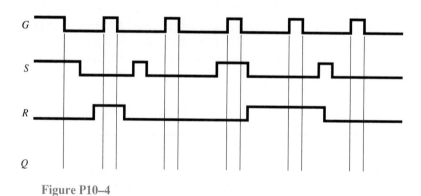

Figure P10–4

Section 10–3

D

10–5. Referring to Figures 10–8 and 10–12, sketch the logic diagram using NORs, ANDs, and inverters that will function as a gated D flip-flop.

10–6. How many integrated-circuit chips will be required to build the gated D flip-flop that you sketched in Problem 10–5?

10–7. Make the necessary connections to a 7402 quad NOR and a 7408 quad AND to form the gated D flip-flop of Problem 10–5.

10–8. Sketch the Q output waveform for the gated D flip-flop of Figure 10–12 given the D and G inputs shown in Figure P10–8.

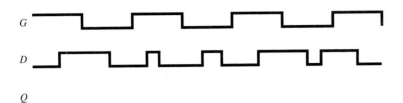

Q

Figure P10–8

10–9. Repeat Problem 10–8 for the G and D inputs shown in Figure P10–9.

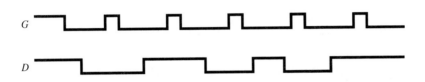

Q

Figure P10–9

Section 10–4

10–10. The logic symbol for one-fourth of a 7475 transparent D latch is given in Figure P10–10. Sketch the Q output waveform given the inputs at E and D.

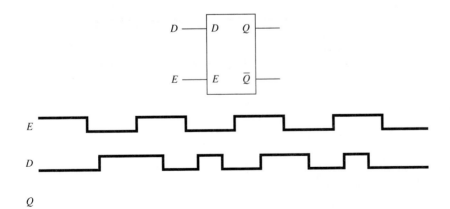

Q

Figure P10–10

10–11. Repeat Problem 10–10 for the waveforms at E and D shown in Figure P10–11.

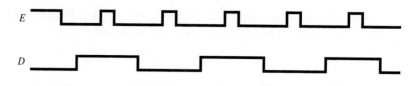

E

D

Q

Figure P10–11

10–12. Explain why the 7475 is called *transparent* and why it is called a *latch*.

10–13. The 7475 is transparent while the E input is _____ (LOW or HIGH) and it is latched while E is _____ (LOW or HIGH).

Section 10–5

10–14 **(a)** What are the asynchronous inputs to the 7474 D flip-flop?

(b) What are the synchronous inputs to the 7474 D flip-flop?

10–15. The logic symbol for one-half of a 7474 dual D flip-flop is given in Figure P10–15(a). Sketch the Q output wave given the inputs at C_p, D, $\overline{S_D}$, and $\overline{R_D}$ shown in Figure P10–15(b).

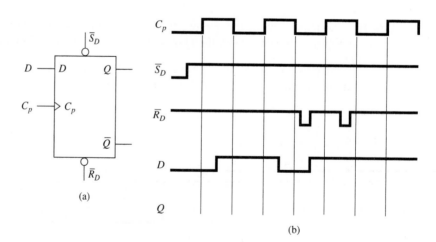

Figure P10–15

10–16. Repeat Problem 10–15 for the input waves shown in Figure P10–16.

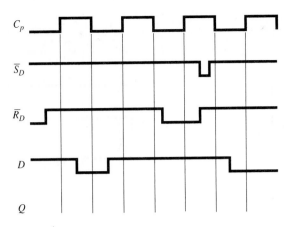

Figure P10–16

10–17. Describe several differences between the 7474 *D* flip-flop and the 7475 *D* latch.

10–18. Describe the differences between the asynchronous inputs and the synchronous inputs of the 7474.

10–19. What does the small triangle on the C_p line of the 7474 indicate?

10–20. To disable the asynchronous inputs to the 7474, should they be connected to a HIGH or a LOW?

D   **10–21.** Using the *universal gate* capability of a NAND gate covered in Section 5–5, redesign Figure 10–18 using only one 74HCT00 NAND IC.

D   **10–22.** Design a circuit similar to Figure 10–18 that can be used as a *negative* edge detector instead of a positive edge detector.

Sections 10–6, 10–7, and 10–8

10–23. What is the one additional synchronous operating mode that the *J-K* flip-flop has that the *S-R* flip-flop does not have?

10–24. What are the asynchronous inputs to the 7476 *J-K* flip-flop? Are they active LOW or active HIGH?

10–25. The 7476 is called a *pulse-triggered master–slave* flip-flop, whereas the 74LS76 is called an *edge-triggered* flip-flop. Describe the differences between them.

C   **10–26.** The logic symbol and input waveforms for both the 7476 and 74LS76 are given in Figure P10–26. Sketch the waveform at each *Q* output.

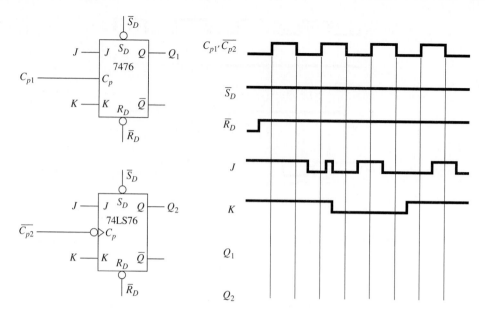

Figure P10–26

C   **10–27.** Repeat Problem 10–26 for the input waveforms shown in Figure P10–27.

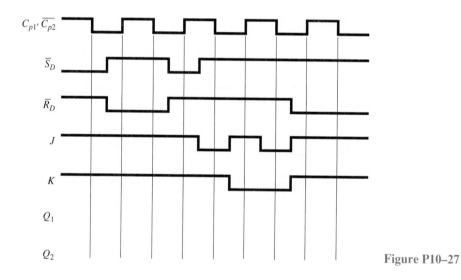

Figure P10–27

10–28. Sketch the output waveform at Q for Figure P10–28.

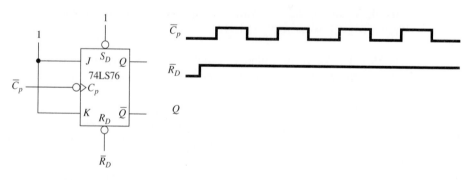

Figure P10–28

10–29. Sketch the output waveform at Q for Figure P10–29.

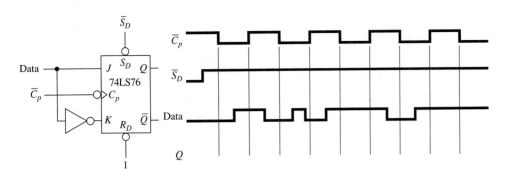

Figure P10–29

C   **10–30.** Sketch the output waveform at Q for Figure P10–30.

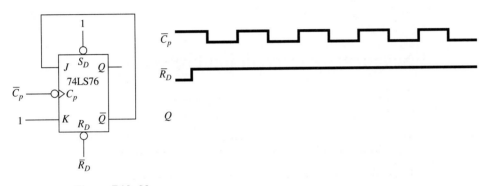

Figure P10–30

C   **10–31.** Sketch the output waveform at Q for Figure P10–31.

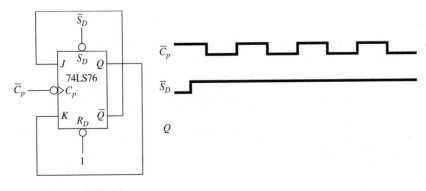

Figure P10–31

C   **10–32.** Sketch the output waveform at Q for Figure P10–32.

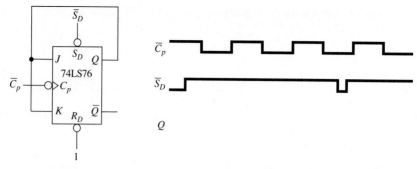

Figure P10–32

Section 10–9

C D   **10–33.** The 74HCT373 (or 74LS373) is an octal transparent latch. Refer to a data book (CMOS or TTL) to review its operation. Discuss why it can or cannot be used to replace the 273 in Figure 10–45.

C T   **10–34.** A designer decides to change the timing pulse increment in Figure 10–45 from 10 s to 10 ms. When she does, the least significant digit always displays the number **8**. Explain why.

Schematic Interpretation Problems

See Appendix G for the schematic diagrams.

S C   **10–35.** Find U1:A of the Watchdog Timer schematic. Assume that initially, WATCHDOG_EN = LOW and /CPU_RESET is pulsed LOW.

       (a) What is the output level of U2:A?

       (b) When WATCHDOG_EN goes HIGH does the output of U2:A go LOW?

       (c) What must happen to U1:A to make the output of U2:A go LOW?

S   **10–36.** In the Watchdog Timer schematic, both U14 flip-flops are Reset when there is a LOW /CPU_RESET _____ (and, or) a LOW $\overline{Q}$ from U14:B.

S   **10–37.** After being Reset, U14:A will be Set as soon as _____.

S   **10–38.** U5 and U6 are octal D flip-flops in the Watchdog Timer schematic. They provide two stages of latching for the 8-bit data bus labeled D(7:0).

       (a) How are they initially Reset? (*Hint:* CLR is the abbreviation for CLEAR, which is the same as Master Reset.)

       (b) What has to happen for the Q-outputs of U5 to receive the value of the data bus?

       (c) What has to happen for the Q-outputs of U6 to receive the value of the U5 outputs?

Answers to Review Questions

10–1. True

10–2. $S = 1, R = 0$

10–3. None

10–4. An S-R flip-flop is asynchronous because the output responds immediately to input changes. The gated S-R is synchronous because it operates sequentially with the control input at the gate.

10–5. False

10–6. $D = 0, G = 1$

10–7. 4

10–8. HIGH

10–9. True

10–10. Transitions at Q will occur only at the edge of the input trigger pulse.

10–11. C_p and D are synchronous. $\overline{S_D}$ and $\overline{R_D}$ are asynchronous.

10–12. False

10–13. Invert, delay

10–14. Because the master will latch onto any HIGH inputs while the clock pulse is HIGH and transfer them to the slave when the clock goes LOW.

10–15. J, K

10–16. True

10–17. It switches the Q and $\overline{Q}$ outputs to their opposite states.

10–18. Apply a LOW on the $\overline{R_D}$ input.

10–19. False

10–20. HIGH

10–21. The uppercase letters mean steady state. The lowercase letters are used for levels one setup time prior to the negative clock edge.

11 Practical Considerations for Digital Design

OUTLINE

11–1 Flip-Flop Time Parameters
11–2 Automatic Reset
11–3 Schmitt Trigger ICs
11–4 Switch Debouncing
11–5 Sizing Pull-Up Resistors
11–6 Practical Input and Output Considerations

Objectives

Upon completion of this chapter, you should be able to

- Describe the causes and effects of a race condition on synchronous flip-flop operation.

- Use manufacturers' data sheets to determine IC operating specifications such as setup time, hold time, propagation delay, and input/output voltage and current specifications.

- Perform worst-case analysis on the time-dependent operations of flip-flops and sequential circuitry.

- Design a series *RC* circuit to provide an automatic power-up reset function.

- Describe the wave-shaping capability and operating characteristics of Schmitt trigger ICs.

- Describe the problems caused by switch bounce and how to eliminate its effects.

- Calculate the optimum size for a pull-up resistor.

Introduction

We now have the major building blocks required to form sequential circuits. There are a few practical time and voltage considerations that we have to deal with first before we connect ICs to form sequential logic.

For instance, ideally a 74LS76 flip-flop switches on the negative edge of the input clock, but actually it could take the output as long as 30 ns to switch. Thirty nanoseconds (30×10^{-9} s) does not sound like much, but when you cascade several flip-flops end to end or any time you have combinational logic with flip-flops that rely on a high degree of accurate timing, the IC delay times could cause serious design problems.

Digital ICs have to keep up with the high speed of the microprocessors used in modern computer systems. For example, a microprocessor operating at a clock frequency of 20 MHz will have a clock period of 50 ns. The slower IC families that have switching speeds of 20 to 30 ns would create timing problems that would lead to misinterpretation of digital levels in a system operating that fast. To ensure reliability in high-speed systems, designers must consider the worst-case timing scenario. They cannot simply play it safe and build in a large margin of error because that would mean that a slower operating system would be developed, which would not compete favorably in the modern marketplace.

In this chapter we look at the *actual* operating characteristics of digital ICs as they relate to output delay times, input setup requirements, and input/output voltage and current levels. With a good knowledge of the practical aspects of digital ICs, we then develop the external circuitry needed to interface to digital logic.

11–1   Flip-Flop Time Parameters

There are several time parameters listed in IC manufacturers' data manuals that require careful analysis. For example, let's look at Figure 11–1, which uses a 74LS76 flip-flop with the J and $\overline{C_p}$ inputs brought in from some external circuit.

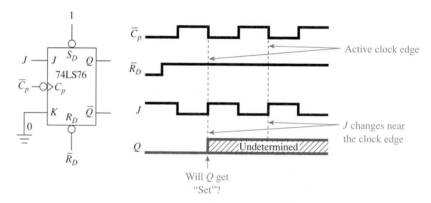

Figure 11–1   A possible race condition on a *J-K* flip-flop creates an undetermined result at Q.

The waveform shown for J and $\overline{C_p}$ will create a **race condition.** *Race* is the term used when the inputs to a triggerable device (like a flip-flop) are changing at the same time that the active trigger edge of the input clock is making its transition. In the case of Figure 11–1, the J waveform is changing from LOW to HIGH exactly at the negative edge of the clock; so what is J at the negative edge of the clock, LOW or HIGH?

Now when you look at Figure 11–1, you should ask the question, Will Q ever get Set? Remember from Chapter 10 that J must be HIGH at the negative edge of $\overline{C_p}$ in order to set the flip-flop. Actually, J must be HIGH one *setup time* prior to the negative edge of the clock.

The **setup time** is the length of time prior to the active clock edge that the flip-flop looks back to determine the levels to use at the inputs. In other words, for Figure 11–1, the flip-flop will look back one setup time prior to the negative clock edge to determine the levels at J and K.

The setup time for the 74LS76 is 20 ns, so we must ask, Were J and K HIGH or LOW 20 ns prior to the negative clock edge? Well, K is tied to ground, so it was LOW, and depending on when J changed from LOW to HIGH, the flip-flop may have Set ($J = 1, K = 0$) or Held ($J = 0, K = 0$).

In a data manual, the manufacturer will provide **ac waveforms** that illustrate the measuring points for all the various time parameters. The illustration for setup time will look something like Figure 11–2.

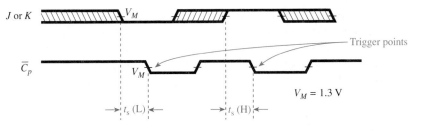

The shaded areas indicate when the input is
permitted to change for predictable output
performance

Figure 11–2   Setup time waveform specifications for a 74LS76.

The active transition (trigger point) of the $\overline{C_p}$ input (clock) occurs when $\overline{C_p}$ goes from above to below the 1.3-V level.

Setup time (LOW), t_s (L), is given as 20 ns. This means that J and K can be changing states 21 ns or more before the active transition of $\overline{C_p}$; but in order to be interpreted as a LOW, they must be 1.3 V *or less* at 20 ns *before* the active transition of $\overline{C_p}$.

Setup time (HIGH), t_s (H), is also given as 20 ns. This means that J and K can be changing states 21 ns or more before the active edge of $\overline{C_p}$; but to be interpreted as a HIGH, they must be 1.3 V *or more* at 20 ns *before* the active transition of $\overline{C_p}$.

Did you follow all that? If not, go back and read it again! Sometimes, material like this has to be read over and over again, carefully, to be fully understood.

Not only does the input have to be set up some definite time *before* the clock edge, but it also has to be *held* for a definite time after the clock edge. This time is called the **hold time** [t_h (L) and t_h (H)].

The hold time for the 74LS76 (and most other flip-flops) is given as 0 ns. This means that the desired levels at J and K must be held 0 ns *after* the **active clock edge.** In other words, the levels do not have to be held beyond the active clock edge for most flip-flops. In the case of the 74LS76, the desired level for J and K must be present from 20 ns before the negative clock edge to 0 ns after the clock edge.

For example, for a 74LS76 to have a LOW level on J and K, the waveforms in Figure 11–3 illustrate the *minimum* setup and hold times allowed to still have the LOW reliably interpreted as a LOW. Figure 11–3 shows us that J and K are allowed to change states any time greater than 20 ns before the negative clock edge, and because the hold time is zero, they are permitted to change immediately after the negative clock edge.

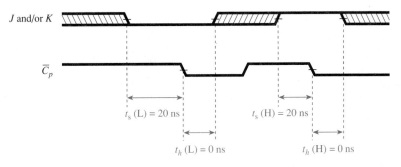

Figure 11–3   Setup and hold parameters for a 74LS76 flip-flop.

Do you notice in Examples 11–1 and 11–2 that the Q output changes *exactly* on the negative clock edge? Do you really think that it will? It won't! Electrical charges that build up inside any digital logic circuit won't allow it to change states instantaneously as the inputs change. This delay from input to output is called **propagation delay.** There are propagation delays from the synchronous inputs to the output and also from the asynchronous inputs to the output.

EXAMPLE 11–1

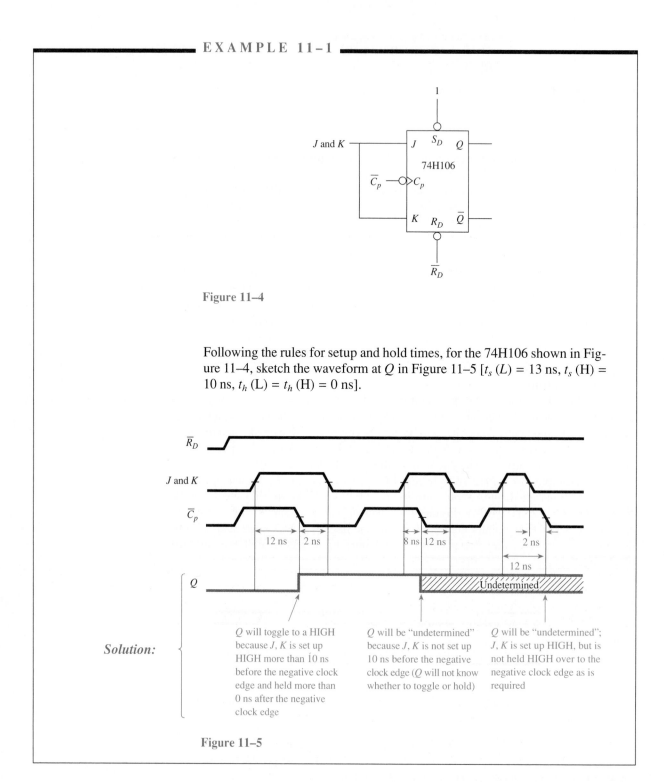

Figure 11–4

Following the rules for setup and hold times, for the 74H106 shown in Figure 11–4, sketch the waveform at Q in Figure 11–5 [t_s (L) = 13 ns, t_s (H) = 10 ns, t_h (L) = t_h (H) = 0 ns].

Solution:

Q will toggle to a HIGH because *J*, *K* is set up HIGH more than 10 ns before the negative clock edge and held more than 0 ns after the negative clock edge

Q will be "undetermined" because *J*, *K* is not set up 10 ns before the negative clock edge (*Q* will not know whether to toggle or hold)

Q will be "undetermined"; *J*, *K* is set up HIGH, but is not held HIGH over to the negative clock edge as is required

Figure 11–5

EXAMPLE 11-2

Sketch the Q output for the 74H106 shown in Figure 11-6, with the input waveforms given in Figure 11-7 [t_s (L) = 13 ns, t_s (H) = 10 ns, t_h (L) = t_h (H) = 0 ns].

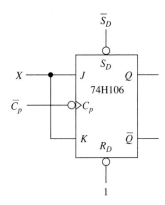

Figure 11-6

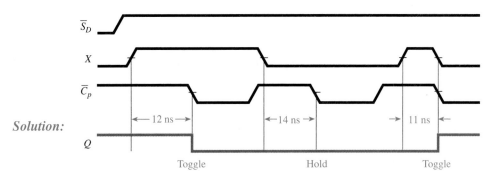

Solution:

Figure 11-7

For example, there is a propagation delay period from the instant that $\overline{R_D}$ or $\overline{S_D}$ goes LOW until the Q output responds accordingly. The data manual shows a *maximum* propagation delay for $\overline{S_D}$ to Q of 20 ns and for $\overline{R_D}$ to Q of 30 ns. Since a LOW on $\overline{S_D}$ causes Q to go *LOW to HIGH*, the propagation delay is abbreviated t_{PLH}. A LOW on $\overline{R_D}$ causes Q to go *HIGH to LOW*; therefore, use t_{PHL} for that propagation delay, as illustrated in Figure 11-8.

The propagation delay from the clock trigger point to the Q output is also called t_{PLH} or t_{PHL}, depending on whether the Q output is going LOW to HIGH or HIGH to LOW. For the 74LS76 clock to output, t_{PLH} = 20 ns and t_{PHL} = 30 ns. Figure 11-9 illustrates the synchronous propagation delays.

Besides setup, hold, and propagation delay times, the manufacturer's data manual will also give

1. Maximum frequency (f_{max}), the maximum frequency allowed at the clock input. Any frequency above this limit will yield unpredictable results.

2. Clock pulse width (LOW) [t_w(L)], the minimum width (in nanoseconds) that is allowed at the clock input during the LOW level for reliable operation.

Some students mistakenly think that the propagation delays in this figure should be additive, yielding a total delay of 80 ns.

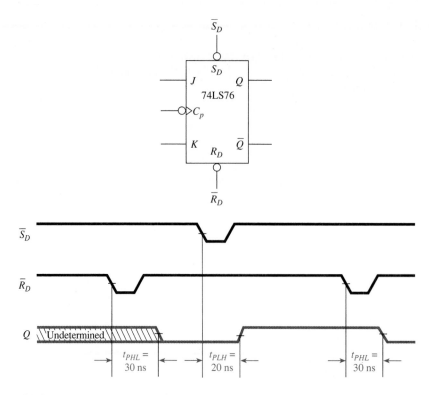

Figure 11–8 Propagation delay for the asynchronous input to Q output for the 74LS76.

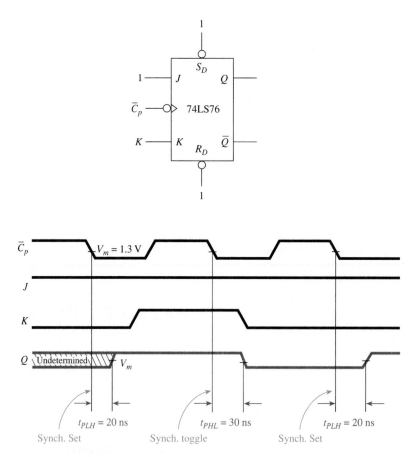

Figure 11–9 Propagation delay for the clock to output of the 74LS76.

CHAPTER 11 | PRACTICAL CONSIDERATIONS FOR DIGITAL DESIGN

3. Clock pulse width (HIGH) [t_w(H)], the minimum width (in nanoseconds) that is allowed at the clock input during the HIGH level for reliable operation.

4. Set or Reset pulse width (LOW) [t_w(L)], the minimum width (in nanoseconds) of the LOW pulse at the Set ($\overline{S_D}$) or Reset ($\overline{R_D}$) inputs.

Figure 11–10 shows the measurement points for these specifications.

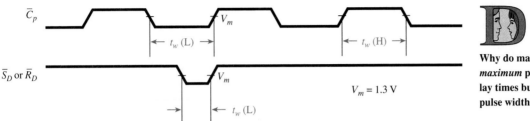

Figure 11–10 Minimum pulse-width specifications.

Complete specifications for the 7476/74LS76 flip-flop are given in Figure 11–11. Can you locate all the specifications that we have discussed so far? If you have your own data manual, look at some of the other flip-flops and see how they compare. In the front of the manual you will find a section that describes all the IC specifications and abbreviations used throughout the data manual.

Now that we understand most of the operating characteristics of digital ICs, let's examine why they are so important and what implications they have on our design of digital circuits.

To get the flip-flop in Example 11–3 on page 360 to toggle, we have to move C_p to the right by at least 10 ns so that J is HIGH 10 ns *before* the positive edge of C_p. By saying "move it to the right," we mean "delay it by at least 10 ns."

One common way to introduce delay is to insert one or more IC gates in the C_p line, as shown in Figure 11–14, so that their combined propagation delay is greater than 10 ns. From the manufacturer's specifications, we can see that the propagation delay for a 7432 input to output is $t_{PHL} = 22$ ns max. and $t_{PLH} = 15$ ns max. [typically, the propagation delay will be slightly less than the maximum (worst case) rating].

Now let's redraw the waveforms as shown in Figure 11–15 with the delayed clock to see if the flip-flop will toggle. The 7432 will delay the LOW-to-HIGH edge of the clock by approximately 15 ns, so J *will* be HIGH one setup time prior to the trigger point on C_p; thus Q *will* toggle.

An important point to be made here is that in Figure 11–14 we are relying on the propagation delay of the 7432 to be 15 ns, which according to the manufacturer is the worst case (maximum) for the 7432. What happens if the *actual* propagation delay is less than 15 ns, let's say only 8 ns? The clock (C_p) would not be delayed far enough to the right for J to be set up in time.

Special *delay-gate* ICs are available that provide exact, predefined delays specifically for the purpose of delaying a particular signal to enable proper time relationships. One such delay gate is shown in Figure 11–16. To use this delay gate, you would connect the signal that you want delayed to the C_p input terminal. You then select the delayed output waveform that suits your needs. The output waveforms are identical to the input except delayed by 5, 10, 15, or 20 ns. Complemented, delayed waveforms are also available at the $\overline{5}$, $\overline{10}$, $\overline{15}$, and $\overline{20}$ outputs. Delay gates with various other delay intervals are also available.

Team Discussion

Why do manufacturers give *maximum* propagation delay times but *minimum* pulse widths?

Helpful Hint

It is very important that you learn how to interpret a data manual. New devices and ICs are constantly being introduced. Often, the only way to learn their operation and features is to study a data sheet.

FLIP-FLOPS

54/7476, LS76

Dual J-K Flip-Flop

Describes the operation and function of the chip.

DESCRIPTION

The '76 is a dual J-K flip-flop with individual J, K, Clock, Set, and Reset inputs. The 7476 is positive pulse-triggered. JK information is loaded into the master while the Clock is HIGH and transferred to the slave on the HIGH-to-LOW Clock transition. The J and K inputs must be stable while the Clock is HIGH for conventional operation.

The 74LS76 is a negative edge-triggered flip-flop. The J and K inputs must be stable only one setup time prior to the HIGH-to-LOW Clock transition.

The Set ($\overline{S}_D$) and Reset ($\overline{R}_D$) are asynchronous active LOW inputs. When LOW, they override the Clock and Data inputs, forcing the outputs to the steady state levels as shown in the Function Table.

TYPE	TYPICAL f$_{MAX}$	TYPICAL SUPPLY CURRENT (Total)
7476	20MHz	10mA
74LS76	45MHz	4mA

Gives part numbers for various package types, V_{CC}, and temperature ranges.

ORDERING CODE

PACKAGES	COMMERCIAL RANGES $V_{CC} = 5V \pm 5\%$; $T_A = 0°C$ to $+70°C$		MILITARY RANGES $V_{CC} = 5V \pm 10\%$; $T_A = -55°C$ to $+125°C$	
Plastic DIP	N7476N	• N74LS76N		
Ceramic Dip			S5476F	• S54LS76F
Flatpack			S5476W	• S54LS76W

2 unit loads means that these inputs draw 2 times the I_I ratings in this family.

INPUT AND OUTPUT LOADING AND FAN-OUT TABLE

PINS	DESCRIPTION	54/74	54/74LS
$\overline{CP}$	Clock input	2ul	2LSul
$\overline{R}_D, \overline{S}_D$	Reset and Set inputs	2ul	2LSul
J, K	Data inputs	1ul	1LSul
Q, $\overline{Q}$	Outputs	10ul	10LSul

NOTE
Where a 54/74 unit load (ul) is understood to be 40µA I_{IH} and -1.6 mA I_{IL} and a 54/74LS unit load (LSul) is 20µA I_{IH} and -0.4mA I_{IL}.

10 unit loads means that these outputs can supply 10 times the amount of current required for a single unit input load (or fan-out = 10).

PIN CONFIGURATION

Notice the overbars indicating active-LOW terminals.

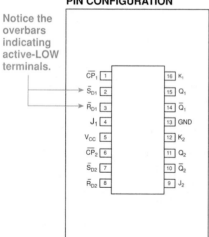

LOGIC SYMBOL

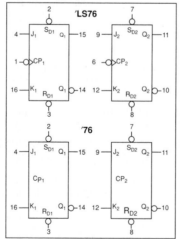

LOGIC SYMBOL (IEEE/IEC)

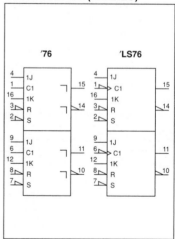

Note: In this figure, the abbreviation for Clock is $\overline{CP}$ instead of $\overline{C_p}$.

Figure 11–11 Typical data sheet for a 7476/74LS76. (Courtesy of Signetics Corporation.)

FLIP-FLOPS

54/7476, LS76

Gives the input requirements and output result for each operating mode.

LOGIC DIAGRAM

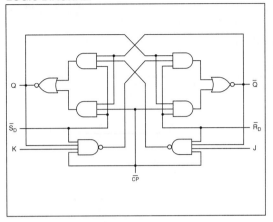

FUNCTION TABLE

OPERATING MODE	INPUTS					OUTPUTS	
	$\overline{S}_D$	$\overline{R}_D$	$\overline{CP}^{(b)}$	J	K	Q	$\overline{Q}$
Asynchronous Set	L	H	X	X	X	H	L
Asynchronous Reset (Clear)	H	L	X	X	X	L	H
Undetermined[a]	L	L	X	X	X	H	H
Toggle	H	H	⌐⌐	h	h	$\overline{q}$	q
Load "0" (Reset)	H	H	⌐⌐	l	h	L	H
Load "1" (Set)	H	H	⌐⌐	h	l	H	L
Hold "no change"	H	H	⌐⌐	l	l	q	$\overline{q}$

H = HIGH voltage level steady state.

h = HIGH voltage level one setup time prior to the HIGH-to-LOW Clock transition.[C]

L = LOW voltage level steady state.

l = LOW voltage level one setup time prior to the HIGH-to-LOW Clock transition.[C]

q = Lowercase letters indicate the state of the referenced output prior to the HIGH-to-LOW Clock transition.

X = Don't care.

⌐⌐ = Positive Clock pulse.

NOTES

a. Both outputs will be HIGH while both $\overline{S}_D$ and $\overline{R}_D$ are LOW, but the output states are unpredictable if $\overline{S}_D$ and $\overline{R}_D$ go HIGH simultaneously.

b. The 74LS76 is edge triggered. Data must be stable one setup time prior to the negative edge of the Clock for predictable operation.

c. The J and K inputs of the 7476 must be stable while the Clock is HIGH for conventional operation.

ABSOLUTE MAXIMUM RATINGS (Over operating free-air temperature range unless otherwise noted.)

	PARAMETER	54	54LS	74	74LS	UNIT
V_{CC}	Supply voltage	7.0	7.0	7.0	7.0	V
V_{IN}	Input voltage	− 0.5 to + 5.5	− 0.5 to + 7.0	− 0.5 to + 5.5	− 0.5 to + 7.0	V
I_{IN}	Input current	− 30 to + 5	− 30 to + 1	− 30 to + 5	− 30 to + 1	mA
V_{OUT}	Voltage applied to output in HIGH ouput state	− 0.5 to + V_{CC}	− 0.5 to + V_{CC}	− 0.5 to + V_{CC}	− 0.5 to + V_{CC}	V
T_A	Operating free-air temperature range	− 55 to + 125		0 to 70		°C

RECOMMENDED OPERATING CONDITIONS

HIGH/LOW input voltage requirements.

Maximum output currents.

	PARAMETER		54/74			54/74LS			UNIT
			Min	Nom	Max	Min	Nom	Max	
V_{CC}	Supply voltage	Mil	4.5	5.0	5.5	4.5	5.0	5.5	V
		Com'l	4.75	5.0	5.25	4.75	5.0	5.25	V
V_{IH}	HIGH-level input voltage		2.0			2.0			V
V_{IL}	LOW-level input voltage	Mil			+ 0.8			+ 0.7	V
		Com'l			+ 0.8			+ 0.8	V
I_{IK}	Input clamp current				− 12			− 18	mA
I_{OH}	HIGH-level output current				− 400			− 400	µA
I_{OL}	LOW-level output current	Mil			16			4	mA
		Com'l			16			8	mA
T_A	Operating free-air temperature	Mil	− 55		+ 125	− 55		+ 125	°C
		Com'l	0		70	0		70	°C

Figure 11–11 *(Continued)*

FLIP-FLOPS

DC ELECTRICAL CHARACTERISTICS (Over recommended operating free-air temperature range unless otherwise noted.)

Typical and worst-case output voltages.

HIGH/LOW input current requirements.

PARAMETER		TEST CONDITIONS[1]		54/7476 Min	54/7476 Typ[2]	54/7476 Max	54/74LS76 Min	54/74LS76 Typ[2]	54/74LS76 Max	UNIT
V_{OH}	HIGH-level output voltage	V_{CC} = MIN, V_{IH} = MIN, V_{IL} = MAX, I_{OH} = MAX	Mil	2.4	3.4		2.5	3.4		V
			Com'l	2.4	3.4		2.7	3.4		V
V_{OL}	LOW-level output voltage	V_{CC} = MIN, V_{IL} = MAX, V_{IH} = MIN I_{OL} = MAX	Mil		0.2	0.4		0.25	0.4	V
			Com'l		0.2	0.4		0.35	0.5	V
		I_{OL} = 4mA	74LS					0.25	0.4	V
V_{IK}	Input clamp voltage	V_{CC} = MIN, I_I = I_{IK}				− 1.5			− 1.5	V
I_I	Input current at maximum input voltage	V_{CC} = MAX	V_I = 5.5 V			1.0				mA
			V_I = 7.0V J, K Inputs						0.1	mA
			$\overline{S}_D, \overline{R}_D$ Inputs						0.3	mA
			$\overline{CP}$ Inputs						0.4	mA
I_{IH}	HIGH-level input current	V_{CC} = MAX	V_I = 2.4V J, K Inputs			40				μA
			$\overline{S}_D, \overline{R}_D$ Inputs			80				μA
			$\overline{CP}$ Inputs			80				μA
			V_I = 2.7V J, K Inputs						20	μA
			$\overline{S}_D, \overline{R}_D$ Inputs						60	μA
			$\overline{CP}$ Inputs						80	μA
I_{IL}	LOW-level input current[5]	V_{CC} = MAX, V_I = 0.4V	J, K Inputs			− 1.6			− 0.4	mA
			$\overline{S}_D, \overline{R}_D$ Inputs			− 3.2			− 0.8	mA
			$\overline{CP}$ Inputs			− 3.2			− 0.8	mA
I_{OS}	Short-circuit output current[3]	V_{CC} = MAX	Mil	− 20		− 57	− 20		− 100	mA
			Com'l	− 18		− 57	− 20		− 100	mA
I_{CC}	Supply current[4] (total)	V_{CC} = MAX			10	40		4	8	mA

NOTES
1. For conditions shown as MIN or MAX, use the appropriate value specified under recommended operating conditions for the applicable type.
2. All typical values are at V_{CC} = 5V, T_A = 25° C.
3. I_{OS} is tested with V_{OUT} = + 0.5V and V_{CC} = V_{CC} MAX + 0.5V. Not more than one output should be shorted at a time and duration of the short circuit should not exceed one second.
4. With the Clock input grounded and all outputs open, I_{CC} is measured with the Q and $\overline{Q}$ outputs HIGH in turn.
5. $\overline{S}_D$ is tested with $\overline{R}_D$ HIGH, and $\overline{R}_D$ is tested with $\overline{S}_D$ HIGH.

Waveforms 1, 2, and 3 on the next page show the measurement points for frequency and propagation delay values.

AC CHARACTERISTICS T_A = 25° C, V_{CC} = 5.0V

PARAMETER		TEST CONDITIONS	54/74 C_L = 15pF, R_L = 400Ω Min	54/74 C_L = 15pF, R_L = 400Ω Max	54/74LS C_L = 15pF, R_L = 2kΩ Min	54/74LS C_L = 15pF, R_L = 2kΩ Max	UNIT
f_{MAX}	Maximum Clock frequency	Waveform 3	15		30		MHz
t_{PLH} t_{PHL}	Propagation delay Clock to output	Waveform 1, 'LS76 Waveform 3, '76		25 40		20 30	ns
t_{PLH} t_{PHL}	Propagation delay $\overline{S}_D$ or $\overline{R}_D$ to output	Waveform 2		25 40		20 30	ns

NOTE
Per industry convention, f_{MAX} is the worst case value of the maixmum device operating frequency with no constraints on t_r, t_f, pulse width, or duty cycle.

Figure 11–11 *(Continued)*

CHAPTER 11 ｜ PRACTICAL CONSIDERATIONS FOR DIGITAL DESIGN

FLIP-FLOPS

AC SETUP REQUIREMENTS $T_A = 25°$ C, $V_{CC} = 5.0V$

Minimum pulse widths

PARAMETER		TEST CONDITIONS	54/74		54/74LS		UNIT
			Min	Max	Min	Max	
$t_W(H)$	Clock pulse width (HIGH)	Waveform 1	20		20		ns
$t_W(L)$	Clock pulse width (LOW)	Waveform 1	47				ns
$t_W(L)$	Reset pulse width (LOW)	Waveform 2	25		25		ns
t_s	Setup time J or K to Clock[C]	Waveform 1	0		20		ns
t_h	Hold time J or K to Clock	Waveform 1	0		0		ns

Setup and hold times

AC WAVEFORMS

These waveforms define time measurement points.

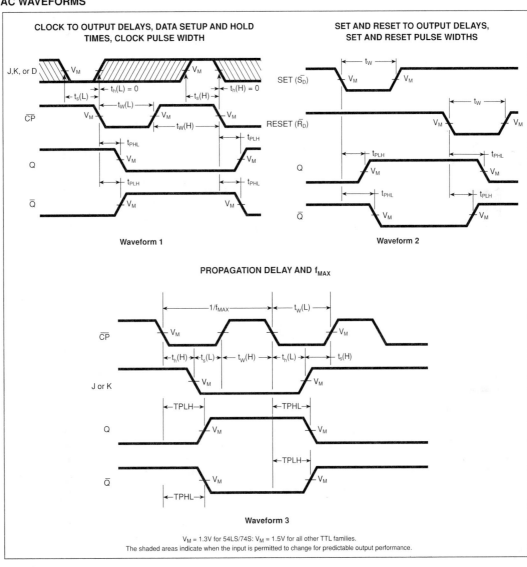

$V_M = 1.3V$ for 54LS/74S; $V_M = 1.5V$ for all other TTL families.
The shaded areas indicate when the input is permitted to change for predictable output performance.

Figure 11–11 *(Continued)*

The 74109 is a positive edge-triggered J-$\overline{K}$ flip-flop. If we attach the J input to the clock as shown in Figure 11–12, will the flip-flop's Q output toggle?

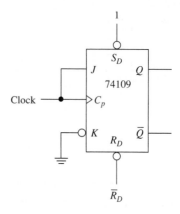

Figure 11–12

Solution: t_s for a 74109 is 10 ns, which means that J must be HIGH and $\overline{K}$ must be LOW 10 ns *before* the positive clock edge. When we draw the waveforms as shown in Figure 11–13, we see that J and C_p are exactly the same. Therefore, because J is not HIGH one setup time prior to the positive clock edge, Q will not toggle.

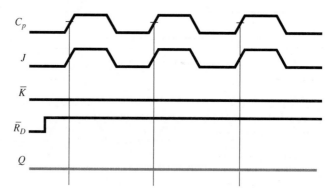

Figure 11–13

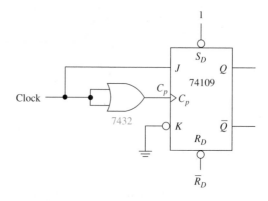

Figure 11–14   Modification of flip-flop in Example 11–3 to allow it to toggle.

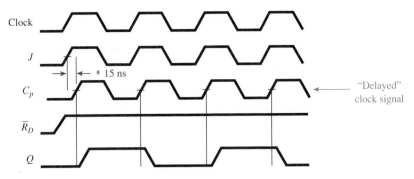

Team Discussion

If manufacturers provided a minimum *and* a maximum propagation delay, which would you use?

Figure 11–15 Timing waveforms for Figure 11–14.

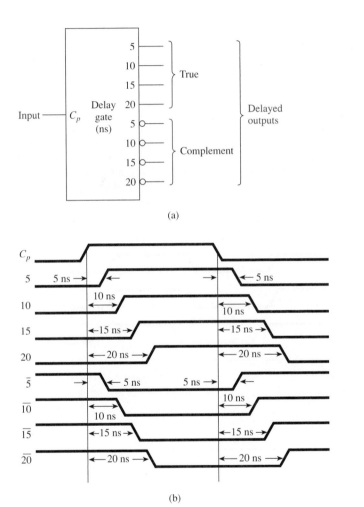

(a)

Helpful Hint

The circuitry inside a typical delay gate is given in Figure 16–10(a).

(b)

Figure 11–16 A 5-ns multitap delay gate: (a) logic symbol; (b) output waveforms.

EXAMPLE 11–4

Use the setup, hold, and propagation delay times from a data manual to determine if the 74109 J-$\overline{K}$ flip-flop in the circuit shown in Figure 11–17 will toggle. (Assume that the flip-flop is initially Reset, and remember that for a toggle, $J = 1$, $\overline{K} = 0$.)

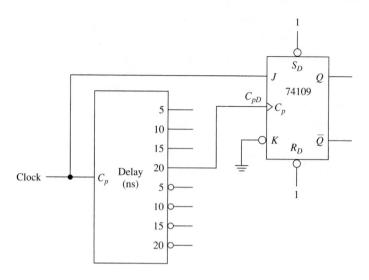

Figure 11–17

Solution: First, draw the waveforms as shown in Figure 11–18. C_{pD}, the delayed clock, makes its LOW-to-HIGH transition and triggers the flip-flop 20 ns *after* J makes its transition. Looking at the waveforms, J *is* HIGH ≥ 10 ns *before* the positive edge of C_{pD} and *is* held HIGH ≥ 6 ns *after* the positive edge of C_{pD}. Therefore, *the flip-flop will toggle* at each positive edge of C_{pD}.

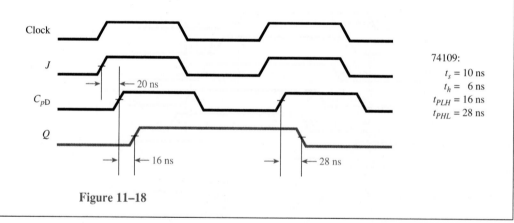

74109:

$t_s = 10$ ns
$t_h = 6$ ns
$t_{PLH} = 16$ ns
$t_{PHL} = 28$ ns

Figure 11–18

EXAMPLE 11–5

Use the specifications from a data manual to determine if the 74LS112 *J-K* flip-flop in the circuit shown in Figure 11–19 will toggle. (Assume that the flip-flop is initially Reset.)

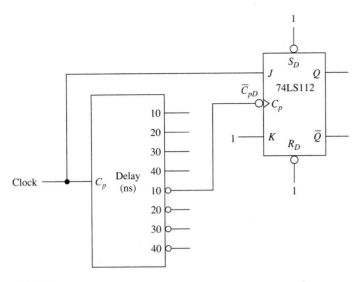

Figure 11–19

Solution: First, draw the waveforms as shown in Figure 11–20. $\overline{C_{pD}}$ is inverted and delayed by 10 ns from the Clock and *J* waveforms. Each negative edge of $\overline{C_{pD}}$ triggers the flip-flop. Looking at the waveforms, the *J* input *is not* set up HIGH ≥ 20 ns before the negative $\overline{C_{pD}}$ edge and therefore *is not* interpreted as a HIGH. The flip-flop output will be *undetermined* from then on because it cannot distinguish if *J* is a HIGH or a LOW at each negative $\overline{C_{pD}}$ edge.

To correct the problem, $\overline{C_{pD}}$ should be connected to the $\overline{30}$-ns tap on the delay gate instead of the $\overline{10}$-ns tap. This way, when the flip-flop "looks back" 20 ns from the negative edge of $\overline{C_{pD}}$, it will see a HIGH on *J*, allowing the toggle operation to occur.

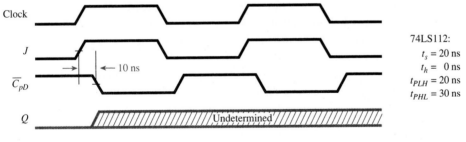

74LS112:
$t_s = 20$ ns
$t_h = 0$ ns
$t_{PLH} = 20$ ns
$t_{PHL} = 30$ ns

Figure 11–20

EXAMPLE 11–6

The repetitive waveforms shown in Figure 11–21(b) are input to the 7474 *D* flip-flop shown in Figure 11–21(a). Because of poor timing, *Q* never goes HIGH. Add a delay gate to correct the timing problem. (Assume that rise, fall, and propagation delay times are 0 ns.)

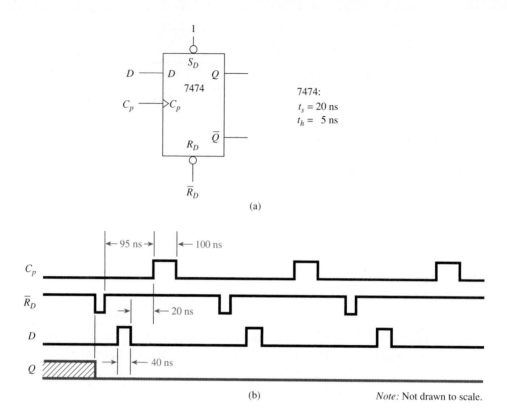

7474:
$t_s = 20$ ns
$t_h = 5$ ns

(a)

95 ns
100 ns

C_p

$\overline{R}_D$
20 ns

D

Q
40 ns

(b)

Note: Not drawn to scale.

Figure 11–21

Team Discussion

To obtain the fastest possible operating speed in a digital system, designers sometimes push the times to the absolute limits specified by the manufacturer. Discuss some of the considerations the designer must watch for and some of the problems that may occur.

Solution: Q never goes HIGH because the 40-ns HIGH pulse on D does not occur at the positive edge of C_p. Delaying the D waveform by 30 ns will move D to the right far enough to fulfill the necessary setup and hold times to allow the flip-flop to get Set at every positive edge of C_p, as shown in Figure 11–22. (D_D is the delayed D waveform, which has been shifted to the right by 30 ns to correct the timing problem.)

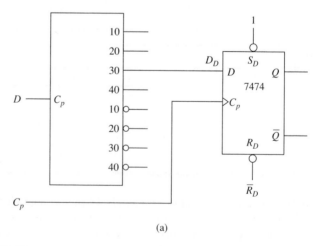

(a)

Figure 11–22

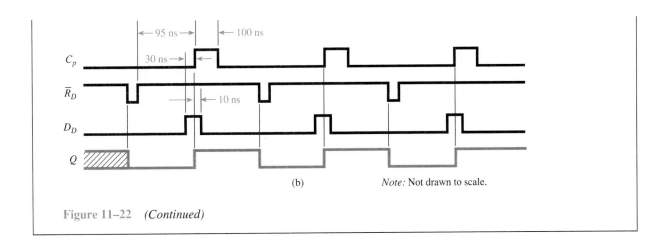

(b)   *Note:* Not drawn to scale.

Figure 11–22   *(Continued)*

Review Questions

11–1. A *race condition* occurs when the Q output of a flip-flop changes at the same time as its clock input. True or false?

11–2. *Setup time* is the length of time that the clock input must be held stable before its active transition. True or false?

11–3. Describe what manufacturers mean when they specify a *hold time* of 0 ns for a flip-flop.

11–4. The abbreviation t_{PHL} is used to specify the _____ of an IC from input to output. The letters *HL* in the abbreviation refer to the _____ (input, output) waveform changing from HIGH to LOW.

11–5. Under what circumstances would a digital circuit design require a delay gate?

11–2   Automatic Reset

Often it is advantageous to automatically Reset (or Set) all resettable (or settable) devices as soon as power is first applied to a digital circuit. In the case of resetting flip-flops, we want a LOW voltage level (0) present at the $\overline{R_D}$ inputs for a short duration immediately following **power-up,** but then after a short time (usually a few microseconds), we want the $\overline{R_D}$ line to return to a HIGH (1) level so that the flip-flops can start their synchronous operations.

To implement such an operation, we might use a series **RC circuit** to charge a capacitor that is initially discharged (0). A short time after the power-up voltage is applied to the *RC* circuit and the flip-flop's V_{CC}, the capacitor will charge up to a value high enough to be considered a HIGH (1) by the $\overline{R_D}$ input.

Basic electronic theory states that in a series *RC* circuit, the capacitor becomes almost fully charged after a time equal to the product $5RC$. This means that in Figure

11–23(a) the capacitor (which is initially discharged via the internal resistance of the $\overline{R_D}$ terminal) will begin to charge toward the 5-V level through R as soon as the switch is closed.

Before the capacitor reaches the HIGH-level threshold of the 74LS76 (approximately 2.0 V), the temporary LOW on the $\overline{R_D}$ terminal will cause the flip-flop to Reset. As soon as the capacitor charges to above 2.0 V, the $\overline{R_D}$ terminal will see a HIGH, allowing the flip-flop to perform its normal synchronous operations. The waveforms that appear on the V_{CC} and $\overline{R_D}$ lines as the power switch is closed and opened are shown in Figure 11–23(b).

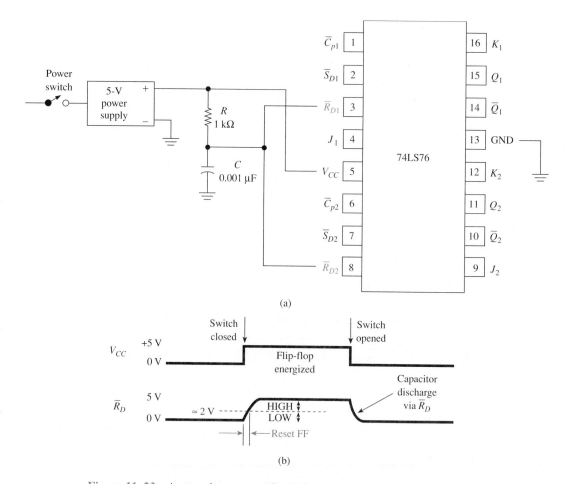

Figure 11–23 Automatic power-up Reset for a *J-K* flip-flop: (a) circuit connections; (b) waveforms.

This automatic resetting scheme can be used in circuits employing single or multiple resettable ICs. Depending on the device being Reset, the length of time that the Reset line is at a LOW level will be approximately 1 μs.

As you add more and more devices to the Reset line, the time duration of the LOW will decrease because of the additional charging paths supplied by the internal circuitry of the ICs. Remember, there is a minimum allowable width for the LOW Reset pulse ($\approx$ 25 ns for a 74LS76). To increase the time, you can increase the capacitor to 0.01 μF, or to eliminate loading effects and create a sharp edge on the $\overline{R_D}$ line, a 7407 buffer could be inserted in series with the $\overline{R_D}$ input. (A silicon diode can also be added from the top of the capacitor up to the V_{CC} line to discharge the capacitor more rapidly when power is removed.) We will utilize the **automatic Reset** feature several times in Chapters 12 and 13.

11–3   Schmitt Trigger ICs

A **Schmitt trigger** is a special type of integrated circuit that is used to transform slowly changing waveforms into sharply defined, jitter-free output signals. They are useful for changing clock edges that may have slow rise and fall times into straight vertical edges.

The Schmitt trigger employs a technique called **positive feedback** internally to speed up the level transitions and also to introduce an effect called **hysteresis.** Hysteresis means that the switching threshold on a positive-going input signal is at a higher level than the switching threshold on a negative-going input signal [see Figure 11–24(b)]. This is useful for devices that have to ignore small amounts of **jitter,** or electrical noise on input signals. Notice in Figure 11–24(a) that when the positive- and negative-going thresholds are exactly the same, as with standard gates, and a small amount of noise causes the input to jitter slightly, the output will switch back and forth several times until the input level is far above the threshold voltage.

Figure 11–24 illustrates the difference in the output waveforms for a standard 7404 inverter and a 7414 Schmitt trigger inverter. As you can see in Figure 11–24(b), the output (V_{out2}) is an inverted, jitter-free pulse. On the other hand, just think if V_{out1} were fed into the $\overline{C_p}$ input of a 74LS76 hooked up as a toggle flip-flop; the flip-flop would toggle three times (three negative edges) instead of once as was intended.

The difference between the positive- and negative-going thresholds is defined as the hysteresis voltage. For the 7414, the positive-going threshold (V_{T+}) is typically 1.7 V and the negative-going threshold (V_{T-}) is typically 0.9 V, yielding a hysteresis voltage (ΔV_T) of 0.8 V. The small box symbol ($\Box$) inside the 7414 symbol is used to indicate that it is a Schmitt trigger inverter instead of a regular inverter.

The most important specification for Schmitt trigger devices is illustrated by use of a **transfer function** graph, which is a plot of V_{out} versus V_{in}. From the transfer function, we can determine the HIGH- and LOW-level output voltages (typically 3.4 V and 0.2 V, the same as for most TTL gates), as well as V_{T+}, V_{T-}, and ΔV_T.

Figure 11–25 shows the transfer function for the 7414. The transfer function graph is produced experimentally by using a variable voltage source at the input to the Schmitt and a voltmeter (VOM) at V_{in} and V_{out}, as shown in Figure 11–26.

As the V_{in} of Figure 11–26 is increased from 0 V up toward 5 V, V_{out} will start out at approximately 3.4 V (1) and switch to 0.2 V (0) when V_{in} exceeds the positive-going **threshold** ($\approx$ 1.7 V). The output transition from HIGH to LOW is indicated in Figure 11–25 by the downward arrow. As V_{in} is increased up to 5 V, V_{out} remains at 0.2 V (0).

As the input voltage is then decreased down toward 0 V, V_{out} will remain LOW until the negative-going threshold is passed ($\approx$ 0.9 V). At that point the output will switch up to 3.4 V (1), as indicated by the upward arrow in Figure 11–25. As V_{in} continues to 0 V, V_{out} remains HIGH at 3.4 V. The hysteresis in this case is 1.7 V − 0.9 V = 0.8 V.

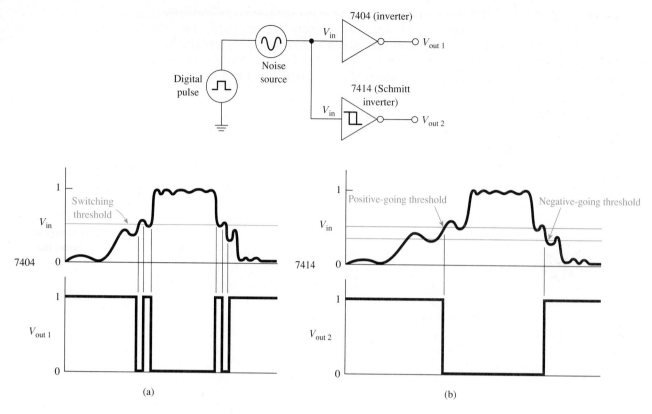

Figure 11–24 Edge-sharpening, jitter-free operation of a Schmitt trigger: (a) regular inverter; (b) Schmitt inverter.

Common Misconception

Using an oscilloscope in the X-Y mode, you can view the transfer function as you apply a 0- to 5-V triangle wave to V_{in}. You may think that it is not working, because the vertical lines are not apparent. They do not show, because the output switches so fast. However, the threshold points are still obvious to see.

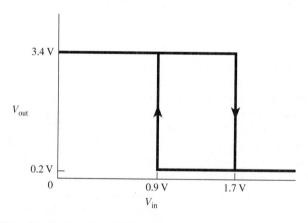

Figure 11–25 Transfer function for a 7414 Schmitt trigger inverter.

Team Discussion

How would the transfer function of a noninverting Schmitt differ from Figure 11–25?

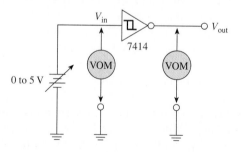

Figure 11–26 Circuit used to experimentally produce a Schmitt trigger transfer function.

EXAMPLE 11-7

Let's use the Schmitt trigger to convert a small-signal sine wave (E_s) into a square wave (V_{out}).

Solution: The diode is used to short the negative 4 V from E_s to ground to protect the Schmitt input, as shown in Figure 11–27(a). The 1-kΩ resistor will limit the current through the diode when it is conducting. [I_{diode} = (4 − 0.7 V)/1 kΩ = 3.3 mA, which is well within the rating of most silicon diodes.]

Helpful Hint

Actually, there will be a slight negative voltage of −0.7 V at V_{in} during the negative cycle.

Team Discussion

Why is the duty cycle of V_{out} always going to be greater than 50%?

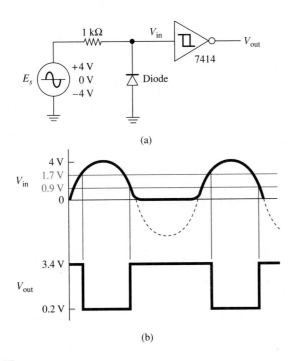

(a)

(b)

Figure 11–27

Also, the HIGH-level input current to the Schmitt (I_{IH}) is only 40 μA, causing a voltage drop of 40 μA × 1 kΩ = 0.04 V when V_{in} is HIGH. (We can assume that 0.04 V is negligible compared to +4.0 V.)

The input to the Schmitt will therefore be a half-wave signal with a 4.0-V peak. The output will be a square wave, as shown in Figure 11–27(b).

EXAMPLE 11-8

The V_{in} waveform to the 74132 Schmitt trigger NAND gate in Figure 11–28 is given in Figure 11–29.

Figure 11–28

74132

V_{in} ⎯ V_{out}

(a) Sketch the V_{out} waveform. (The 74132 has the same voltage specifications as the 7414.)
(b) Determine the duty cycle of the output waveform; the **duty cycle** is defined as

$$\frac{\text{time HIGH}}{\text{time HIGH} + \text{time LOW}} \times 100\%$$

Solution:

(a) The V_{out} waveform is shown in Figure 11–29.

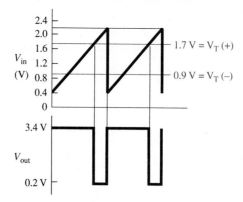

Figure 11–29

(b) V_{out} stays HIGH while V_{in} increases from 0.4 to 1.7 V, for a change of 1.3 V. V_{out} stays LOW while V_{in} increases from 1.7 to 2.2 V, for a change of 0.5 V. Since the input voltage increases linearly with respect to time, the change in V_{in} is proportional to time duration, so

$$\text{duty cycle} = \frac{1.3\ \text{V}}{1.3\ \text{V} + 0.5\ \text{V}} \times 100\% = 72.2\%$$

E X A M P L E   1 1 – 9

Sketch V_{out} of the 7414 in Figure 11–30 given the V_{in} waveform shown in Figure 11–31.

Figure 11–30

Solution:

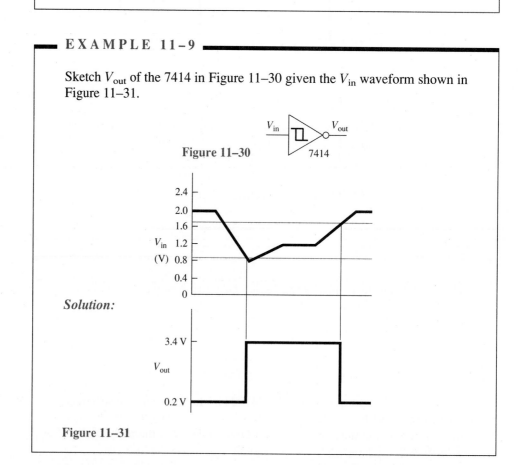

Figure 11–31

EXAMPLE 11–10

Draw and completely label the V_{out} versus V_{in} transfer function for the Schmitt trigger device whose V_{in} and V_{out} waveforms are given in Figure 11–32.

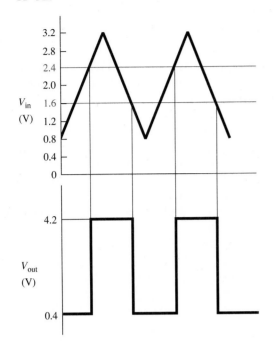

Figure 11–32

Solution: The transfer function is shown in Figure 11–33.

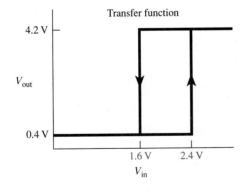

Figure 11–33

Common Misconception

Students often draw this transfer function backward, thinking that it is an *inverting* Schmitt.

Review Questions

11–6. In an automatic power-up Reset RC circuit, the voltage across the _____ (resistor, capacitor) provides the initial LOW to the $\overline{R_D}$ inputs.

11–7. The input voltage to a Schmitt trigger IC has to cross two different switching points called the _____ and the _____. The voltage differential between these two switching points is called the _____.

11–8. A Schmitt trigger IC is capable of "cleaning up" a square wave that may have a small amount of noise on it. Briefly explain how it works.

11–9. The transfer function of a Schmitt trigger device graphically shows the relationship between the _____ and _____ voltages.

11–4 Switch Debouncing

Often, mechanical switches are used in the design of digital circuits. Unfortunately, however, most switches exhibit a phenomenon called **switch bounce.** Switch bounce is the action that occurs when a mechanical switch is opened or closed. For example, when the contacts of a switch are closed, the electrical and mechanical connection is first made, but due to a slight springing action of the contacts, they will bounce back open, then close, then open, then close repeatedly until they finally settle down in the closed position. This bouncing action will typically take place for as long as 50 ms.

A typical connection for a single-pole, single-throw (**SPST**) **switch** is shown in Figure 11–34. This is a poor design because, if we except the toggle to operate only once when we close the switch, we will be out of luck because of switch bounce. Why do we say that? Let's look at the waveform at $\overline{C_p}$ to see what actually happens when a switch is closed.

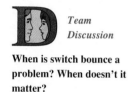

Team Discussion

When is switch bounce a problem? When doesn't it matter?

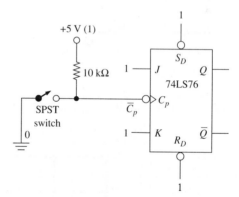

Figure 11–34 Switch used as a clock input to a toggle flip-flop.

Figure 11–35 shows that $\overline{C_p}$ will receive several LOW pulses each time the switch is closed instead of the single pulse that we expect. The 10-kΩ pull-up resistor in Figure 11–34 is necessary to hold the voltage level at $\overline{C_p}$ up close to +5 V while the switch is open. If the pull-up resistor were not used, the voltage at $\overline{C_p}$ with the switch open would be undetermined; but *with* the 10 kΩ (and realizing that the current into the $\overline{C_p}$ terminal is negligible), the level at the $\overline{C_p}$ terminal will be held at approximately +5 V while the switch is open.

There are several ways to eliminate the effects of switch bounce. If you need to debounce a single-pole, single-throw switch or push button, the Schmitt trigger scheme shown in Figure 11–36 can be used. With the switch open, the capacitor will be charged to +5 V (1), keeping $V_{out} = 0$. When the switch is closed, the capacitor will discharge rapidly to zero via the 100-Ω current-limiting resistor, making V_{out} equal to 1. Then, as the switch bounces, the capacitor will repeatedly try to charge slowly back up to a HIGH and then discharge rapidly to zero. The *RC* charging time constant (10 kΩ × 0.47 μF) is long enough that the capacitor will not get the chance to charge up high enough (above V_{T+}) before the switch bounces back to the closed position. This keeps V_{out} equal to 1.

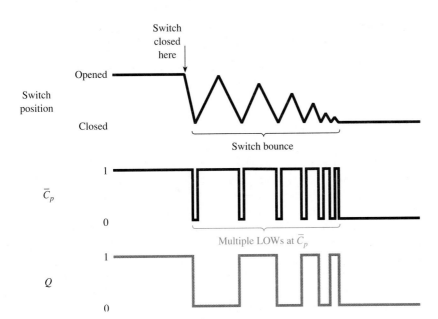

Team Discussion

Why are the LOW pulses on $\overline{C_p}$ drawn so narrow?

Figure 11–35 Waveform at point $\overline{C_p}$ for Figure 11–34.

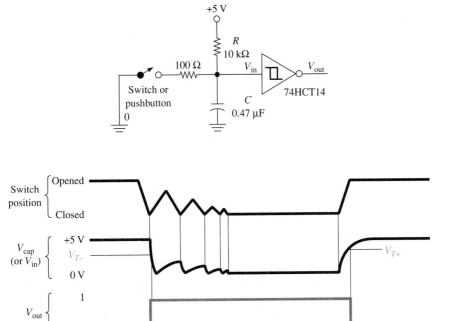

Figure 11–36 *D* flip-flop method of debouncing a single-pole, double-throw switch.

When the switch is reopened, the capacitor is allowed to charge all the way up to +5 V. When it crosses V_{T+}, V_{out} will switch to 0, as shown in Figure 11–36. The result is that by closing the switch or push button once *you will get only a single pulse at the output even though the switch is bouncing.*

To debounce single-pole, double-throw switches, a different method is required, as illustrated in Figures 11–37 and 11–38. The single-pole, double-throw switch shown in Figure 11–37(a) actually has three positions: (1) position A, (2) in between position A and position B while it is making its transition, and (3) position B. The cross-NAND debouncer works very similarly to the cross-NAND *S-R* flip-flop presented in Figure 10–2.

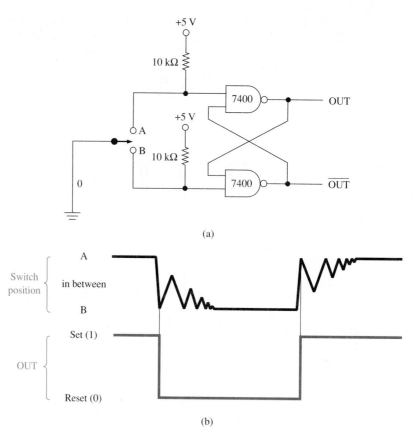

(a)

(b)

Figure 11–37 (a) Cross-NAND method of debouncing a single-pole, double-throw switch; (b) waveforms for part (a).

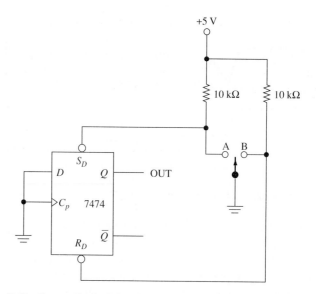

Figure 11–38 *D* flip-flop method of debouncing a single-pole, double-throw switch.

When the switch is in position A, OUT will be Set (1). When the switch is moved to position B, it bounces, causing OUT to Reset, Hold, Reset, Hold, Reset, Hold repeatedly until the switch stops bouncing and settles into position B. From the time the switch first touched position B until it is returned to position A, OUT will be Reset even though the switch is bouncing.

When the switch is returned to position A, it will bounce, causing OUT to be Set, Hold, Set, Hold, Set, Hold repeatedly until the switch stops bouncing. In this case, OUT will be Set and remain Set from the moment the switch first touched position A, even though the switch is bouncing.

Figure 11–38 shows another way to debounce a single-pole, double-throw switch using a 7474 D flip-flop. (Actually, any flip-flop with asynchronous $\overline{S_D}$ and $\overline{R_D}$ inputs can be used.)

The waveforms created from Figure 11–38 will look the same as Figure 11–37(b). As the switch is moved to position A but is still bouncing, the flip-flop will Set, Hold, Set, Hold, Set, Hold repeatedly until the switch settles into position A, keeping the flip-flop Set. When it is moved to position B, the flip-flop will Reset, Hold, Reset, Hold, and so on, until it settles down, keeping the flip-flop Reset.

11–5 Sizing Pull-Up Resistors

By the way, how do we know what size **pull-up resistors** to use in circuits like the one in Figure 11–38? Remember, the object of the pull-up resistor is to keep a terminal at a HIGH level when it would normally be at a **float** (not 1 or 0) level. In Figure 11–38, when the switch is *between* points A and B, current will flow down through the 10-kΩ resistor to $\overline{S_D}$. I_{IH} for $\overline{S_D}$ is 80 μA. This causes a voltage drop of 80 μA × 10 kΩ = 0.8 V, leaving 4.2 V at $\overline{S_D}$, which is well within the HIGH specifications of the 7474.

You may ask: Why not just make the pull-up resistor very small to minimize the voltage drop across it? Well, when the switch is in position A or B, we have a direct connection to ground. If the resistor is too small, we will have excessive current and high power consumption. However, a 10-kΩ or larger resistor would work just fine. So check the I_{IH} and V_{IH} values in a data book and keep within their ratings. Usually, a 10-kΩ pull-up resistor is a good size for most digital circuits.

When you have to provide a **pull-down resistor** (to keep a floating terminal LOW), a much smaller resistor is required because I_{IL} is typically much higher than I_{IH}. For example, if $I_{IL} = -1.6$ mA and a 100-Ω pull-down resistor is used, the voltage across the resistor is 0.160 V, which will be interpreted as a LOW. One concern of using a pull-down resistor is the high power dissipation of the resistor.

EXAMPLE 11–11

Determine the power dissipation of the 10-kΩ pull-up resistors used in Figure 11–37(a). Also, determine the HIGH-level voltage at the input to the NAND gates.

Solution: The specs for a 7400 NAND from a TTL data manual are

$$I_{IL} = 1.6 \text{ mA max.}$$
$$I_{IH} = 40 \text{ }\mu\text{A max.}$$
$$V_{IL} = 0.8 \text{ V max.}$$
$$V_{IH} = 2.0 \text{ V min.}$$

You can review these terms in Chapter 9.

When the switch is *between* positions A and B, I_{IH} will flow from the +5 V, through the 10-kΩ resistor, into the NAND. The power dissipation is

$$P = I^2 \times R$$
$$= (40 \text{ }\mu\text{A})^2 \times 10 \text{ k}\Omega$$
$$= 16 \text{ }\mu\text{W}$$

The high-level input voltage

$$V = V_{CC} - I_{IH} \times R$$
$$= 5\ V - 40\ \mu A \times 10\ k\Omega$$
$$= 4.6\ V$$

The 4.6-V HIGH-level input voltage is about the 2.0-V V_{IH} limit given in the specs, and 16 μW is negligible for most applications.

When the switch is moved to either A or B, the power dissipation in the resistor is

$$P = \frac{E^2}{R}$$
$$= \frac{5\ V^2}{10\ k\Omega}$$
$$= 2.5\ mW$$

The value of 2.5 mW is still negligible for most applications. If not, increase the size of the 10-kΩ pull-up resistor and recalculate for the HIGH-level input voltage and power dissipation. As long as you keep the HIGH-level input voltage above the specified limit of 2.0 V, the circuit will operate properly.

Team Discussion

Sizing pull-up resistors for open-collector outputs requires that you also consider the size of the load resistor. For example, if the load is 10 kΩ, what happens if you use a 10-kΩ pull up?

11–6 Practical Input and Output Considerations

Before designing and building the practical digital circuits in the next few chapters, let's study some circuit designs for (1) a simple 5-V power supply, (2) a clock to drive synchronous trigger inputs, and (3) circuit connections for LED interfacing to the outputs of integrated-circuit chips.

A 5-V Power Supply

Helpful Hint

This is an inexpensive but very useful circuit for you to build for yourself. With the additional expense of a breadboard and a few ICs, you can test out several of the textbook circuits at home.

For now, we limit our discussion to the TTL family of integrated circuits. From the data manual we can see that TTL requires a constant supply voltage of 5.0 V $\pm$ 5%. Also, the total supply current requirement into the V_{CC} terminal ranges from 20 to 100 mA for most TTL ICs.

For the power supply, the 78XX series of integrated-circuit **voltage regulators** is inexpensive and easy to use. To construct a regulated 5.0-V supply, we use the 7805 (the 05 designates 5 V; a 7808 would designate an 8.0-V supply). The 7805 is a three-terminal device (input, ground, output) capable of supplying 5.0 V $\pm$ 0.2% at 1000 mA. Figure 11–39 shows how a 7805 voltage regulator is used in conjunction with an ac-to-dc **rectifier** circuit.

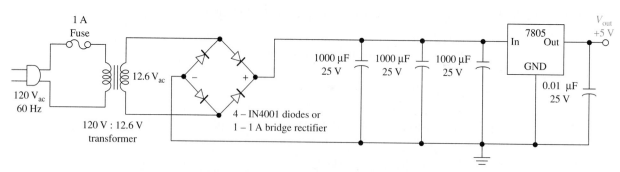

Figure 11–39   Complete 5-V, 1-A TTL power supply.

In Figure 11–39 the 12.6-V ac rms is rectified by the diodes (or a four-terminal bridge rectifier) into a full-wave dc of approximately 20 V. The 3000 μF of capacitance is required to hold the dc level into the 7805 at a high, steady level. The 7805 will automatically decrease the 20-V dc input to a solid, **ripple**-free 5.0-V dc output.

The 0.01-μF capacitor is recommended by TTL manufacturers for *decoupling* the power supply. Tantalum capacitors work best and should be mounted as close as possible to the V_{CC}-to-ground pins on every TTL IC used in your circuit. Their size should be between 0.01 and 0.1 μF with a voltage rating ≥ 5 V. The purpose of the capacitor is to eliminate the effects of voltage spikes created from the internal TTL switching and electrostatic noise generated on the power and ground lines.

The 7805 will get very hot when your circuit draws more than 0.5 A. In that case, it should be mounted on a heat sink to help dissipate the heat.

A 60-Hz Clock

Figure 11–40 shows a circuit design for a simple 60-Hz TTL-level (0 to 5 V) clock that can be powered from the same transformer used in Figure 11–39 and used to drive the clock inputs to our synchronous ICs. Our electric power industry supplies us with accurate 60-Hz ac voltages. It is a simple task to reduce the voltage to usable levels and still maintain a 60-Hz [60-pulse-per-second (pps)] signal.

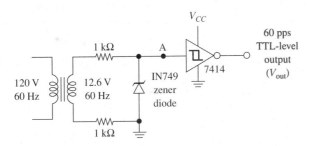

Helpful Hint

This circuit can be attached to the same transformer used in the power supply circuit.

Figure 11–40 Accurate 60-Hz, TTL-level clock pulse generator.

From analog electronics courses, you may remember that a zener diode will conduct normally in the forward-biased direction, and in the reverse-biased direction it will start conducting when a voltage level equal to its *reverse* **zener breakdown** rating is reached (4.3 V for the IN749).

The 1-kΩ resistors are required to limit the zener current to reasonable levels. Figure 11–41 shows the waveform that will appear at point A and V_{out} of Figure 11–40. The

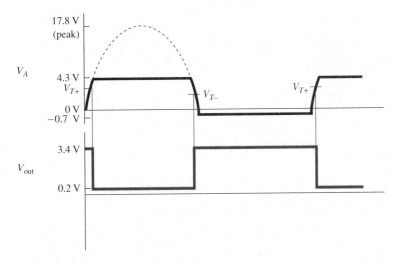

Figure 11–41 Voltage waveform at point A and V_{out} of Figure 11–40.

zener breaks down at 4.3 V, which is high enough for a one-level input to the Schmitt trigger but not too high to burn out the chip. The V_{out} waveform will be an accurate 60-pulse-per-second, approximately 50% duty cycle square wave. As we will see in Chapter 12, this frequency can easily be divided down to 1 pulse per second by using toggle flip-flops. One pulse per second is handy because it is slow enough to see on visual displays (like LEDs) and accurate enough to use as a trigger pulse on a digital clock.

Driving Light-Emitting Diodes

Light-emitting diodes (LEDs) are good devices to visually display a HIGH (1) or LOW (0) digital state. A typical red LED will drop 1.7 V cathode to anode when forward biased (positive anode-to-cathode voltage) and will illuminate with 10 to 20 mA flowing through it. In the reverse-biased direction (zero or negative anode-to-cathode voltage), the LED will block current flow and not illuminate.

Because it takes 10 to 20 mA to illuminate an LED, we may have trouble driving it with a TTL output. From the TTL data manual, we can determine that most ICs can sink (0-level output) a lot more current than they can source (1-level output). Typically, the maximum sink current, I_{OL}, is 16 mA and the maximum source current, I_{OH}, is only 0.4 mA. Therefore, we had better use a LOW level (0) to turn on our LED instead of a HIGH level.

Figure 11–42 shows how we can drive an LED from the output of a TTL circuit (a *J-K* flip-flop in this case). The *J-K* flip-flop is set up in the toggle mode so that *Q* will flip states once each second.

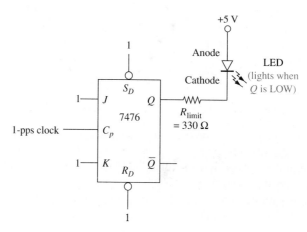

Team Discussion

If we want the light to come on when *Q* is HIGH, could we connect the circuit to $\overline{Q}$ instead?

Figure 11–42 Driving an LED.

When *Q* is LOW (0 V), the LED is forward biased and current will flow through the LED and resistor and sink into the *Q* output. The 330-Ω resistor is required to limit the series current to 10 mA [$I = (5\ V - 1.7\ V)/330\ \Omega = 10\ mA$], and 10 mA into the LOW-level *Q* output will not burn out the flip-flop. If, however, we were trying to turn the LED on with a HIGH-level output, we would turn the LED around and connect the cathode to ground. But 10 mA would exceed the limit of I_{OH} on the 7476 and either burn it out or just not illuminate the LED.

Phototransistor Input to a Latching Alarm System

A **phototransistor** is made to turn off and on by shining light on its base region. It is encased in clear plastic and is turned on when light strikes its base region or if the base connection is forward biased with an external voltage. The resistance from collector to emitter for an OFF transistor is typically 1 to 10 MΩ. An ON transistor will range from 1000 Ω to as low as 10 Ω depending on the light intensity.

The circuit of Figure 11–43 uses a phototransistor in an alarm system. The phototransistor could be placed in a doorway and positioned so that light is normally striking it. This will keep its resistance low and the voltage at point A low. When a person walks through the doorway, the light is interrupted, making the voltage at point A momentarily high. The 74HCT14 Schmitt inverters will react by outputting a LOW-to-HIGH pulse. This creates the clock trigger to the D flip-flop, which will latch HIGH, turning on the alarm. The alarm will remain on until the Reset pushbutton is pressed.

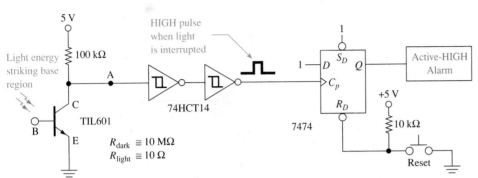

Team Discussion

Could a toggle flip-flop be used in place of the D flip-flop? How would the operation change?

Figure 11–43   Phototransistor used as an input to a latching alarm system.

Using an Optocoupler for Level Shifting

An **optocoupler** (or optoisolator) is an integrated circuit with an LED and phototransistor encased in the same package. The phototransistor has a very high OFF resistance (dark) and a low ON resistance (light), which are controlled by the amount of light striking its base from the LED. The terms *optocoupler* and *optoisolator* come from the fact that the output side of the device is electrically *isolated* from the input side and can therefore be used to *couple* one circuit to another without being concerned about incompatible or harmful voltage levels.

Figure 11–44 shows how an optocoupler can be used to transmit TTL-level data to another circuit having 25-V logic levels. The 7408 is used to sink current through the optocoupler's LED each time the input signal goes LOW. Each time the LED illuminates, the phototransistor will exhibit a low resistance, making V_{out} LOW. Therefore, the output signal will be in phase with the input signal, but its HIGH/LOW levels

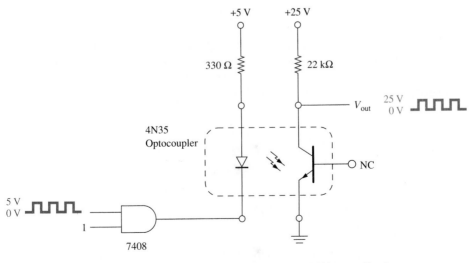

Figure 11–44   An optocoupler provides isolation in a level-shifting application.

will be approximately 25 V/0 V. Notice that the 25-V circuit is totally separate from the 5-V TTL circuit, providing complete isolation from the potentially damaging higher voltage.

A Power MOSFET Used to Drive a Relay and AC Motor

The output drive capability of digital logic severely limits the size of the load that can be connected. An LS-TTL buffer such as the 74LS244 can sink up to 24 mA, and the BiCMOS 74ABT244 can sink up to 64 mA, but this is still far below the current requirements of some loads. A common way to boost the current capability is to use a *power MOSFET,* which is particularly well suited for these applications. This is a transistor specifically designed to have a very high input impedance to limit current draw into its gate and also be capable of passing a high current through its drain to source. (See Section 9–5 to review MOSFET operation.)

Figure 11–45 shows a circuit that could be used to drive a $\frac{1}{3}$-hp ac motor from a digital logic circuit. Because the starting current of a motor can be extremely high, we will use a relay with a 24-V dc coil and a 50-A contact rating. A relay of this size may require as much as 200 mA to energize the coil to pull in the contacts. A MOSFET such as the IRF130 can pass up to 12 A through its drain to source, so it can easily handle this relay coil requirement.

Team Discussion

The relay and the optocoupler are common means used to interface digital logic to the outside world. List several devices that might be driven from a digital or microprocessor circuit.

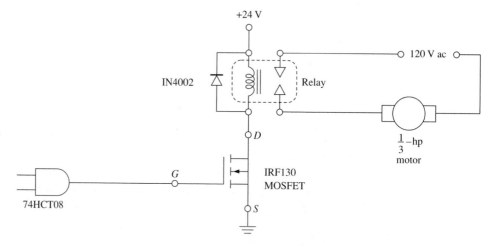

Figure 11–45 Using a power MOSFET to interface digital logic to high-power ac circuitry.

When the 74HCT08 outputs a HIGH (5 V) to the gate of the MOSFET, the drain to source becomes a short (approximately 0.2 Ω). This allows current to flow through the relay coil, creating the magnetic flux required to pull in the contacts. The motor will start. The 1N4002 diode provides arc protection across the coil when it is deenergized. (See Section 2–6 for a review of relays.)

Review Questions

11–10. What is the cause of switch bounce, and why is it harmful in digital circuits?

11–11. What size resistor is better suited for a pull-up resistor, 10 kΩ or 100 Ω?

11–12. What 78XX series IC voltage regulator could be used to build an inexpensive 12-V power supply?

11–13. The zener diode serves two purposes in the pulse generator design in Figure 11–40. What are those purposes?

11–14. Why are LEDs usually connected as active-LOW indicator lights instead of active HIGH?

11–15. Is the phototransistor alarm in Figure 11–43 better suited for a home burglar alarm or as an alarm to announce when a customer has entered a store? Why?

11–16. How would the operation of the alarm in Figure 11–43 change if only one Schmitt inverter was used instead of two?

11–17. Why is an optocoupler sometimes referred to as an optoisolator?

11–18. In Figure 11–45, why can't you drive the relay directly with 74HCT08 instead of using the MOSFET?

Summary

In this chapter we have learned that

1. Unpredictable results on IC logic can occur if strict timing requirements are not met.

2. A setup time is required to ensure that the input data to a logic circuit is present some definite time prior to the active clock edge.

3. A hold time is required to ensure that the input data to a logic circuit is held for some definite time after the active clock edge.

4. The propagation delay is the length of time it takes for the output of a logic circuit to respond to an input stimulus.

5. Delay gates are available to purposely introduce time delays when required.

6. The charging voltage on a capacitor in a series *RC* circuit can be used to create a short delay for a power-up reset.

7. The two key features of Schmitt trigger ICs are that they output extremely sharp edges and they have two distinct input threshold voltages. The difference between the threshold voltages is called the hysteresis voltage.

8. Mechanical switches exhibit a phenomenon called switch bounce, which can cause problems in most kinds of logic circuits.

9. Pull-up resistors are required to make a normally floating input act like a HIGH. Pull-down resistors are required to make a normally floating input act like a LOW.

10. A practical, inexpensive 5-V power supply can be made with just a transformer, four diodes, some capacitors, and a voltage regulator.

11. A 60-pulse-per-second clock oscillator can be made using the power supply's transformer and a few additional components.

12. The resistance from collector to emitter of a phototransistor changes from about 10 MΩ down to about 1000 Ω when light shines on its base region.

13. An optocoupler provides electrical isolation from one part of a circuit to another.

14. Power MOSFETs are commonly used to increase the output drive capability of IC logic from less than 100 mA to more than 1 A.

Glossary

Active Clock Edge: A clock edge is the point in time where the waveform is changing from HIGH to LOW (negative edge) or LOW to HIGH (positive edge). The *active* clock edge is the edge (either positive or negative) used to trigger a synchronous device to accept input digital states.

AC Waveforms: Test waveforms that are supplied by IC manufacturers for design engineers to determine timing sequence and measurement points for such quantities as setup, hold, and propagation delay times.

Automatic Reset: A scheme used to automatically Set or Reset all storage ICs (usually flip-flops) to a Set or Reset condition when power is first applied to them so that their starting condition can always be determined.

Duty Cycle: The ratio of the length of time a periodic wave is HIGH versus the total period of the wave.

Float: A condition in which an input or output line in a circuit is neither HIGH nor LOW because it is not directly connected to a high or low voltage level.

Hold Time: The length of time *after* the active clock edge that the input data to be recognized (usually *J* and *K*) must be held stable to ensure recognition.

Hysteresis: In digital Schmitt trigger ICs, hysteresis is the difference in voltage between the positive-going switching threshold and the negative-going switching threshold at the input.

Jitter: A term used in digital electronics to describe a waveform that has some degree of electrical noise on it, causing it to rise and fall slightly between and during level transitions.

Optocoupler: A device having an LED and a phototransistor encased in the same package. Illuminating the LED turns the transistor on, providing optical coupling and isolation between two circuits.

Phototransistor: An optically sensitive transistor that is turned on when light strikes its base region.

Positive Feedback: A technique employed by Schmitt triggers that involves taking a small sample of the output of a circuit and feeding it back into the input of the same circuit to increase its switching speed and introduce hysteresis.

Power-Up: The term used to describe the initial events or states that occur when power is first applied to an IC or digital system.

Propagation Delay: The length of time that it takes for an input level change to pass through an IC and appear as a level change at the output.

Pull-Down Resistor: A resistor with one end connected to a LOW voltage level and the other end connected to an input or output line so that, when that line is in the float condition (not HIGH or LOW), the voltage level on that line will, instead, be pulled down to a LOW state.

Pull-Up Resistor: A resistor with one end connected to a HIGH voltage level and the other end connected to an input or output line so that, when that line is in a float condition (not HIGH or LOW), the voltage level on that line will instead be pulled up to a HIGH state.

Race Condition: The condition that occurs when a digital input level (1 or 0) is changing states at the same time as the active clock edge of a synchronous device, making the input level at that time undetermined.

RC Circuit: A simple series circuit consisting of a resistor and a capacitor used to provide time delay.

Rectifier: An electronic device used to convert an ac voltage into a dc voltage.

Ripple: A small fluctuation in the output voltage of a power supply that is the result of poor filtering and regulation.

Schmitt Trigger: A circuit used in digital electronics to provide ultrafast level transitions and introduce hysteresis for improving jittery or slowly rising waveforms.

Setup Time: The length of time *prior to* the active clock edge that the input data to be recognized (usually J and K) must be held stable to ensure recognition.

SPST Switch: The abbreviation for single pole, single throw. A SPST switch is used to make or break contact in a single electrical line.

Switch Bounce: An undesirable characteristic of most switches when they physically make and break contact several times (bounce) each time they are opened or closed.

Threshold: The exact voltage level at the input to a digital IC that causes it to switch states. In Schmitt trigger ICs, there are two different threshold levels: the positive-going threshold (LOW to HIGH) and the negative-going threshold (HIGH to LOW).

Transfer Function: A plot of V_{out} versus V_{in} that is used to graphically determine the operating specifications of a Schmitt trigger.

Voltage Regulator: An electronic device or circuit that is used to adjust and control a voltage to remain at a constant level.

Zener Breakdown: The voltage across the terminals of a zener diode when it is conducting current in the reverse-biased direction.

Problems

Section 11–1

11–1. Sketch the Q output waveform for a 74LS76 given the input waveforms shown in Figure P11–1 [use t_s (L) = 20 ns, t_s (H) = 20 ns, t_h (L) = 0 ns, t_h (H) = 0 ns, t_{PLH} = 0 ns, t_{PHL} = 0 ns].

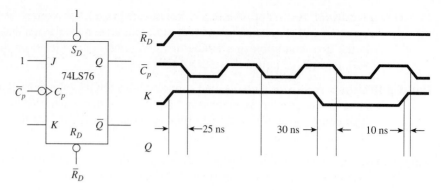

Figure P11–1

11–2. Repeat Problem 11–1 for the waveforms shown in Figure P11–2.

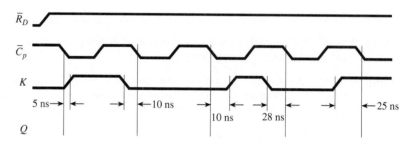

Figure P11–2

11–3. Using actual specifications for a 74LS76, label the propagation delay times on the waveforms shown in Figure P11–3.

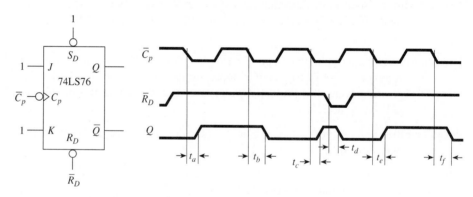

Figure P11–3

11–4. Repeat Problem 11–3 for the waveforms shown in Figure P11–4. Use specifications for a 74109 in the toggle mode ($J = 1$, $\overline{K} = 0$).

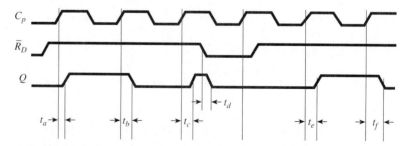

Figure P11–4

11–5. Describe the problem that may arise when using the 7432 OR gate to delay the clock signal into the flip-flop circuit of Figure 11–14.

C   **11–6.** Sketch the output at $\overline{C_{pD}}$ and Q for the flip-flop circuit shown in Figure P11–6. (Ignore propagation delays in the 74LS76.)

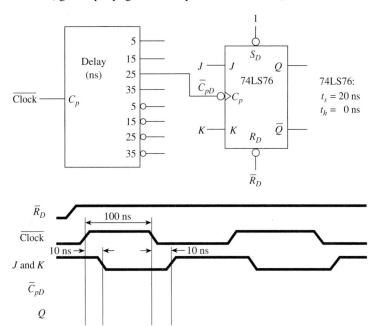

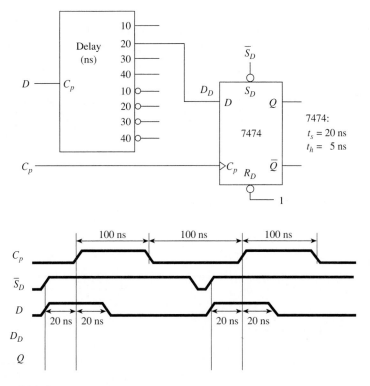

Figure P11–6

C   **11–7.** Redraw the waveforms given in Problem 11–6 if the 35-ns delay tap is used instead of the 25-ns tap.

Figure P11–8

C   **11–8.** (a) Sketch the output at D_D and Q for the flip-flop circuit shown in Figure P11–8. (Ignore propagation delays in the 7474.) (b) Connect D_D to the 30-ns tap and repeat part (a). See Figure P11–8 on page 385.

Sections 11–2 and 11–3

11–9. One particular Schmitt trigger inverter has a positive-going threshold of 1.9 V and a negative-going threshold of 0.7 V. Its V_{OH} (typical) is 3.6 V and V_{OL} (typical) is 0.2 V. Sketch the transfer function (V_{out} versus V_{in}) for this Schmitt trigger.

11–10. If the input waveform (V_{in}) shown in Figure P11–10 is fed into the Schmitt trigger described in Problem 11–9, sketch its output waveform (V_{out}).

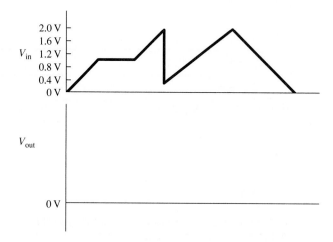

Figure P11–10

11–11. If the waveform shown in Figure P11–11 is fed into a 7414 Schmitt trigger inverter, sketch V_{out} and determine the duty cycle of V_{out}.

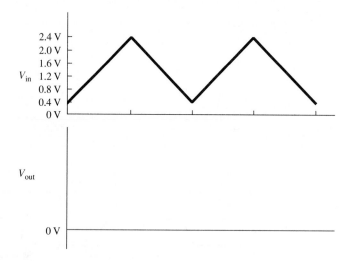

Figure P11–11

11–12. If the V_{in} and V_{out} waveforms shown in Figure P11–12 are observed on a Schmitt trigger device, determine its characteristics and sketch the transfer function (V_{out} versus V_{in}).

CHAPTER 11 I PRACTICAL CONSIDERATIONS FOR DIGITAL DESIGN

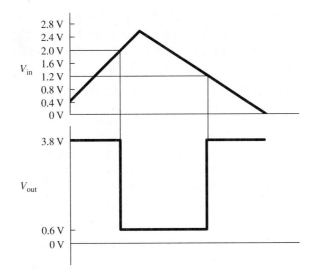

Figure P11–12

Section 11–4

T   **11–13.** The Q output of the 74LS76 in Figure 11–34 is used to drive an LED. Sometimes when the switch is closed, the LED toggles to its opposite state, but sometimes it does not. Discuss the probable cause and a solution to the problem.

Section 11–5

D   **11–14.** Occasionally, instead of using a pull-up resistor, a pull-down resistor is required because a floating connection must be held LOW as shown in Figure P11–14. Why can't a 10-kΩ resistor be used for this purpose? Could a 100-Ω resistor be used? How would the use of a 74HCT74 improve the situation?

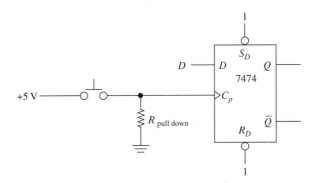

Figure P11–14

Section 11–6

T   **11–15.** A problem arises in a digital system that you have designed. Using a multimeter you find that none of your ICs is receiving 5-V V_{CC} power. Your 5-V power supply is the one given in Figure 11–39. Outline a procedure that you would follow to troubleshoot your power supply.

D   **11–16.** Design a 60-pulse-per-second TTL-level pulse generator similar to Figure 11–40 using an optocoupler instead of the zener diode.

11–17. In Figure 11–43, assume that the phototransistor has an ON resistance of 1 kΩ and an OFF resistance of 1 MΩ. Determine the voltage at point A when the light is striking, and then not striking, the phototransistor.

C T **11–18.** You are asked to troubleshoot the alarm circuit in Figure 11–43, which is not working. You find that the voltage at C_p is stuck at 0.2 V for both light and dark conditions. The voltage at point A is also stuck at about 1.0 V. You then take the inverters out of the circuit, test them, and determine that they are working. While the inverters are out, you notice that the voltage at point A jumped up to 4.8 V. When you shine a light on the phototransistor, it drops to 0.2 V! Looking further, you notice that a 7414 was substituted for the 74HCT14. What is the problem?

D **11–19.** A good choice for an alarm in Figure 11–43 is a 5-V piezo buzzer. The problem is that it takes about 10 mA to operate the buzzer and the 7474 can only source 0.4 mA. Any ideas?

11–20. Assume that the ON and OFF resistances of the phototransistor in a 4N35 optocoupler are 1 kΩ ON, 1 MΩ OFF. Determine the actual values of V_{out} in Figure 11–44.

11–21. If the relay used in Figure 11–45 has a coil resistance of 100 Ω, determine the coil current when the MOSFET is ON. (Assume that $R_{DS(ON)} = 0.2$ Ω.)

Schematic Interpretation Problems

See Appendix G for the schematic diagrams.

S **11–22.** Find the section of the watchdog timer schematic that shows U14:A, U15:A, and U14:B. Assume that pins 2 and 13 of U15:A are both HIGH and U14:A is initially reset. Apply a positive pulse on the line labeled WATCHDOG_SEL.

(a) Discuss the possible setup time problems that may occur with U14:B.

(b) Discuss how the situation changes if pin 1 of U15:A is already HIGH and the positive pulse comes in on pin 2 instead.

S D **11–23.** On a separate piece of paper draw the circuit connections to add a bank of eight LEDs with current-limiting resistors to the octal D flip-flop, U5, in the 4096/4196 schematic.

S C D **11–24.** On a separate piece of paper draw the connections to input the following values to port PA of the 68HC11 microcontroller in the HC11D0 master board schematic.

(a) Monitor light/no light conditions by using a light-sensitive phototransistor connected to PA1.

(b) Interface the 0-V/15-V (LOW/HIGH) levels from a 4050B CMOS buffer to PA3 via an optocoupler.

S C **11–25.** S2 in grid location B-1 in the HC11D0 schematic is a set of seven 10-kΩ pull-up resistors contained in a single DIP. They all have a common connection to V_{CC}, as shown. Explain their purpose as they relate to the U12 DIP-switch package and the MODA, MODB inputs to the 68HC11 microcontroller.

S D   **11–26.** On a separate piece of paper, add the circuitry to provide +5 V on sheet 2 of the 4096/4196 schematic. (Tap off of the +UNREG signal provided.)

Answers to Review Questions

11–1. False

11–2. False

11–3. It means that the input levels don't have to be held beyond the active clock edge.

11–4. Propagation delay, output

11–5. To enable proper setup and hold times

11–6. Capacitor

11–7. Positive-going threshold, negative-going threshold, hysteresis

11–8. The switching threshold on a positive-going input signal is at a higher level than the switching threshold on a negative-going input signal. This is called hysteresis. The output is steady as long as the input noise does not exceed the hysteresis voltage.

11–9. Input, output

11–10. It is caused by the springing action of the contacts, and it can cause false triggering of a digital circuit.

11–11. 10 kΩ

11–12. 7812

11–13. It cuts off the negative cycle of the sine wave and limits the positive cycle to 4.3 V.

11–14. Because the ICs to which they are connected can sink more current than they can source

11–15. Home alarm, because the alarm is latched in the ON state until the flip-flop is manually reset

11–16. The clock signal would be normally HIGH and drop LOW when interrupted, and the flip-flop would latch when the signal returned HIGH.

11–17. The output side is electrically isolated from the input side.

11–18. The current capability of the output is too low to trigger the relay.

12 Counter Circuits and Applications

OUTLINE

12–1 Analysis of Sequential Circuits
12–2 Ripple Counters
12–3 Design of Divide-by-N Counters
12–4 Ripple Counter Integrated Circuits
12–5 System Design Applications
12–6 Seven-Segment LED Display Decoders
12–7 Synchronous Counters
12–8 Synchronous Up/Down-Counter ICs
12–9 Applications of Synchronous Counter ICs

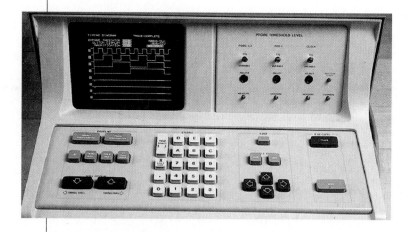

Objectives

Upon completion of this chapter, you should be able to

- Use timing diagrams for the analysis of sequential logic circuits.

- Design any modulus ripple counter and frequency divider using *J-K* flip-flops.

- Describe the difference between ripple counters and synchronous counters.

- Solve various counter design applications using 4-bit counter ICs and external gating.

- Connect seven-segment LEDs and BCD decoders to form multidigit numeric displays.

- Cascade counter ICs to provide for higher counting and frequency division.

Introduction

Now that we understand the operation of flip-flops and latches, we can apply our knowledge to the design and application of sequential logic circuits. One common application of sequential logic arrives from the need to count events and time the duration of various processes. These applications are called **sequential** because they follow a predetermined sequence of digital states and are triggered by a timing pulse or clock.

To be useful in digital circuitry and microprocessor systems, counters normally count in binary and can be made to stop or recycle to the beginning at any time. In a recycling counter, the number of different binary states defines the **modulus** (MOD) of the counter. For example, a counter that counts from 0 to 7 is called a MOD-8 counter. For a counter to count from 0 to 7, it must have three binary outputs and one clock trigger input, as shown in Figure 12–1.

Normally, each binary output will come from the *Q* output of a flip-flop. Flip-flops are used because they can hold, or remember, a binary state until the next clock or trigger pulse comes along. The count sequence of a 0 to 7 binary counter is shown in Table 12–1 and Figure 12–2.

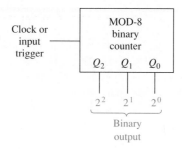

Figure 12–1   Simplified block diagram of a MOD-8 binary counter.

Table 12–1   Binary Count Sequence of a MOD-8 Binary Counter

Q_2	Q_1	Q_0	Count	
0	0	0	0	
0	0	1	1	
0	1	0	2	
0	1	1	3	Eight different
1	0	0	4	binary states
1	0	1	5	
1	1	0	6	
1	1	1	7	
0	0	0	0	
0	0	1	1	
0	1	0	2	
0	1	1	3	
and so on				

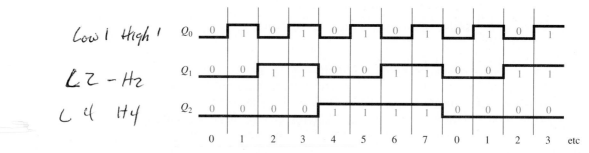

Low 1 High 1

L2 - H2

L 4 H4

Figure 12–2   Waveforms for a MOD-8 binary counter.

Before studying counter circuits, let's analyze some circuits containing logic gates with flip-flops to get a feeling for the analytical process involved in determining the output waveforms of sequential circuits.

12–1   Analysis of Sequential Circuits

To get our minds thinking in terms of sequential analysis, let's look at an example that combines regular logic gates with flip-flops and whose operation is dictated by a specific sequence of input waveforms, as shown in Example 12–1.

The 7474 shown in Figure 12–3 is a positive edge-triggered *D* flip-flop. The waveforms shown in Figure 12–4 are applied to the inputs at *A* and C_p. Sketch the resultant waveform at *D*, *Q*, $\overline{Q}$, and *X*.

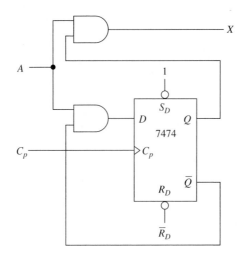

Figure 12–3

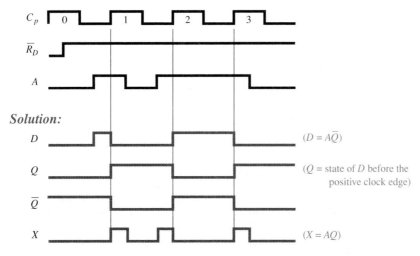

Solution:

Figure 12–4

1. $Q = 0$, $\overline{Q} = 1$ during the 0 period because of $\overline{R_D}$.

2. *D* is equal to $A\overline{Q}$; *X* is equal to AQ (therefore, the level at *D* and *X* will change whenever the inputs to the AND gates change, regardless of the state of the input clock).

3. At the positive edge of pulse 1, *D* is HIGH, so *Q* will go HIGH and $\overline{Q}$ will go LOW and remain there until the positive edge of pulse 2.

4. During period 1, *D* will equal $A\overline{Q}$ and *X* will equal AQ, as shown.

5. At the positive edge of pulse 2, *D* is LOW, so the flip-flop will Reset ($Q = 0$; $\overline{Q} = 1$) and remain there until the positive edge of pulse 3.

6. At the positive edge of pulse 3, *D* is HIGH, so the flip-flop will Set ($Q = 1$; $\overline{Q} = 0$).

The timing analysis in Examples 12–1 and 12–2 was done by observing the level on D prior to the positive clock edge and realizing that D follows the level of $A\overline{Q}$.

EXAMPLE 12–2

Using the same circuit of Example 12–1, sketch the waveforms at D, Q, $\overline{Q}$, and X, given the input waves shown in Figure 12–5.

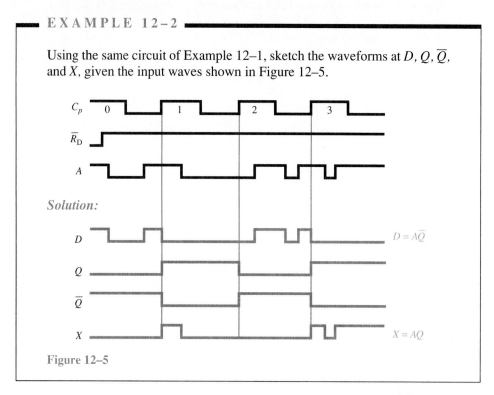

Figure 12–5

When a J-K flip-flop is used, we have to consider the level at J and K at the active clock edge as well as any asynchronous operations that may be taking place. Examples 12–3 and 12–4 illustrate the timing analysis of sequential circuits utilizing J-K flip-flops.

EXAMPLE 12–3

The 74LS76 shown in Figure 12–6 is a negative edge-triggered flip-flop. The waveforms in Figure 12–7 are applied to the inputs at A and $\overline{C_{p0}}$. Sketch the resultant waveforms at J_1, K_1, Q_0, and Q_1. Notice that the clock input to the second flip-flop comes from Q_0. Also, $J_1 = AQ_1$, $K_1 = AQ_0$.

Common Misconception

Students often think that *both* flip-flops are triggered from $\overline{C_{p0}}$.

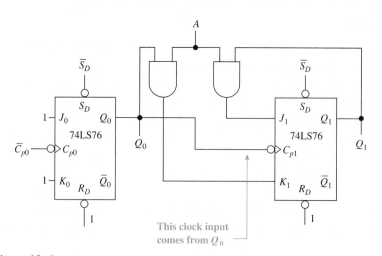

Figure 12–6

CHAPTER 12 I COUNTER CIRCUITS AND APPLICATIONS

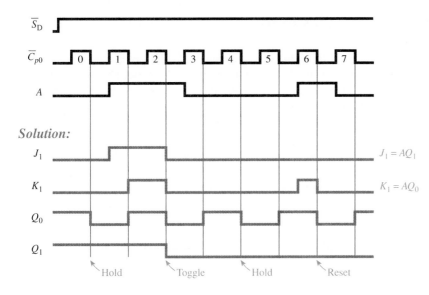

Figure 12–7

1. Because $J_0 = 1$ and $K_0 = 1$, then Q_0 will *toggle* at each negative edge of $\overline{C_{p0}}$.

2. The second flip-flop will be triggered at each negative edge of the Q_0 line.

3. The levels at J_1 and K_1 just prior to the negative edge of the Q_0 line will determine the synchronous operation of the second flip-flop.

4. After Q_0 and Q_1 are determined for each period, the new levels for J_1 and K_1 can be determined from $J_1 = AQ_1$ and $K_1 = AQ_0$.

EXAMPLE 12–4

Repeat Example 12–3 for the waveforms shown in Figure 12–8.

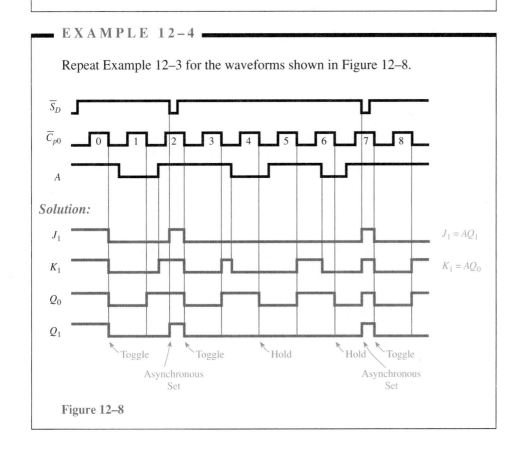

Figure 12–8

12–2   Ripple Counters

Flip-flops can be used to form binary counters. The counter output waveforms discussed in the beginning of this chapter (Figure 12–2) could be generated by using three flip-flops cascaded together (**cascade** means to connect the Q output of one flip-flop to the clock input of the next). Three flip-flops are needed to form a 3-bit counter (each flip-flop will represent a different power of 2: $2^2, 2^1, 2^0$). With three flip-flops we can produce 2^3 different combinations of binary outputs ($2^3 = 8$). The eight different binary outputs from a 3-bit binary counter will be 000, 001, 010, 011, 100, 101, 110, and 111.

If we have a 4-bit binary counter, we would count from 0000 up to 1111, which is 16 different binary outputs. As it turns out, we can determine the number of different binary output states (modulus) by using the following formula:

$$\text{modulus} = 2^N, \qquad \text{where } N = \text{number of flip-flops}$$

To form a 3-bit binary counter, we cascade three *J-K* flip-flops, each operating in the *toggle mode* as shown in Figure 12–9. The clock input used to increment the binary count comes into the $\overline{C_p}$ input of the first flip-flop. Each flip-flop will toggle every time its clock input receives a HIGH-to-LOW edge.

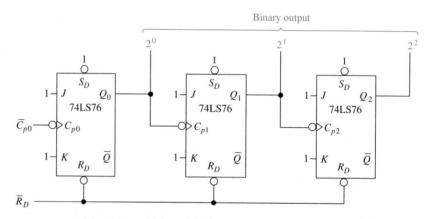

Figure 12–9   Three-bit binary ripple counter.

Now, with the knowledge that we have gained by analyzing the sequential circuits in Section 12–1, it should be easy to determine the output waveforms of the 3-bit binary **ripple counter** of Figure 12–9.

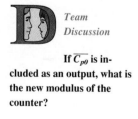

Team Discussion

If $\overline{C_{p0}}$ is included as an output, what is the new modulus of the counter?

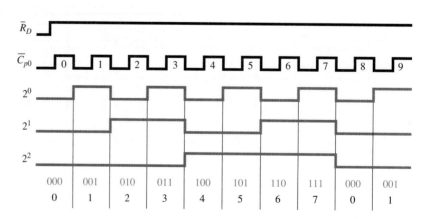

Figure 12–10   Waveforms generated from the 3-bit binary ripple counter.

When we analyze the circuit and waveforms, we see that Q_0 toggles at each negative edge of $\overline{C_{p0}}$, Q_1 toggles at each negative edge of Q_0, and Q_2 toggles at each negative edge of Q_1. The result is that the outputs will "count" repeatedly from 000 up to 111 and then 000 to 111, as shown in Figure 12–10. The term *ripple* is derived from the fact that the input clock trigger is not connected to each flip-flop directly but instead has to propagate down through each flip-flop to reach the next.

For example, look at clock pulse 7. The negative edge of $\overline{C_{p0}}$ causes Q_0 to toggle LOW . . . which causes Q_1 to toggle LOW . . . which causes Q_2 to toggle LOW. There will definitely be a propagation delay between the time that $\overline{C_{p0}}$ goes LOW until Q_2 finally goes LOW. Because of this delay, ripple counters are called **asynchronous counters,** meaning that each flip-flop is not triggered at exactly the same time.

Synchronous counters can be formed by driving each flip-flop's clock by the same clock input. Synchronous counters are more complicated, however, and will be covered after we have a thorough understanding of asynchronous ripple counters.

The propagation delay inherent in ripple counters places limitations on the maximum frequency allowed by the input trigger clock. The reason is that if the input clock has an active trigger edge before the previous trigger edge has propagated through all the flip-flops, you will get an erroneous binary output.

Let's look at the 3-bit counter waveforms in more detail, now taking into account the propagation delays of the 74LS76 flip-flops. In reality, the 2^0 waveform will be delayed to the right (**skewed**) by the propagation of the first flip-flop. The 2^1 waveform will be skewed to the right from the 2^0 waveform, and the 2^2 waveform will be skewed to the right from the 2^1 waveform. This is a cumulative effect that causes the 2^2 waveform to be skewed to the right of the original $\overline{C_{p0}}$ waveform by three propagation delays. (Remember, however, that the propagation delay for most flip-flops is in the 20-ns range, which will not hurt us until the input clock period is very short, 100 to 200 ns (5 to 10 MHz).) Figure 12–11 illustrates the effect of propagation delay on the output waveform.

From Figure 12–11 we can see that the length of time that it takes to change from binary 011 to 100 (3 to 4) will be

$$t_{\text{PHL1}} + t_{\text{PHL2}} + t_{\text{PLH3}} = 30\ \text{ns} + 30\ \text{ns} + 20\ \text{ns} = 80\ \text{ns}$$

Team Discussion

For 80 ns during period 4, the output is invalid. How does this affect the minimum clock period that can be used?

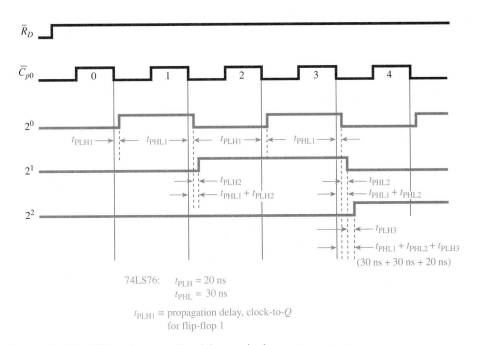

Figure 12–11 Effect of propagation delay on ripple counter outputs.

As we cascade more and more flip-flops to form higher-modulus counters, the cumulative effect of the propagation delay becomes more of a problem.

A MOD-16 ripple counter can be built using four ($2^4 = 16$) flip-flops. Figures 12–12 and 12–13 show the circuit design and waveforms for a MOD-16 ripple counter. From the waveforms we can see that the 2^1 line toggles at every negative edge of the 2^0 line, the 2^2 line toggles at every negative edge of the 2^1 line, and so on, down through each successive flip-flop. When the count reaches 15 (1111), the next negative edge of $\overline{C_{p0}}$ causes all four flip-flops to toggle and changes the count to 0 (0000). Figure 12–13(b) is a photograph of the MOD-16 waveforms displayed on an eight-trace logic analyzer.

Team Discussion

On paper, connect a 4-input NOR gate to the 2^0–2^3 outputs in Figure 12–12. Sketch the output of the NOR gate, including *glitches* (short-duration error pulses) that occur due to propagation delay.

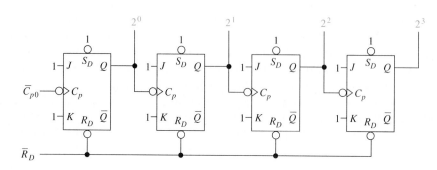

Figure 12–12   MOD-16 ripple counter.

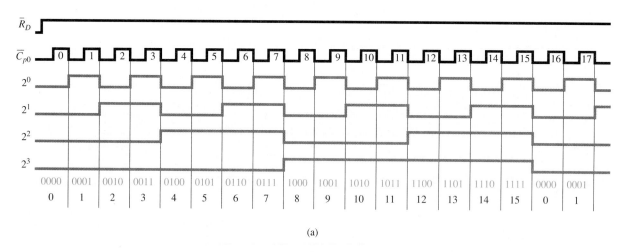

(a)

Figure 12–13   MOD-16 ripple counter waveforms: (a) theoretical counter waveforms;

Down-Counters

On occasion there is a need to count down in binary instead of counting up. To form a down-counter, simply take the binary outputs from the $\overline{Q}$ outputs instead of the Q outputs, as shown in Figure 12–14. The down-counter waveforms are shown in Figure 12–15.

When you compare the waveforms of the up-counter of Figure 12–10 to the down-counter of Figure 12–15, you can see that they are exact complements of each other. That is easy to understand because the binary output is taken from $\overline{Q}$ instead of Q.

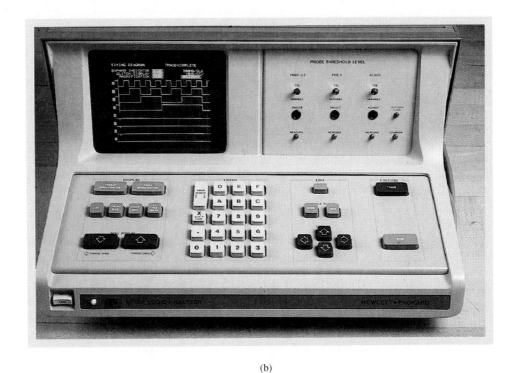

(b)

Figure 12–13   *(Continued)* (b) actual counter output display on a logic analyzer.

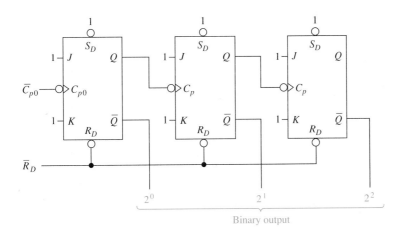

Figure 12–14   MOD-8 ripple down-counter.

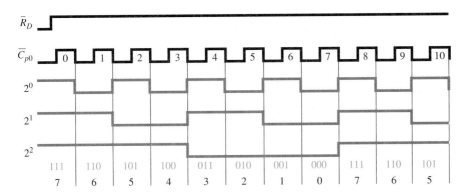

Figure 12–15   MOD-8 down-counter waveforms.

Review Questions

12–1. When analyzing digital circuits containing basic gates combined with sequential logiclike flip-flops, you must remember that gate outputs can change at any time, whereas sequential logic only changes at the active clock edges. True or false?

12–2. For a binary ripple counter to function properly, all J and K inputs must be tied _____ (HIGH, LOW), and all $\overline{S_D}$ and $\overline{R_D}$ inputs must be tied _____ (HIGH, LOW) to count.

12–3. What effect does propagation delay have on ripple counter outputs?

12–4. How can a ripple up-counter be converted to a down-counter?

12–3 Design of Divide-by-N Counters

Counter circuits are also used as frequency dividers to reduce the frequency of periodic waveforms. For example, if we study the waveforms generated by the MOD-8 counter of Figure 12–10, we can see that the frequency of the 2^2 output line is one-eighth of the frequency of the $\overline{C_{p0}}$ input clock line. This concept is illustrated in the block diagram of Figure 12–16, assuming that the input frequency is 24 kHz. So, as it turns out, a MOD-8 counter can be used as a divide-by-8 frequency divider and a MOD-16 can be used as a divide-by-16 frequency divider. Notice that the duty cycle of each of the outputs in Figures 12–10 and 12–13 is 50%.

Clock in ($f = 24$ kHz)   MOD-8 counter (divide-by-8)   Output ($f = 3$ kHz)

Figure 12–16 Block diagram of a divide-by-8 counter.

Common Misconception

As a frequency divider, this circuit has a single input and a single output. As a MOD-5 counter, all three outputs are used.

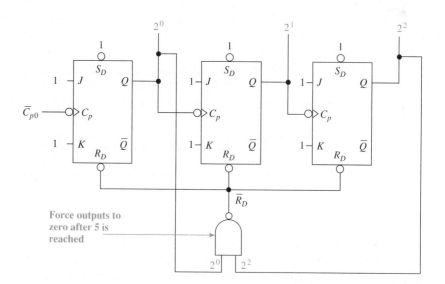

Figure 12–17 Connections to form a divide-by-5 (MOD-5) binary counter.

What if we need a divide-by-5 (MOD-5) counter? We can modify the MOD-8 counter so that when it reaches the number 5 (101) all flip-flops will be Reset. The new count sequence will be 0–1–2–3–4–0–1–2–3–4–0–, and so on. To get the counter to Reset at number 5 (binary 101), you will have to monitor the 2^0 and 2^2 lines and, when they are both HIGH, put out a LOW Reset pulse to all flip-flops. Figure 12–17 shows a circuit that can do this for us.

As you can see, the inputs to the NAND gate are connected to the 2^0 and 2^2 lines, so when the number 5 (101) comes up, the NAND puts out a LOW level to Reset all flip-flops. The waveforms in Figure 12–18 illustrate the operation of the MOD-5 counter of Figure 12–17.

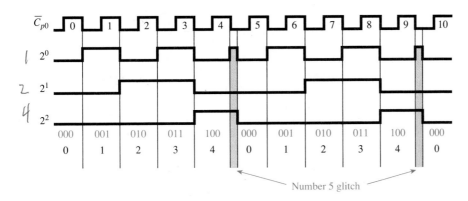

Team Discussion

Besides the 2^2 output, what other output can provide a divide-by-5 signal?

Figure 12–18 Waveforms for the MOD-5 counter.

As we can see in Figure 12–18, the number 5 will appear at the outputs for a short duration, just long enough to Reset the flip-flops. The resulting short pulse on the 2^0 line is called a **glitch.** Do you think you could determine how long the glitch is? (Assume that the flip-flop is a 74LS76 and the NAND gate is a 7400.)

Because t_{PHL} of the NAND gate is 15 ns, it takes that long just to drive the $\overline{R_D}$ inputs LOW. But then it also takes 30 ns (t_{PHL}) for the LOW on $\overline{R_D}$ to Reset the Q output to LOW. Therefore, the total length of the glitch is 45 ns. If the input clock period is in the microsecond range, then 45 ns is insignificant, but at extremely high clock frequencies, that glitch could give us erroneous results. Also notice that the duty cycle of each of the outputs is not 50% anymore.

Any modulus counter (**divide-by-N** counter) can be formed by using external gating to Reset at a predetermined number. The following examples illustrate the design of some other divide-by-N counters.

EXAMPLE 12–5

Design a MOD-6 ripple up-counter that can be manually Reset by an external push button.

Solution: The ripple up-counter is shown in Figure 12–19. The count sequence will be 0–1–2–3–4–5. When 6 (binary 110) is reached, the output of the AND gate will go HIGH, causing the NOR gate to put a LOW on the $\overline{R_D}$ line, resetting all flip-flops to zero.

Team
Discussion

Why do we need such a small value for the pull-down resistor? What if we use a 10-kΩ resistor?

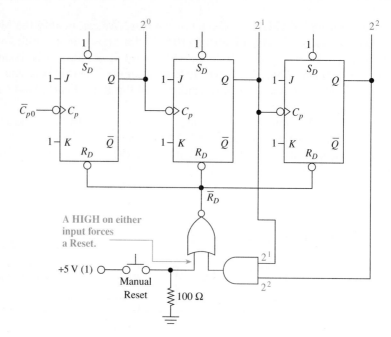

Figure 12–19

As soon as all outputs return to zero, the AND gate will go back to a LOW output, causing the NOR and $\overline{R_D}$ to return to a HIGH, allowing the counter to count again.

This cycle continues to repeat until the manual Reset push button is pressed. The HIGH from the push button will also cause the counter to Reset. The 100-Ω pull-down resistor will keep the input to the NOR gate LOW when the push button is in the open position. [I_{IL} (NOR) = −1.6 mA, $V_{100\,\Omega}$ = 1.6 mA × 100 Ω = 0.160 V ≡ LOW.]

EXAMPLE 12–6

Design a MOD-10 ripple up-counter with a manual push button Reset.

Solution: The ripple up-counter is shown in Figure 12–20. Four flip-flops are required to give us a possibility of $2^4 = 16$ binary states ($2^3 = 8$ would not be enough). We want to stop the count and automatically Reset when 10 (binary 1010) is reached. This is taken care of by the AND gate feeding into the NOR, making the $\overline{R_D}$ line go LOW when 10 is reached. The count sequence will be 0–1–2–3–4–5–6–7–8–9–0–1–, and so on, which is a MOD-10 up-counter. (The number 10 is only a glitch and is not considered a part of the output count.)

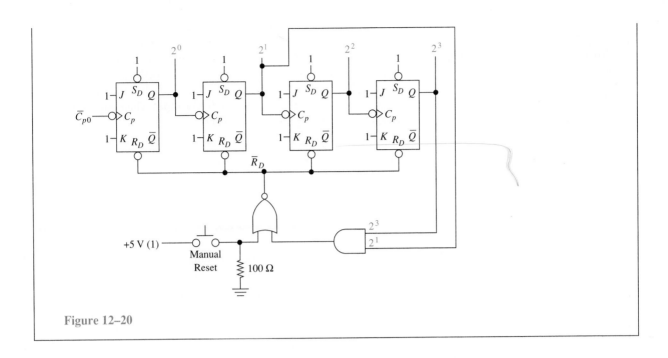

Figure 12–20

EXAMPLE 12–7

Design a MOD-6 down-counter with a manual push button Reset (the count sequence should be 7–6–5–4–3–2–7–6–5–, and so on).

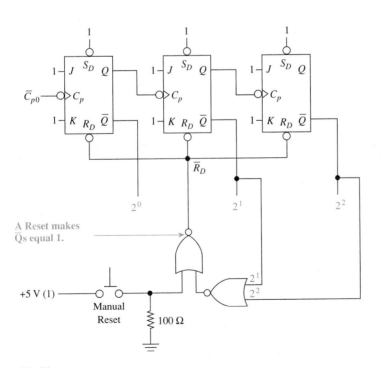

A Reset makes Qs equal 1.

Figure 12–21

Common Misconception

A LOW on $\overline{R_D}$ normally makes the outputs 000, but in this case it makes them 111.

Solution: The down-counter is shown in Figure 12–21. First, by pressing the manual Reset push button, all flip-flops will Reset, making the counter outputs, taken from the $\overline{Q}$'s, to be 1 1 1. The count sequence that we want is 7–6–5–4–3–2, then Reset to 7 again when 1 is reached (binary 001). When 1 is reached, that is the first time that 2^1 and 2^2 are both LOW. The NOR gate connected to 2^1 and 2^2 will give a HIGH output when both of its inputs are LOW, causing the $\overline{R_D}$ line to go LOW.

EXAMPLE 12–8

Design a MOD-5 up-counter that counts in the sequence 6–7–8–9–10–6–7–8–9–10–6–, and so on.

Solution: The up-counter is shown in Figure 12–22. By pressing the manual Preset push button, the 2^1 and 2^2 flip-flops get Set while the 2^0 and 2^3 flip-flops get Reset. This will give the number 6 (binary 0110) at the output. In the count mode, when the count reaches 11 (binary 1011), the output of the AND gates goes HIGH, causing the $\overline{Preset}$ line to go LOW and recycling the count to 6 again.

Helpful Hint

It is useful to list out the bit configurations for the numbers 6 through 11.

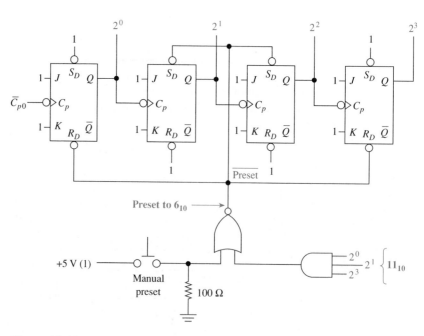

Figure 12–22

EXAMPLE 12–9

Design a down-counter that counts in the sequence 6–5–4–3–2–6–5–4–3–2–6–5–, and so on.

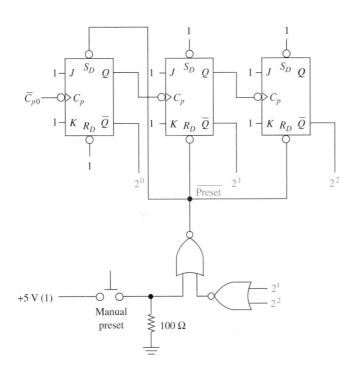

Figure 12–23

Solution: The down-counter is shown in Figure 12–23. When the $\overline{\text{Preset}}$ line goes LOW, the 2^0 flip-flop is Set and the other two flip-flops are Reset (this gives a 6 at the $\overline{Q}$ *outputs*). As the counter counts down toward zero, the 2^1 and 2^2 will both go LOW at the count of 1 (binary 001) and the $\overline{\text{Preset}}$ line will then go LOW again, starting the cycle over again.

EXAMPLE 12–10

Design a counter that counts 0–1–2–3–4–5 and then stops and turns on an LED. The process is initiated by pressing a start push button.

Solution: The required counter is shown in Figure 12–24. When power is first applied to the circuit (power-up), the capacitor will charge up toward 5 V. It starts out at a zero level, however, which causes the 7474 to Reset ($Q_D = 0$). The LOW at Q_D will remain there until the start button is pressed. With a LOW at Q_D, the three counter flip-flops are all held in the Reset state (binary 000). The output of the NAND gate is HIGH, so the LED is OFF.

When the start button is pressed, Q_D goes HIGH and stays HIGH after the button starts bouncing and is released. With Q_D HIGH, the counter begins counting: 0–1–2–3–4–5. When 5 is reached, the output of the NAND gate goes LOW, turning on the LED. The current through the LED will be (5 V − 1.7 V)/330 Ω = 10 mA. The NAND gate can sink a maximum of 16 mA (I_{OL} = 16 mA), so 10 mA will not burn it out.

The LOW output of the NAND gate is also fed to the input of the AND gate, which will disable the clock input. Since the clock cannot get through the AND gate to the first flip-flop, the counter stays at 5 and the LED stays lit.

If you want to Reset the counter to zero again, you could put a push button in parallel across the capacitor so that, when it is pressed, Q_D will go LOW and stay LOW until the start button is pressed again.

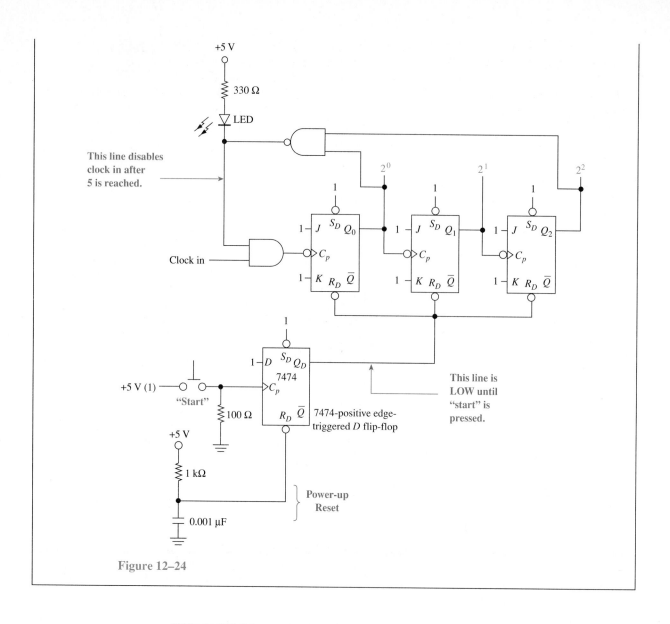

Figure 12–24

Review Questions

12–5. A MOD-16 counter can function as a divide-by-16 frequency divider by taking the output from the _____ output.

12–6. To convert a 4-bit MOD-16 counter to a MOD-12 counter, the flip-flops must be Reset when the counter reaches the number _____ (11, 12, 13).

12–7. Briefly describe the operation of the Manual Reset push-button circuitry used in the MOD-*N* counters in this section.

12–4 Ripple Counter Integrated Circuits

Four-bit binary ripple counters are available in a single integrated-circuit package. The most popular are the 7490, 7492, and 7493 TTL ICs.

Figure 12–25 shows the internal logic diagram for the 7493 4-bit binary ripple counter. The 7493 has 4 *J-K* flip-flops in a single package. It is divided into two sections: a divide-by-2 and a divide-by-8. The first flip-flop provides the divide-by-2 with its $\overline{C_{p0}}$ input and Q_0 output. The second group has three flip-flops cascaded to each other and provides the divide-by-8 via the $\overline{C_{p1}}$ input and $Q_1Q_2Q_3$ outputs. To get a divide-by-16 you can *externally* connect Q_0 to $\overline{C_{p1}}$ so that all four flip-flops are cascaded end to end, as shown in Figure 12–26. Notice that two Master Reset inputs (MR_1, MR_2) are provided to asynchronously Reset all four flip-flops. When MR_1 and MR_2 are both HIGH, all Q's will be Reset to 0. (MR_1 or MR_2 must be held LOW to enable the count mode.)

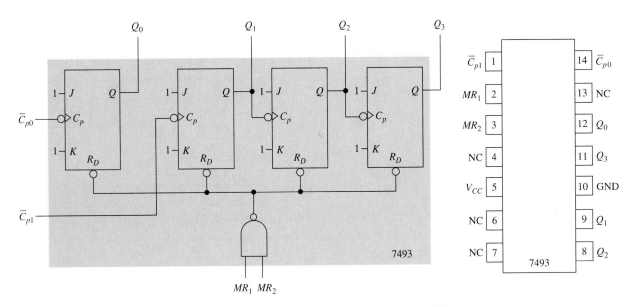

Figure 12–25 Logic diagram and pin configuration for a 7493 4-bit ripple counter IC.

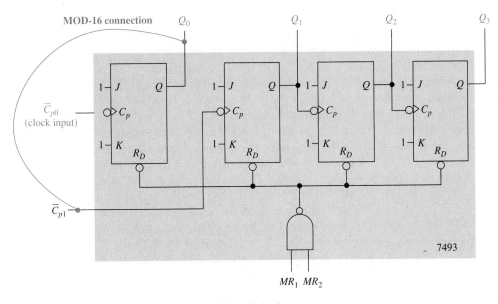

Figure 12–26 A 7493 connected as a MOD-16 ripple counter.

SECTION 12–4 I RIPPLE COUNTER INTERGRATED CIRCUITS

407

With the MOD-16 connection, the frequency output at Q_0 is equal to one-half the frequency input at $\overline{C_{p0}}$. Also, $f_{Q1} = \frac{1}{4}f\overline{C_{p0}}, f_{Q2} = \frac{1}{8}f\overline{C_{p0}}$, and $f_{Q3} = \frac{1}{16}f\overline{C_{p0}}$.

The 7493 can be used to form any modulus counter less than or equal to MOD-16 by utilizing the MR_1 and MR_2 inputs. For example, to form a MOD-12 counter, simply make the external connections shown in Figure 12–27.

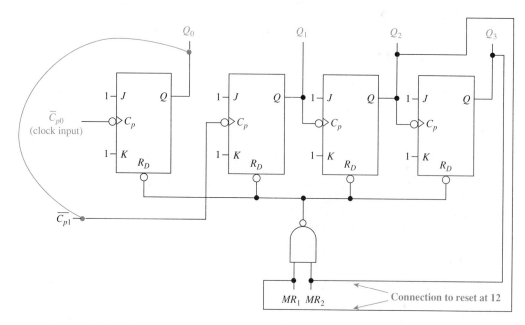

Figure 12–27 External connections to a 7493 to form a MOD-12 counter.

The count sequence of the MOD-12 counter will be 0–1–2–3–4–5–6–7–8–9–10–11–0–1–, and so on. Each time 12 (1100) tries to appear at the outputs, a HIGH–HIGH is placed on MR_1–MR_2 and the counter resets to zero.

Two other common ripple counter ICs are the 7490 and 7492. They both have four internal flip-flops like the 7493, but through the application of internal gating, they automatically recycle to zero after 9 and 11, respectively.

The 7490 is a 4-bit ripple counter consisting of a divide-by-2 section and a divide-by-5 section (see Figure 12–28). The two sections can be cascaded together to form a divide-by-10 (decade or BCD) counter by connecting Q_0 to $\overline{C_{p1}}$ externally. The 7490 is most commonly used for applications requiring a decimal (0 to 9) display.

Notice in Figure 12–28 that, besides having Master Reset inputs (MR_1–MR_2), the 7490 also has Master Set inputs (MS_1–MS_2). When both MS_1 and MS_2 are made HIGH, the clock and MR inputs are overridden, and the Q outputs will be asynchronously Set to a 9 (1001). This is a very useful feature because, if used, it ensures that *after* the first active clock transition the counter will start counting from 0.

The 7492 is a 4-bit ripple counter consisting of a divide-by-2 section and a divide-by-6 section (see Figure 12–29). The two sections can be cascaded together to form a divide-by-12 (MOD-12) by connecting Q_0 to $\overline{C_{p1}}$ and using $\overline{C_{p0}}$ as the clock input. The 7492 is most commonly used for applications requiring MOD-12 and MOD-6 frequency dividing, such as in digital clocks. You can get a divide-by-6 frequency divider simply by ignoring the $\overline{C_{p0}}$ input of the first flip-flop and, instead, bringing the clock input into $\overline{C_{p1}}$, which is the input to the divide-by-6 section. (One peculiarity of the 7492 is that, when connected as a MOD-12, it does *not* count sequentially from 0 to 11. Instead, it counts from 0 to 13, skipping 6 and 7, but still functions as a divide-by-12.)

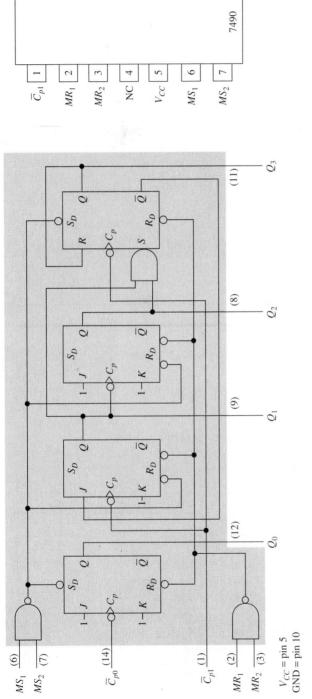

Figure 12–28 Logic diagram and pin configuration for a 7490 decade counter.

( ) = pin numbers
V_{CC} = pin 5
GND = pin 10

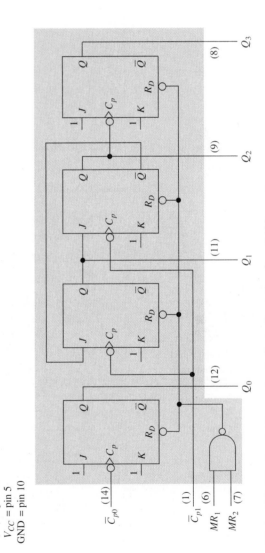

Figure 12–29 Logic diagram and pin configuration for a 7492 counter.

409

EXAMPLE 12–11

Make the necessary external connections to a 7490 to form a MOD-10 counter.

Solution: The MOD-10 counter is shown in Figure 12–30.

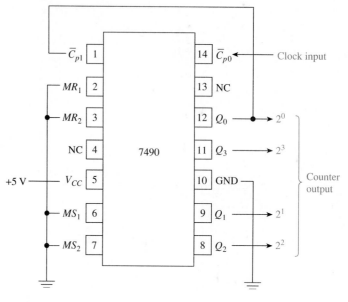

Figure 12–30

EXAMPLE 12–12

Make the necessary external connections to a 7492 to form a divide-by-6 frequency divider ($f_{out} = \frac{1}{6}f_{in}$).

Solution: The frequency divider is shown in Figure 12–31.

Team Discussion

Sketch the f_{in} and f_{out} waveforms that you would observe on a dual-trace oscilloscope.

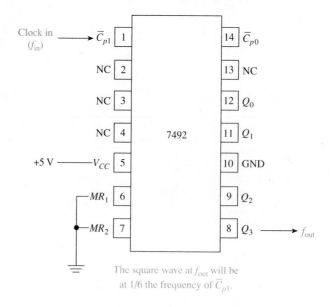

The square wave at f_{out} will be at 1/6 the frequency of $\overline{C}_{p1}$.

Figure 12–31

EXAMPLE 12–13

Make the necessary external connections to a 7490 to form a MOD-8 counter (0 to 7). Also, upon initial power-up, set the counter at 9 so that after the first active input clock edge the output will be 0 and the count sequence will proceed from there.

Solution: The MOD-8 counter is shown in Figure 12–32. The output of the 7414 Schmitt inverter will initially be HIGH when power is first turned on because the capacitor feeding its input is initially discharged to zero. This HIGH on MS_1 and MS_2 will Set the counter to 9. Then, as the capacitor charges up above 1.7 V, the Schmitt will switch to a LOW output, allowing the counter to start its synchronous counting sequence. The inverter on MR_1 is necessary to keep the counter from Resetting when the outputs are at 9 (Q_0, $Q_3 = 1$, 1).

Q_0 is connected to $\overline{C_{p1}}$ so that all four flip-flops are cascaded. When the counter reaches 8 (1000), the MR_1 and MR_2 lines will equal 1–1, causing the counter to Reset to 0. The counter will continue to count in the sequence 0–1–2–3–4–5–6–7–0–1–2–, and so on, continuously.

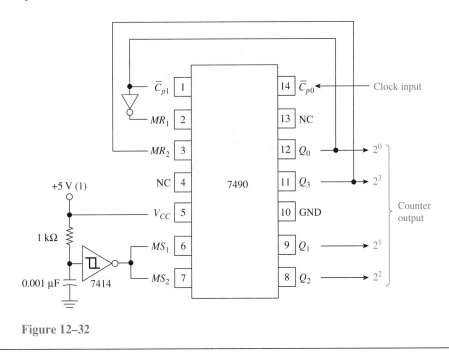

Figure 12–32

Review Questions

12–8. What is the highest modulus of each of the following counter ICs: 7490, 7492, 7493?

12–9. Why does the 7493 counter IC have *two* clock inputs?

12–10. What happens to the Q-outputs of the 7490 counter when you put 1's on the *MS* inputs?

12–5   System Design Applications

Integrated-circuit counter chips are used in a multitude of applications dealing with timing operations, counting, sequencing, and frequency division. To implement a complete system application, output devices such as LED indicators, seven-segment LED displays, relay drivers, and alarm buzzers must be configured to operate from the counter outputs. The synchronous and asynchronous inputs can be driven by such devices as a clock oscillator, a push-button switch, the output from another digital IC, or control signals provided by a microprocessor.

APPLICATION 12–1

For example, let's consider an application that requires an LED indicator to illuminate for 1 s once every 13 s to signal an assembly line worker to perform some manual operation.

Solution: To solve this design problem, we first have to come up with a clock oscillator that produces 1 pulse per second (pps).

The first part of Figure 12–33(a), which is used to produce the 60-pps clock, was described in detail in Section 11–6. To divide the 60 pps down to 1 pps, we can cascade a MOD-10 counter with a MOD-6 counter to create a divide-by-60 circuit.

The 7490 connected as a MOD-10 is chosen for the divide-by-10 section. If you study the output waveforms of a MOD-10 counter, you can see that Q_3 will oscillate at a frequency one-tenth of the frequency at $\overline{C_{p0}}$.

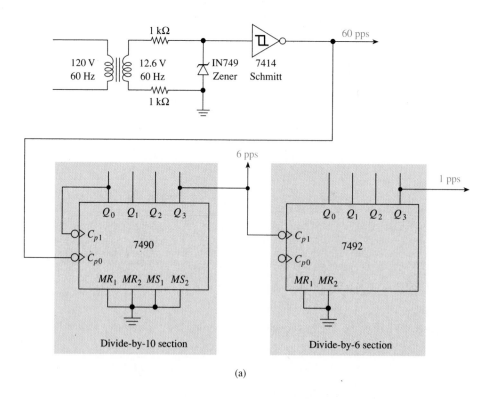

(a)

Figure 12–33   (a) Circuit used to produce 1 pps;

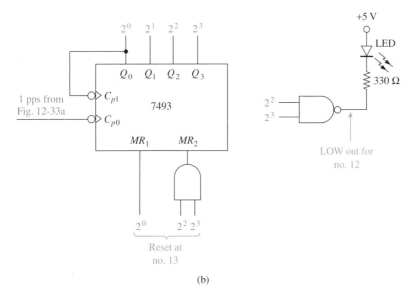

Figure 12–33 *(Continued)* (b) circuit used to illuminate an LED once every 13 s.

Then, if we use Q_3 to trigger the input clock of the divide-by-6 section, the overall effect will be a divide-by-60. [The 7492 is used for the divide-by-6 section simply by using $\overline{C_{p1}}$ as the input and taking the 1-pps output from Q_3, as shown in Figure 12–33(a).]

The next step in the system design is to use the 1-pps clock to enable a circuit to turn on an LED for 1 s once every 13 s. It sounds like we need a MOD-13 counter (0 to 12) and a gating scheme that turns on an LED when the count is on the number 12. A 7493 can be used for a MOD-13 counter and a NAND gate can be used to sink the current from an LED when the number 12 ($Q_2 = 1$, $Q_3 = 1$) occurs. Figure 12–33(b) shows the necessary circuit connections.

Notice in Figure 12–33(b) that a MOD-13 is formed by connecting Q_0 to $\overline{C_{p1}}$ and resetting the counter when the number 13 is reached, resulting in a count of 0 to 12. Also, when the number 12 is reached, the NAND gate's output goes LOW, turning on the LED. [$I_{\text{LED}} = (5 \text{ V} - 1.7 \text{ V})/ 330 \text{ }\Omega = 10 \text{ mA}$].

APPLICATION 12–2

Design a circuit to turn on an LED for 20 ms once every 100 ms. Assume that you have a 50-Hz (50-pps) clock available.

Solution: Because 20 ms is one-fifth of 100 ms, we should use a MOD-5 counter such as the one available in the 7490 IC. To determine which outputs to use to drive the LED, let's look at the waveforms generated by a 7490 connected as a MOD-5 counter.

Remember that the second section of a 7490 is a MOD-5 counter (0 to 4). If the input frequency is 50 Hz, each count will last for 20 ms ($\frac{1}{50}$ Hz = 20 ms), as shown in Figure 12–34(a).

Notice that the Q_3 line goes HIGH for 20 ms once every 100 ms. So if we just invert the Q_3 line and use it to drive the LED, we have the solution to our problem! Figure 12–34(b) shows the final solution.

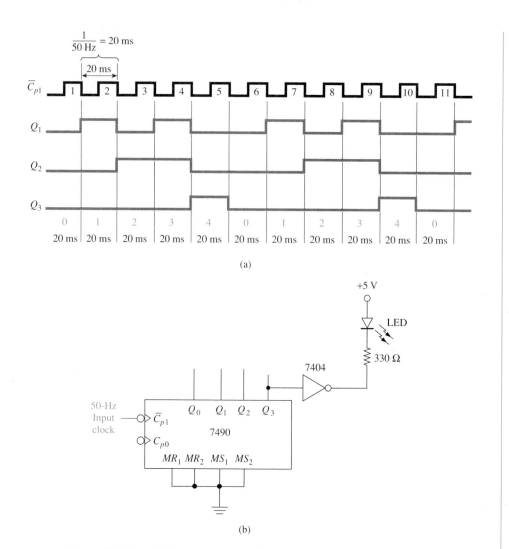

Figure 12–34   (a) Output waveforms from a MOD-5 counter driven by a 50-Hz input clock; (b) solution to Application 12–2.

APPLICATION 12–3

Design a three-digit decimal counter that can count from 000 to 999.

Solution: We have already seen that a 7490 is a single-digit decimal (0 to 9) counter. If we cascade three 7490s together and use the low-order counter to trigger the second digit counter and the second digit counter to trigger the high-order-digit counter, they will count from 000 up to 999. (Keep in mind that the outputs will be binary-coded decimal in groups of 4. In Section 12–6 we will see how we can convert the BCD outputs into actual decimal digits.)

If you review the output waveforms of a 7490 connected as a MOD-10 counter, you can see that at the end of the cycle, when the count changes from 9 (1001) to 0 (0000), the 2^3 output line goes from HIGH to LOW. When cascading counters, you can use that HIGH-to-LOW transition to trigger the input to the next-highest-order counter. That will work out great, because we want the next-highest-order decimal digit to increment by 1 each time the lower-order digit has completed its 0-through-9 cycle

Helpful Hint

Review MOD-10 waveforms to see why the 2^3 signal is used to clock each successive BCD digit.

(that is, the transition from 009 to 010). The complete circuit diagram for a 000 to 999 BCD counter is shown in Figure 12-35.

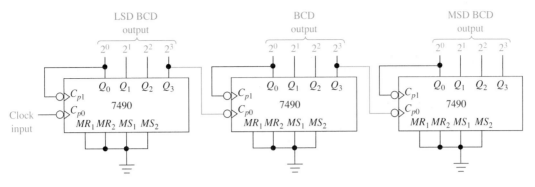

Figure 12–35   Cascading 7490s to form a 000-999 BCD output counter.

APPLICATION 12–4

Design and sketch a block diagram of a digital clock capable of displaying hours, minutes, and seconds.

Solution: First, we have to design a 1-pps clock to feed into the least significant digit of the seconds counter. The seconds will be made up of two cascaded counters that count 00 to 59. When the seconds change from 59 to 00, that transition will be used to trigger the minutes digits to increment by 1. The minutes will also be made up of two cascaded counters that count from 00 to 59. When the minutes change from 59 to 00, that transition will be used to trigger the hours digits to increment by 1. Finally, when the hours reach 12, all counters should be Reset to 0. The digital clock will display the time from 00:00:00 to 11:59:59.

Figure 12–36 is the final circuit that could be used to implement a digital clock. A 1-pps clock (similar to the one shown in Figure 12–33) is used as the initial clock trigger into the least significant digit (LSD) counter of the seconds display. This counter is a MOD-10 constructed from a 7490 IC. Each second this counter will increment. When it changes from 9 to 0, the HIGH-to-LOW edge on the 2^3 line will serve as a clock pulse into the most significant digit (MSD) counter of the seconds display. This counter is a MOD-6 constructed from a 7492 IC.

After 59 s, the 2^2 output of the MOD-6 counter will go HIGH to LOW [once each minute (1 ppm)], triggering the MOD-10 of the minutes section. When the minutes exceed 59, the 2^2 output of that MOD-6 counter will trigger the MOD-10 of the hours section.

The MOD-2 of the hours section is just a single toggle flip-flop having a 1 or 0 output. The hours section is set up to count from 0 to 11. When 12 is reached, the AND gate resets both hours counters. The clock display will therefore be 00:00:00 to 11:59:59.

If you want the clock to display 1:00:00 to 12:59:59 instead, you will have to check for a 13 in the hours section instead of 12. When 13 is reached, you will want to Reset the MOD-2 counter and *Preset* the MOD-10 counter to a 1. Presettable counters such as the 74192 are used in a case like this. (Presettable counters are covered later in this chapter.)

The decoders are required to convert the BCD from the counters into a special code that can be used by the actual display device. Digit displays and decoders are discussed in Section 12–6.

Team Discussion

Discuss ways that you might modify the clock generator circuit to provide a "fast-forward" feature for setting the time.

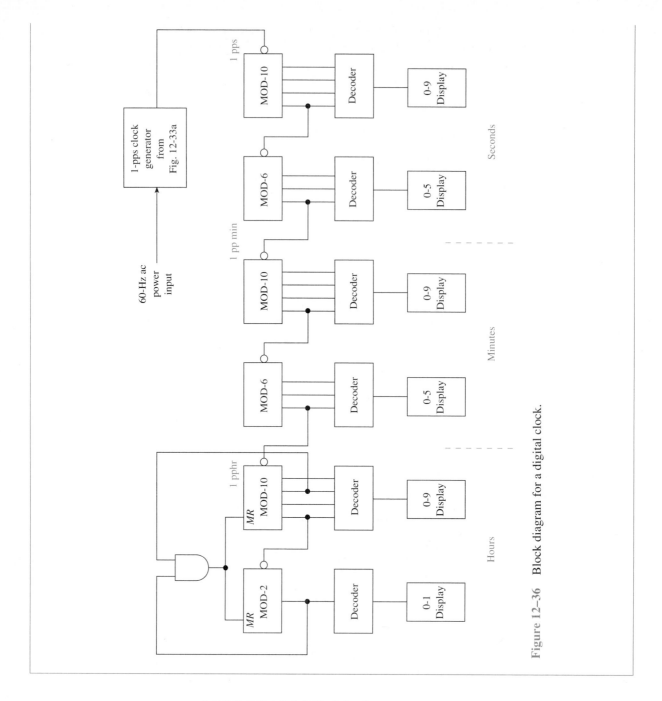

Figure 12–36 Block diagram for a digital clock.

APPLICATION 12-5

Design an egg-timer circuit. The timer will be started when you press a push button. After 3 min, a 5-V, 10-mA dc piezoelectric buzzer will begin buzzing.

Solution: The first thing to take care of is to divide the 1-pps clock previously designed in Figure 12–33(a) down to a 1-ppm clock. At 1 ppm, when the count reaches 3, the buzzer should be enabled and the input clock disabled. An automatic power-up Reset is required on all the counters so that the minute counter will start at zero. A *D* latch can be utilized for the push-button starter so that, after the push button is released, the latch remembers and will keep the counting process going.

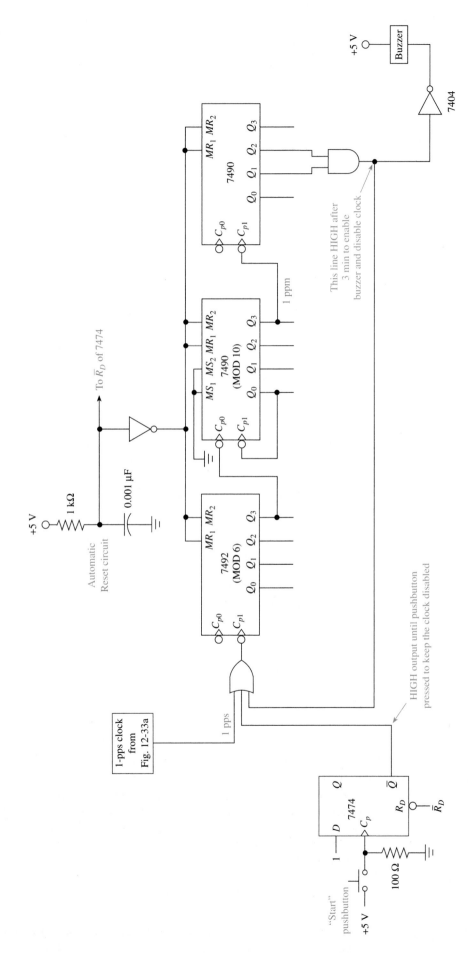

Figure 12–37 Egg-timer circuit design for Application 12–5.

The circuit of Figure 12–37 can be used to implement this design. When power is first turned on, the automatic Reset circuit will Reset all counter outputs and Reset the 7474, making $\overline{Q} = 1$. With $\overline{Q}$ HIGH, the OR gate will stay HIGH, disabling the clock from getting through to the first 7492.

When the start push button is momentarily depressed, $\overline{Q}$ will go LOW, allowing the 1-pps clock to reach $\overline{C_{p1}}$. The first two counters are connected as a MOD-6 and a MOD-10 to yield a divide-by-60, so we have 1 ppm available for the last counter, which serves as a minute counter. When the count reaches 3 in the last 7490, the AND gate goes HIGH, disabling the clock input. This causes the 7404 to go LOW, providing sink current for the buzzer to operate. The buzzer is turned off by turning off the main power supply.

Review Questions

12–11. How could you form a divide-by-60 using two IC counters?

12–12. When cascading several counter ICs end to end, which Q-output drives the clock input to each successive stage?

12–6   Seven-Segment LED Display Decoders

In Section 12–5 we discussed counter circuits that are used to display decimal (0 to 9) numbers. If a counter is to display a decimal number, the count on each 4-bit counter cannot exceed 9 (1001). In other words, the counters must be outputting binary-coded decimal (BCD). As described in Chapter 2, BCD is a 4-bit binary string used to represent the 10 decimal digits. To be useful, however, the BCD must be decoded by a decoder into a format that can be used to drive a decimal numeric display. The most popular display technique is the **seven-segment LED** display.

A seven-segment LED display is actually made up of seven separate light-emitting diodes in a single package. The LEDs are oriented so as to form an ⊟. Most seven-segment LEDs have an eighth LED used for a decimal point.

The job of the decoder is to convert the 4-bit BCD code into a seven-segment code that will turn on the appropriate LED segments to display the correct decimal digit. For instance, if the BCD is 0111 (7), the decoder must develop a code to turn on the top segment and the two right segments (⌐|).

Common-Anode LED Display

The physical layout of a seven-segment LED display is shown in Figure 12–38. This figure shows that the anode of each LED (segment) is connected to the +5-V supply. Now, to illuminate an LED, its cathode must be grounded through a series-limiting resistor, as shown in Figure 12–39. The value of the limiting resistor can be found by knowing that the voltage drop across an LED is 1.7 V and that it takes approximately 10 mA to illuminate it. Therefore,

$$R_{\text{limit}} = \frac{5.0 \text{ V} - 1.7 \text{ V}}{10 \text{ mA}} = 330 \ \Omega$$

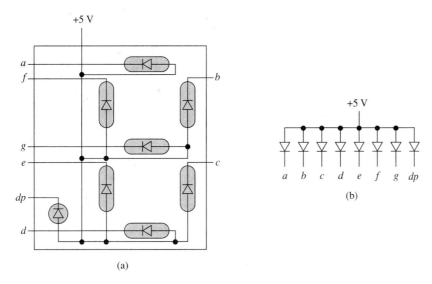

(a)

(b)

Figure 12–38 Seven-segment common-anode LED display: (a) physical layout; (b) schematic.

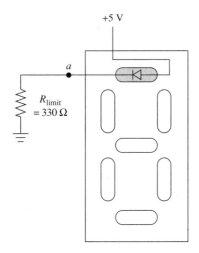

Figure 12–39 Illuminating the *a* segment.

Each segment in the display unit is illuminated in the same way. Figure 12–40 shows the numerical designations for the 10 allowable decimal digits.

Common-anode displays are *active-LOW* (LOW-enable) devices because it takes a LOW to turn on (illuminate) a segment. Therefore, the decoder IC used to drive a **common-anode LED** must have active-LOW outputs.

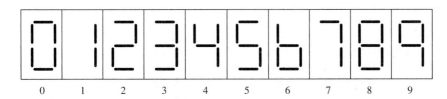

| 0 | 1 | 2 | 3 | 4 | 5 | 6 | 7 | 8 | 9 |

Figure 12–40 Numerical designations for a seven-segment LED.

Common-cathode LEDs and decoders are also available, but they are not as popular because they are *active-HIGH*, and ICs typically cannot *source* (1 output) as much current as they can *sink* (0 output).

BCD-to-Seven-Segment Decoder/Driver ICs*

The 7447 is the most popular common-anode decoder/LED driver. Basically, the 7447 has a 4-bit BCD input and seven individual active-LOW outputs (one for each LED segment). As shown in Figure 12–41, it also has a *lamp test ($\overline{LT}$)* input for testing *all* segments, and it also has **ripple blanking** input and output.

To complete the connection between the 7447 and the seven-segment LED, we need seven 330-Ω resistors (eight if the decimal point is included) for current limiting. Dual-in-line package (DIP) *resistor networks* are available and simplify the wiring process because all seven (or eight) resistors are in a single DIP.

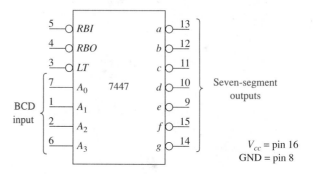

Figure 12–41 Logic symbol for a 7447 decoder.

Figure 12–42 shows typical decoder–resistor–DIP–LED connections. As an example of how Figure 12–42 works, if a MOD-10 counter's outputs are connected to the BCD input and the count is at six (0110_{BCD}), the following will happen:

1. The decoder will determine that a 0110_{BCD} must send the $\overline{c}, \overline{d}, \overline{e}, \overline{f}, \overline{g}$ outputs LOW ($\overline{a}, \overline{b}$ will be HIGH for ⊔).

2. The LOW on those outputs will provide a path for the sink current in the appropriate LED segments via the 330-Ω resistors (the 7447 can sink up to 40 mA at each output).

3. The decimal number ⊔ will be illuminated, together with the decimal point if the dp switch is closed.

A complete three-digit decimal display system is shown in Figure 12–43. The three counters in the figure are connected as MOD-10 counters with the input clock oscillator connected to the least significant counter. The three counters are cascaded by connecting the Q_3 output of the first to the $\overline{C_{p0}}$ of the next, and so on.

Notice that the decimal point of the LSD is always on, so the counters will therefore count from .0 up to 99.9. If the clock oscillator is set at 10 pps, the LSD will indicate tenths of seconds. Also notice that the ripple blanking inputs and outputs ($\overline{RBI}$ and $\overline{RBO}$) are used in this design. They are active LOW and are used for *leading-zero suppression*. For example, if the display output is at 1.4, would you like it to read 01.4 or 1.4? To suppress the leading zero and make it a blank, ground the $\overline{RBI}$ terminal of the MSD decoder. How about if the output is at .6; would you like it to read 00.6, 0.6, or .6? To suppress the second zero when the MSD is blank, simply connect the $\overline{RBO}$ of the MSD decoder to the $\overline{RBI}$ of the second digit decoder. The way this works is that if

---

*Very versatile CMOS seven-segment decoders are the 4543 and its high-speed version, the 74HCT4543. The 4543 provides active-HIGH *or* active-LOW outputs and can drive LED displays as well as **liquid-crystal displays (LCDs).** LCDs are used in low-power battery applications such as calculators and watches. (See Application 16–2.) Their segments don't actually emit light, but instead the individual liquid-crystal segments will polarize to become either opaque (black) or transparent (white) to external light.

the MSD is blank (zero suppressed) the MSD decoder puts a LOW out at $\overline{RBO}$. This LOW is connected to the $\overline{RBI}$ of the second decoder, which forces a blank output (zero suppression) if its BCD input is zero.

The $\overline{RBI}$ and $\overline{RBO}$ can also be used for zero suppression of trailing zeros. For example, if you have an eight-digit display, the $\overline{RBI}$s and $\overline{RBO}$s could be used to automatically suppress the number 0046.0910 to be displayed as 46.091.

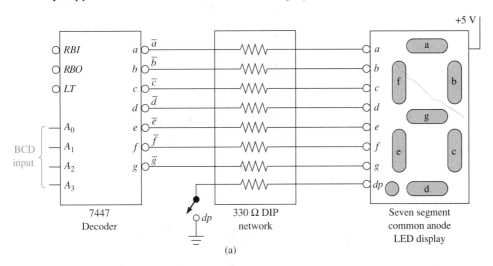

(a)

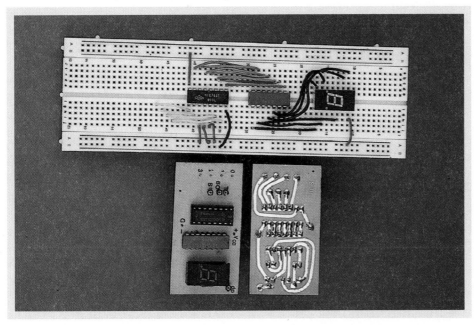

(b)

Figure 12–42 Driving a seven-segment LED display: (a) logic circuit connections; (b) photo of the actual circuit on a breadboard and a printed circuit.

Intelligent LED displays are also available. These displays contain an integrated logic circuit in the same package with the light-emitting diodes. For example, the TIL306 has a built-in BCD counter, a 4-bit data latch, a BCD-to-seven-segment decoder, and the drive circuitry along with the display LEDs. For displaying hexadecimal digits, the TIL311 can be used. It doesn't contain a counter like the TIL306, but it does have a latch, decoder, and driver along with the LED display. It accepts a 4-bit hex input and displays the 16 digits 0 through F.

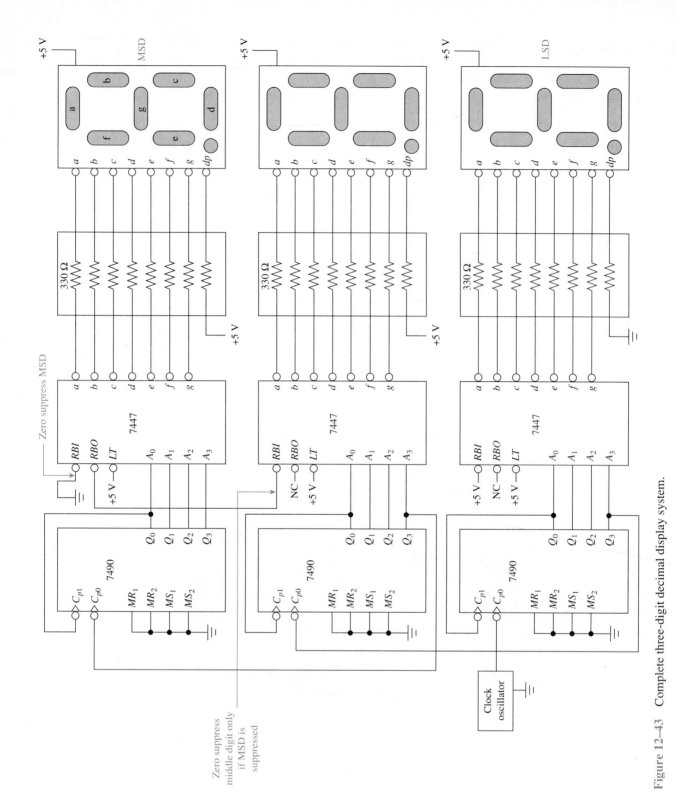

Figure 12–43 Complete three-digit decimal display system.

Team Discussion

Which of the following numbers would be zero-suppressed, 00.7, 10.7, 01.5, or 00.0?

Driving a Multiplexed Display with a Microcontroller

Multidigit LED or LCD displays are commonly used in microprocessor systems. To drive each digit of a six-digit display using separate, dedicated drivers would require six 8-bit I/O ports. Instead, a *multiplexing* scheme is usually used. Using the multiplexing technique, up to eight digits can be driven by using only two output ports. One output port is used to select which *digit* is to be active, whereas the other port is used to drive the appropriate *segments* within the selected digit. Figure 12–44 shows how two I/O ports of an 8051 microcontroller can be used to drive a six-digit multiplexed display.

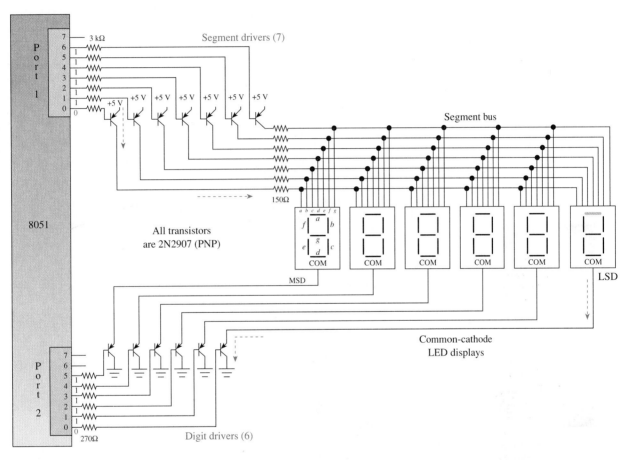

Figure 12–44 Multiplexed six-digit display with the *a* segment of the LSD illuminated. (From *Digital and Microprocessor Fundamentals,* by William Kleitz, Prentice Hall, Englewood Cliffs, N.J., 1990.)

Team Discussion

What codes must be sent to Port 1 and Port 2 to display the number 7 in the MSD? How about 77 in the MSDs?

The displays used in Figure 12–44 are common-cathode LEDs. To enable a digit to work, the connection labeled COM must be grounded. The individual segments are then illuminated by supplying +5 V via a 150-Ω limiting resistor to the appropriate segment.

It takes about 10 mA to illuminate a single segment. If all segments in one digit are on, as with the number 8, the current in the COM line will be 70 mA. The output ports of the 8051 can only sink 1.6 mA. This is why we need the *PNP* transistors set up as current buffers. When port 1 outputs a 0 on bit 0, the first *PNP* turns on, shorting the emitter to collector. This allows current to flow from the +5-V supply through the 150-Ω limiting resistor, to the *a* segments. None of the *a* segments will illuminate unless one of the digits' COM lines is brought LOW. To enable the LSD, port 2 will output a 0 on bit 0, which shorts the emitter to collector of that transistor. This provides a path for current to flow from the COM on the LSD to ground.

Notice that enabling both the segment and the digit requires an active-LOW signal. To drive all six digits, we have to *scan* the entire display repeatedly with the appropriate numbers to be displayed. For example, to display the number 123456, we need to turn on the segments for the number 1 (*b* and *c*) and then turn on the MSD. We then turn off all digits, turn on the segments for the number 2 (*a*, *b*, *g*, *e*, and *d*), and turn on the next digit. We then turn off all digits, turn on the segments for the number 3, and turn on the next digit. This process repeats until all six digits have been flashed on once. At that point, the MSD is cycled back on, followed by each of the next digits. By repeating this cycle over and over again, the number 123456 appears to be on all the time. This process of decoding the segments and scanning the digits is performed by software instructions written for the 8051 output ports.

Review Questions

12–13. Seven-segment displays are either common anode or common cathode. What does this mean?

12–14. List the active segments, by letter, that form the following digits on a seven-segment display: 5, 0.

12–15. Why are series resistors required when driving a seven-segment LED display?

12–16. The 7447 IC is used to convert _____ data into _____ data for common-_____ LEDs.

12–17. Liquid crystal displays (LCDs) use more power but are capable of emitting a brighter light than LEDs. True or false?

12–18. What is the advantage of using a multiplexing scheme for multi-digit displays like the one shown in Figure 12–44?

12–7 Synchronous Counters

Remember the problems we discussed with ripple counters due to the accumulated propagation delay of the clock from flip-flop to flip-flop? (See Figure 12–11.) Well, synchronous counters eliminate that problem, because all the clock inputs ($\overline{C_p}$'s) are tied to a common clock input line, so each flip-flop will be triggered at the same time (thus any Q output transitions will occur at the same time).

If we want to design a 4-bit synchronous counter, we need four flip-flops, giving us a MOD-16 (2^4) binary counter. Keep in mind that, because all the $\overline{C_p}$ inputs receive a trigger at the same time, we must hold certain flip-flops from making output transitions until it is their turn. To design the connection scheme for the synchronous counter, let's first study the output waveforms of a 4-bit binary counter to determine which flip-flops are to be held from toggling, and when.

From the waveforms in Figure 12–45(a), we can see that the 2^0 output is a continuous toggle off the clock input line. The 2^1 output line toggles on every negative edge of the 2^0 line, but since the 2^1's $\overline{C_{p0}}$ input is also connected to the clock input, it must be held from toggling until the 2^0 line is HIGH. This can be done simply by tying the J and K inputs to the 2^0 line, as shown in Figure 12–45(b).

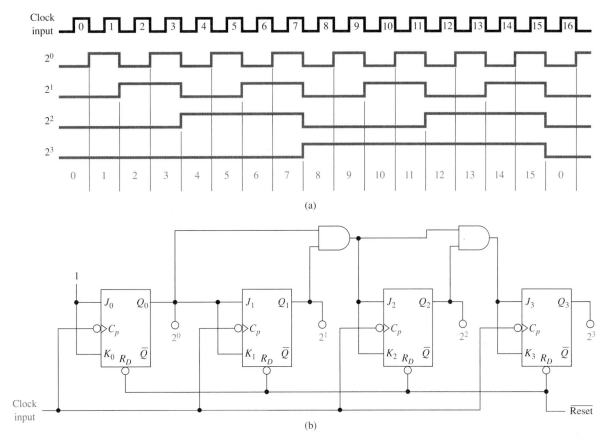

Figure 12–45 Four-bit MOD-16 synchronous counter: (a) output waveforms; (b) circuit connections.

The same logic follows through for the 2^2 and 2^3 output lines. The 2^2 line must be held from toggling until the 2^0 *and* 2^1 lines are both HIGH. Also, the 2^3 line must be held from toggling until the 2^0 *and* 2^1 *and* 2^2 lines are all HIGH.

To keep the appropriate flip-flops in the *hold* condition or *toggle* condition, their J and K inputs are tied together and, through the use of additional AND gates, as shown in Figure 12–45(b), the *J-K* inputs will be both 0 or 1, depending on whether they are to be in the hold or toggle mode.

From Figure 12–45(b) we can see that the same clock input is driving all four flip-flops. The 2^1 flip-flop will be in the hold mode ($J_1 = K_1 = 0$) until the 2^0 output goes HIGH, which will force J_1-K_1 HIGH, allowing the 2^1 flip-flop to toggle when the next negative clock edge comes in.

Now, observe the output waveforms [Figure 12–45(a)] while you look at the circuit design [Figure 12–45(b)] to determine the operation of the last two flip-flops. From the waveforms we see that the 2^2 output must not be allowed to toggle until 2^0 *and* 2^1 are both HIGH. Well, the first AND gate in Figure 12–45(b) takes care of that by holding J_2-K_2 LOW. The same method is used to keep the 2^3 output from toggling until the 2^0 *and* 2^1 *and* 2^2 outputs are *all* HIGH.

As you can see, the circuit is more complicated, but the cumulative effect of propagation delays through the flip-flops is not a problem as it was in ripple counters, because all output transitions will occur at the same time, because all flip-flops are triggered from the same input line. (There *is* a propagation delay through the AND gates, but it will not affect the Q outputs of the flip-flops.)

As with ripple counters, synchronous counters can be used as down-counters by taking the output from the $\overline{Q}$ outputs and can form any modulus count by resetting the count to zero after some predetermined binary number has been reached.

EXAMPLE 12–14

Design a MOD-6 synchronous binary up-counter.

Solution: A MOD-6 counter will count 0–1–2–3–4–5–0–1–, and so on. To count to 5, we will need three flip-flops and will have to Reset the count to zero when the number 6 (110_2) is reached, as shown in the circuit in Figure 12–46.

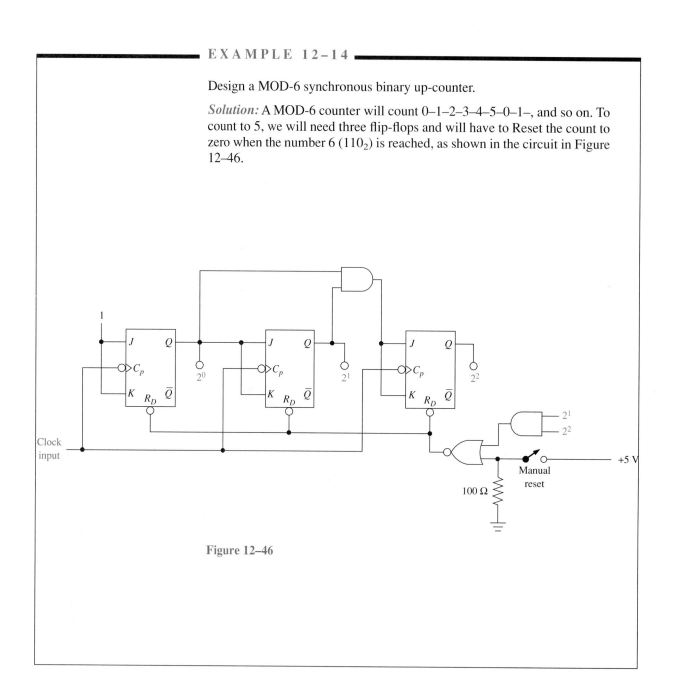

Figure 12–46

System Design Application

Synchronous binary counters have many applications in the timing and sequencing of digital systems. The following design will illustrate one of these applications.

APPLICATION 12–6

Let's say that your company needs a system that will count the number of hours of darkness each day. The senior design engineer for your company will be connecting his microcontroller-based system to your counter outputs after you are sure that your system is working correctly. After the counter outputs are read, the microcontroller will issue a LOW Reset pulse to your counter to Reset it to all zeros sometime before sunset.

Solution: You decide to use a synchronous counter but realize that it may be dark outside for as many as 18 h per day. A 4-bit counter will not count high enough, so first you have to come up with the 5-bit synchronous counter design that is shown in Figure 12–47. That was not hard; you just had to add one more AND gate and a flip-flop to a 4-bit counter.

From analog electronics you remembered that a phototransistor has varying resistance from collector to emitter, depending on how much light strikes it. The **phototransistor** that you decide to use has a resistance of 10 MΩ when it is in the dark and 10 Ω when it is in the daylight. Your final circuit design is shown in Figure 12–47.

Explanation of Figure 12–47: Let's start with the 5-bit synchronous counter. With the addition of the last AND gate and flip-flop, it will be capable of counting from 0 up to 31 (MOD-32). The manual Reset push button, when depressed, will Reset the counter to zero.

The phototransistor collector-to-ground voltage will be almost zero during daylight because the collector-to-emitter resistance acts almost like a short. (Depending on the transistor used, the ON resistance may be as low as 10 Ω.) The Schmitt inverters are used to give a sharp HIGH-to-LOW and LOW-to-HIGH at sunset and sunrise to eliminate any false clock switching. Schmitt triggers are most commonly available as inverting functions, so two of them are necessary so that a LOW at the collector will come through as a LOW at point A.

The LOW at point A during the daylight will hold all the MOD counters (divide-by-N's) at zero so that at the beginning of sunset the waveform at point B will start out LOW and take one full hour before it goes HIGH to LOW, triggering the first transition at Q. During the nighttime, point B will **oscillate** at 1 pulse per hour, incrementing the counter once each hour. At sunrise, point A goes LOW, forcing all MODs LOW and disabling the clock. The counter outputs at Q_0 to Q_4 will be read by the microcontroller during the day and then Reset.

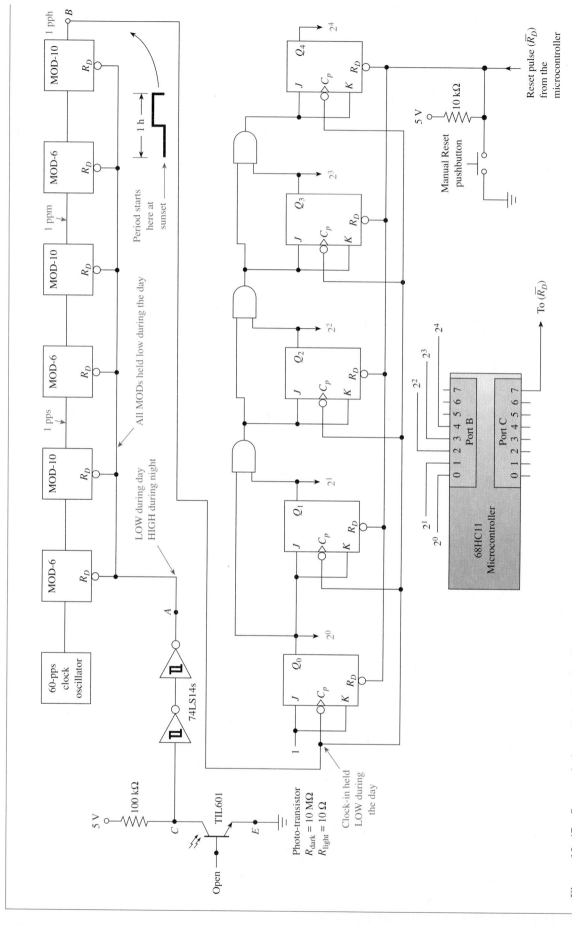

Figure 12–47 System design solution for the "hours of darkness counter."

Review Questions

12–19. What advantage do synchronous counters have over ripple counters?

12–20. Because each flip-flop in a synchronous counter is driven by the same clock input, what keeps *all* flip-flops from toggling at each active clock edge?

12–21. The 5-bit synchronous counter in Figure 12–47 counts continuously, day and night, but is ignored by the microcontroller during the day. True or false?

12–8 Synchronous Up/Down-Counter ICs

Four-bit synchronous binary counters are available in a single integrated-circuit (IC) package. Two popular synchronous IC counters are the 74192 and 74193. They both have some features that were not available on the ripple counter ICs. They can count *up or down* and can be *preset* to any count that you desire. The 74192 is a BCD decade **up/down-counter,** and the 74193 is a 4-bit binary up/down-counter. The logic symbol used for both counters is shown in Figure 12–48.

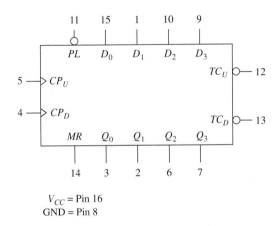

V_{CC} = Pin 16
GND = Pin 8

Figure 12–48 Logic symbol for the 74192 and 74193 synchronous counter ICs.

There are two separate clock inputs: C_{pU} for counting up and C_{pD} for counting down. One clock must be held HIGH while counting with the other. The binary output count is taken from Q_0 to Q_3, which are the outputs from four internal *J-K* flip-flops. The Master Reset *(MR)* is an active-HIGH Reset for resetting the Q outputs to zero.

The counter can be preset by placing any binary value on the parallel data inputs (D_0 to D_3) and then driving the Parallel Load $(\overline{PL})$ line LOW. The parallel load operation will change the counter outputs regardless of the conditions of the clock inputs.

The **Terminal Count** Up $(\overline{TC_U})$ and Terminal Count Down $(\overline{TC_D})$ are normally HIGH. The $\overline{TC_U}$ is used to indicate that the maximum count is reached and the count is about to recycle to zero (carry condition). The $\overline{TC_U}$ line goes LOW for the 74193 when the count reaches 15 *and* the input clock (C_{pU}) goes HIGH to LOW. $\overline{TC_U}$ remains LOW until C_{pU} returns HIGH. This LOW pulse at $\overline{TC_U}$ can be used as a clock input to the next-higher-order stage of a multistage counter.

The $\overline{TC_U}$ output for the 74192 is similar except that it goes LOW at 9 *and* a LOW C_{pU} (see Figure 12–49 on page 431). The Boolean equations for $\overline{TC_U}$, therefore, are as follows:

$$\text{LOW at } \overline{TC_U} = Q_0 Q_1 Q_2 Q_3 \overline{C_{pU}} \qquad (74193)$$

$$\text{LOW at } \overline{TC_U} = Q_0 Q_3 \overline{C_{pU}} \qquad (74192)$$

The Terminal Count Down ($\overline{TC_D}$) is used to indicate that the minimum count is reached and the count is about to recycle to the maximum (15 or 9) count (borrow condition). Therefore, $\overline{TC_D}$ goes LOW when the down-count reaches zero and the input clock (C_{pD}) goes LOW (see Figure 12–51). The Boolean equation at $\overline{TC_D}$ is

$$\text{LOW at } \overline{TC_D} = \overline{Q_0}\ \overline{Q_1}\ \overline{Q_2}\ \overline{Q_3}\ \overline{C_{pD}} \qquad (74192 \text{ and } 74193)$$

The function table shown in Table 12–2 can be used to show the four operating modes (Reset, Load, Count up, and Count down) of the 74192/74193.

The best way to illustrate how these chips operate is to exercise all their functions and observe the resultant waveforms, as shown in the following examples.

Table 12–2 Function Table for the 74192/74193 Synchronous Counter IC[a]

Operating Mode	Inputs								Outputs					
	MR	$\overline{PL}$	C_{pU}	C_{pD}	D_0	D_1	D_2	D_3	Q_0	Q_1	Q_2	Q_3	$\overline{TC_U}$	$\overline{TC_D}$
Reset	H	×	×	L	×	×	×	×	L	L	L	L	H	L
	H	×	×	H	×	×	×	×	L	L	L	L	H	H
Parallel load	L	L	×	L	L	L	L	L	L	L	L	L	H	L
	L	L	×	H	L	L	L	L	L	L	L	L	H	H
	L	L	L	×	H	H	H	H	H	H	H	H	L	H
	L	L	H	×	H	H	H	H	H	H	H	H	H	H
Count up	L	H	↑	H	×	×	×	×	Count up				H	H
Count down	L	H	H	↑	×	×	×	×	Count down				H	H

[a]H = HIGH voltage level; L = LOW voltage level; × = don't care; ↑ = LOW-to-HIGH clock transition.

EXAMPLE 12–15

Common Misconception

Students often mistakenly draw the LOW at $\overline{TC}$ for the entire terminal count clock period instead of only while the clock is LOW.

Draw the input and output timing waveforms for a 74192 that goes through the following sequence of operation:

1. Reset all outputs to zero.
2. Parallel load a 7 (0111).
3. Count up five counts.
4. Count down five counts.

Solution: The timing waveforms are shown in Figure 12–49.

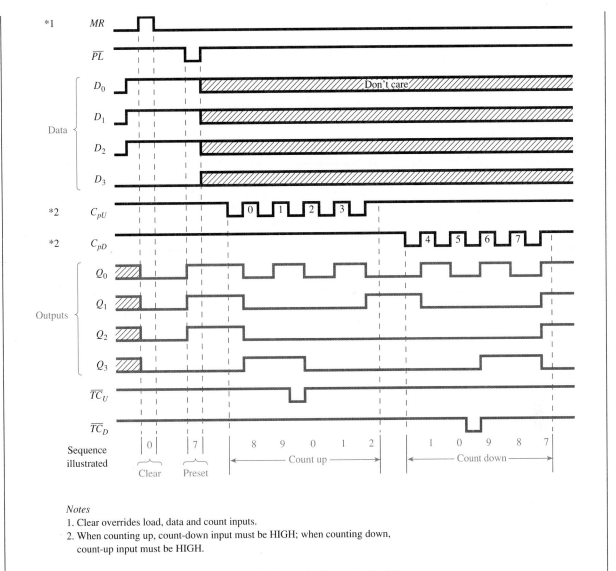

Figure 12–49 Timing waveforms for the 74192 used in Example 12–15.

Notes
1. Clear overrides load, data and count inputs.
2. When counting up, count-down input must be HIGH; when counting down, count-up input must be HIGH.

EXAMPLE 12–16

Draw the output waveforms for the 74193 shown in Figure 12–50, given the waveforms shown in Figure 12–51. (Initially, set $D_0 = 1$, $D_1 = 0$, $D_2 = 1$, $D_3 = 1$, and $MR = 0$.)

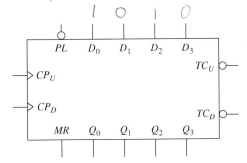

Figure 12–50 Circuit connections for Example 12–16.

Solution:

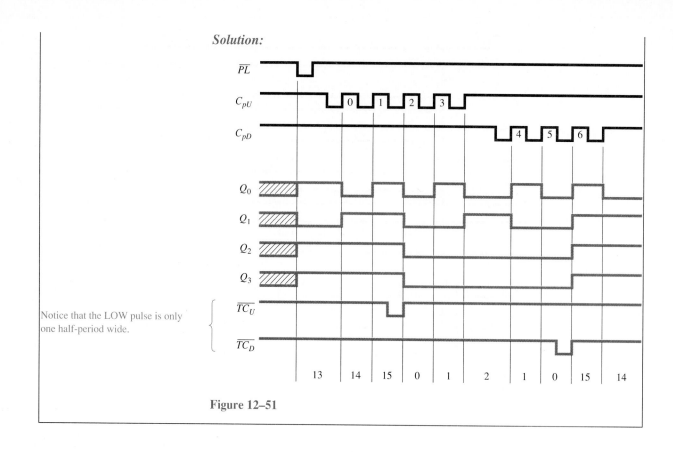

Notice that the LOW pulse is only one half-period wide.

Figure 12–51

EXAMPLE 12-17

Design a decimal counter that will count from 00 to 99 using two 74192 counters and the necessary drive circuitry for the two-digit display. (Display circuitry was explained in Section 12–5.)

Solution: The 74192s can be used to form a multistage counter by connecting the $\overline{TC_U}$ of the first counter to the C_{pU} of the second counter. $\overline{TC_U}$ will go LOW, then HIGH, when the first counter goes from 9 to 0 (carry). This LOW-to-HIGH edge can be used as the clock input to the second stage, as shown in Figure 12–52.

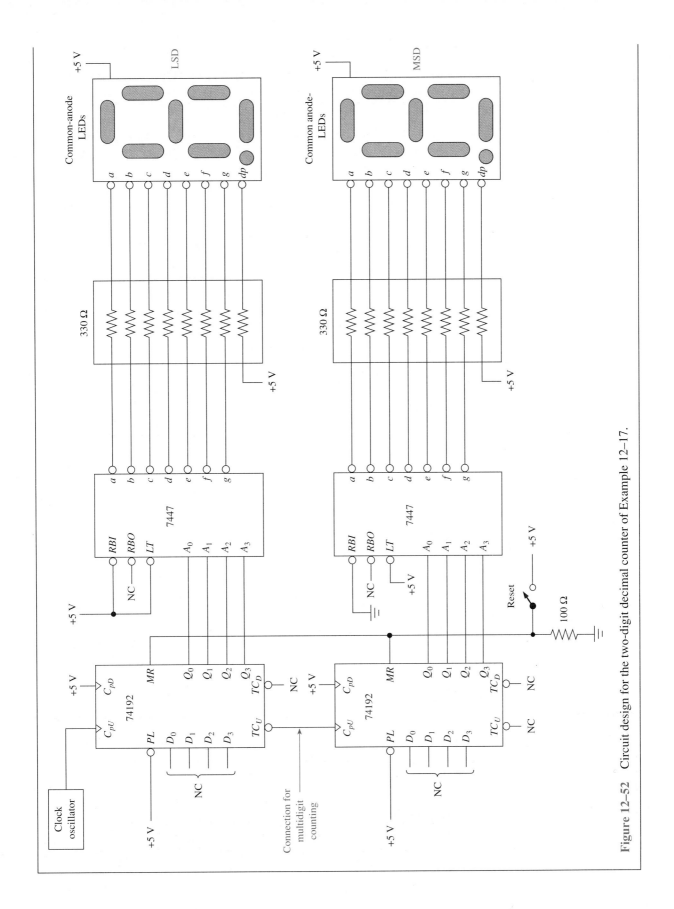

Figure 12–52 Circuit design for the two-digit decimal counter of Example 12–17.

74190/74191 Synchronous Counter ICs

Other forms of synchronous counters are the 74190 and 74191. The 74190 is a BCD counter (0 to 9) and the 74191 is a 4-bit counter (0 to 15). They have some different features and input/output pins, as shown in Figure 12–53.

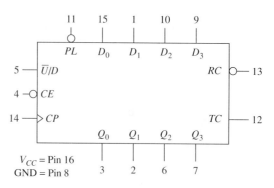

Figure 12–53   Logic symbol for the 74190/74191 synchronous counters.

The 74190/74191 can be preset to any count by using the **Parallel Load** *(PL)* operation. It can count up or down by using the $\overline{U}/D$ input. With $\overline{U}/D = 0$ it will count up, and with $\overline{U}/D = 1$ it will count down. The Count Enable input *(CE)* is an active-LOW input used to enable/inhibit the counter. With $\overline{CE} = 0$, the counter is enabled. With $\overline{CE} = 1$, the counter stops and holds the current states of the Q_0 to Q_3 outputs.

The Terminal Count output *(TC)* is normally LOW, but it goes HIGH when the counter reaches zero in the count-down mode and 15 (or 9) in the count-up mode. The ripple clock output $(\overline{RC})$ follows the input clock (C_p) whenever *TC* is HIGH. In other words, in the count-down mode, when zero is reached, $\overline{RC}$ will go LOW when C_p goes LOW. The $\overline{RC}$ output can be used as a clock input to the next higher stage of a multistage counter, just the way that the $\overline{TC}$ outputs of the 74192/74193 were used. In either case, however, the multistage counter will not be truly synchronous because of the small propagation delay from C_p to $\overline{RC}$ of each counter.

For a multistage counter to be *truly* synchronous, the C_p of each stage must be connected to the *same* clock input line. The 74190/74191 counters enable you to do this by using the *TC* output to inhibit each successive stage from counting until the previous stage is at its Terminal Count. Figure 12–54 shows how three 74191s can be connected to form a true 12-bit binary synchronous counter.

In Figure 12–54, we can see that each counter stage is driven by the same clock, making it truly synchronous. The second stage is inhibited from counting until the first stage reaches 15. The second stage will then increment by 1 at the next positive clock edge. Stage 1 will then inhibit stage 2 via the *TC*-to-$\overline{CE}$ connection while stage 1 is counting up to 15 again. The same operation between stages 2 and 3 also keeps stage 3 from incrementing until stages 1 and 2 both reach 15.

74160/61/62/63 Synchronous Counter ICs

Finally, another type of counter allows you to perform true synchronous counting without using external gates, as we had to in Figure 12–54. The 74160/74161/74162/74163 synchronous counter ICs have *two* Count Enable inputs *(CEP* and *CET)* and a Terminal Count output to facilitate high-speed synchronous counting. The logic symbol is given in Figure 12–55. From the logic symbol we can see that this counter is similar to the previous synchronous counters, except that it has two active-HIGH Count Enable inputs *(CEP* and *CET)* and an active-HIGH Terminal Count *(TC)* output. (There are other differences between this and other synchronous counters, but we will leave it up to you to determine those from reading your TTL data manual.)

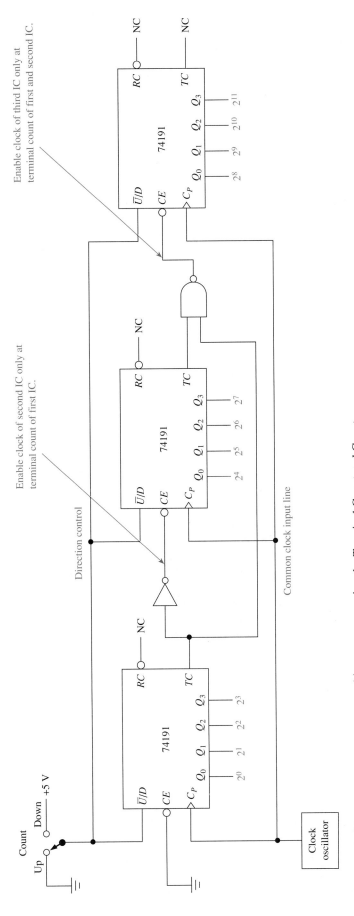

Figure 12–54 Twelve-bit synchronous binary counter using the Terminal Count and Count Enable features.

435

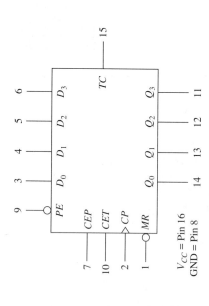

Figure 12–55 Logic symbol for the 74160/74161/74162/74163 synchronous counter.

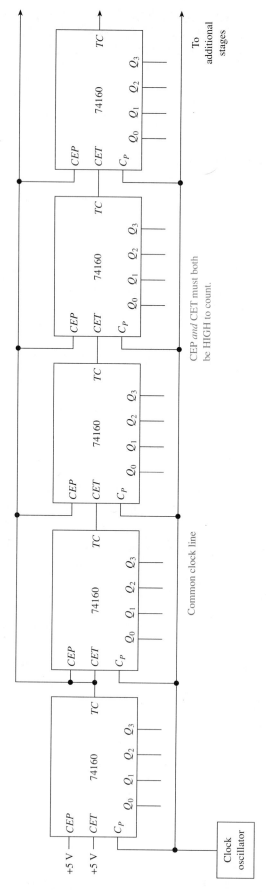

Figure 12–56 High-speed multistage synchronous counter.

Both count enables *(CEP* and *CET)* must be HIGH to count. The Terminal Count output *(TC)* will go HIGH when the highest count is reached. *TC* will be forced LOW, however, when *CET* goes LOW, even though the highest count may be reached. This is an important feature that enables the multistage counter of Figure 12–56 to operate properly.

Review Questions

12–22. What is the function of the $\overline{TC_U}$ and $\overline{TC_D}$ output pins on the 74193 synchronous counter IC?

12–23. How do you change the 74190 from an up-counter to a down-counter?

12–24. The $\overline{CE}$ input to the 74190 synchronous counter is the *Chip Enable* used to enable/disable the *Q*-outputs. True or false?

12–9 Applications of Synchronous Counter ICs

The following applications will explain some useful design strategy and circuit operations using synchronous counter ICs.

APPLICATION 12–7

Design a counter that will count up 0 to 9, then down 9 to 0, then up 0 to 9 repeatedly using a synchronous counter and various gates.

Solution: Because the count is 0 to 9, a BCD counter will work. Also, we want to go up, then down, then up, and so on, so it would be easy if we had a reversible counter like the 74190 and just toggled the $\overline{U}/D$ terminal each time the Terminal Count is reached. Figure 12–57 could be used to implement this circuit. When power is first applied, the 74190 will be Parallel

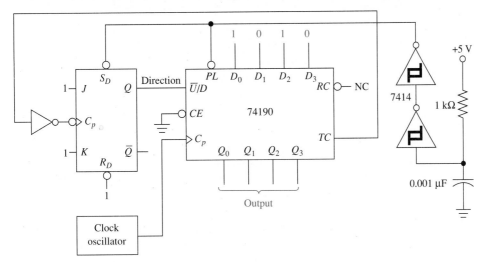

Figure 12–57 Self-reversing BCD counter (solution to Application 12–7).

Loaded with a 5 (0101) and the direction line will be 1. (5 is chosen arbitrarily because it is somewhere between the terminal counts 0 and 9.) The counter will count down to 0, at which time TC will go HIGH, causing the flip-flop to toggle and changing the direction to 0. With the clock oscillator still running, the counter will reverse and start counting up. When 9 is reached, TC goes HIGH, again changing the direction and the cycle repeats.

APPLICATION 12–8

Design and sketch the timing waveforms for a divide-by-9 frequency divider using a 74193 counter.

Solution: We can use the Parallel Load feature of the 74193 to set the counter at some initial value and then count down to zero. When we reach zero, we will have to Parallel Load the counter to its initial value and count down again, making sure the repetitive cycle repeats once every nine clock periods. Figure 12–58 could be used to implement such a circuit. $\overline{TC_D}$ is fed back into $\overline{PL}$. This means that, when the Terminal Count is reached, the LOW out of $\overline{TC_D}$ will enable the Parallel Load, making the outputs equal to the D_0 to D_3 inputs (1001).

Helpful Hint

This circuit configuration works fine as a frequency divider but not as a counter, because the 0 and 9 appear for only one-half of a period.

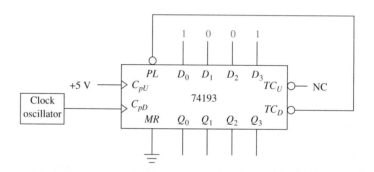

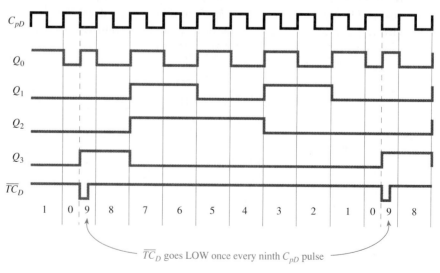

$\overline{TC_D}$ goes LOW once every ninth C_{pD} pulse

Figure 12–58 Circuit design and timing waveforms for a divide-by-9 frequency divider.

The time waveforms arbitrarily start at 1 and count down. Notice at zero (Terminal Count) that $\overline{TC_D}$ goes LOW when C_{pD} goes LOW (remember that a LOW at $\overline{TC_D} = \overline{Q_0}\,\overline{Q_1}\,\overline{Q_2}\,\overline{Q_3}\,\overline{C_{pD}}$). As soon as $\overline{TC_D}$ goes LOW, the outputs return to 9, thus causing $\overline{TC_D}$ to go back HIGH again. Therefore, $\overline{TC_D}$ is a narrow pulse just long enough to perform the Parallel Load operation.

The down counting resumes until zero is reached again, which causes the Parallel Load of 9 to occur again. The $\overline{TC_D}$ pulse occurs once every ninth C_{pD} pulse; thus we have a divide-by-9. (A different duty-cycle divide-by-9 can be gotten from the Q_3 or Q_2 outputs.)

APPLICATION 12–9

Design a divide-by-200 using synchronous counters.

Solution: The number 200 exceeds the maximum count of a single 4-bit counter. Two 4-bit counters can be cascaded together to form an 8-bit counter capable of counting 256 states ($2^8 = 256$).

The 74193 is a logical choice for a 4-bit counter. We can cascade two of them together to form an 8-bit down-counter. If we preload with the binary equivalent of the number 200 and count down to zero, we can use the borrow output ($\overline{TC_{D2}}$) to drive the Parallel Load *(PL)* line LOW to recycle back to 200. Figure 12–59 shows the circuit connections to form this 8-bit divide-by-200 counter. The two 74193 counters will start out at some unknown value and start counting down toward zero. The borrow-out ($\overline{TC_{D2}}$) line will go LOW when the count reaches zero and C_{pD2} is LOW. As soon as $\overline{TC_{D2}}$ goes LOW, a Parallel Load of number 200 takes place, making $\overline{TC_{D2}}$ go back HIGH again. Therefore, $\overline{TC_{D2}}$ is just a short glitch, and the number zero will appear at the outputs for just one-half of a clock period, and the number 200 will appear the other one-half of the same clock period. The remainder of the numbers will follow a regular counting sequence (199 down to 1), giving us 200 complete clock pulses between the LOW pulses on $\overline{TC_{D2}}$. If the short glitch on $\overline{TC_{D2}}$ is not wide enough as a divide-by-200 output, it could be widened to produce any duty cycle without affecting the output frequency by using a one-shot multivibrator pulse stretcher. (One shots are discussed in Chapter 14.)

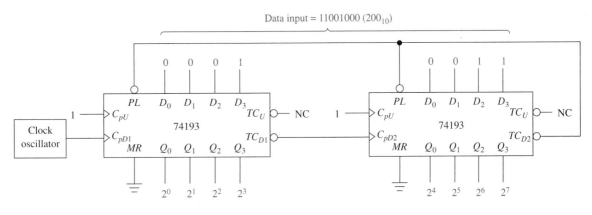

Figure 12–59 Eight-bit divide-by-200 counter.

Use a 74163 to form a MOD-7 synchronous up-counter. Sketch the timing waveforms.

Solution: The 74163 has a *synchronous* Reset feature—that is, a LOW level at the Master Reset *($\overline{MR}$)* input will Reset all flip-flops (Q_0 to Q_3) at the next positive clock *(C_p)* edge. Therefore, what we can do is bring the Q_1 and Q_2 (binary 6) lines into a NAND gate to drive the $\overline{MR}$ line LOW when the count is at 6. The next positive C_p edge would normally increase the count to 7, but instead will Reset the count to 0. The result is a count from 0 to 6, which is a MOD-7.

Remember that with previous MOD-*N* counters we would look for the number that was one greater than the last number to be counted, and when we reached it we would Reset the count to zero, because we went beyond the modulus required. That method of resetting after the fact works, but it lets a short-duration glitch (unwanted state) through to the outputs before resetting (see Figure 12–18). With the 74163, using a synchronous Reset, as shown in Figure 12–60, we can avoid that problem.

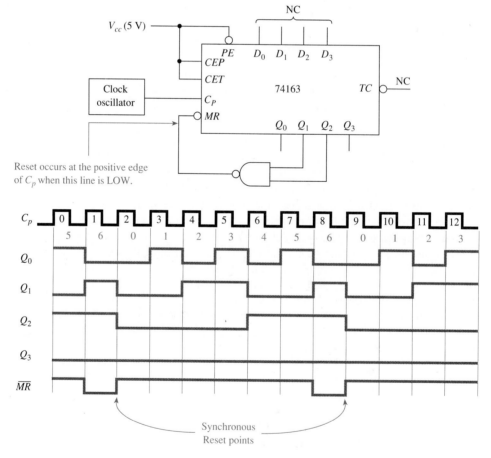

Figure 12–60   Using the synchronous Reset on the 74163 to form a glitch-free MOD-7 up-counter.

Summary

In this chapter we have learned that

1. Toggle flip-flops can be cascaded end to end to form ripple counters.

2. Ripple counters cannot be used in high-speed circuits because of the problem they have with the accumulation of propagation delay through all the flip-flops.

3. A down-counter can be built by taking the outputs from the $\overline{Q}$'s of a ripple counter.

4. Any modulus (or divide-by) counter can be formed by resetting the basic ripple counter when a specific count is reached.

5. A glitch is an unwanted level transition that may appear on some of the output bits of a ripple counter.

6. Ripple counter ICs such as the 7490, 7492, and 7493 have four flip-flops integrated into a single package providing four-bit counter operations.

7. Four-bit counter ICs can be cascaded end to end to form counters with higher than MOD-16 capability.

8. Seven-segment LED displays choose between seven separate LEDs (plus a decimal point LED) to form the 10 decimal digits. They are constructed with either the anodes or the cathodes connected to a common pin.

9. LED displays require a decoder/driver IC such as the 7447 to decode BCD data into a seven-bit code to activate the appropriate segments to illuminate the correct digit.

10. Synchronous counters eliminate the problem of accumulated propagation delay associated with ripple counters by driving all four flip-flops with a common clock.

11. The 74192 and 74193 are 4-bit synchronous counter ICs. They have a count-up/count-down feature and can accept a 4-bit parallel load of binary data.

12. The 74190 and 74191 synchronous counter ICs are similar to the 74192/74193 except they are better for constructing multistage counters of more than 4 bits. The 74160 series goes one step further and allows for truly synchronous high-speed multistage counting.

Glossary

Asynchronous Counter: See *ripple counter.*

Cascade: In multistage systems, when the output of one stage is fed directly into the input of the next.

Common-Anode LED: A seven-segment LED display whose LED anodes are all connected to a common point and supplied with +5 V. Each LED segment is then turned on by supplying a LOW level (via a limiting resistor) to the appropriate LED cathode.

Divide-by-*N*: The Q outputs in counter operations will oscillate at a frequency that is at some multiple *(N)* of the input clock frequency. For example, in a divide-by-8 (MOD-8) counter the output frequency of the highest-order Q (Q_2) is one-eighth the frequency of the input clock.

Glitch: A short-duration-level change in a digital circuit.

Liquid Crystal Display (LCD): A low-power display technology that creates an image by selectively making sections of its crystalline structure either opaque or transparent to incident light. This forms dark and light segments that together create alphanumeric images.

Modulus: In a digital counter the modulus is the number of different counter steps.

Oscillate: Change digital states repeatedly (HIGH–LOW–HIGH–LOW–, and so on).

Parallel Load: A feature on some counters that allows you to load all 4 bits of a counter at the same time, asynchronously.

Phototransistor: A transistor whose collector-to-emitter current and resistance vary, depending on the amount of light shining on its base junction.

Ripple Blanking: A feature supplied with display decoders to enable the suppression of leading and trailing zeros.

Ripple Counter: (Asynchronous counter) A multibit counter whose clock input trigger is not connected to each flip-flop but instead has to propagate through each flip-flop to reach the input of the next. The fact that the clock has to "ripple" through from stage to stage tends to decrease the maximum operational frequency of the ripple counter.

Sequential: Operations that follow a predetermined sequence of digital states triggered by a timing pulse or clock.

Seven-Segment LED: Seven light-emitting diodes fabricated in a single package. By energizing various combinations of LED segments, the 10 decimal digits can be displayed.

Skewed: A skewed waveform or pulse is one that is offset to the right or left with respect to the time axis.

Synchronous Counter: A multibit counter whose clock input trigger is connected to each flip-flop so that each flip-flop will operate in step with the same input clock transition.

Terminal Count: The highest (or lowest) count in a multibit counting sequence.

Up/Down-Counter: A counter that is capable of counting up or counting down.

Problems

Section 12–1

12–1. How are sequential logic circuits different from combinational logic gate circuits?

12–2. The waveforms shown in Figure P12–2 are applied to the inputs at A, $\overline{R_D}$, and C_p. Sketch the resultant waveforms at D, Q, $\overline{Q}$, and X.

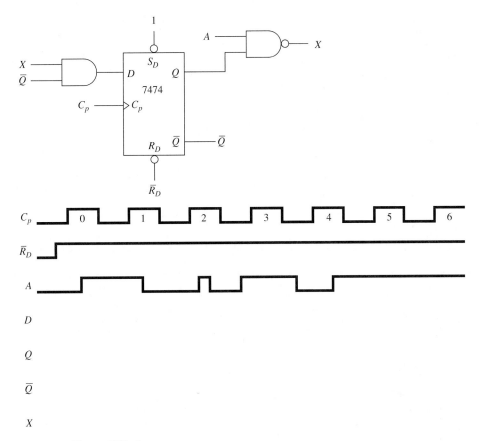

Figure P12–2

12–3. Repeat Problem 12–2 for the input waveforms shown in Figure P12–3.

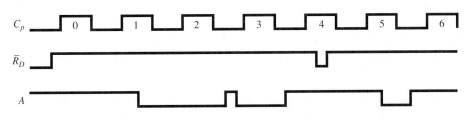

Figure P12–3

12–4. The waveforms shown in Figure P12–4 on page 444 are applied to the inputs at A, $\overline{C_p}$, and $\overline{R_D}$. Sketch the resultant waveforms at J, K, Q, and $\overline{Q}$.

12–5. Repeat Problem 12–4 for the input waveforms shown in Figure P12–5 on page 444.

Section 12–2

12–6. What is the modulus of a counter whose output counts from

(**a**) 0 to 7? (**d**) 10 to 0?

(**b**) 0 to 18? (**e**) 2 to 15?

(**c**) 5 to 0? (**f**) 7 to 3? -5

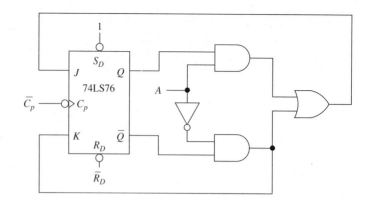

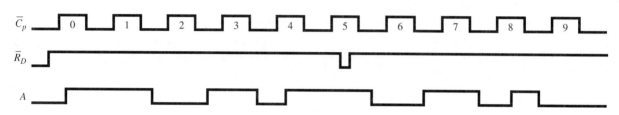

Figure P12–4

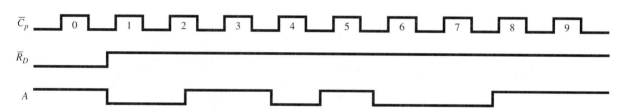

Figure P12–5

12–7. How many *J-K* flip-flops are required to construct the following counters?

(a) MOD-7   (d) MOD-20

(b) MOD-8   (e) MOD-33

(c) MOD-2   (f) MOD-15

12–8. If the input frequency to a 6-bit counter is 10 MHz, what is the frequency at the following output terminals?

(a) 2^0 (d) 2^3

(b) 2^1 (e) 2^4

(c) 2^2 (f) 2^5

12–9. Draw the timing waveforms at $\overline{C_p}$, 2^0, 2^1, and 2^2 for a 3-bit binary up-counter for 10 clock pulses.

12–10. Repeat Problem 12–9 for a binary down-counter.

12–11. What is the highest binary number that can be counted using the following number of flip-flops?

(a) 2   (c) 7

(b) 4   (d) 1

12–12. In a 5-bit counter the frequency at the following output terminals is what fraction of the input clock frequency?

(a) 2^0 (d) 2^3

(b) 2^1 (e) 2^4

(c) 2^2

12–13. How many flip-flops are required to form the following divide-by-N frequency dividers?

(a) Divide-by-4 (c) Divide-by-12

(b) Divide-by-15 (d) Divide-by-18

12–14. Explain why the propagation delay of a flip-flop affects the maximum frequency at which a ripple counter can operate.

C **12–15.** Sketch the $\overline{C_p}$, 2^0, 2^1, and 2^2 output waveforms for the counter shown in Figure P12–15. (Assume that flip-flops are initially Reset.)

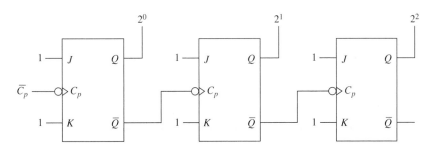

Figure P12–15

12–16. Is the counter of Problem 12–15 an up- or down-counter, and is it a MOD-8 or MOD-16?

12–17. Sketch the connections to a 3-bit ripple up-counter that can be used as a divide-by-6 frequency divider.

D **12–18.** Design a circuit that will convert a 2-MHz input frequency into a 0.4-MHz output frequency.

D **12–19.** Design and sketch a MOD-11 ripple up-counter that can be manually Reset by an external push button.

C D **12–20.** Design and sketch a MOD-5 ripple down-counter with a manual Reset push button. (The count sequence should be 7–6–5–4–3–7–6–5–, and so on.)

C D **12–21.** Repeat Problem 12–20 for a count sequence of 10–9–8–7–6–10–9–8–, and so on.

C D **12–22.** Design a MOD-4 ripple up-counter that counts in the sequence 10–11–12–13–10–11–12–, and so on.

C D **12–23.** Redesign the Reset circuitry of Figure 12–19 using a 7401 open-collector NAND and an active-LOW push button with a 10-kΩ pull-up resistor. (Use no other gates.)

C T **12–24.** When you test your design for Problem 12–23, it works fine until you depress the push button. After that, it becomes a MOD-8 counter.

When you check the ICs you find that a 7400 was used instead of the 7401. What caused the MOD-6 counter to turn into a MOD-8?

C D   **12–25.** The circuit in Figure P12–25 is being considered as a replacement for the Reset circuitry of Figure 12–19. Do you think that it will work? Why?

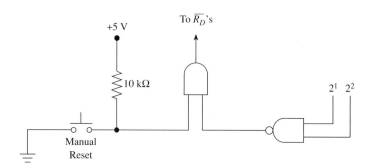

Figure P12–25

Section 12–4

12–26. Describe the major differences among the 7490, 7492, and the 7493 TTL ICs.

12–27. Assume that you have one 7490 and one 7492. Show the external connections that are required to form a divide-by-24.

12–28. Repeat Problem 12–27 using two 7492s to form a divide-by-36.

12–29. Using as many 7492s and 7490s as you need, sketch the external connections required to divide a 60-pps clock down to one pulse per day.

12–30. Make the necessary external connections to a 7493 to form a MOD-10 counter.

Section 12–5

D   **12–31.** Design a ripple counter circuit that will flash an LED ON for 40 ms and OFF for 20 ms (assume that a 100-Hz clock oscillator is available). (*Hint:* Study the output waveforms from a MOD-6 counter.)

C D   **12–32.** Design a circuit that will turn on an LED 6 s after you press a momentary push button. (Assume that a 60-pps clock is available.)

12–33. What modification to the egg timer circuit of Figure 12–37 could be made to allow you to turn off the buzzer without shutting off the power?

Section 12–6

12–34. Calculate the size of the series current-limiting resistor that could be used in Figure 12–39 to limit the LED current to 15 mA instead of 10 mA.

T   **12–35.** In Figure 12–42, instead of using a resistor dip network, some designers use a single limiting resistor in series with the 5-V supply and connect the 7447 outputs directly to the LED inputs to save money. It works, but the display does not look as good; can you explain why?

12–36. What advantage does a synchronous counter have over a ripple counter?

12–37. Sketch the waveforms at $\overline{C_p}$, 2^0, 2^1, and 2^2 for 10 clock pulses for the 3-bit synchronous counter shown in Figure P12–37.

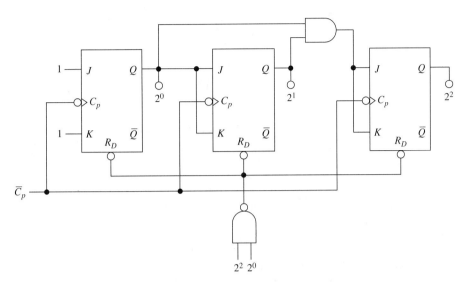

Figure P12–37

12–38. The duty cycle of a square wave is defined as the time the wave is HIGH, divided by the total time for one period. From the waveforms that you sketched for Problem 12–37, find the duty cycle for the 2^2 output wave.

Sections 12–8 and 12–9

C   **12–39.** Sketch the timing waveforms at $\overline{TC_D}$, $\overline{TC_U}$, Q_0, Q_1, Q_2, and Q_3 for the 74192 counter shown in Figure P12–39.

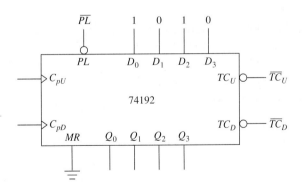

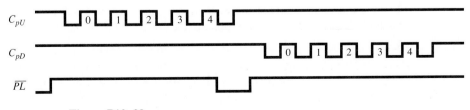

Figure P12–39

C   **12–40.** Sketch the timing waveforms at $\overline{RC}$, TC, Q_0, Q_1, Q_2, and Q_3 for the 74191 counter shown in Figure P12–40.

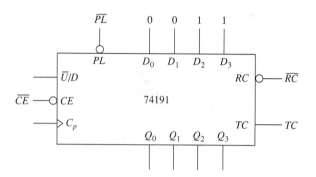

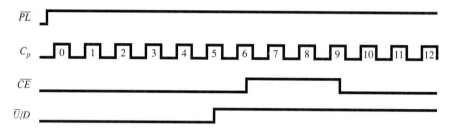

Figure P12–40

C D   **12–41.** Make all the necessary pin connections to a 74193 without using external gating to form a divide-by-4 frequency divider. Make it an up-counter and show the waveforms at C_{pU}, $\overline{TC_U}$, Q_0, Q_1, Q_2, and Q_3.

C D   **12–42.** Using the synchronous Reset feature of the 74163 counter, make the necessary connections to form a glitch-free MOD-12 up-counter.

Schematic Interpretation Problems

See Appendix G for the schematic diagrams.

S   **12–43.** The 74161s in the Watchdog Timer schematic are used to form an 8-bit counter.

(a) Which is the HIGH-order and which is the LOW-order counter?

(b) Is the parallel-load feature being used on these counters?

(c) How are the counters reset in this circuit?

S D   **12–44.** On a separate piece of paper, redesign the counter section of the Watchdog Timer schematic by replacing the 74161s with 74193s.

S C D   **12–45.** The 68HC11 microcontroller in the HC11D0 master board schematic provides a clock output signal at the pin labeled E. This clock signal is used as the input to the LCD controller, M1 (grid location E-7). The frequency of this signal is 9.8304, as dictated by the crystal on the 68HC11. To experiment with different clock speeds on the LCD controller, you want to divide that frequency by 2, 4, 8, and 16 before inputting it to pins 6 and 10. Design a circuit using a 4-bit counter IC connected as a frequency divider and a multiplexer IC to select which counter output is sent to the LCD controller for its clock signal.

Answers to Review Questions

12–1. True

12–2. HIGH, HIGH

12–3. It places limitations on the maximum frequency allowed by the input trigger clock because each output is delayed from the previous one.

12–4. By taking the binary output from the $\overline{Q}$ outputs

12–5. 2^3

12–6. 12

12–7. When the push button is pressed, 5 V is applied to the NOR gate, driving its output LOW. The 100-Ω pull-down resistor will keep the input LOW when the push button is in the open position.

12–8. MOD-10, MOD-12, MOD-16

12–9. One is for the divide-by-2 section, and the other is for the divide-by-8 section.

12–10. $Q_0 = 1, Q_1 = 0, Q_2 = 0, Q_3 = 1$

12–11. By cascading a divide-by-10 with a divide-by-6

12–12. The Q associated with the most significant bit

12–13. This means that all of the anodes or cathodes of the LED segments are connected together.

12–14. a c d f g, a b c d e f

12–15. To limit the current flowing through the LED segments

12–16. Active-HIGH, active-LOW, anode

12–17. False

12–18. It reduces both the number of I/O ports and data paths needed.

12–19. They don't have the problem of accumulated propagation delay.

12–20. J and K inputs are tied together and are controlled through the use of an AND gate.

12–21. False

12–22. They are the terminal count pins, which are used to indicate when the terminal count is reached, and the count is about to recycle.

12–23. $\overline{U}/D = 1$

12–24. False

13 Shift Registers

OUTLINE

13–1 Shift Register Basics
13–2 Parallel-to-Serial Conversion
13–3 Recirculating Register
13–4 Serial-to-Parallel Conversion
13–5 Ring Shift Counter and Johnson Shift Counter
13–6 Shift Register ICs
13–7 System Design Applications for Shift Registers
13–8 Driving a Stepper Motor with a Shift Register
13–9 Three-State Buffers, Latches, and Transceivers

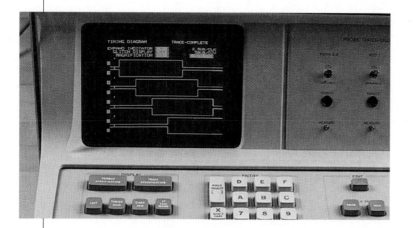

Objectives

Upon completion of this chapter, you should be able to

- Connect *J-K* flip-flops as serial or parallel-in to serial or parallel-out multi-bit shift registers.

- Draw timing waveforms to illustrate shift register operation.

- Explain the operation and application of ring and Johnson shift counters.

- Make external connections to MSI shift register ICs to perform conversions between serial and parallel data formats.

- Explain the operation and application of three-state output buffers.

- Discuss the operation of circuit design applications that employ shift registers.

Introduction

Registers are required in digital systems for the temporary storage of a group of bits. **Data bits** (1's and 0's) traveling through a digital system sometimes have to be temporarily stopped, copied, moved, or even shifted to the right or left one or more positions.

A shift register facilitates this movement and storage of data bits. Most shift registers can handle parallel movement of data bits, as well as serial movement, and can also be used to convert from parallel to serial and serial to parallel.

13–1   Shift Register Basics

Let's take a look at the contents of a 4-bit shift register as it receives 4 bits of parallel data and shifts them to the right four positions into some other digital device. The timing for the shift operations is provided by the input clock. The data bits will shift to the right by one position for each input clock pulse, as shown in Figure 13–1.

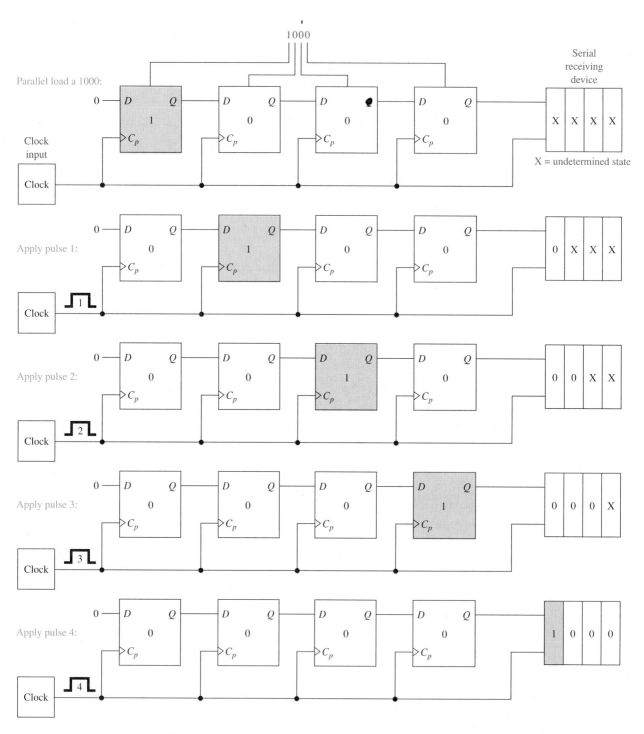

Figure 13–1   Block diagram of a 4-bit shift register for parallel-to-serial conversion.

In the figure, the group of four boxes is four D flip-flops comprising the 4-bit shift register. The first step is to parallel load the register with a 1–0–0–0. *Parallel load* means to load all four flip-flops at the same time. This is done by momentarily enabling the appropriate asynchronous Set ($\overline{S_D}$) and Reset ($\overline{R_D}$) inputs.

Next, the first clock pulse causes all bits to shift to the right by 1 because the input to each flip-flop comes from the Q output of the flip-flop to its left. Each successive pulse causes all data bits to shift one more position to the right.

At the end of the fourth clock pulse, all data bits have been shifted all the way across, and now all four original data bits appear, in the correct order, in the serial receiving device. The connections between the fourth flip-flop and the serial receiving device could be a three-conductor serial transmission cable (serial data, clock, and ground).

Figure 13–1 illustrated a parallel-to-serial conversion. Shift registers can also be used for serial-to-parallel, parallel-to-parallel, and serial-to-serial, as well as shift-right, shift-left operations as indicated in Figure 13–2(a). Each of these configurations and an explanation of the need for a *recirculating line* are explained in upcoming sections.

Figure 13–2(b) shows how shift registers are commonly used in data communications systems. Computers operate on data internally in a *parallel format*. To communicate over a serial cable like the one used by the RS232 standard or a telephone line, the data must first be converted (**data conversion**) to the *serial format*. For example, for computer A to send data to computer B, computer A will parallel load 8 bits of data into shift register A and then apply eight clock pulses. The 8 data bits output

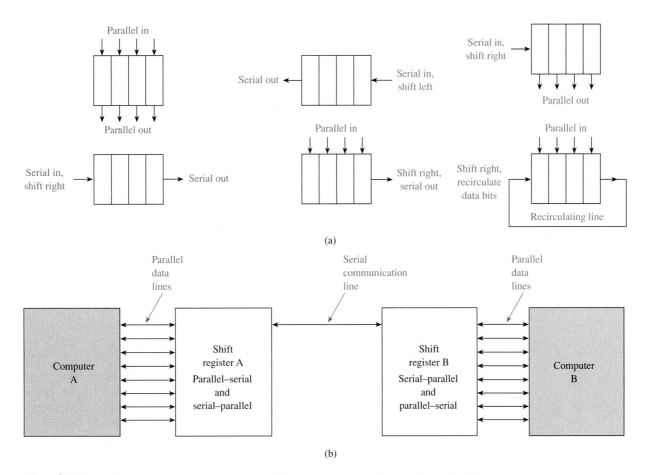

Figure 13–2   Shift register operations: (a) types of data movement and conversion; (b) using shift registers to provide serial communication between computers having parallel data.

from shift register A will travel across the serial communication line to shift register B, which is concurrently loading the 8 bits. After shift register B has received all 8 data bits, it will output them on its parallel output lines to computer B. This is a simplification of the digital communication* that takes place between computers, but it illustrates the heart of the system, the **shift register.**

13–2   Parallel-to-Serial Conversion

Now let's look at the actual circuit connections for a shift register. The data storage elements can be D flip-flops, S-R flip-flops, or J-K flip-flops. We are familiar with J-K flip-flops, so let's stick with them. Most J-Ks are negative edge triggered (like the 74LS76) and will have an active-LOW asynchronous Set $(\overline{S_D})$ and Reset $(\overline{R_D})$.

Figure 13–3 shows the circuit connections for a 4-bit parallel-in, serial-out shift register that is first Reset and then parallel loaded with an active-LOW 7 (1000), and then shifted right four positions.

Notice in Figure 13–3(a) that all $\overline{C_p}$ inputs are fed from a common clock input. Each flip-flop will respond to its J-K inputs at every negative clock input edge. Because every J-K input is connected to the preceding stage output, then at each negative clock edge each flip-flop will change to the state of the flip-flop to its left. In other words, all data bits will be shifted one position to the right.

Now, looking at the timing diagram, in the beginning of period 1, $\overline{R_D}$ goes LOW, resetting Q_0 to Q_3 to zero. Next, the parallel data are input (parallel loaded) via the D_0 to D_3 input lines. (Because the $\overline{S_D}$ inputs are active LOW, the complement of the number to be loaded must be used. The $\overline{S_D}$ inputs must be returned HIGH before shifting can be initiated.)

At the first negative clock edge,

Q_0 takes on the value of Q_1

Q_1 takes on the value of Q_2

Q_2 takes on the value of Q_3

Q_3 is Reset by $J = 0, K = 1$

In effect, the bits have all shifted one position to the right. Next, the negative edges of periods 2, 3, and 4 will each shift the bits one more position to the right.

The serial output data comes out of the right-end flip-flop (Q_0). Because the LSB was parallel loaded into the rightmost flip-flop, the LSB will be shifted out first. The order of the parallel input data bits could have been reversed and the MSB would have come out first. Either case is acceptable. It is up to the designer to know which is first, MSB or LSB, and when to sample (or read) the serial output data line.

13–3   Recirculating Register

Recirculating the rightmost data bits back into the beginning of the register can be accomplished by connecting Q_0 back to J_3 and $\overline{Q_0}$ back to K_3. This way, the original parallel-loaded data bits will never be lost. After every fourth clock pulse, the Q_3 to Q_0 outputs will contain the original 4 data bits. Therefore, with the addition of the recirculating lines to Figure 13–3(a), the register becomes a parallel-in, serial, *and* parallel-out.

---

*An integrated circuit called the UART (universal asynchronous receiver transmitter) is used to perform the shift register operation and to create the other control signals necessary for computer communication. That circuitry is used in a computer modem for communication over telephone lines.

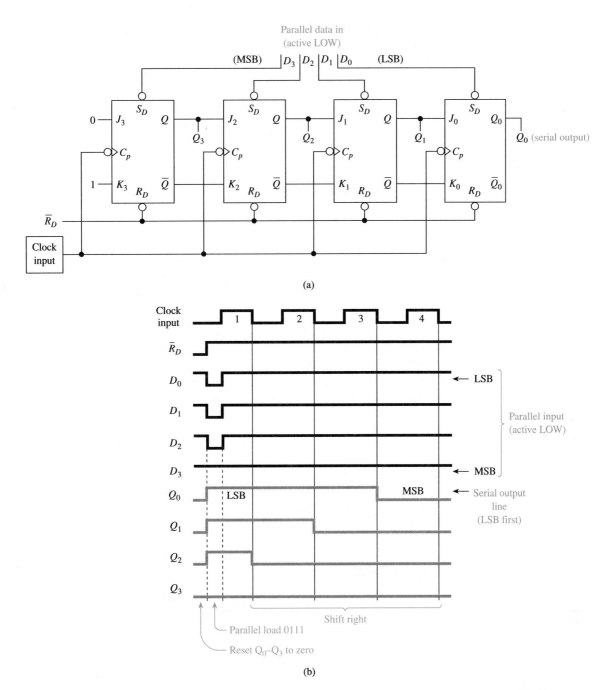

Parallel data in
(active LOW)

(a)

(b)

Figure 13–3 (a) Four-bit parallel-in, serial-out shift register using 74LS76 *J-K* flip-flops; (b) waveforms produced by parallel loading a 7 (0 1 1 1) and shifting right by four clock pulses.

Helpful Hint

Can you appreciate the story about the imaginary "bit bucket" that is used to catch the bits that drop out of the Q_0 output if the clock is allowed to continue?

Review Questions

13–1. All flip-flops within a shift register are driven by the same clock input. True or false?

13–2. What connections allow data to pass from one flip-flop to the next in a shift register?

13–3. How are data parallel loaded into a shift register constructed from *J-K* flip-flops?

13–4. If a hexadecimal C is parallel loaded into the shift register of Figure 13–3 and four clock pulses are applied, what is the state of the Q outputs? If recirculating lines are connected and the same operation occurs, what is the state of the Q outputs?

13–4   Serial-to-Parallel Conversion

Serial-in, parallel-out shift registers can also be made up of *J-K* flip-flop storage and a shift-right operation. The idea is to put the serial data in on the serial input line, LSB first (or MSB first, depending on the direction of the shift) and clock the shift register four times (for a 4-bit register), stop, and then read the parallel data from the Q_0 to Q_3 outputs. Figure 13–4(a) shows a 4-bit serial-to-parallel shift register converter. The serial data are coming in on the left at D_S. The flip-flops are connected in a shift-right fashion. The inverter at D_S is required to ensure that if $D_S = 1$ then $J = 1$, $K = 0$, and the first flip-flop will Set. Each of the other flip-flops takes on the value of the flip-flop to its left at each negative clock edge.

Each bit of the serial input must be present on the D_S line before the corresponding negative clock edge. After four clock pulses, all 4 serial data bits have been shifted into their appropriate flip-flop. At that time the parallel output data can be read by some other digital device.

If the clock were allowed to continue beyond four pulses, the data bits would continue shifting out of the right end of the register and be lost if you tried to read them again later. This problem is corrected by the **Strobe** line. It is used to Enable-then-Disable the clock at the appropriate time so that the shift-right process will stop. Using a Strobe signal is a popular technique in digital electronics to Enable or Disable some function during a specific time period.

13–5   Ring Shift Counter and Johnson Shift Counter

Two common circuits that are used to create sequential control waveforms for digital systems are the ring and Johnson **shift counters.** They are similar to a synchronous counter because the clock input to each flip-flop is driven by the same clock input. Their outputs do not count in true binary, but instead provide a repetitive sequence of digital output levels. These shift counters are used to control a sequence of events in a digital system **(digital sequences).**

In the case of a 4-bit *ring shift counter,* the output at each flip-flop will be HIGH for one clock period, then LOW for the next three, and then repeat, as shown in Figure 13–5(b). To form the ring shift counter of Figure 13–5(a), the Q-$\overline{Q}$ output of each stage is fed to the *J-K* input of the next stage, and the Q-$\overline{Q}$ output of the last stage is fed back to the *J-K* input of the first stage. Before applying clock pulses, the shift counter is preset with a 1–0–0–0.

Ring Shift Counter Operation

The *RC* circuit connected to the power supply will provide a LOW-then-HIGH as soon as the power is turned on, forcing a HIGH–LOW–LOW–LOW at Q_0–Q_1–Q_2–Q_3, which is the necessary preset condition for a ring shift counter. At the first negative clock input edge, Q_0 will go LOW because just before the clock edge J_0 was LOW

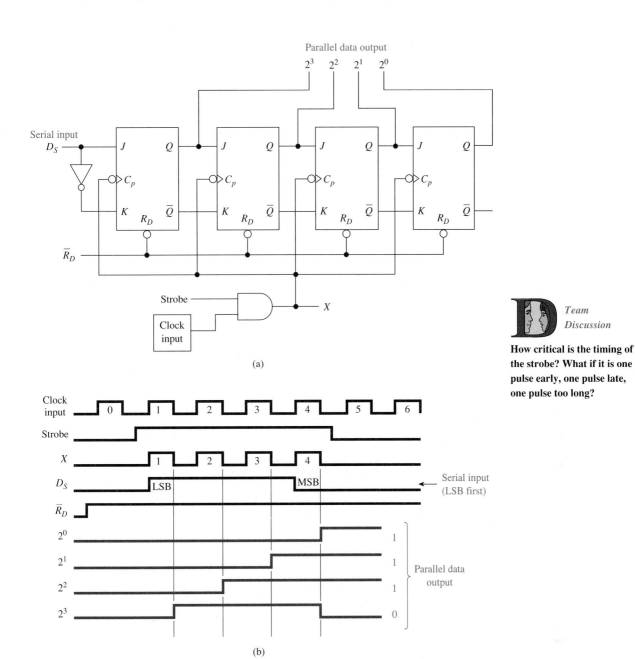

Parallel data output

Team
Discussion

How critical is the timing of
the strobe? What if it is one
pulse early, one pulse late,
one pulse too long?

(a)

(b)

Figure 13–4   (a) Four-bit serial-to-parallel shift register; (b) waveforms produced by a
serial-to-parallel conversion of the binary number 0 1 1 1.

(from Q_3) and K_0 was HIGH (from $\overline{Q_3}$). At that same clock edge, Q_1 will go HIGH because its J-K inputs are connected to Q_0-$\overline{Q_0}$, which were 1–0. The Q_2 and Q_3 flip-flops
will remain Reset (LOW) because their J-K inputs see a 0–1 from the previous flip-
flops.

Now, the ring shift counter is outputting a 0–1–0–0 (period 2). At the negative
edge of period 2, the flip-flop outputs will respond to whatever levels are present at
their J-K inputs, the same as explained in the preceding paragraph. That is, because
J_2-K_2 are looking back at (connected to) Q_1-$\overline{Q_1}$ (1–0), then Q_2 will go HIGH. All other
flip-flops are looking back at a 0–1, so they will Reset (LOW). This cycle repeats continuously. The system acts like it is continuously "pushing" the initial HIGH level at
Q_0 through the four flip-flops.

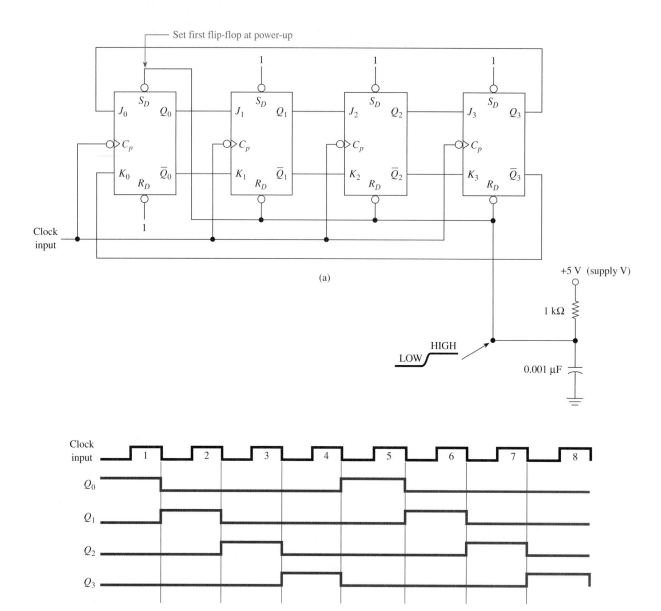

(a)

(b)

Figure 13–5 Ring shift counter: (a) circuit connections; (b) output waveforms.

Team Discussion

How would you implement this circuit using 7474 *D* flip-flops?

The *Johnson shift counter* circuit is similar to the ring shift counter except that the output lines of the last flip-flop are crossed (thus an alternative name is *twisted ring counter*) before feeding back to the input of the first flip-flop and *all* flip-flops are initially Reset as shown in Figure 13–6.[*]

Johnson Shift Counter Operation

The *RC* circuit provides an automatic Reset to all four flip-flops, so the initial outputs will all be Reset (LOW). At the first negative clock edge, the first flip-flop will Set (HIGH) because J_0 is connected to $\overline{Q_3}$ (HIGH) and K_0 is connected to Q_3 (LOW). The Q_1, Q_2, and Q_3 outputs will follow the state of their preceding flip-flop because of their direct connection *J*-to-*Q*. Therefore, during period 2, the output is 1–0–0–0.

---

[*]See Appendix E for a PLD design of Figure 13–6.

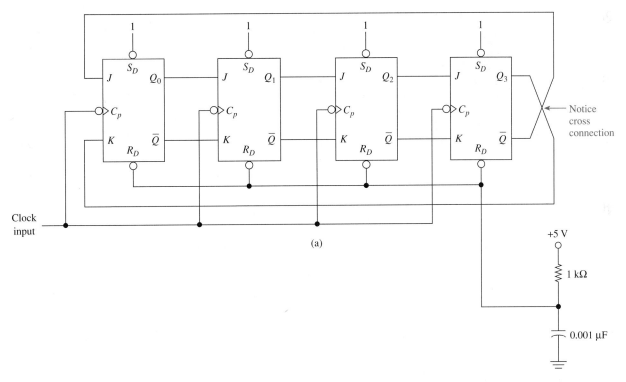

(a)

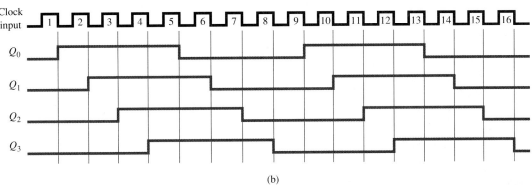

(b)

Figure 13–6 Johnson shift counter: (a) circuit connections; (b) output waveforms.

Team Discussion

To see if you really understand the circuit operation, try to sketch the waveforms if the cross-connection is made between the third and the fourth flip-flop instead of at the end.

At the next negative clock edge, Q_0 remains HIGH because it takes on the *opposite* state of Q_3, Q_1 goes HIGH because it takes on the *same* state as Q_0, Q_2 stays LOW, and Q_3 stays LOW. Now the output is 1–1–0–0.

The sequence continues as shown in Figure 13–6. Notice that, during period 5, Q_3 gets Set HIGH. At the end of period 5, Q_0 gets Reset LOW because the outputs of Q_3 are crossed, so Q_0 takes on the opposite state of Q_3.

Review Questions

13–5. What happens to the initial parallel-loaded data in the shift register of Figure 13–4 if the *Strobe* line is never disabled?

13–6. To operate properly, a ring shift counter must be parallel loaded with _____ and a Johnson shift counter must be parallel loaded with _____.

13–6 Shift Register ICs

Four-bit and 8-bit shift registers are commonly available in integrated-circuit packages. Depending on your needs, practically every possible load, shift, and conversion operation is available in a shift register IC.

Let's look at four popular shift register ICs to get familiar with using our data manuals and understanding the terminology and procedure for performing the various operations.

The 74164 8-Bit Serial-In, Parallel-Out Shift Register

By looking at the logic symbol and logic diagram for the 74164 (Figure 13–7), we can see that it saves us the task of wiring together eight D flip-flops. The 74164 has two serial input lines (D_{Sa} and D_{Sb}), synchronously read in by a positive edge-triggered clock (C_p). The logic diagram [Figure 13–7(b)] shows both D_S inputs feeding into an AND gate. Therefore, either input can be used as an active-HIGH enable for data entry through the other input. Each positive edge clock pulse will shift the data bits one position to the right. Therefore, the first data bit entered (either LSB or MSB) will end up in the far right D flip-flop (Q_7) after eight clock pulses. The $\overline{MR}$ is an active-LOW Master Reset that resets all eight flip-flops when pulsed LOW.

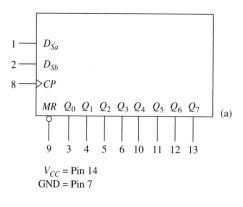

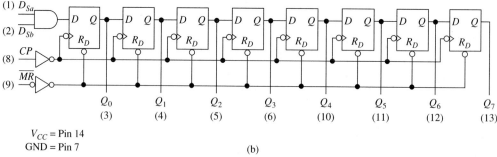

Figure 13–7 The 74164 shift register IC: (a) logic symbol; (b) logic diagram. [(b) Courtesy of Signetics Corporation]

Draw the circuit connections and timing waveforms for the serial-to-parallel conversion of the binary number 11010010 using a 74164 shift register.

Solution: The serial-to-parallel conversion circuit and waveforms are shown in Figure 13–8. First, the register is cleared by a LOW on $\overline{MR}$, making Q_0–Q_7 = 0. The Strobe line is required to make sure that we get only eight clock pulses. The serial data are entered on the D_{Sb} line, MSB first. After eight clock pulses, the 8 data bits can be read at the parallel output pins (MSB at Q_7 and LSB at Q_0).

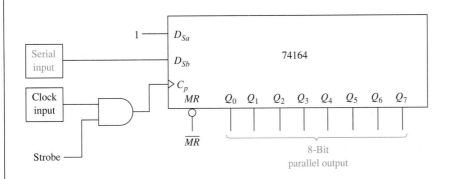

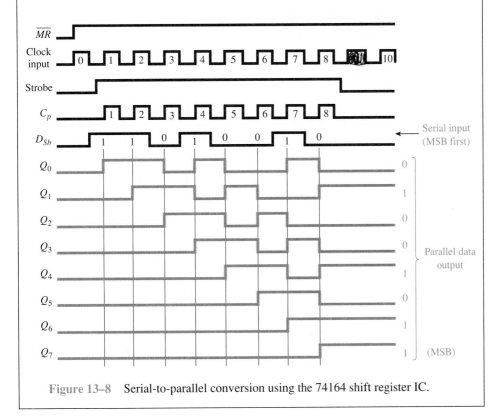

Figure 13–8   Serial-to-parallel conversion using the 74164 shift register IC.

The 74165

The next IC to consider is the 74165 8-bit serial *or* parallel-in, serial-out shift register. The logic symbol for the 74165 is given in Figure 13–9.

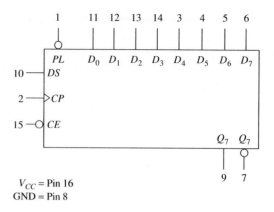

Figure 13–9 Logic symbol for the 74165 8-bit serial or parallel-in, serial-out register.

Just by looking at the logic symbol, you should be able to determine the operation of the 74165. The $\overline{PL}$ is an active-LOW terminal for performing a parallel load of the 8 parallel input data bits. The $\overline{CE}$ is an active-LOW clock enable for starting/stopping (shifting/holding) the shift operation by enabling/disabling the clock (same function as the Strobe in Example 13–1).

The clock input (C_p) is positive edge-triggered, so after each positive edge the data bits are shifted one position to the right. The serial output (Q_7) and its complement ($\overline{Q_7}$) are available from the rightmost flip-flop's outputs.

The 74194

Another shift register IC is the 74194 4-bit bidirectional universal shift register. It is called universal because it has a wide range of applications, including serial or parallel input, serial or parallel output, shift left or right, hold, and asynchronous Reset. The logic symbol for the 74194 is shown in Figure 13–10.

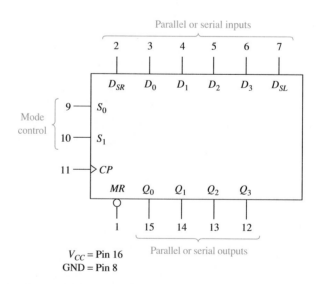

Figure 13–10 Logic symbol for the 74194 universal shift register.

CHAPTER 13 | SHIFT REGISTERS

The major differences with the 74194 are that there are separate serial inputs for shifting left or shifting right, and the operating mode is determined by the digital states of the **mode control** inputs, S_0 and S_1. S_0 and S_1 can be thought of as receiving a 2-bit binary code representing one of four possible operating modes ($2^2 = 4$ combinations). The four operating modes are shown in Table 13–1.

Table 13–1 Operating Modes of the 74194		
Operating Mode	S_1	S_0
Hold	0	0
Shift left	1	0
Shift right	0	1
Parallel load	1	1

A complete mode select-function table for the 74194 is shown in Table 13–2. Table 13–2 can be used to determine the procedure and expected outcome of the various shift register operations.

Table 13–2 Mode Select-Function Table for the 74194[a]

	Inputs							Outputs			
Operating Mode	C_p	$\overline{MR}$	S_1	S_0	D_{SR}	D_{SL}	D_n	Q_0	Q_1	Q_2	Q_3
Reset (clear)	×	L	×	×	×	×	×	L	L	L	L
Hold (do nothing)	×	H	l[b]	l[b]	×	×	×	q_0	q_1	q_2	q_3
Shift left	H	h	l[b]	×	1	×	q_1	q_2	q_3	L	
($Q_N \leftarrow Q_{N+1}$, $Q_3 \leftarrow D_{SL}$)	↑	H	h	l[b]	×	h	×	q_1	q_2	q_3	H
Shift right	↑	H	l[b]	h	1	×	×	L	q_0	q_1	q_2
($D_{SR} \rightarrow Q_∅$, $Q_N \rightarrow Q_{N+1}$)	↑	H	l[b]	h	h	×	×	H	q_0	q_1	q_2
Parallel load	↑	H	h	h	×	×	d_n	d_0	d_1	d_2	d_3

Courtesy of Signetics Corporation

[a]H = HIGH voltage level; h = HIGH voltage level one setup time prior to the LOW-to-HIGH clock transition; L = LOW voltage level; l = LOW voltage level one setup time prior to the LOW-to-HIGH clock transition; $d_n(q_n)$ = lowercase letters indicate the state of the referenced input (or output) one setup time prior to the LOW-to-HIGH clock transition; × = don't care; ↑ = LOW-to-HIGH clock transition.

[b]The HIGH-to-LOW transition of the S_0 and S_1 inputs on the 74194 should take place only while C_p is HIGH for conventional operation.

From the function table we can see that a LOW input to the Master Reset ($\overline{MR}$) asynchronously resets Q_0 to Q_3 to 0. A parallel load is accomplished by making S_0, S_1 both HIGH and placing the parallel input data on D_0 to D_3. The register will then be parallel loaded synchronously by the first positive clock (C_p) edge. The 4 data bits can then be shifted to the right or left by making S_0–S_1 1–0 or 0–1 and applying an input clock to C_p.

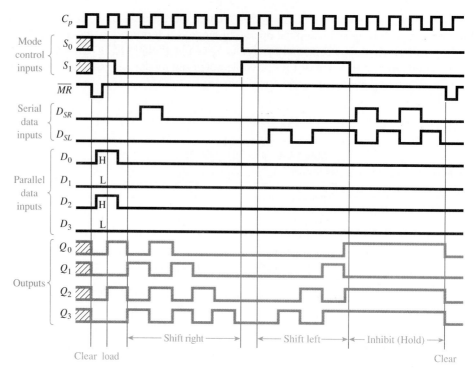

Figure 13–11 Typical clear–load–shift right–shift left–inhibit–clear sequence for a 74194.

A **recirculating** shift-right register can be set up by connecting Q_3 back into D_{SR} and applying a clock input (C_p) with $S_1 = 0$, $S_0 = 1$. Also, a recirculating shift-left register can be set up by connecting Q_0 into D_{SL} and applying a clock input (C_p) with $S_1 = 1$, $S_0 = 0$.

The best way to get a feel for the operation of the 74194 is to study the timing waveforms for a typical sequence of operations. Figure 13–11 shows the input data, control waveforms, and the output (Q_0 to Q_3) waveforms generated by a clear–load–shift right–shift left–inhibit–clear sequence. Study these waveforms carefully until you thoroughly understand the setup of the mode controls and the reason for each state change in the Q_0 to Q_3 outputs.

EXAMPLE 13–2

Draw the circuit connection and timing waveforms for a recirculating shift-right register. The register should be loaded initially with a hexadecimal D (1101).

Solution: The shift-right register is shown in Figure 13–12. First, the S_0 and S_1 mode controls are set to 1–1 for parallel loading D_0 to D_3. When the first positive clock edge (pulse 0) comes in, the data present on D_0 to D_3 are loaded into Q_0 to Q_3. Next, S_0 and S_1 are made 1–0 to perform shift-right operations. At the positive edge of each successive clock edge, the data are shifted one position to the right ($Q_0 \rightarrow Q_1$, $Q_1 \rightarrow Q_2$, $Q_2 \rightarrow Q_3$, $Q_3 \rightarrow D_{SR}$, $D_{SR} \rightarrow Q_0$). The recirculating connection from Q_3 back to D_{SR} keeps the data from being lost. After each fourth clock pulse, the circulating data are back in their original position.

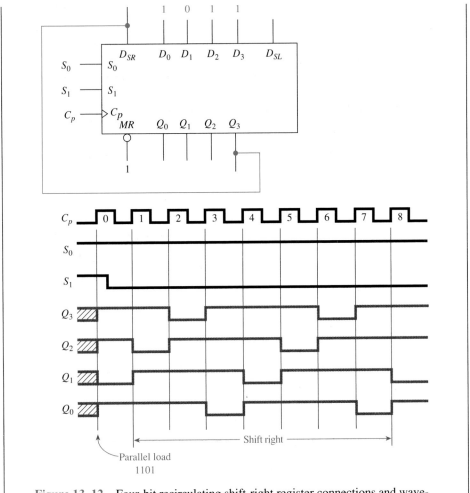

Figure 13–12   Four-bit recirculating shift-right register connections and waveforms using the 74194.

Three-State Outputs

Another valuable feature available on some shift registers is a **three-state output.** The term *three-state* (or *tristate*) is derived from the fact that the output can have one of three levels: HIGH, LOW, or float. The symbol and function table for a three-state output buffer are shown in Figure 13–13.

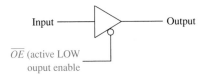

$\overline{OE}$ (active LOW
ouput enable

Input	$\overline{OE}$	Output
1	0	1
0	0	0
1	1	Float
0	1	Float

Figure 13–13   Three-state output buffer symbol and function table.

From Figure 13–13, we can see that the circuit acts like a straight buffer (output = input) when $\overline{OE}$ is LOW (active-LOW Output Enable). When the output is disabled ($\overline{OE}$ = HIGH), the output level is placed in the **float**, or **high-impedance state.** In the high-impedance state, the output looks like an open circuit to anything else connected to it. In other words, in the float state the output is neither HIGH nor LOW and cannot sink nor source current.

Three-state outputs are necessary when you need to connect the output of more than one register to the same points. For example, if you have two 4-bit registers, one containing data from device 1 and the other containing different data from device 2, and you want to connect both sets of outputs to the same receiving device, one device must be in the float condition while the other's output is enabled, and vice versa. This way, only one set of outputs is connected to the receiving device at a time to avoid a conflict.

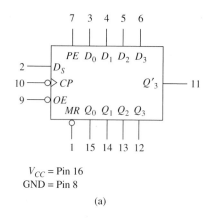

V_{CC} = Pin 16
GND = Pin 8

(a)

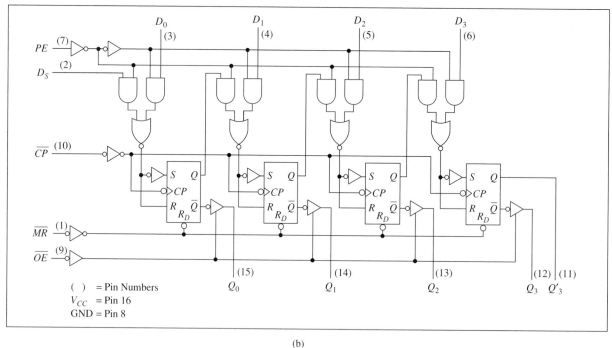

(b)

Figure 13–14 The 74395A 4-bit shift register with three-state outputs: (a) logic symbol; (b) logic circuit. [(b) Courtesy of Signetics Corporation]

The 74395A

To further illustrate the operation of three-state buffers, let's look at a register that has three-state outputs and discuss a system design example that uses two 4-bit registers to feed a single receiving device. The 74395A is a 4-bit shift (right) register with three-state outputs as shown in Figure 13–14. From the logic circuit diagram [Figure 13–14(b)], we can see that the Q_0 and Q_3 outputs are "three-stated" and will not be allowed to pass data unless a LOW is present at the Output Enable pin ($\overline{OE}$). Also, a non-three-stated output, Q_3', is made available to enable the user to cascade with another register and shift data bits to the cascaded register whether the regular outputs (Q_0 to Q_3) are enabled or not (that is, to cascade two 4-bit registers, Q_3' would be connected to D_S of the second stage).

Otherwise, the chip's operation is similar to previously discussed shift registers. The **Parallel Enable** (PE) input is active-HIGH for enabling the parallel data input (D_0 to D_3) to be synchronously loaded on the negative clock edge. D_S is the serial data input line for synchronously loading serial data, and $\overline{MR}$ is an active-LOW Master Reset.

EXAMPLE 13–3

Sketch the circuit connections for a two-register system that alternately feeds one register and then the other into a 4-bit receiving device. Upon power-up, load register 1 with 0111 and register 2 with 1101.

Solution: The two-register system is shown in Figure 13–15. Notice that both sets of outputs go to a common point. The three-state outputs allow us to do this by only enabling one set of outputs at a time so that there is no conflict between HIGHs and LOWs. One way to alternately enable one register and then the other is to use a toggle flip-flop and feed the Q output

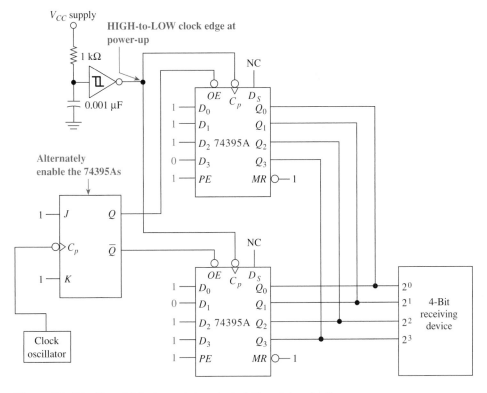

Figure 13–15 Two 4-bit, three-state output shift registers feeding a common receiving device.

to the upper Output Enable ($\overline{OE}$) and the $\overline{Q}$ to the lower Output Enable, as shown in Figure 13–15.

Also, upon initial power-up we want to parallel load both 4-bit registers. To do this, the D_0 to D_3 inputs contain the proper bit string to be loaded, and the Parallel Enable (PE) is held HIGH. Since the parallel-load function is synchronous (needs a clock trigger), we will supply a HIGH-then-LOW pulse to $\overline{C_p}$ via the RC Schmitt circuit.

Review Questions

13–7. To input serial data into the 74164 shift register, one D_S input must be held _____ (HIGH, LOW) while the other receives the serial data.

13–8. What is the function of the $\overline{CE}$ input to the 74165 shift register?

13–9. Why is the 74194 IC sometimes called a "universal" shift register?

13–10. List the steps that you would follow to parallel load a hexadecimal B into a 74194 shift register.

13–11. To make the 74194 act as a shift-left *recirculating* register, a connection must be made from _____ to _____ and S_0 and S_1 must be _____ _____.

13–12. How does the operation of the Parallel Enable (PE) on the 74395A shift register differ from the Parallel Load ($\overline{PL}$) of the 74165?

13–13. The outputs of the 74395A shift register are disabled by making $\overline{OE}$ _____ (HIGH, LOW), which makes Q_0–Q_3 _____ (HIGH, LOW, float).

13–7 System Design Applications for Shift Registers

Shift registers have many applications in digital sequencing, storage, and transmission of serial and parallel data. The following designs will illustrate some of these applications.

━━━ **APPLICATION 13–1** ━━━

Using a ring shift counter as a sequencing device, design a traffic light controller that goes through the following sequence: green, 20 s; yellow, 10 s; red, 20 s. Also, at night, flash the yellow light on and off continuously.

Solution: By studying the waveforms from Figure 13–5, you will notice that, if we add one more flip-flop to that 4-bit ring shift counter and use a clock input of 1 pulse per 10 s, we could tap off the Q outputs using OR gates to get a 20–10–20 sequence. Also, we could use a phototransistor to determine night from day. During the night we want to stop the ring shift counter and flash the yellow light. Figure 13–16 shows a 5-bit ring counter that could be used as this traffic light controller. First, let's make sure that the green–yellow–red sequence will work properly during the daytime. During the daytime, outdoor light shines on the phototransistor, making its collector-to-emitter resistance LOW, placing a low voltage at the input to the first Schmitt inverter, and causing a LOW input to OR gate 4. The

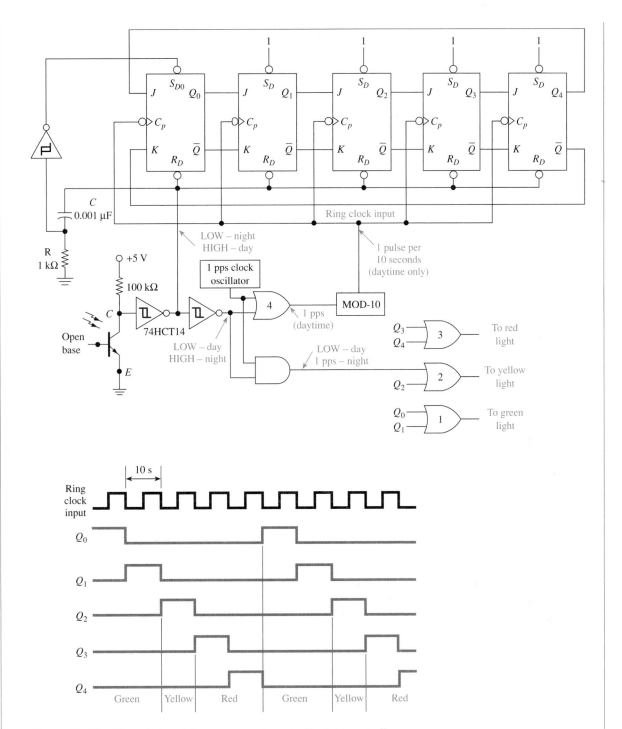

Figure 13–16 Five-bit ring shift counter used as a traffic light controller.

1-pps clock oscillator will pass through OR gate 4 into the MOD-10, which divides the frequency down to one pulse per 10 s. The output from the MOD-10 is used to drive the clock input to the 5-bit ring shift counter, which will circulate a single HIGH level down through each successive flip-flop for 10 s at each Q output, as shown in the timing waveforms.

OR gates 1, 2, and 3 are connected to the ring counter outputs in such a way that the green light will be on if Q_0 or Q_1 is HIGH, which occurs for 20 s. The yellow light will come on next for 10 s due to Q_2 being on, and then the red light will come on for 20 s because either Q_3 or Q_4 is HIGH.

At nighttime the phototransistor changes to a high resistance, placing a HIGH at the input to the first Schmitt inverter, which places a HIGH at OR gate 4. This makes its output HIGH, stopping the clock input oscillations to the ring counter.

Also at nighttime, the LOW output from the first Schmitt inverter is connected to the ring counter Resets, holding the Q outputs at 0. The HIGH output from the second Schmitt inverter enables the AND gate to pass the 1-pps clock oscillator on to OR gate 2, causing the yellow light to flash.

At sunrise the output from the first Schmitt inverter changes from a LOW to a HIGH, allowing the ring counter to start again. This LOW-to-HIGH transition causes an instantaneous surge of current to flow through the RC circuit. That current will cause a HIGH at the input of the third Schmitt inverter, which places a LOW at $\overline{S_{D0}}$, setting Q_0 HIGH. When the surge current has passed (a few microseconds), $\overline{S_{D0}}$ returns to a HIGH and the ring counter will proceed to rotate the HIGH level from Q_0 to Q_1 to Q_2 to Q_3 to Q_4 continuously, all day, as shown in the timing waveforms.

APPLICATION 13–2

Design a 16-bit serial-to-parallel converter.

Solution: First, we have to look through a TTL data manual to see what is available. The 74164 is an 8-bit serial-in, parallel-out shift register. Let's cascade two of them together to form a 16-bit register. Figure 13–17 shows that the Q_7 output is fed into the serial input of the second 8-bit register. This way, as the data bits are shifted through the register, when they reach Q_7 the next shift will pass the data into Q_0 of the second register (via D_{Sa}), making it a 16-bit register. The second serial input (D_{Sb}) of each stage is internally ANDed with the D_{Sa} input so that it serves as an active-HIGH enable input.

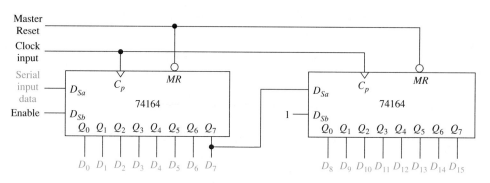

Figure 13–17 Sixteen-bit serial-to-parallel converter.

APPLICATION 13–3

Design a circuit and provide the input waveforms required to perform a parallel-to-serial conversion. Specifically, a hexadecimal B (1011) is to be parallel loaded and then transmitted repeatedly to a serial device LSB first.

Solution: By controlling the mode control inputs (S_0–S_1) of a 74194, we can perform a parallel load and then shift right repeatedly. The serial output data are taken from Q_3, as shown in Figure 13–18. The 74194 universal shift register is connected as a recirculating parallel-to-serial converter. Each time the Q_3 serial output level is sent to the serial device, it is also recirculated back into the left end of the shift register.

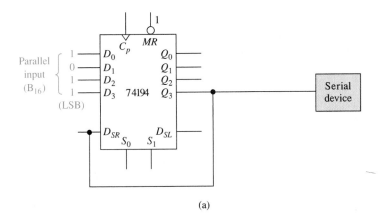

(a)

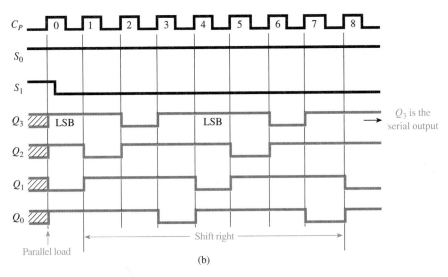

(b)

Figure 13–18 Four-bit parallel-to-serial converter: (a) circuit connections; (b) waveforms.

First, at the positive edge of clock pulse 0, the register is parallel loaded with a 1011 (B_{16}) because the mode controls (S_0–S_1) are HIGH–HIGH. (D_3 is loaded with the LSB because it will be the first bit out when we shift right.)

Next, the mode controls (S_0–S_1) are changed to HIGH–LOW for a shift-right operation. Now, each successive positive clock edge will shift the data bit one position to the right. The Q_3 output will continuously have the levels 1101–1101–1101–, and so on, which is a backward hexadecimal B (LSB first).

APPLICATION 13–4

Design an interface to an 8-bit serial printer. Sketch the waveforms required to transmit the single ASCII code for an asterisk (*). *Note:* ASCII is the 7-bit code that was given in Chapter 1. The ASCII code for an asterisk is 010 1010. Let's make the unused eighth bit (MSB) a zero.

Solution: The circuit design and waveforms are shown in Figure 13–19. The 74165 is chosen for the job because it is an 8-bit register that can be parallel loaded and then shifted synchronously by the clock input to provide the serial output to the printer.

During pulse 0, the register is loaded with the ASCII code for an asterisk (the LSB is put into D_7, because we want it to come out first). The clock input is then enabled by a LOW on $\overline{CE}$ (**Clock Enable).** Each positive pulse on C_p from then on will shift the data bits one position to the right. After the eighth clock pulse (0 to 7), the printer will have received all 8 serial data bits. Then the $\overline{CE}$ line is brought HIGH to disable the synchronous clock input. To avoid any racing problems, the printer will read the Q_7 line at each negative edge of C_p so that the level will definitely be a stable HIGH or LOW, as shown in Figure 13–19.

At this point you may be wondering how, practically, we are going to electronically provide the necessary signals on the $\overline{CE}$ and $\overline{PL}$ lines. An exact degree of timing must be provided on these lines to ensure that the register–printer interface communicates properly. These signals will be provided by a microprocessor and are called the **handshaking** signals.

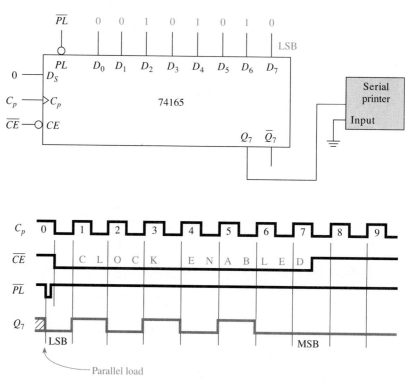

Figure 13–19   Circuit design and waveforms for the transmission of an ASCII character to a serial printer.

Microprocessor theory and programming are advanced digital topics and are not discussed in this book. For now, it is important for us to realize that these signals are required and to be able to sketch their timing diagrams.

13–8   Driving a Stepper Motor with a Shift Register

A **stepper motor** makes its rotation in steps instead of a smooth continuous motion as with conventional motors. Typical **stepping angles** are 15° or 7.5° per step, requiring 24 or 48 steps, respectively, to complete one revolution. The stepping action is controlled by digital levels that energize magnetic coils within the motor.

Because they are driven by sequential digital signals, it is a good application for shift registers. For example, a shift register circuit could be developed to cause the stepper motor to rotate at 100 rpm for 32 revolutions and then stop. This is useful for applications requiring exact positioning control without the use of **closed-loop feedback** circuitry to monitor the position. Typical applications are floppy disk Read/Write head positioning, printer type head and line feed control, and robotics.

There are several ways to construct a motor to achieve this digitally controlled stepping action. One such way is illustrated in Figure 13–20. This particular stepper motor construction uses four **stator** (stationary) **coils** set up as four **pole pairs.** Each stator pole is offset from the previous one by 45°. The directions of the windings are such that energizing any one coil will develop a north field at one pole and a south field at the opposite pole. The north and south poles created by energizing coil 1 are shown in Figure 13–20. The rotating part of the motor (the **rotor**) is designed with three ferromagnetic pairs spaced 60° apart from each other. (A **ferromagnetic** material is one that is attracted to magnetic fields.) Because the stator poles are spaced 45° apart, this makes the next stator-to-rotor 15° out of alignment.

In Figure 13–20, the rotor has aligned itself with the **flux lines** created by the north–south stator poles of coil 1. To step the rotor 15° clockwise, coil 1 is deenergized

Common Misconception

Before you learned the operation of a stepper motor, you may have thought that it was limited to 16 steps per rotation, because it has only four binary inputs ($2^4 = 16$).

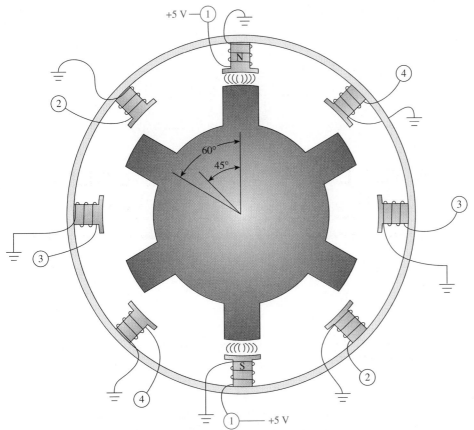

Figure 13–20   A four-coil stepper motor with stator coil 1 energized. (From *Digital and Microprocessor Fundamentals* by William Kleitz, Prentice Hall, Englewood Cliffs, NJ, 1990.)

and coil 2 is energized. The closest rotor pair to coil 2 will now line up with stator pole pair 2's flux lines. The next 15° steps are made by energizing coil 3, then 4, then 1, then 2, and so on, for as many steps as you require. Figure 13–21 shows the stepping action achieved by energizing each successive coil six times. Table 13–3 shows the digital codes that are applied to the stator coils for 15° clockwise and 15° counterclockwise rotation. Figure 13–22 shows a stepper motor with the motor removed to expose two of the stator coils.

Figure 13–22   Stepper motor with the rotor removed to expose two of the stator coils.

Figure 13–21   Coil energizing sequence for 15° clockwise steps.

Table 13–3 Digital Codes for 15° Clockwise and Counterclockwise Rotation	
Clockwise Coil 1 2 3 4	Counterclockwise Coil 1 2 3 4
1 0 0 0	0 0 0 1
0 1 0 0	0 0 1 0
0 0 1 0	0 1 0 0
0 0 0 1	1 0 0 0
1 0 0 0	0 0 0 1
0 1 0 0	0 0 1 0
and so on	and so on

The amount of current required to energize a coil pair is much higher than the capability of a 74194, so we will need some current-buffering circuitry similar to that shown in Figure 13–23.

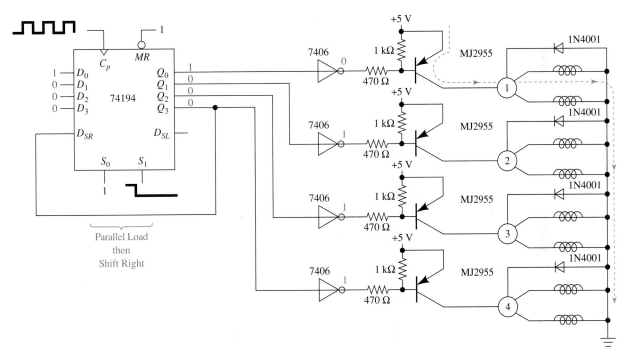

Figure 13–23   Drive circuitry for a four-coil stepper motor showing the number 1 coils energized.

Team Discussion

Sketch the 74194 waveforms for the parallel load and 8 pulses.

The output of the upper 7406 inverting buffer in Figure 13–23 is LOW, forward biasing the base-emitter of the MJ2955 *PNP* power transistor. This causes the collector–emitter to short, allowing the large current to flow through the number 1 coils to ground. The IN4001 diodes protect the coils from arcing over when the current is stopped.

The 74194 is first parallel loaded with 0001 and then changed to a shift-right operation by making S_0, $S_1 = 1$, 0. Each positive clock edge shifts the ON bit one position to the right. The Q outputs will follow the clockwise pattern shown in Table 13–3, causing the motor to rotate. The speed of revolution is dictated by the period of C_p.

Review Questions

13–14. The traffic light controller of Figure 13–16 flashes the yellow light at night because the level at the collector of the phototransistor is _____ (LOW, HIGH), which _____ (enables, disables) the AND gate.

13–15. What circuitry is responsible for parallel loading a 1 into the first flip-flop of Figure 13–16 at the beginning of each day?

13–16. The serial-to-parallel converter in Figure 13–17 could also be used for serial in to serial out. True or false?

13–17. How would the waveforms change in Figure 13–18 if Q_3 were connected to D_{SL} instead of D_{SR}?

13–18. How is the $\overline{CE}$ input used on the 74165 in Figure 13–19?

13–19. Is the stepper motor in Figure 13–23 turning clockwise or counter-clockwise? How would you change its direction?

13–9 Three-State Buffers, Latches, and Transceivers

When we start studying microprocessor hardware, we'll see a need for transmitting a number of bits simultaneously as a group (**data transmission**). A single flip-flop will not suffice. What we need is a group of flip-flops, called a **register,** to facilitate the movement and temporary storage of binary information. The most commonly used registers are 8 bits wide and function as either a buffer, latch, or transceiver.

Three-State Buffers

In microprocessor systems, several input and output devices must share the same data lines going into the microprocessor IC. (These shared data lines are called the **data bus**). For example, if an 8-bit microprocessor interfaces with four separate 8-bit input devices, we must provide a way to enable just one of the devices to place its data on the data bus and disable the other three. One way this procedure can be accomplished is to use three-state octal buffers.

In Figure 13–24, the second buffer is enabled, which allows the 8 data bits from Input Device 2 to reach the data bus. The other three buffers are disabled, thus keeping their outputs in a *float* condition.

A buffer is simply a device that, when enabled, passes a digital level from its input to its output unchanged. It provides isolation, or a **buffer,** between the input device and the data bus. A buffer also provides the sink or source current required by any devices connected to its output without loading down the input device. An octal buffer IC has eight individual buffers within a single package.

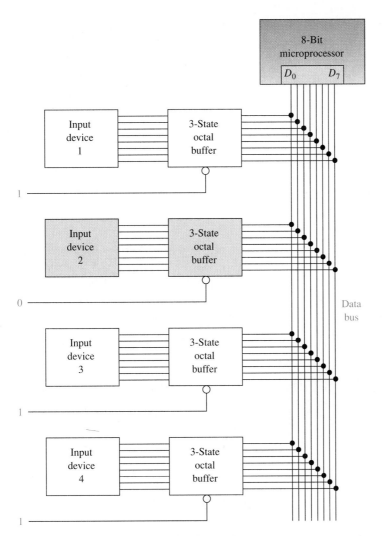

Figure 13–24 Using a three-state octal buffer to pass 8 data bits from input device 2 to the data bus.

A popular three-state octal buffer is the 74LS244 shown in Figure 13–25. Notice that the buffers are configured in two groups of four. The first group (group a) is controlled by $\overline{OE_a}$ and the second group (group b) is controlled by $\overline{OE_b}$. OE is an abbreviation for **Output Enable** and is active-LOW, meaning that it takes a LOW to allow data to pass from the inputs (I) to the outputs (Y). Other features of the 74LS244 are that it has Schmitt trigger hysteresis and very high sink and source current capabilities (24 and 15 mA, respectively).

Octal Latches/Flip-Flops

In microprocessor systems we need latches and flip-flops to remember digital states that a microprocessor issues before it goes on to other tasks. Take, for example, a microprocessor system that drives two separate 8-bit output devices, as shown in Figure 13–26.

To send information to output device 1, the microprocessor first sets up the data bus (D_0–D_7) with the appropriate data and then issues a LOW-to-HIGH pulse on line C_1. The positive edge of the pulse causes the data at D_0–D_7 of the flip-flop to be stored at Q_0–Q_7. Because $\overline{OE}$ is tied LOW, its data are sent on to output device 1. (The

**It is instructive for you to
see the ´244 in a practical
microprocessor application,
such as the one shown in
Figure 17–6.**

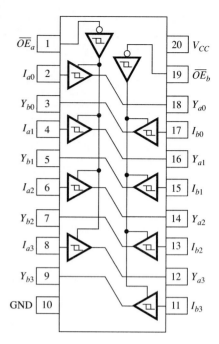

Figure 13–25   Pin configuration for the
74LS244 three-state octal buffer.

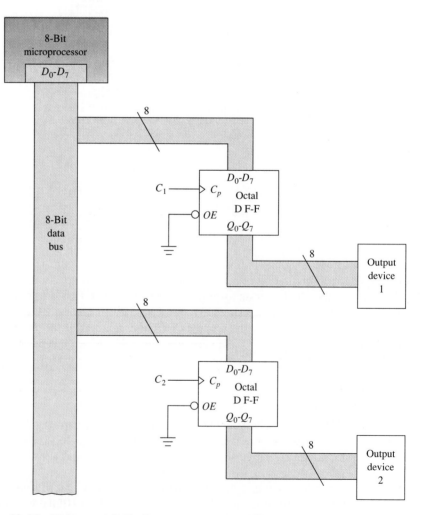

Figure 13–26   Using octal *D* flip-flops to capture data that appear momentarily on a microprocessor data bus.

diagonal line with the number 8 above it is a shorthand method used to indicate eight separate lines or conductors.)

Next, the microprocessor sets up the data bus with data for output device 2 and issues a LOW-to-HIGH pulse on C_2. Now the second **octal** D flip-flop is loaded with valid data. The outputs of the D flip-flops will remain at those digital levels, thus allowing the microprocessor to go on to perform other tasks.

Earlier in this text we studied the 7475 **transparent latch** and the 7474 D flip-flop. The 74LS373 and 74LS374 shown in Figure 13–27 operate similarly, except that they were developed to handle 8-bit data operations.

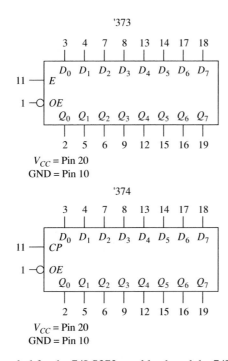

Helpful Hint

The ´374 is used in the microprocessor application in Figure 17–6.

Figure 13–27 Logic symbol for the 74LS373 octal latch and the 74LS374 octal D flip-flop.

Transceivers

Another way to connect devices to a shared data bus is to use a **transceiver** (transmitter/receiver). The transceiver differs from a buffer or latch because it is **bidirectional** This capability is necessary for interfacing devices that are used for *both input and output* to a microprocessor. Figure 13–28 shows a common way to connect an I/O device to a data bus via a transceiver.

To make input/output device 1 the active interface, the $\overline{CE}$ (Chip Enable) line must first be made LOW. If $\overline{CE}$ is HIGH, the transceiver disconnects the I/O device from the bus by making the connection float.

After making $\overline{CE}$ LOW, the microprocessor then issues the appropriate level on the $S/\overline{R}$ line depending on whether it wants to *send data to* the I/O device or *receive data from* the I/O device. If $S/\overline{R}$ is made HIGH, the transceiver allows data to pass to the I/O device (from A to B). If $S/\overline{R}$ is made LOW, the transceiver allows data to pass to the microprocessor data bus (from B to A).

To see how a transceiver is able to both send and receive data, study the internal logic of the 74LS245 shown in Figure 13–29.

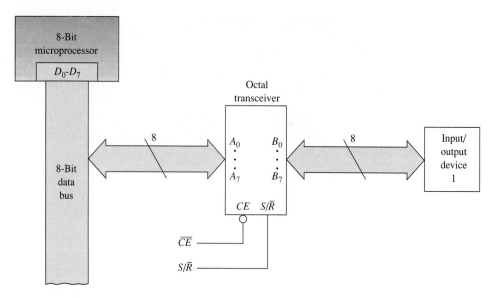

Figure 13–28 Using an octal transceiver to interface an input/output device to an 8-bit data bus.

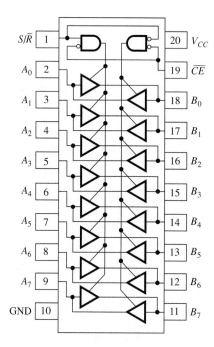

Figure 13–29 Pin configuration and internal logic of the 74LS245 octal three-state transceiver.

Review Questions

13–20. The 74LS244 provides buffering for a total of _____ signals. The outputs are all forced to their *float* state by making _____ and _____ HIGH.

13–21. The 74LS374 octal *D* flip-flop is a _____ device, whereas the 74LS244 octal buffer is a _____ device (synchronous, asynchronous).

13–22. A transceiver like the 74LS245 is *bidirectional*, allowing data to flow in either direction through it. True or false?

Summary

In this chapter we have learned that

1. Shift registers are used for serial-to-parallel and parallel-to-serial conversions.

2. One common form of digital communication is for a sending computer to convert its data from parallel to serial and then transmit over a telephone line to a receiving computer, which converts back from serial to parallel.

3. Simple shift registers can be constructed by connecting the *Q*-outputs of one *J-K* flip-flop into the *J-K* inputs of the next flip-flop. Several flip-flops can be cascaded together this way, driven by a common clock, to form multibit shift registers.

4. The ring and Johnson shift counters are two specialized shift registers used to create sequential control waveforms.

5. Several multibit shift register ICs are available for the designer to choose from. They generally have four or eight internal flip-flops and are designed to shift either left or right and perform either serial-to-parallel or parallel-to-serial conversions.

6. The 74194 is called a universal 4-bit shift register, because it can shift in either direction and can receive and convert to either format.

7. Three-state outputs are used on ICs that must have their outputs go to a common point. They are capable of the normal HIGH/LOW levels but can also output a float (or high-impedance) state.

8. The rotation of a stepper motor is made by taking small angular steps. This is controlled by sequential digital strings often generated by a recirculating shift register.

9. Three-state buffers, latches, and transceivers are an integral part of microprocessor interface circuitry. They allow the microprocessor system to have external control of 8-bit groups of data. The *buffer* can be used to allow multiple input devices to feed a common point or to provide high output current to a connected load. The *latch* can be used to remember momentary data from the microprocessor that needs to be held for other devices in the system. The *transceiver* provides bidirectional (input or output) control of interface circuitry.

Glossary

Bidirectional: Allowing data to flow in either direction.

Buffer: A logic device connected between two digital circuits, providing isolation, high sink, and source current, and usually three-state control.

Clock Enable: A separate input pin included on some ICs, used to enable or disable the clock input signal.

Closed-Loop Feedback: A system that sends information about an output device back to the device that is driving the output device, to keep track of the particular activity.

Data Bit: A single binary representation (0 or 1) of digital information.

Data Bus: A group of eight lines or electrical conductors usually connected to a microprocessor and shared by a number of other devices connected to it.

Data Conversion: Transformation of digital information from one format to another (for example, serial-to-parallel conversion).

Data Transmission: The movement of digital information from one location to another.

Digital Sequencer: A system (like a shift counter) that can produce a specific series of digital waveforms to drive another device in a specific sequence.

Ferromagnetic: A material in which magnetic flux lines can easily pass (high permeability).

Float: A digital output level that is neither HIGH nor LOW but instead is in a *high-impedance state*. In this state, the output acts like a high impedance with respect to ground and will float to any voltage level that happens to be connected to it.

Flux Lines: The north-to-south magnetic field set up by magnets is made up of flux lines.

Handshaking: The communication between a data sending device and receiving device that is necessary to determine the status of the transmitted data.

High-Impedance State: *See* float.

Mode Control: Input pins available on some ICs used to control the operating functions of that IC.

Octal: When referring to an IC, octal means that a single package contains *eight* logic devices.

Output Enable: An input pin on an IC that can be used to enable or disable the outputs. When disabled, the outputs are in the float condition.

Parallel Enable: An IC input pin used to enable or disable a synchronous parallel load of data bits.

Pole Pair: Two opposing magnetic poles situated opposite each other in a motor housing and energized concurrently.

Recirculating: In a shift register, instead of letting the shifting data bits drop out of the end of the register, a recirculating connection can be made to pass the bits back into the front end of the register.

Register: Two or more flip-flops (or storage units) connected as a group and operated simultaneously.

Rotor: The rotating part of the stepper motor.

Shift Counter: A special-purpose shift register with modifications to its connections and preloaded with a specific value to enable it to output a special sequence of digital waveforms. It does not count in true binary, but instead is used for special sequential waveform generation.

Shift Register: A storage device containing two or more data bits, capable of moving the data to the left or right and performing conversions between serial and parallel.

Stator Coil: A stationary coil, mounted on the inside of the motor housing.

Step Angle: The number of degrees that a stepper motor rotates for each change in the digital input signal (usually $15°$ or $7.5°$).

Stepper Motor: A motor whose rotation is made in steps that are controlled by a digital input signal.

Strobe: A connection used in digital circuits to enable or disable a particular function.

Three-State Output: A feature on some ICs that allows you to connect several outputs to a common point. When one of the outputs is HIGH or LOW, all others will be in the float condition (the three output levels are HIGH, LOW, and float).

Transceiver: A data transmission device that is bidirectional, allowing data to flow through it in either direction.

Transparent Latch: An asynchronous device whose outputs hold onto the most recent digital state of the inputs. The outputs immediately follow the state of the inputs (transparent) while the trigger input is active and then latch onto that information when the trigger is removed.

Problems

Sections 13–1 Through 13–4

13–1. In Figure P13–1, will the data bits be shifted right or left with each clock pulse? Will they be shifted on the positive or negative clock edge?

13–2. If the register of Figure P13–1 is initially parallel loaded with $D_3 = 0$, $D_2 = 1$, $D_1 = 0$, and $D_0 = 0$, what will the output at Q_3 to Q_0 be after two clock pulses? After four clock pulses?

13–3. Repeat Problem 13–2 for $J_3 = 1$, $K_3 = 0$.

13–4. Change Figure P13–1 to a recirculating shift register by connecting Q_0 back to J_3 and $\overline{Q_0}$ back to K_3. If the register is initially loaded with $D_3–D_0 = 1001$, what is the output at Q_3 to Q_0:

(a) After two clock pulses?   **(b)** After four clock pulses?

13–5. Outline the steps that you would take to parallel load the binary equivalent of a hex B into the register of Figure P13–1.

Parallel data input

D_3 D_2 D_1 D_0

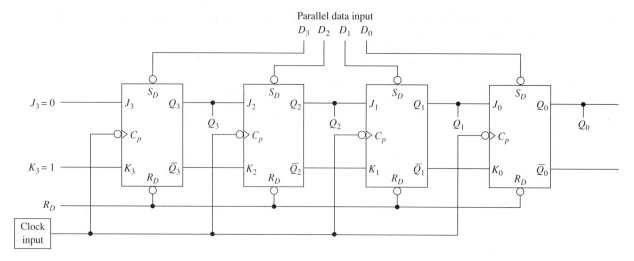

Figure P13–1

13–6. To use Figure P13–1 as a parallel-to-serial converter, where are the data input line(s) and data output line(s)?

13–7. Repeat Problem 13–6 for a serial-to-parallel converter.

Section 13–5

13–8. What changes have to be made to the circuit of Figure P13–1 to make it a Johnson shift counter?

13–9. How many flip-flops are required to produce the waveform shown in Figure P13–9 at the Q_0 output of a ring shift counter?

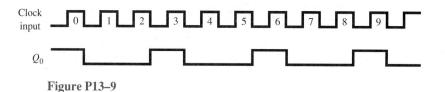

Figure P13–9

13–10. Repeat Problem 13–9 for the waveforms shown in Figure P13–10.

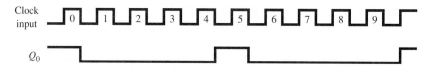

Figure P13–10

13–11. Which flip-flop(s) of a 4-bit ring shift counter must be initially Set to produce the waveform shown in Figure P13–11 at Q_0?

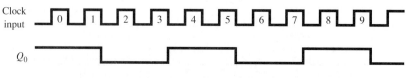

Figure P13–11

CHAPTER 13 | SHIFT REGISTERS

C   **13–12.** Sketch the waveforms at Q_2 for the first seven clock pulses generated by the circuit shown in Figure P13–12.

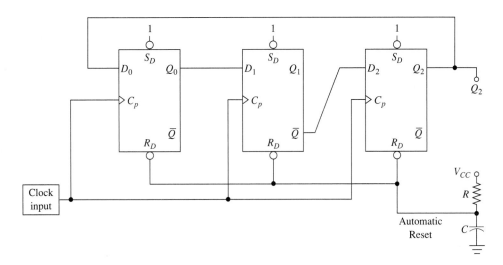

Figure P13–12

C   **13–13.** In Figure P13–12, connect the automatic Reset line to the three $\overline{S_D}$ inputs instead of the three $\overline{R_D}$ inputs and sketch the waveforms at Q_2 for the first seven clock pulses.

C   **13–14.** Sketch the waveforms at C_p, Q_0, Q_1, and Q_2 for seven clock pulses for the ring shift counter shown in Figure P13–14.

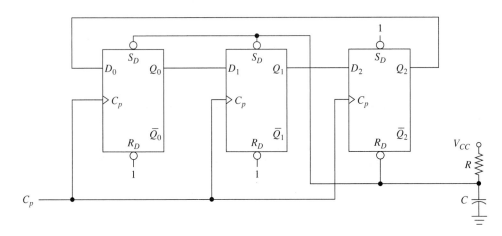

Figure P13–14

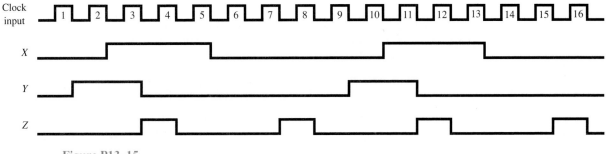

Figure P13–15

13–15. Using the Johnson shift counter output waveforms in Figure 13–6, add some logic gates to produce the waveforms at *X*, *Y*, and *Z* shown in Figure P13–15 on page 485.

D   **13–16.** Redesign the Johnson shift counter of Figure 13–6(a) using 7474 *D* flip-flops in place of the *J-K* flip-flops.

Sections 13–6 and 13–7

D   **13–17.** What modification could be made to the circuit in Figure 13–16 to cause the yellow light to flash all day on Sundays? (Assume that someone will throw a switch at the beginning and end of each Sunday.)

13–18. Sketch the output waveforms at Q_0 to Q_3 for the 74194 circuit shown in Figure P13–18. Also, list the operating mode at each positive clock edge.

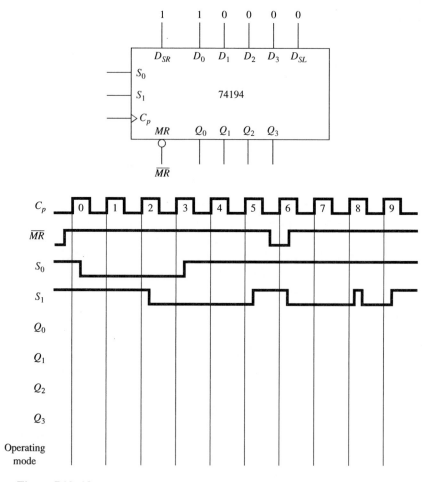

Figure P13–18

13–19. Repeat Problem 13–18 for the input waveforms shown in Figure P13–19.

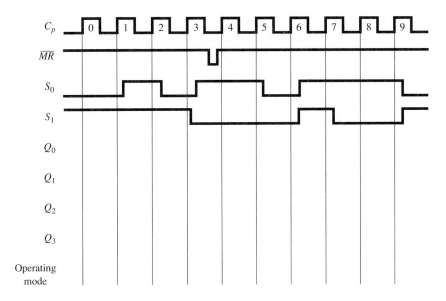

Figure P13–19

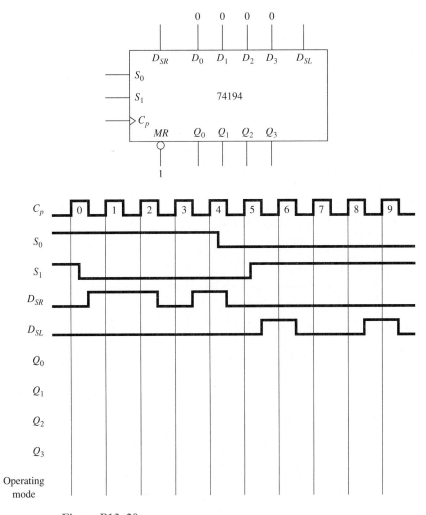

Figure P13–20

13–20. Sketch the output waveforms at Q_0 to Q_3 for the 74194 circuit shown in Figure P13–20 on page 487. Also, list the operating mode at each positive clock edge.

13–21. Repeat Problem 13–20 for the waveforms shown in Figure P13–21.

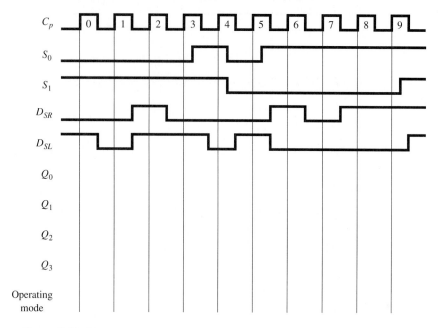

Figure P13–21

13–22. Draw the timing waveforms (similar to Figure 13–8) for a 74164 used to convert the serial binary number 10010110 into parallel.

13–23. Draw the circuit connections and timing waveforms for a 74165 used to convert the parallel binary number 1001 0110 into serial, MSB first.

13–24. Using your TTL data manual, describe the differences between the 74195 and the 74395A.

13–25. Using your TTL data manual, describe the differences between the 74164 and the 74165.

13–26. Describe how the procedure for parallel loading the 74165 differs from parallel loading a 74166.

C D   **13–27.** Design a system that can be used to convert an 8-bit serial number LSB first into an 8-bit serial number MSB first. Show the timing waveforms for 16 clock pulses and any control pulses that may be required for the binary number 10110100.

Section 13–8

13–28. How many clock pulses are required at C_p to cause the stepper motor to make one revolution in Figure 13–23?

C   **13–29.** Sketch the waveforms at C_p, S_0, S_1, Q_0, Q_1, Q_2, and Q_3 for six steps of the motor in Figure 13–23.

C   **13–30.** What must the clock frequency be in Figure 13–23 to make the stepper motor revolve at 600 rpm (rotations per minute)?

13–31. Describe the difference between a buffer and a latch and between a buffer and a transceiver.

13–32. Why is it important to use devices with three-state outputs when interfacing to a microprocessor data bus?

Schematic Interpretation Problems

See Appendix G for the schematic diagrams.

S **13–33.** Identify the following ICs on the 4096/4196 schematic (sheets 1 and 2):

(a) The three-state octal buffers

(b) The three-state octal D flip-flops

(c) The three-state octal transceivers

(d) The three-state octal latches

S **13–34.** Describe the operation of U6 in the 4096/4196 schematic. Use the names of the input/output labels provided on the IC for your discussion.

S C **13–35.** Refer to sheet 2 of the 4096/4196 schematic. Describe the sequence of operations that must take place to load the 8-bit data string labeled IA0–IA7 and the 8-bit data string labeled ID0–ID7. Include reference to U30, U32, U23, U13:A, U1:F, and U33.

Answers to Review Questions

13–1. True

13–2. The output of a flip-flop connected to the input of the next flip-flop (Q to J, $\overline{Q}$ to K)

13–3. By using the active-LOW asynchronous set ($\overline{S_D}$)

13–4. $Q_3 = 0, Q_2 = 0, Q_1 = 0, Q_0 = 0$; $Q_3 = 1, Q_2 = 1, Q_1 = 0, Q_0 = 0$

13–5. The data would continue shifting out the register and would be lost.

13–6. 1000, 0000

13–7. HIGH

13–8. It's an active-LOW clock enable for starting/stopping the clock.

13–9. Because the data can be input or output, serial or parallel, shifted left or right, held, and reset

13–10. $\overline{MR} = 1, S_1 = 1, S_0 = 1, D_0 = 1, D_1 = 1, D_2 = 0, D_3 = 1$. Data are loaded on the rising edge of C_p.

13–11. D_{SL}, Q_0, 0 1

13–12. Parallel enable (PE) enables the data to be loaded synchronously on the negative clock edge, and parallel load (PL) loads the data asynchronously.

13–13. HIGH, float

13–14. HIGH, enables

13–15. RC circuit and the Schmitt trigger

13–16. True

13–17. Instead of recycling, the parallel loaded information would be lost, because D_{SL} is ignored.

13–18. It is used as a strobe to enable the clock.

13–19. Clockwise. Do a shift-left in the 74194.

13–20. 8, $\overline{OE_a}\ \overline{OE_b}$

13–21. Synchronous, asynchronous

13–22. True

14 Multivibrators and the 555 Timer

OUTLINE

14–1 Multivibrators
14–2 Capacitor Charge and Discharge Rates
14–3 Astable Multivibrators
14–4 Monostable Multivibrators
14–5 IC Monostable Multivibrators
14–6 Retriggerable Monostable Multivibrators
14–7 Astable Operation of the 555 IC Timer
14–8 Monostable Operation of the 555 IC Timer
14–9 Crystal Oscillators

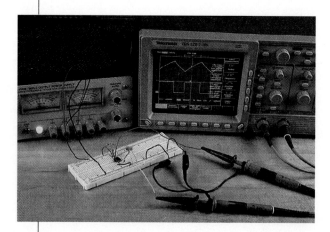

Objectives

Upon completion of this chapter, you should be able to

- Calculate capacitor charging and discharging rates in series *RC* timing circuits.

- Sketch the waveforms and calculate voltage and time values for astable and monostable multivibrators.

- Connect integrated-circuit monostable multivibrators to output a waveform with a specific pulse width.

- Explain the operation of the internal components of the 555 IC timer.

- Connect a 555 IC timer as an astable multivibrator and as a monostable multivibrator.

- Discuss the operation and application of crystal oscillator circuits.

Introduction

We have seen that timing is very important in digital electronics. Clock oscillators, used to drive counters and shift registers, must be designed to oscillate at a specific frequency. Specially designed **pulse-stretching** and time-delay circuits are also required to produce specific pulse widths and delay periods.

14–1   Multivibrators

Multivibrator circuits have been around for years, designed from various technologies, to fulfill electronic circuit timing requirements. A **multivibrator** is a circuit that changes between the two digital levels on a continuous, free-running basis or on demand from some external trigger source. Basically, there are three types of multivibrators: bistable, astable, and monostable.

The *bistable* multivibrator is triggered into one of the two digital states by an external source and stays in that state until it is triggered into the opposite state. The *S-R* flip-flop is a bistable multivibrator; it is in either the Set or Reset state.

The *astable* multivibrator is a free-running oscillator that alternates between the two digital levels at a specific frequency and duty cycle.

The *monostable* multivibrator, also known as a *one shot,* provides a single output pulse of a specific time length when it is triggered from an external source.

The bistable multivibrator (*S-R* flip-flop) was discussed in detail in Chapter 10. The astable and monostable multivibrators discussed in this chapter can be built from basic logic gates or from special ICs designed specifically for timing applications. In either case, the charging and discharging rate of a capacitor is used to provide the specific time durations required for the circuits to operate.

14–2   Capacitor Charge and Discharge Rates

Because the capacitor is so critical in determining the time durations, let's briefly discuss the capacitor charge and discharge formulas. We will use Figure 14–1 to determine the voltages on the capacitor at various periods of time after the switch is closed. In Figure 14–1, with the switch in position 1, conventional current will flow clockwise from the *E* source through the *RC* circuit. The capacitor will charge at an exponential rate, toward the value of the *E* source. The rate that the capacitor charges is dependent on the product of *R* times *C*:

$$\Delta v = E(1 - e^{-t/RC}) \tag{14-1}$$

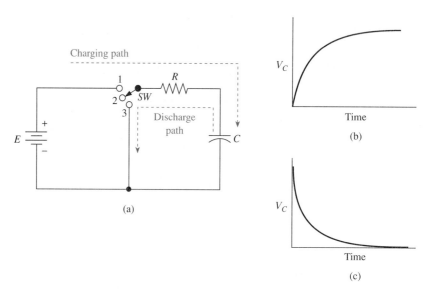

Figure 14–1   Basic *RC* charging/discharging circuit: (a) *RC* circuit; (b) charging curve; (c) discharging curve.

where $\Delta v \equiv$ change in capacitor voltage over a period of time t

$E \equiv$ voltage difference between the initial voltage on the capacitor and the total voltage that it is trying to reach

$e \equiv$ base of the natural log (2.718)

$t \equiv$ time that the capacitor is allowed to charge

$R \equiv$ resistance, ohms

$C \equiv$ capacitance, farads

In some cases the capacitor is initially discharged, and Δv is equal to the final voltage on the capacitor. But with astable multivibrator circuits, the capacitor usually is not fully discharged, and Δv is equal to the final voltage minus the starting voltage. If you think of the y axis in the graph of Figure 14–1(b) as a distance that the capacitor voltage is traveling through, the variables in Equation 14–1 take on new meaning, as follows:

$\Delta v \equiv$ distance that the capacitor voltage travels

$E \equiv$ total distance that the capacitor voltage is trying to travel

Using these new definitions, Equation 14–1 can be used whether the capacitor is charging *or* discharging (a discharging capacitor can be thought of as *charging to a lower voltage*).

When the switch in Figure 14–1 is thrown to position 3, the capacitor discharges counterclockwise through the *RC* circuit. The values for the variables in Equation 14–1 are determined the same way as they were for the charging condition, except that the voltage on the capacitor is decreasing **exponentially,** as shown in Figure 14–1(c).

Transposing the Capacitor Charging Formula to Solve for t

Often in the design of timing circuits it is necessary to solve for t,[*] given Dv, E, R, and C. To make life easy for ourselves, let's develop a new equation by rearranging Equation 14–1 to solve for t instead of Dv.

$$\Delta v = E(1 - e^{-t/RC})$$

$$\frac{\Delta v}{E} = 1 - e^{-t/RC} \qquad \text{Divide both sides by } E.$$

$$\frac{\Delta v}{E} - 1 = -e^{-t/RC} \qquad \text{Subtract 1 from both sides.}$$

$$1 - \frac{\Delta v}{E} = e^{-t/RC} \qquad \text{Multiply both sides by } (-1).$$

$$\frac{1}{1 - \Delta v/E} = \frac{1}{e^{-t/RC}} \qquad \text{Take the reciprocal of both sides.}$$

$$\frac{1}{1 - \Delta v/E} = e^{t/RC} \qquad \frac{1}{e^{-x}} = e^x$$

$$\ln\left(\frac{1}{1 - \Delta v/E}\right) = \ln e^{t/RC} \qquad \text{Take the natural logarithm of both sides.}$$

$$\ln\left(\frac{1}{1 - \Delta v/E}\right) = \frac{t}{RC} \qquad \ln e^x = x$$

$$t = RC \ln\left(\frac{1}{1 - \Delta v/E}\right) \qquad\qquad (14\text{–}2)$$

---

[*] A practical assumption often used by technicians is that a capacitor is 99% charged after a time equal to $5 \times R \times C$.

The following examples illustrate the use of Equations 14–1 and 14–2 for solving capacitor timing problems.

EXAMPLE 14–1

The capacitor in Figure 14–2 is initially discharged. Determine the voltage on the capacitor 0.5 ms after the switch is moved from position 2 to position 1.

Team Discussion

Would the capacitor voltage be more than or less than 3.27 V if the resistor is doubled? If the capacitor is doubled?

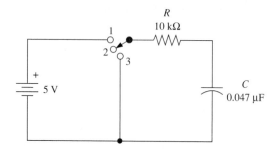

Figure 14–2 Circuit for Examples 14–1 through 14–4.

Solution: E, the total distance that the capacitor voltage is trying to charge to, is 5 V. Using Equation 14–1 yields

$$\Delta v = E(1 - e^{-t/RC})$$
$$= 5.0 \text{ V}(1 - e^{-0.5 \text{ ms}/(10 \text{ k}\Omega \times 0.047 \text{ }\mu\text{F})})$$
$$= 5.0(1 - e^{-1.06})$$
$$= 5.0(1 - 0.345)$$
$$= 5.0(0.655)$$
$$= 3.27 \text{ V} \quad \textit{Answer}$$

Thus the distance the capacitor voltage traveled in 0.5 ms is 3.27 V. Because it started at 0 V, $V_{\text{cap}} = 3.27$ V.

EXAMPLE 14–2

The capacitor in Figure 14–2 is initially discharged. How long after the switch is moved from position 2 to position 1 will it take for the capacitor to reach 3 V?

Solution: Δv, the distance that the capacitor voltage travels through, is 3 V. E, the total distance that the capacitor voltage is trying to travel, is 5 V. Using Equation 14–2, we obtain

$$t = RC \ln\left(\frac{1}{1 - \Delta v/E}\right)$$
$$= (10 \text{ k}\Omega)(0.047 \text{ }\mu\text{F}) \ln\left(\frac{1}{1 - 3/5}\right)$$
$$= 0.00047 \ln\left(\frac{1}{0.4}\right)$$
$$= 0.00047 \ln (2.5)$$
$$= 0.00047(0.916)$$
$$= 0.431 \text{ ms} \quad \textit{Answer}$$

EXAMPLE 14-3

For this example, let's assume that the capacitor in Figure 14–2 is initially charged to 1 V. How long after the switch is thrown from position 2 to position 1 will it take for the capacitor to reach 3 V?

Solution: Δv, the distance through which the capacitor voltage travels, is 2 V (3 V − 1 V). *E,* the total distance that the capacitor voltage is trying to travel, is 4 V (5 V − 1 V). Using Equation 14–2 gives us

$$t = RC \ln\left(\frac{1}{1 - \Delta v/E}\right)$$

$$= (10 \text{ k}\Omega)(0.047 \text{ }\mu\text{F}) \ln\left(\frac{1}{1 - 2/4}\right)$$

$$= 0.326 \text{ ms} \quad \textit{Answer}$$

The graph of the capacitor voltage is shown in Figure 14–3.

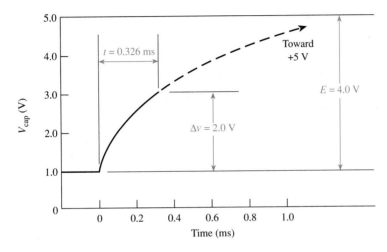

Figure 14–3   Graphical illustration of the capacitor voltage for Example 14–3.

EXAMPLE 14-4

The capacitor in Figure 14–2 is initially charged to 4.2 V. How long after the switch is thrown from position 2 to position 3 will it take to drop to 1.5 V?

Solution: Equation 14–2 can be used to solve for *t* by thinking of the capacitor as *charging to a lower voltage.* Δv, the distance that the capacitor voltage travels through, is 2.7 V (4.2 V − 1.5 V). *E,* the total distance that the capacitor voltage is trying to travel, is 4.2 V (4.2 V − 0 V). Using Equation 14–2 yields

$$t = RC \ln\left(\frac{1}{1 - \Delta v/E}\right)$$

$$= (10 \text{ k}\Omega)(0.047 \text{ }\mu\text{F}) \ln\left(\frac{1}{1 - 2.7 \text{ V}/4.2 \text{ V}}\right)$$

$$= 0.484 \text{ ms} \quad \textit{Answer}$$

The graph of the capacitor voltage is shown in Figure 14–4.

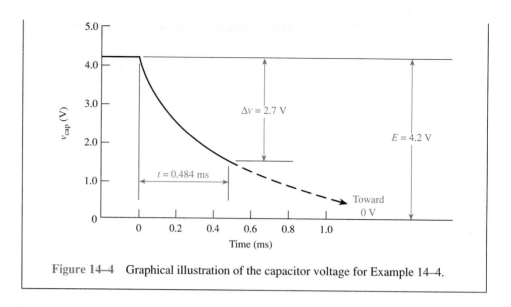

Figure 14–4   Graphical illustration of the capacitor voltage for Example 14–4.

Review Questions

14–1. The *astable* multivibrator, also known as a *one shot,* produces a single output pulse after it is triggered. True or false?

14–2. The voltage on a charging capacitor will increase _____ (faster, slower) if its series resistor is increased.

14–3. A 1-μF capacitor with a 10-kΩ series resistor will have the same charging rate as a 10-μF capacitor with a 1-kΩ series resistor. True or false?

14–3   Astable Multivibrators

A very simple astable multivibrator (free-running oscillator) can be built from a single Schmitt trigger inverter and an *RC* circuit, as shown in Figure 14–5. The oscillator of Figure 14–5 operates as follows:

Team Discussion

How would the operation of this circuit change if a 7414 were used in place of the 74HC14? (Consider I_I's and V_O's.)

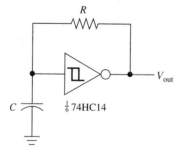

Figure 14–5   Schmitt trigger astable multivibrator.

1. When the IC supply power is first turned on, V_{cap} is 0 V, so V_{out} will be HIGH ($\approx$ 5.0 V for high-speed CMOS).

2. The capacitor will start charging toward the 5 V at V_{out}.

3. When V_{cap} reaches the positive-going threshold (V_{T+}) of the Schmitt trigger, the output of the Schmitt will change to a LOW (≈ 0 V).

4. Now, with $V_{out} \approx 0$ V, the capacitor will start discharging toward 0 V.

5. When V_{cap} drops below the negative-going threshold (V_{T-}), the output of the Schmitt will change back to a HIGH.

6. The cycle repeats now, with the capacitor charging back up to V_{T+}, then down to V_{T-}, then up to V_{T+}, and so on. (The waveform at V_{out} will be a square wave oscillating between V_{OH} and V_{OL}, as shown in Figure 14–6.)

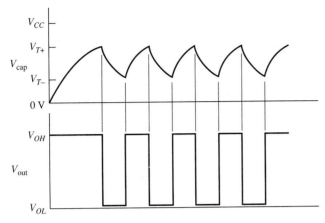

Figure 14–6 Waveforms from the oscillator circuit of Figure 14–5.

EXAMPLE 14–5

(a) Sketch and label the waveforms for the Schmitt RC oscillator of Figure 14–5, given the following specifications for a 74HC14 high-speed CMOS Schmitt inverter ($V_{CC} = 5.0$ V).

$$V_{OH} = 5.0 \text{ V}, \qquad V_{OL} = 0.0 \text{ V}$$
$$V_{T+} = 2.75 \text{ V}, \qquad V_{T-} = 1.67 \text{ V}$$

(b) Calculate the time HIGH (t_{HI}), time LOW (t_{LO}), duty cycle, and frequency if $R = 10$ kΩ and $C = 0.022$ μF.

Solution: (a) The waveforms for the oscillator are shown in Figure 14–7.

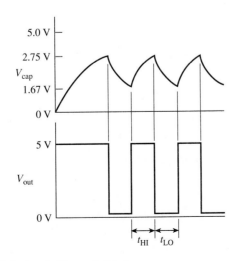

Figure 14–7 Solution to Example 14–5.

(b) To solve for t_{HI}:

$$\Delta V = 2.75 - 1.67 = 1.08 \text{ V}$$

$$E = 5.00 - 1.67 = 3.33 \text{ V}$$

$$t_{HI} = RC \ln\left(\frac{1}{1 - \Delta v/E}\right)$$

$$= (10 \text{ k}\Omega)(0.022 \text{ } \mu\text{F}) \ln\left(\frac{1}{1 - 1.08 \text{ V}/3.33 \text{ V}}\right)$$

$$= 86.2 \text{ } \mu\text{s}$$

To solve for t_{LO}:

$$\Delta v = 2.75 - 1.67 = 1.08 \text{ V}$$

$$E = 2.75 - 0 = 2.75 \text{ V}$$

$$t_{LO} = RC \ln\left(\frac{1}{1 - \Delta v/E}\right)$$

$$= (10 \text{ k}\Omega)(0.022 \text{ } \mu\text{F}) \ln\left(\frac{1}{1 - 1.08 \text{ V}/2.75 \text{ V}}\right)$$

$$= 110 \text{ } \mu\text{s}$$

To solve for duty cycle: **Duty cycle** is a ratio of the length of time a square wave is HIGH to the total period:

$$D = \frac{t_{HI}}{t_{HI} + t_{LO}}$$

$$= \frac{86.2 \text{ } \mu\text{s}}{86.2 \text{ } \mu\text{s} + 110 \text{ } \mu\text{s}}$$

$$= 0.439 = 43.9\%$$

To solve for frequency:

$$f = \frac{1}{t_{HI} + t_{LO}}$$

$$= \frac{1}{86.2 \text{ } \mu\text{s} + 110 \text{ } \mu\text{s}}$$

$$= 5.10 \text{ kHz}$$

Review Questions

14–4. The capacitor voltage levels in a Schmitt trigger astable multivibrator are limited by _____ and the output voltage is limited by _____.

14–5. One way to increase the frequency of a Schmitt trigger astable multivibrator is to _____ (increase, decrease) the resistor.

14–4 Monostable Multivibrators

The block diagram and I/O waveforms for a monostable multivibrator (commonly called a one shot) are shown in Figure 14–8. The one shot has one *stable state*, which

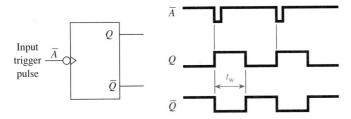

Figure 14–8 Block diagram and input/output waveforms for a monostable multivibrator.

is Q = LOW and $\overline{Q}$ = HIGH. The outputs switch to their opposite state for a length of time t_w only when a trigger is applied to the $\overline{A}$ input. $\overline{A}$ is a negative edge trigger in this case (other one shots use a positive edge trigger or both). The input/output waveforms in Figure 14–8 show the effect that $\overline{A}$ has on the Q output. Q is LOW until the HIGH-to-LOW edge of $\overline{A}$ causes Q to go HIGH for the length of time t_w. The output pulse width (t_w) is determined by the discharge rate of a capacitor in an RC circuit.

A simple monostable multivibrator can be built from NAND gates and an RC circuit as shown in Figure 14–9. The operation of Figure 14–9 is as follows:

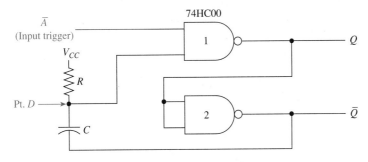

Figure 14–9 Two-gate monostable multivibrator.

1. When power is first applied, make the following assumptions: $\overline{A}$ is HIGH, Q is LOW, $\overline{Q}$ is HIGH, C is discharged. Therefore, point D is HIGH.

2. When a negative-going pulse is applied at $\overline{A}$, Q is forced HIGH, which forces $\overline{Q}$ LOW.

3. Since the capacitor voltage cannot change instantaneously, point D will drop to 0 V.

4. The 0 V at point D will hold one input to gate 1 LOW, even if the $\overline{A}$ trigger goes back HIGH. Therefore, Q stays HIGH, $\overline{Q}$ stays LOW.

5. Meanwhile, the capacitor is charging toward V_{CC}. When the capacitor voltage at point D reaches the HIGH-level input voltage rating (V_{IH}) of gate 1, Q will switch to a LOW, making $\overline{Q}$ HIGH.

6. The circuit is back in its stable state, awaiting another trigger signal from $\overline{A}$. The capacitor will discharge back to ≈ 0 V ($\approx V_{CC}$ on each side).

The waveforms in Figure 14–10 show the input/output characteristics of the circuit and will enable us to develop an equation to determine t_w. In the stable state ($\overline{Q}$ = HIGH), the voltage at point D will sit at V_{CC} ① because the capacitor is discharged (it has V_{CC} on both sides of it). When $\overline{Q}$ goes LOW due to an input trigger at $\overline{A}$, point D will follow $\overline{Q}$ LOW ② because the capacitor is still discharged. Now, the capacitor will start charging toward V_{CC}. When point D reaches V_{IH} ③, $\overline{Q}$ will switch back HIGH. The capacitor still has V_{IH} volts across it: $\overset{\longrightarrow)}{}\Big|\!\!\underset{V_{IH}}{\overset{+}{}}$. The capacitor voltage is added to the HIGH-level output of $\overline{Q}$, which causes point D to shoot up to $V_{CC} + V_{IH}$ ④. As the capacitor voltage discharges back to 0 V, point D drops back to the V_{CC} level ⑤ and awaits the next trigger.

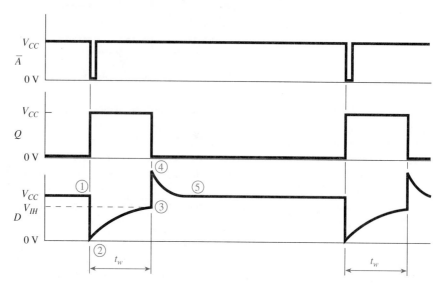

Figure 14–10   Input/output waveforms for the circuit of Figure 14–9.

EXAMPLE 14–6

(a) Sketch and label the waveforms for the monostable multivibrator of Figure 14–9, given the input waveform at $\overline{A}$ and the following specifications for a 74HC00 high-speed CMOS NAND gate ($V_{CC} = 5.0$ V).

$$V_{OH} = 5.0 \text{ V}, \qquad V_{OL} = 0.0 \text{ V}$$
$$V_{IH} = 3.5 \text{ V}, \qquad V_{IL} = 1.0 \text{ V}$$

(b) Calculate the output pulse width (t_w) for $R = 4.7$ kΩ and $C = 0.0047$ μF.

Solution:

(a) The waveforms for the multivibrator are shown in Figure 14–11.

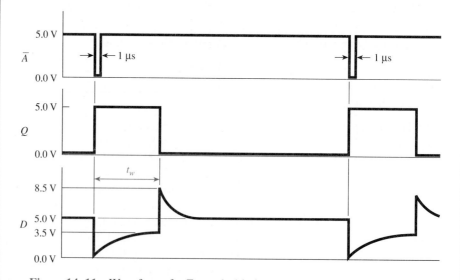

Figure 14–11   Waveforms for Example 14–6.

CHAPTER 14 | MULTIVIBRATORS AND THE 555 TIMER

(b) To solve for t_w using Equation 14–2:

$$\Delta v = 3.5 \text{ V} - 0 \text{ V} = 3.5 \text{ V}$$

$$E = 5.0 \text{ V} - 0 \text{ V} = 5.0 \text{ V}$$

$$t_w = RC \ln \left(\frac{1}{1 - \Delta v/E} \right)$$

$$= (4.7 \text{ k}\Omega)(0.0047 \text{ } \mu\text{F}) \ln \left(\frac{1}{1 - 3.5 \text{ V}/5.0 \text{ V}} \right)$$

$$= 26.6 \text{ } \mu\text{s} \quad \textit{Answer}$$

14–5   IC Monostable Multivibrators

Monostable multivibrators are available in an integrated-circuit package. Two popular ICs are the 74121 (nonretriggerable) and the 74123 (**retriggerable**) monostable multivibrators. To use these ICs, you need to connect the RC timing components to achieve the proper pulse width. The 74121 provides for two active-LOW and one active-HIGH trigger inputs $(\overline{A_1}, \overline{A_2}, B)$ and true and complemented outputs $(Q, \overline{Q})$. Figure 14–12 shows the 74121 block diagram and function table that we can use to figure out its operation.

To trigger the multivibrator at point T in Figure 14–12, the inputs to the Schmitt AND gate must both be HIGH. To do that, you need B with $(\overline{A_1}$ or $\overline{A_2})$. Holding $\overline{A_1}$ or $\overline{A_2}$ LOW and bringing the input trigger in on B is useful if the trigger signal is slow rising or if it has noise on it, because the Schmitt input will provide a definite trigger point.

Common Misconception

Students sometimes mistakenly sketch the output at Q as a negative-going pulse.

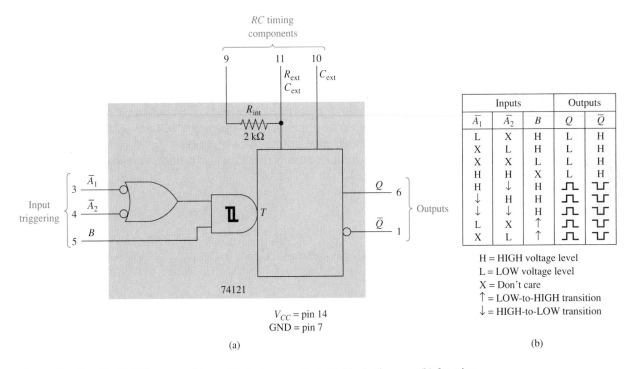

	Inputs		Outputs	
$\overline{A_1}$	$\overline{A_2}$	B	Q	$\overline{Q}$
L	X	H	L	H
X	L	H	L	H
X	X	L	L	H
H	H	X	L	H
H	↓	H	⊓	⊔
↓	H	H	⊓	⊔
↓	↓	H	⊓	⊔
L	X	↑	⊓	⊔
X	L	↑	⊓	⊔

H = HIGH voltage level
L = LOW voltage level
X = Don't care
↑ = LOW-to-HIGH transition
↓ = HIGH-to-LOW transition

V_{CC} = pin 14
GND = pin 7

(a)

(b)

Figure 14–12   The 74121 monostable multivibrator one shot: (a) block diagram; (b) function table. (Courtesy of Signetics Corporation)

The RC timing components are set up on pins 9, 10, and 11. If you can use the 2-kΩ *internal* resistor, just connect pin 9 to V_{CC} and put a timing capacitor between pins 10 and 11. An *external* timing resistor can be used instead by placing it between pin 11 and V_{CC} and putting the timing capacitor between pins 10 and 11. If the external timing resistor is used, pin 9 must be left open. The allowable range of R_{ext} is 1.4 to 40 kΩ and C_{ext} is 0 to 1000 μF. If an electrolytic capacitor is used, its positive side must be connected to pin 11.

The formula that the IC manufacturer gives for determining the output pulse width is

$$t_w = R_{ext}C_{ext} \ln 2 \tag{14-3}$$

(Substitute 2 kΩ for R_{ext} if the internal timing resistor is used.)

For example, if the external timing RC components are 10 kΩ and 0.047 μF, then t_w will equal 10 kΩ $\times$ 0.047 μF $\times$ ln 2, which works out to be 326 μs. Using the maximum allowed values of $R_{ext}C_{ext}$, the maximum pulse width is almost 28 s (40 kΩ $\times$ 1000 μF $\times$ ln 2).

The function table in Figure 14–12 shows that the Q output is LOW and the $\overline{Q}$ output is HIGH as long as the $\overline{A}_1$, $\overline{A}_2$, B inputs do not provide a HIGH–HIGH to the Schmitt-AND inputs. But by holding B HIGH and applying a HIGH-to-LOW edge to $\overline{A}_1$ or $\overline{A}_2$, the outputs will produce a pulse. Also, the function table shows, in its last two entries, that a LOW-to-HIGH edge at input B will produce an output pulse as long as either $\overline{A}_1$ or $\overline{A}_2$ is held LOW.

The following examples illustrate the use of the 74121 for one-shot operation.

EXAMPLE 14–7

Design a circuit using a 74121 to convert a 50-kHz, 80% duty cycle square wave to a 50-kHz, 50% duty cycle square wave. (In other words, stretch the negative-going pulse to cover 50% of the total period.)

Solution. First, let's draw the original square wave [Figure 14–13(a)] to see what we have to work with ($t = 1/50$ kHz = 20 μs, $t_{HI} = 80\% \times 20$ μs = 16 μs).

Now, we want to stretch the 4-μs negative pulse out to 10 μs to make the duty cycle 50%. If we use the HIGH-to-LOW edge on the negative pulse to trigger the $\overline{A}_1$ input to a 74121 and set the output pulse width (t_w) to 10 μs, we should have the solution. The output will be taken from $\overline{Q}$ because it provides a negative pulse when triggered.

Using the formula given in the IC specifications, we can calculate an appropriate R_{ext}, C_{ext} to yield 10 μs.

$$t_w = R_{ext}C_{ext} \ln (2)$$
$$10 \ \mu s = R_{ext}C_{ext}(0.693)$$
$$R_{ext}C_{ext} = 14.4 \ \mu s$$

Pick $C_{ext} = 0.001$ μF (1000 pF); then

$$R_{ext} = \frac{14.4 \ \mu s}{0.001 \ \mu F}$$

$\qquad = 14.4$ kΩ $\qquad$ (Use a 10-kΩ fixed resistor with a 5-kΩ potentiometer.)

The value 0.001 μF is a good choice for C_{ext} because it is much larger than any stray capacitance that might be encountered in a typical circuit. Values of capacitance less than 100 pF (0.0001 μF) may be unsuitable, because it is not uncommon for there to be 50 pF of stray capacitance between traces in a printed-circuit board. Also, resistances in the kilohm range are a good choice because they are big enough to limit current flow, but not so big to be susceptible to electrostatic noise. The final circuit design and waveforms are given in Figure 14–13(b) and (c).

Common Misconception

Students are often surprised to measure a 5% to 10% error in the output pulse width in the lab. Even if the pot is set at exactly 14.4 kΩ, there is still a capacitor tolerance to consider as well as stray wiring capacitance and internal IC inaccuracies.

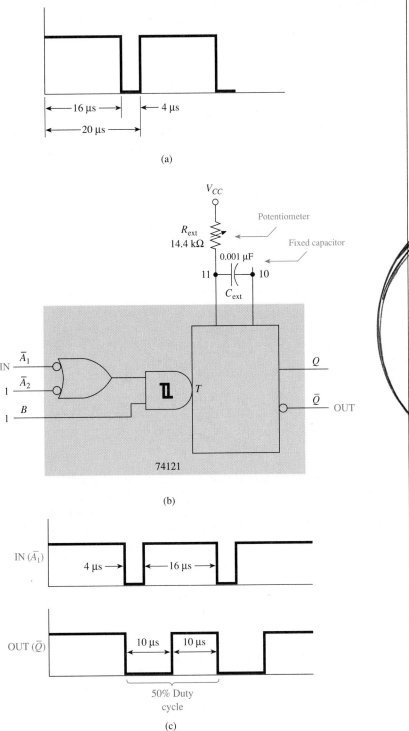

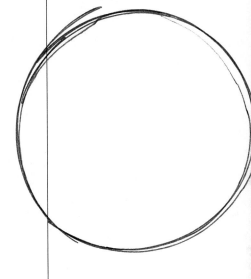

Figure 14–13   (a) Original square wave for Example 14–7; (b) monostable multivibrator circuit connections; (c) input/output waveforms.

EXAMPLE 14–8

In microprocessor systems, most control signals are active-LOW, and often one shots are required to introduce delays for certain devices to wait for other, slower devices to respond. For example, to read from a memory device, a line called $\overline{READ}$ goes LOW to enable the memory device. Most systems have to introduce a delay after the memory device is enabled (to allow for internal propagation delays) before the microprocessor actually reads the data. Design a system using two 74121s to output a 200-ns LOW pulse (called $\overline{Data\text{-}Ready}$) 500 ns after the $\overline{READ}$ line goes LOW.

Solution: The first 74121 will be used to produce the 500-ns delay pulse as soon as the $\overline{READ}$ line goes LOW (see Figure 14–14). The second 74121 will be triggered by the end of the 500-ns delay pulse and will output its own 200-ns LOW pulse for the $\overline{Data\text{-}Ready}$ line. (The 74121s are edge triggered, so they will trigger only on a HIGH-to-LOW or LOW-to-HIGH *edge*.)

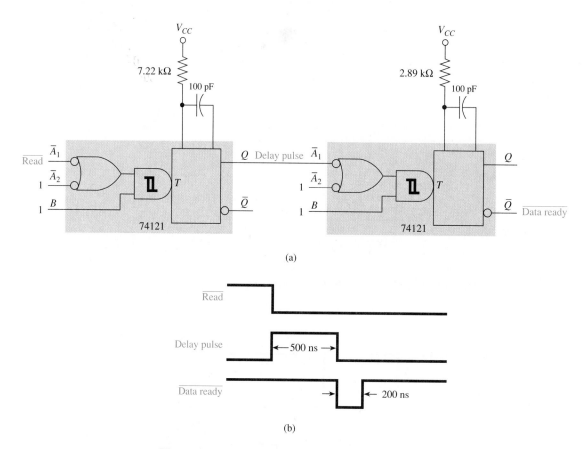

(a)

(b)

Figure 14–14 Solution to Example 14–8: (a) circuit connections for creating a delayed pulse; (b) input/output waveforms.

For $t_w = 500$ ns (output for first 74121):

$$t_w = R_{ext}C_{ext}\ln(2)$$
$$500\text{ ns} = R_{ext}C_{ext}(0.693)$$
$$R_{ext}C_{ext} = 0.722\ \mu s$$

Pick $C_{ext} = 100$ pF; then

$$R_{\text{ext}} = \frac{0.722 \ \mu s}{0.0001 \ \mu F} = 7.22 \ k\Omega$$

For $t_w = 200$ ns (output for second 74121):

$$t_w = R_{\text{ext}}C_{\text{ext}} \ln (2)$$
$$200 \text{ ns} = R_{\text{ext}}C_{\text{ext}}(0.693)$$
$$R_{\text{ext}}C_{\text{ext}} = 0.289 \ \mu s$$

Pick $C_{\text{ext}} = 100$ pF; then

$$R_{\text{ext}} = 2.89 \ k\Omega$$

14–6 Retriggerable Monostable Multivibrators

Have you wondered what might happen if a second input trigger came in before the end of the multivibrator's timing cycle? With the 74121 (which is nonretriggerable), any triggers that come in before the end of the timing cycle are ignored.

Retriggerable monostable multivibrators (such as the 74123) are available, which will start a new timing cycle each time a new trigger is applied. Figure 14–15 illustrates the differences between the retriggerable and nonretriggerable types, assuming that $t_w = 500$ ns and a negative edge-triggered input.

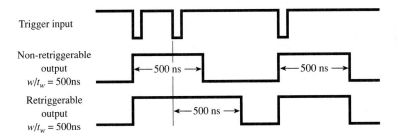

Figure 14–15 Comparison of retriggerable and nonretriggerable one-shot outputs.

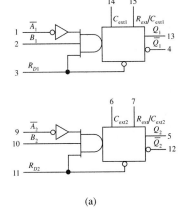

Input			Output	
R_D	$\overline{A}$	B	Q	$\overline{Q}$
L	X	X	L	H
X	H	X	L	H
X	X	L	L	H
H	L	↑	⊓	⊔
H	↓	H	⊓	⊔
↑	L	H	⊓	⊔

H = HIGH voltage level
L = LOW voltage level
X = Don't care
↑ = LOW-to-HIGH transition
↓ = HIGH-to-LOW transition
⊓ = One HIGH-level pulse
⊔ = One LOW-level pulse

(b)

Figure 14–16 The 74123 retriggerable monostable multivibrator: (a) logic symbol; (b) function table. (Courtesy of Signetics Corporation)

As you can see in Figure 14–15, the retriggerable device starts its timing cycle all over again when the second (or subsequent) input trigger is applied. The nonretriggerable device ignores any additional triggers until it has completed its 500-ns timing pulse.

The logic symbol and function table for the 74123 retriggerable monostable multivibrator are given in Figure 14–16.

Besides being retriggerable, some of the other important differences of the 74123 are as follows:

1. It is a *dual* multivibrator (two multivibrators in a single IC package).

2. It has an active-LOW Reset (R_D), which terminates all timing functions by forcing Q LOW, $\overline{Q}$ HIGH.

3. It has no internal timing resistor.

4. It uses a different method for determining the output pulse width, as explained next.

Output Pulse Width of the 74123

If $C_{\text{ext}} > 1000$ pF, the output pulse width is determined by the formula

$$t_w = 0.28 R_{\text{ext}} C_{\text{ext}} \left(1 + \frac{0.7}{R_{\text{ext}}} \right) \tag{14–4}$$

If $C_{\text{ext}} \leq 1000$ pF, the timing chart shown in Figure 14–17 must be used to find t_w.

For example, let's say that we need an output pulse width of 200 ns. Using the chart in Figure 14–17, one choice for the timing components would be $R_{\text{ext}} = 10$ kΩ, $C_{\text{ext}} = 30$ pF, or a better choice might be $R_{\text{ext}} = 5$ kΩ, $C_{\text{ext}} = 90$ pF. (With such small capacitances, required for nanosecond delays, we must be careful to minimize stray capacitance by using proper printed-circuit layout and component placement techniques.)

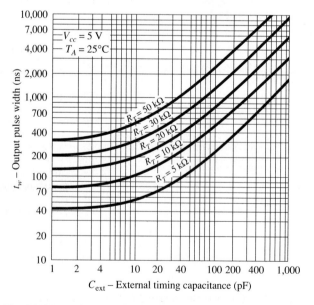

Figure 14–17   The 74123 timing chart for determining t_w when $C_{\text{ext}} \leq 1000$ pF.

CHAPTER 14 I MULTIVIBRATORS AND THE 555 TIMER

EXAMPLE 14–9

To the multivibrator circuit of Figure 14–18, apply the trigger input waveforms shown in Figure 14–19. Determine the output at Q_1.

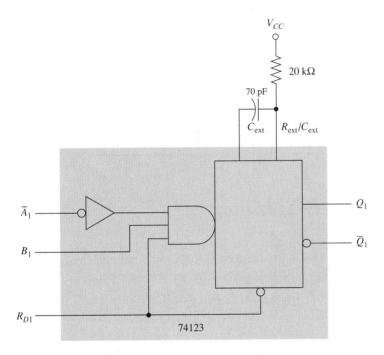

Figure 14–18 The 74123 circuit.

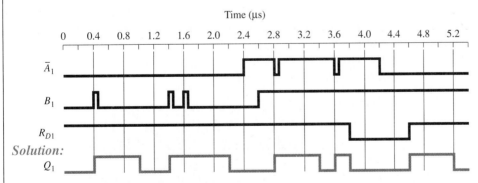

Figure 14–19 Waveforms for Example 14–9.

Explanation: t_w is determined from the timing chart for the 74123 (Figure 14–17). With $C_{ext} = 70$ pF and $R_{ext} = 20$ kΩ, $t_w = 0.6$ μs (600 ns).

0 to 0.4 μs Q is in its stable state (LOW); no trigger has been applied.

0.4 to 1.0 μs Q is triggered HIGH for 0.6 μs by the pulse on B_1.

1.0 to 1.4 μs Q returns to its stable state.

1.4 to 2.2 μs Q is triggered HIGH by B_1 at 1.4 μs; then Q is retriggered at 1.6 μs.

2.2 to 2.8 μs Q returns LOW; the conditions on $A_1 B_1 R_{D1}$ are not right to create a trigger.

2.8 to 3.4 μs Q is triggered HIGH by the LOW pulse on A_1 while $B_1 =$ HIGH and $R_{D1} =$ HIGH.

3.4 to 3.6 μs Q returns LOW.

3.6 to 3.8 μs Q is triggered HIGH by $\overline{A_1}$, but the output pulse is terminated by a LOW on R_{D1}.

3.8 to 4.6 μs Q is held LOW by R_{D1} no matter what the other inputs are doing.

4.6 to 5.2 μs Q is triggered HIGH by the LOW-to-HIGH edge of R_{D1} while $\overline{A_1} =$ LOW, $B_1 =$ HIGH.

5.2 to 5.4 μs Q returns LOW.

Review Questions

14–6. The output of a monostable multivibrator has a predictable pulse width based on the width of the input trigger pulse. True or false?

14–7. To trigger a 74121 one-shot IC, A_1 _____ (and, or) A_2 must be made _____ (LOW, HIGH) _____ (and, or) B must be made _____ (HIGH, LOW).

14–8. Which of the three trigger inputs of the 74121 would you use to trigger from a falling edge of a pulse? What would you do with the other two inputs?

14–9. When the 74121 receives a trigger, the Q output goes _____ (HIGH, LOW) for a time duration t_w.

14–10. The 74123 one-shot IC is *retriggerable*, whereas the 74121 is not. What does this statement mean?

14–7   Astable Operation of the 555 IC Timer

The 555 is a very popular, general-purpose timer IC. It can be connected as a one shot or an astable oscillator, as well as being used for a multitude of custom designs. Figure 14–20 shows a block diagram of the chip with its internal components and the external components that are required to set it up as an astable oscillator.

The 555 got its name from the three 5-kΩ resistors. They are set up as a voltage divider from V_{CC} to ground. The top of the lower 5 kΩ is at $\frac{1}{3}V_{CC}$ and the top of the middle 5 kΩ is at $\frac{2}{3}V_{CC}$. For example, if V_{CC} is 6 V, each resistor will drop 2 V.

The triangle-shaped symbols represent **comparators.** A comparator simply outputs a HIGH or LOW based on a comparison of the analog voltage levels at its input. If the + input is *more positive* than the − input, it outputs a HIGH. If the + input is *less positive* than the − input, it outputs a LOW.

The *S-R flip-flop* is driven by the two comparators. It has an active-LOW Reset and its output is taken from the $\overline{Q}$.

The *discharge transistor* is an NPN, which is used to short pins 7 to 1 when $\overline{Q}$ is HIGH.

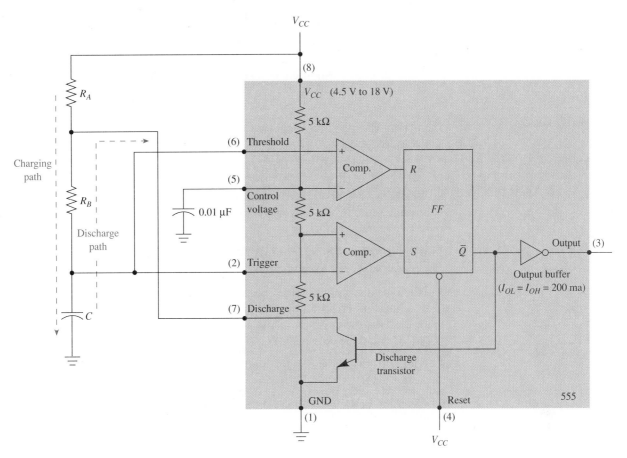

Figure 14–20 Simplified block diagram of a 555 timer with the external timer components to form an astable multivibrator.

The operation and function of the 555 pins are as follows:

Pin 1 (ground): System ground.

Pin 2 (trigger): Input to the lower comparator, which is used to Set the flip-flop. When the voltage at pin 2 crosses from above to below $\frac{1}{3}V_{CC}$, the comparator switches to a HIGH, setting the flip-flop.

Pin 3 (output): The output of the 555 is driven by an inverting buffer capable of sinking or sourcing 200 mA. The output voltage levels are dependent on the output current, but are approximately $V_{OH} = V_{CC} - 1.5$ V and $V_{OL} = 0.1$ V.

Pin 4 (Reset): Active-LOW Reset, which forces $\overline{Q}$ HIGH and pin 3 (output) LOW.

Pin 5 (control): Used to override the $\frac{2}{3}V_{CC}$ level, if required. Usually, it is connected to a grounded 0.01-μF capacitor to bypass noise on the V_{CC} line.

Pin 6 (threshold): Input to the upper comparator, which is used to Reset the flip-flop. When the voltage at pin 6 crosses from below to above $\frac{2}{3}V_{CC}$, the comparator switches to a HIGH, resetting the flip-flop.

Pin 7 (discharge): Connected to the open collector of the NPN transistor. It is used to short pin 7 to ground when $\overline{Q}$ is HIGH (pin 3 LOW), which will discharge the external capacitor.

Pin 8 (V_{CC}): Supply voltage. V_{CC} can range from 4.5 to 18 V.

The operation of the 555 connected in the astable mode shown in Figure 14–20 is explained as follows.

1. When power is first turned on, the capacitor is discharged, which places 0 V at pin 2, forcing the lower comparator HIGH. This sets the flip-flop ($\overline{Q}$ = LOW, output = HIGH).

2. With the output HIGH ($\overline{Q}$ LOW), the discharge transistor is open, which allows the capacitor to charge toward V_{CC} via $R_A + R_B$.

3. When the capacitor voltage exceeds $\frac{1}{3}V_{CC}$, the lower comparator goes LOW, which has no effect on the S-R flip-flop; but when the capacitor voltage exceeds $\frac{2}{3}V_{CC}$, the upper comparator goes HIGH, resetting the flip-flop, forcing $\overline{Q}$ HIGH and the output LOW.

4. With $\overline{Q}$ HIGH, the transistor shorts pin 7 to ground, which discharges the capacitor via R_B.

5. When the capacitor voltage drops below $\frac{1}{3}V_{CC}$, the lower comparator goes back HIGH again, setting the flip-flop and making $\overline{Q}$ LOW, output HIGH.

6. Now, with $\overline{Q}$ LOW, the transistor opens again, allowing the capacitor to start charging up again.

7. The cycle repeats, with the capacitor charging up to $\frac{2}{3}V_{CC}$ and then discharging down to $\frac{1}{3}V_{CC}$ continuously. While the capacitor is charging, the output is HIGH, and when the capacitor is discharging, the output is LOW.

The theoretical waveforms depicting the operation of the 555 as an astable oscillator are shown in Figure 14–21(a). Figure 14–21(b) shows the actual circuit and oscilloscope waveforms at VC and Vout.

The formulas for the time durations t_{LO} and t_{HI} can be derived using the theory presented in Section 14–2 and Equation 14–2. The time duration t_{LO} is determined by realizing that the capacitor voltage (Δv) travels a distance of $\frac{1}{3}V_{CC}$ (from $\frac{2}{3}V_{CC}$ to $\frac{1}{3}V_{CC}$) and the total path that it is trying to travel (E) is equal to $\frac{2}{3}V_{CC}$ (from $\frac{2}{3}V_{CC}$ to 0 V). The path of the discharge current is through R_B and C, so the time constant, τ (tau), is $R_B \times C$. Therefore, the equation for t_{LO} is derived as follows:

$$t_W = RC \ln\left(\frac{1}{1 - \Delta v/E}\right) \tag{14–2}$$

$$t_{LO} = R_B C \ln\left(\frac{1}{1 - \frac{1}{3}V_{CC}/\frac{2}{3}V_{CC}}\right)$$

$$= R_B C \ln\left(\frac{1}{1 - 0.5}\right)$$

$$= R_B C \ln(2)$$

$$= 0.693 R_B C \tag{14–5}$$

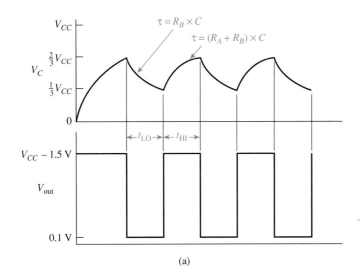

Team Discussion

Why can't the duty cycle of V_{out} ever be less than 50% in this configuration?

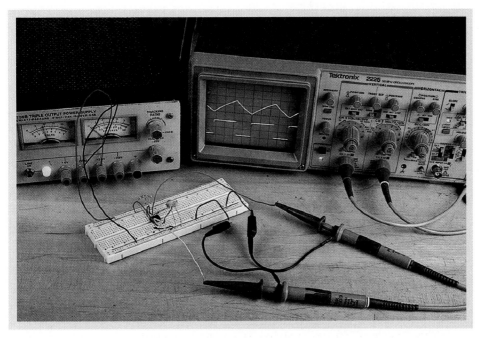

Figure 14–21 The 555 astable multivibrator: (a) theoretical V_c and V_{out} versus time waveforms; (b) actual breadboarded circuit, power supply, and oscilloscope displaying the measured V_c and V_{out} waveforms.

To derive the equation for t_{HI}:

Δv = distance the capacitor voltage travels = $\frac{1}{3}V_{CC}$ (that is, $\frac{2}{3}V_{CC} - \frac{1}{3}V_{CC}$)

E = total distance that the capacitor voltage is trying to travel = $\frac{2}{3}V_{CC}$ (that is, $V_{CC} - \frac{1}{3}V_{CC}$)

τ = **time constant,** or the path that the charging current flows through = $(R_A + R_B) \times C$

$$t_{HI} = (R_A + R_B)C \ln\left(\frac{1}{1 - \frac{1}{3}V_{CC}/\frac{2}{3}V_{CC}}\right)$$

$$= (R_A + R_B)C \ln \left(\frac{1}{1 - 0.5} \right)$$
$$= (R_A + R_B)C \ln (2)$$
$$= 0.693(R_A + R_B)C \tag{14–6}$$

EXAMPLE 14–10

Determine t_{HI}, t_{LO}, duty cycle, and frequency for the 555 astable multivibrator circuit of Figure 14–22.

Team Discussion

If you want to increase the frequency of this oscillator without affecting the duty cycle, do both resistors need adjusting?

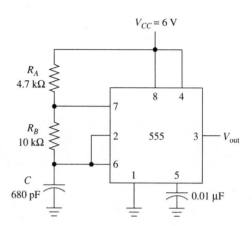

Figure 14–22 The 555 astable connections for Example 14–10.

Solution:

$$t_{LO} = 0.693R_BC$$
$$= 0.693(10 \text{ k}\Omega)680 \text{ pF}$$
$$= 4.71 \ \mu s$$
$$t_{HI} = 0.693(R_a + R_B)C$$
$$= 0.693(4.7 \text{ k}\Omega + 10 \text{ k}\Omega)680 \text{ pF}$$
$$= 6.93 \ \mu s$$
$$\text{duty cycle} = \frac{t_{HI}}{t_{HI} + t_{LO}}$$
$$= \frac{6.93 \ \mu s}{6.93 \ \mu s + 4.71 \ \mu s}$$
$$= 59.5\%$$
$$\text{frequency} = \frac{1}{t_{HI} + t_{LO}}$$
$$= \frac{1}{6.93 \ \mu s + 4.71 \ \mu s}$$
$$= 85.9 \text{ kHz}$$

50% Duty Cycle Astable Oscillator

By studying Figure 14–21 you should realize that to get a 50% duty cycle t_{LO} must equal t_{HI}. You should also realize that for this to occur the capacitor charging time constant (τ) must equal the discharging time constant. But with the astable circuits that we have seen so far, this can never be true because the resistance in one case is just R_B, but in the other case it is R_B *plus* R_A. You cannot just make R_A equal to 0 Ω because that would put V_{CC} directly on pin 7.

However, if we make $R_A = R_B$ and short R_B with a diode during the capacitor charging cycle, we can achieve a 50% duty cycle. The circuit for a 50% duty cycle is shown in Figure 14–23.

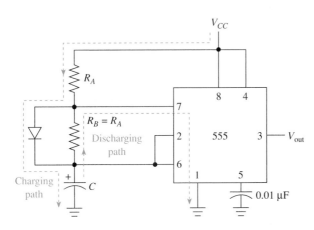

Figure 14–23 The 555 astable multivibrator set up for a 50% duty cycle.

The charging time constant in Figure 14–23 is $R_A \times C$, and the discharging time constant is $R_B \times C$. The formulas therefore become

$$t_{HI} = 0.693R_A C \qquad (14\text{–}7)$$

$$t_{LO} = 0.693R_B C \qquad (14\text{–}8)$$

If R_B is set equal to R_A, t_{HI} will equal t_{LO} and the duty cycle will be 50%. Also, duty cycles of *less than* 50% can be achieved by using the diode and making R_A less than R_B.

14–8 Monostable Operation of the 555 IC Timer

Another common use for the 555 is as a monostable multivibrator, as shown in Figure 14–24. The one shot of Figure 14–24 operates as follows:

1. Initially (before the trigger is applied), V_{out} is LOW, shorting pin 7 to ground and discharging C.

2. Pin 2 is normally held HIGH by the 10-kΩ pull-up resistor. To trigger the one shot, a negative-going pulse (less than $\frac{1}{3}V_{CC}$) is applied to pin 2.

3. The trigger forces the lower comparator HIGH (see Figure 14–20), which sets the flip-flop, making V_{out} HIGH and opening the discharge transistor (pin 7).

4. Now the capacitor is free to charge from 0 V up toward V_{CC} via R_A.

5. When V_c crosses the threshold of $\frac{2}{3}V_{CC}$, the upper comparator goes HIGH, resetting the flip-flop, making V_{out} LOW, and shorting the discharge transistor.

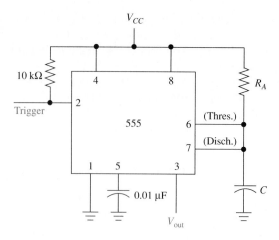

Figure 14–24   The 555 connections for one-shot operation.

6. The capacitor discharges rapidly to 0 V, and the one shot is held in its stable state (V_{out} = LOW) until another trigger is applied.

The waveforms that are generated for the one-shot operation are shown in Figure 14–25.

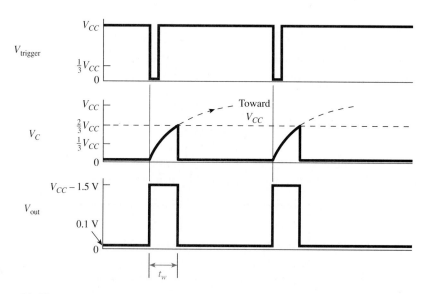

Figure 14–25   Waveforms for the one-shot circuit of Figure 14–24.

To derive the equation for t_w of the one shot, we start with the same capacitor-charging formula (Equation 14–2):

$$t_w = RC \ln \left(\frac{1}{1 - \Delta v / E} \right) \qquad (14\text{–}2)$$

where R = R_A

Δv = distance that the capacitor voltage travels = $\frac{2}{3} V_{CC}$ (from 0 V up to $\frac{2}{3} V_{CC}$)

E = distance that the capacitor voltage is trying to travel = V_{CC} (from 0 V up to V_{CC})

Substitution yields

$$t_w = R_A C \ln \left(\frac{1}{1 - \frac{2}{3}V_{CC}/V_{CC}} \right)$$

$$= R_A C \ln \left(\frac{1}{1 - 0.667} \right)$$

$$= R_A C \ln (3)$$

$$= 1.10 R_A C \qquad (14\text{–}9)$$

EXAMPLE 14–11

Design a circuit using a 555 one shot that will stretch a 1-μs negative-going pulse that occurs every 60 μs into a 10-μs negative-going pulse.

Solution: To set the output pulse width to 10 μs:

$$t_w = 1.10 R_A C$$

$$10 \ \mu s = 1.10 R_A C$$

$$R_A C = 9.09 \ \mu s$$

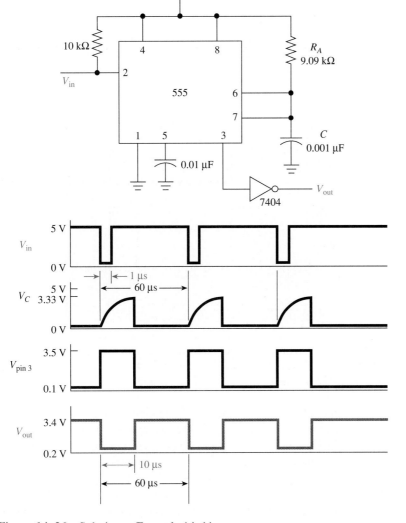

Figure 14–26 Solution to Example 14–11.

Pick $C = 0.001 \; \mu F$; then

$$R_A = 9.09 \; k\Omega$$

Also, since the 555 outputs a positive-going pulse, an inverter must be added to change it to a negative-going pulse. (7404 inverter: $V_{OH} = 3.4$ V, $V_{OL} = 0.2$ V). The final circuit design and waveforms are shown in Figure 14–26.

14–9 Crystal Oscillators

None of the *RC* oscillators or one shots presented in the previous sections are extremely stable. In fact, the standard procedure for building those timing circuits is to prototype them based on the *R* and *C* values calculated using the formulas and then "tweak" (or make adjustments to) the resistor values while observing the time period on an oscilloscope. Normally, standard values are chosen for the capacitors, and potentiometers are used for the resistors.

However, even after a careful calibration of the time period, changes in the components and IC occur as the devices age and as the ambient temperature varies. To partially overcome this problem, some manufacturers will allow their circuits to **burn in,** or age for several weeks before the final calibration and shipment.

Instead of using *RC* components, another timing component is available to the design engineer when extremely critical timing is required. This highly stable and accurate timing component is the *quartz* **crystal** (see Figure 14–27). A piece of quartz crystal is cut to a specific size and shape to vibrate at a specific frequency, similar to an *RLC* resonant circuit. Its frequency is typically in the range 10 kHz to 10 MHz. Accuracy of more than *five significant digits* can be easily achieved using this method.

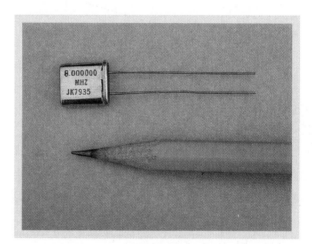

Figure 14–27   Photograph of an 8.000000-MHz quartz crystal.

Crystal **oscillators** are available as an integrated-circuit package or can be built using an external quartz crystal in circuits such as those shown in Figure 14–28. The circuits shown in the figure will oscillate at a frequency dictated by the crystal chosen. In Figure 14–28(a), the 100-kΩ pot may need adjustment to start oscillation.

The 74S124 TTL chip in Figure 14–28(b) is a **voltage-controlled oscillator,** set up to generate a specific frequency at V_{out}. By changing the crystal to a capacitor, the output frequency will vary, depending on the voltage level at the frequency control (pin 1) and frequency range (pin 14) inputs. Using specifications presented in the manufacturer's data manual, you can determine the output frequency of the VCO based on the voltage level applied to its inputs.

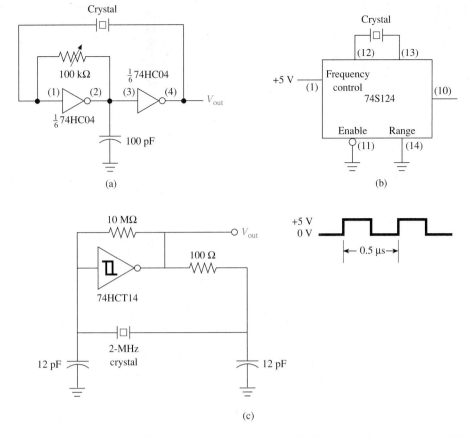

Figure 14–28 Crystal oscillator circuits: (a) high-speed CMOS oscillator; (b) Schottky-TTL oscillator; (c) Schmitt-inverter oscillator.

The oscillator in Figure 14–28(c) uses a 2-MHz crystal and a single HCT-Schmitt to create the highly accurate waveform shown.

Review Questions

14–11. The comparators inside the 555 IC timer will output a LOW if their (−) input is more positive than their (+) input. True or false?

14–12. The discharge transistor inside the 555 shorts pin 7 to ground when the output at pin 3 is _____ (LOW, HIGH).

14–13. When pin 6 (Threshold) of the 555 IC exceeds _____ (1/3 V_{CC}, 2/3 V_{CC}), the flip-flop is _____ (Reset, Set), making the output at pin 3 _____ (LOW, HIGH).

14–14. The 555 is connected as an astable multivibrator in Figure 14–20. V_{out} is HIGH while the capacitor charges through resistor(s) _____, and V_{out} is LOW while the capacitor discharges through resistor(s) _____.

14–15. The 555 astable multivibrator in Figure 14–20 will always have a duty cycle _____ (greater than, less than) 50% because _____.

14–16. When using the 555 as a monostable (one-shot) multivibrator, a _____ (LOW, HIGH) trigger is applied to pin 2, which forces V_{out} _____ (LOW, HIGH) and initiates the capacitor to start _____ (charging, discharging).

14–17. What advantage does a quartz crystal have over an *RC* circuit when used in timing applications?

Summary

In this chapter we have learned that

1. Multivibrator circuits are used to produce free-running clock oscillator waveforms or to produce a timed digital level change triggered by an external source.

2. Capacitor voltage charging and discharging rates are the most common way to produce predictable time duration for oscillator and timing operations.

3. An astable multivibrator is a free-running oscillator whose output oscillates between two voltage levels at a rate determined by an attached *RC* circuit.

4. A monostable multivibrator is used to produce an output pulse that starts when the circuit receives an input trigger and lasts for a length of time dictated by the attached *RC* circuit.

5. The 74121 is an IC monostable multivibrator with two active-LOW and one active-HIGH input trigger sources and an active-HIGH and an active-LOW pulse output terminal.

6. Retriggerable monostable multivibrators allow multiple input triggers to be acknowledged even if the output pulse from the previous trigger had not expired.

7. The 555 IC is a general-purpose timer that can be used to make astable and monostable multivibrators and perform any number of other timing functions.

8. Crystal oscillators are much more accurate and stable than *RC* timing circuits. They are used most often for microprocessor and digital communication timing.

Glossary

Burn-In: A step near the end of the production process in which a manufacturer exercises the functions of an electronic circuit and ages the components before the final calibration step.

Comparator: As used in a 555 timer, it compares the analog voltage level at its two inputs and outputs, a HIGH or a LOW, depending on which input was higher. (If the voltage level on the + input is higher than the voltage level on the − input, the output is HIGH; otherwise, it is LOW.)

Crystal: A material, usually made from quartz, that can be cut and shaped to oscillate at a very specific frequency. It is used in highly accurate clock and timing circuits.

Duty Cycle: A ratio of the lengths of time that a digital signal is HIGH versus its total period:

$$\text{duty cycle} = \frac{t_{HI}}{t_{HI} + t_{LO}}$$

Exponential Charge/Discharge: An exponential rate of charge or discharge is nonlinear, meaning that the rate of change of capacitor voltage is greater in the beginning and then slows down toward the end.

Multivibrator: An electronic circuit or IC used in digital electronics to generate HIGH and LOW logic states. The *bistable* multivibrator is an *S-R* flip-flop triggered into its HIGH or LOW state. The *astable* multivibrator is a free-running oscillator that continuously alternates between its HIGH and LOW states. The *monostable* multivibrator is a one shot that, when triggered, outputs a single pulse of a specific time duration.

Oscillator: An electronic circuit whose output continuously alternates between HIGH and LOW states at a specific frequency.

Pulse Stretching: Increasing the time duration of a pulse width.

Retriggerable: A device that is capable of reacting to a second or subsequent trigger before the action initiated by the first trigger is complete.

Time Constant (tau, τ): τ is equal to the product of resistance times capacitance and is used to determine the *rate* of charge or discharge in a series *RC* circuit. (1τ is equal to the number of seconds that it takes for a capacitor's voltage to reach 63% of its final value.)

Voltage-Controlled Oscillator (VCO): An oscillator whose output frequency is dependent on the analog voltage level at its input.

Problems

Sections 14–1 and 14–2

14–1. Which type of multivibrator is also known as a:

(a) One-shot?

(b) *S-R* flip-flop?

(c) Free-running oscillator?

14–2.

(a) For the *RC* circuit of Figure P14–2, determine the voltage on the capacitor 50 μs after the switch is moved from position 2 to position 1. (Assume that $V_c = 0$ V initially.)

(b) Repeat part (a) for 100 μs.

(c) Repeat part (a) for 150 μs.

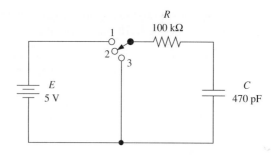

Figure P14–2

(d) Sketch and label a graph of capacitor voltage versus time for the values that you found in parts (a), (b), and (c).

14–3. The capacitor in Figure P14–2 is initially discharged. How long after the switch is moved from position 2 to position 1 will it take for the capacitor to reach 4 V?

14–4. Assume that the capacitor in Figure P14–2 is initially charged to 2 V. How long after the switch is moved from position 2 to position 1 will it take for the capacitor voltage to reach 4 V?

14–5. Assume that the capacitor in Figure P14–2 is initially charged to 4 V. How long after the switch is moved from position 2 to position 3 will it take for the voltage to drop to 2 V?

14–6. If you were successful at solving for the time in Problems 14–4 and 14–5, you will notice that it takes longer for the capacitor voltage to go from 2 V to 4 V than it did to go from 4 V to 2 V. Why is that true?

Section 14–3

14–7. Why is a Schmitt trigger inverter used for the astable multivibrator circuit of Figure 14–5 instead of a regular inverter like a 74HC04?

14–8. In a Schmitt trigger astable multivibrator, if the hysteresis voltage (V_{T+} minus V_{T-}) decreases due to a temperature change, what happens to:

(a) The output frequency?   **(b)** The output voltage?

14–9. Specifications for the 74HC14 Schmitt inverter when powered from a 6-V supply are as follows: $V_{OH} = 6.0$ V, $V_{OL} = 0.0$ V, $V_{T+} = 3.3$ V, and $V_{T-} = 2.0$ V.

(a) Sketch and label the waveforms for V_{cap} and V_{out} in the astable multivibrator circuit of Figure 14–5. (Use $R = 68$ kΩ and $C = 0.0047$ μF.)

(b) Calculate t_{HI}, t_{LO}, duty cycle, and frequency.

Sections 14–4 and 14–5

D   **14–10.** Design a monostable multivibrator using two 74HC00 NAND gates similar to Figure 14–9. Determine the values for R and C such that a negative-going, 2-μs input trigger will create a 50-μs positive-going output pulse.

D   **14–11.** Make the external connections to a 74121 monostable multivibrator to convert a 100-kHz, 30% duty cycle square wave to a 100-kHz, 50% duty cycle square wave.

CHAPTER 14 | MULTIVIBRATORS AND THE 555 TIMER

C D   **14–12.** Use two 74121s as a delay line to reproduce the waveforms shown in Figure P14–12. (The output pulse will look just like the input pulse but delayed by 30 μs.)

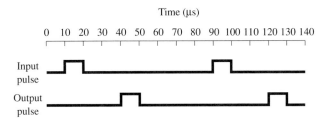

Figure P14–12

Section 14–6

C D   **14–13.** The NPRO microprocessor control line is supposed to issue a 10-μs LOW pulse every 150 μs as long as a certain process is running smoothly. Design a "missing pulse detector" using a 74123 that will normally output a HIGH but will output a LOW if a single pulse on the NPRO line is skipped. (*Hint:* The 74123 will be retriggered by NPRO every 150 μs.)

14–14. Using the timing chart in Figure 14–17 for a 74123, determine a good value for R_{ext} and C_{ext} to give an output pulse width of 400 ns.

14–15. Sketch the output waveforms at Q_1 and Q_2 of the multivibrators given in Figure P14–15 if the $\overline{A_1}$ and B waveforms are as shown.

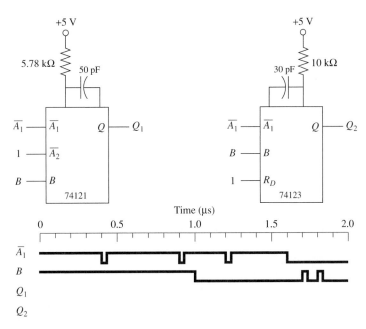

Figure P14–15

14–16. Sketch and label the waveforms at V_{out} for the 555 circuit of Figure P14–16 with the potentiometer set at 0 Ω.

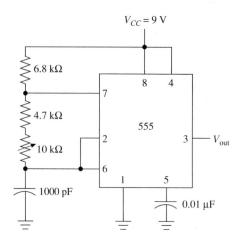

Figure P14–16

C   **14–17.** Determine the maximum and minimum frequency and the maximum and minimum duty cycle that can be achieved by adjusting the potentiometer in Figure P14–16.

C   **14–18.** Derive formulas for duty cycle and frequency in terms of R_A, R_B, and C for a 555 astable multivibrator. (Test your formulas by resolving Problem 14–17.)

C D   **14–19.** Using a 555, design an astable multivibrator that will oscillate at 50 kHz, 60% duty cycle. (So that we all get the same answer, let's pick $C = 0.0022\ \mu$F.)

C D   **14–20.** Design a circuit that will produce a 100-kHz square wave using:

(**a**) A 74HC14   (**c**) A 74123

(**b**) Two 74121s   (**d**) A 555

14–21. Sketch and label the waveforms at $V_{trigger}$, V_{cap}, and V_{out} for the 555 one-shot circuit of figure 14–24. Assume that $V_{trigger}$ is a 5-μs negative-going pulse that occurs every 100 μs and $V_{CC} = 5$ V, $R_A = 47$ kΩ, and $C = 1000$ pF.

Schematic Interpretation Problems

See Appendix G for the schematic diagrams.

S D   **14–22.** The 68HC11 microcontroller in the HC11D0 Master Board schematic provides a clock output signal at the pin labeled E. This clock signal is used as the input to the LCD controller, M1 (grid location E-7). The frequency of this signal is 9.8304, as dictated by the crystal on the 68HC11. To experiment with different clock speeds on the LCD controller, you want to design a variable frequency oscillator that can scan the frequency range of 100 kHz to 1 MHz. Design this oscillator using a 555 that will output its signal to pins 6 and 10 of the LCD controller.

S C D **14–23.** Design a "missing pulse detector" similar to the one designed in Problem 14–13. It will be used to monitor the DAV input line in the 4096/4196 schematic. Assume that the DAV line is supposed to provide a 2-μs HIGH pulse every 100 μs. Monitor the DAV line with a 74123 monostable multivibrator. Have the 74123 output a HIGH to port 1, bit 7 (P1.7) of the 8031 microcontroller if a missing pulse is detected.

Answers to Review Questions

14–1. False

14–2. Slower

14–3. True

14–4. V_{T+} and V_{T-}, V_{OH} and V_{OL}

14–5. Decrease

14–6. False

14–7. Or, LOW, and, HIGH

14–8. $\overline{A_1}$ or $\overline{A_2}$, HIGH-HIGH

14–9. HIGH

14–10. It means that input triggers are ignored during the timing cycle of a 74121, but not for the 74123. A new timing cycle is started each time a trigger is applied to a 74123.

14–11. True

14–12. LOW

14–13. $\frac{2}{3}V_{CC}$, Reset, LOW

14–14. $R_A + R_B$, R_B

14–15. Greater than; the charging path consists of $R_A + R_B$, whereas the discharging path consists of only R_B.

14–16. LOW, HIGH, charging

14–17. It has greater stability and accuracy than an RC circuit.

15 Interfacing to the Analog World

OUTLINE

15–1 Digital and Analog Representations

15–2 Operational Amplifier Basics

15–3 Binary-Weighted Digital-to-Analog Converters

15–4 *R/2R* Ladder Digital-to-Analog Converters

15–5 Integrated-Circuit Digital-to-Analog Converters

15–6 IC Data Converter Specifications

15–7 Parallel-Encoded Analog-to-Digital Converters

15–8 Counter-Ramp Analog-to-Digital Converters

15–9 Successive-Approximation Analog-to-Digital Conversion

15–10 Integrated-Circuit Analog-to-Digital Converters

15–11 Data Acquisition System Application

15–12 Transducers and Signal Conditioning

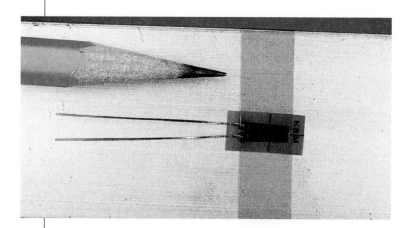

Objectives

Upon completion of this chapter, you should be able to

- Perform the basic calculations involved in the analysis of operational amplifier circuits.

- Explain the operation of binary-weighted and $R/2R$ digital-to-analog converters.

- Make the external connections to a digital-to-analog IC to convert a numeric binary string into a proportional analog voltage.

- Discuss the meaning of the specifications for converter ICs as given in a manufacturer's data manual.

- Explain the operation of parallel-encoded, counter-ramp, and successive-approximation analog-to-digital converters.

- Make the external connections to an analog-to-digital converter IC to convert an analog voltage to a corresponding binary string.

- Discuss the operation of a typical data acquisition system.

Introduction

Most physical quantities that we deal with in this world are *analog* in nature. For example, temperature, pressure, and speed are not simply 1's and 0's but instead take on an infinite number of possible values. To be understood by a digital system, these values must be converted into a binary string representing their value; thus we have the need for *analog-to-digital* conversion. Also, it is important when we need to use a computer to control analog devices to be able to convert from *digital to analog*.

Devices that convert physical quantities into electrical quantities are called **transducers.** Transducers are readily available to convert such quantities as temperature, pressure, velocity, position, and direction into a proportional analog voltage or current. For example, a common transducer for measuring temperature is a thermistor. A **thermistor** is simply a temperature-sensitive resistor. As its temperature changes, so does its resistance. If we send a constant current

Team Discussion

What are some scientific and commercial applications for analog-to-digital conversion and digital-to-analog conversion?

through the thermistor and then measure the voltage across it, we can determine its resistance *and* temperature.

15–1   Digital and Analog Representations

For *analog-to-digital* (**A/D**) or *digital-to-analog* (**D/A**) **converters** to be useful, there has to be a meaningful representation of the analog quantity as a digital representation and the digital quantity as an analog representation. If we choose a convenient range of analog levels such as 0 to 15 V, we could easily represent each 1-V step as a unique digital code, as shown in Figure 15–1.

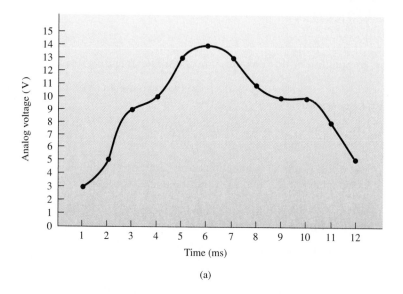

(a)

| | Representation | |
Time (ms)	Analog	Digital
1	3	0011
2	5	0101
3	9	1001
4	10	1010
5	13	1101
6	14	1110
7	13	1101
8	11	1011
9	10	1010
10	10	1010
11	8	1000
12	5	0101

(b)

Figure 15–1   Analog and digital representations: (a) voltage versus time; (b) representations at 1-ms intervals.

Figure 15–1 shows that for each analog voltage we can determine an equivalent digital representation. Using four binary positions gives us 4-bit **resolution,** which allows us to develop 16 different representations, with the increment between each being 1 part in 16. If we need to represent more than just 16 different analog levels, we would have to use a digital code with more than four binary positions. For example, a D/A converter with 8-bit resolution will provide increments of 1 part in 256, which provides much more precise representations.

15–2 Operational Amplifier Basics

Most A/D and D/A circuits require the use of an **op amp** for signal conditioning. Three characteristics of op amps make them an almost *ideal amplifier:* (1) very high input impedance, (2) very high voltage gain, and (3) very low output impedance. In this section we gain a basic understanding of how an op amp works, and in future sections we see how it is used in the conversion process. A basic op-amp circuit is shown in Figure 15–2.

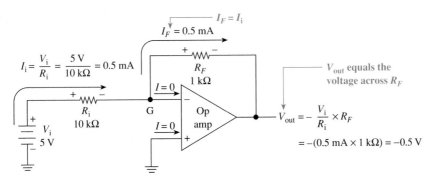

Figure 15–2 Basic op-amp operation.

The symbol for the op amp is the same as that for a comparator, but when it is connected as shown in Figure 15–2, it provides a much different function. The basic theory involved in the operation of the op-amp circuit in Figure 15–2 is as follows:

1. The impedance looking into the + and − input terminals is assumed to be infinite; therefore, I_{in} (+), (−) = 0 A.

2. Point G is assumed to be at the same potential as the + input; therefore, point G is at 0 V, called **virtual ground.** (Virtual means "in effect" but not actual. It is at 0 V, but it cannot sink current.)

3. With point G at 0 V, there will be 5 V across the 10-kΩ resistor, causing 0.5 mA to flow.

4. The 0.5 mA cannot flow into the op amp; therefore, it flows up through the 1-kΩ resistor.

5. Because point G is at virtual ground, and because V_{out} is measured with respect to ground, V_{out} is equal to the voltage across the 1-kΩ resistor, which is −0.5 V.

EXAMPLE 15–1

Find V_{out} in Figure 15–3.

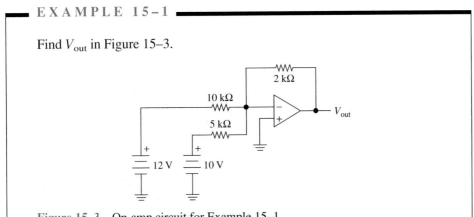

Figure 15–3 Op-amp circuit for Example 15–1.

Solution:

$$I_{10\,k\Omega} = \frac{12\ V}{10\ k\Omega} = 1.2\ mA$$

$$I_{5\,k\Omega} = \frac{10\ V}{5\ k\Omega} = 2\ mA$$

$$I_{2\,k\Omega} = 1.2\ mA + 2\ mA = 3.2\ mA$$

$$V_{out} = -(3.2\ mA \times 2\ k\Omega) = -6.4\ V$$

Review Questions

15–1. Transducers are devices that convert physical quantities like pressure and temperature into electrical quantities. True or false?

15–2. An 8-bit A/D converter is capable of producing how many unique digital output codes?

15–3. The input impedance to an operational amplifier is assumed to be _____. The voltage difference between the (+) input and (−) input is approximately _____ volts.

15–3 Binary-Weighted Digital-to-Analog Converters

A basic D/A converter can be built by expanding on the information presented in Section 15–2. Example 15–1 showed us that the 2-kΩ resistor receives the *sum* of the currents heading toward the op amp from the two input resistors. If we scale the input resistors with a **binary weighting** factor, each input can be made to provide a binary-weighted amount of current, and the output voltage will represent a sum of all the binary-weighted input currents, as shown in Figure 15–4.

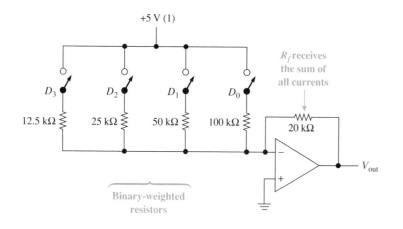

D_3	D_2	D_1	D_0	V_{out} (−V)
0	0	0	0	0
0	0	0	1	1
0	0	1	0	2
0	0	1	1	3
0	1	0	0	4
0	1	0	1	5
0	1	1	0	6
0	1	1	1	7
1	0	0	0	8
1	0	0	1	9
1	0	1	0	10
1	0	1	1	11
1	1	0	0	12
1	1	0	1	13
1	1	1	0	14
1	1	1	1	15

Figure 15–4 Binary weighted D/A converter.

In Figure 15–4 the 20-kΩ resistor *sums* the currents that are provided by closing any of switches D_0 to D_3. The resistors are scaled in such a way as to provide a binary-weighted amount of current to be summed by the 20-kΩ resistor. Closing D_0 causes 50 μA to flow through the 20 kΩ, creating -1.0 V at V_{out}. Closing each successive switch creates *double* the amount of current of the previous switch. Work through several of the switch combinations presented in Figure 15–4 to prove its operation.

If we were to expand Figure 15–4 to an 8-bit D/A converter, the resistor for D_4 would be one-half of 12.5 kΩ, which is 6.25 kΩ. Each successive resistor is one-half of the previous one. Using this procedure, the resistor for D_7 would be 0.78125 kΩ!

Coming up with accurate resistances over such a large range of values is very difficult. This limits the practical use of this type of D/A converter for any more than 4-bit conversions.

EXAMPLE 15–2

Determine the voltage at V_{out} in Figure 15–4 if the binary equivalent of 10_{10} is input on switches D_3 to D_0.

Solution: $10_{10} = 1010_2$ (switches D_3 and D_1 are closed)

$$I_3 = \frac{5 \text{ V}}{12.5 \text{ k}\Omega} = 0.4 \text{ mA}$$

$$I_1 = \frac{5 \text{ V}}{50 \text{ k}\Omega} = 0.1 \text{ mA}$$

$$V_{out} = -[(0.4 \text{ mA} + 0.1 \text{ mA}) \times 20 \text{ k}\Omega] = -10 \text{ V}$$

15–4 *R/2R* Ladder Digital-to-Analog Converters

The method for D/A conversion that is most often used in integrated-circuit D/A converters is known as the *R/2R* ladder circuit. In this circuit, only two resistor values are required, which lends itself nicely to the fabrication of ICs with a resolution of 8, 10, or 12 bits, and higher. Figure 15–5 shows a 4-bit D/A *R/2R* converter. To form converters with higher resolution, all that needs to be done is to add more *R/2R* resistors and switches to the left of D_3. Commercially available D/A converters with resolutions of 8, 10, and 12 bits are commonly made this way.

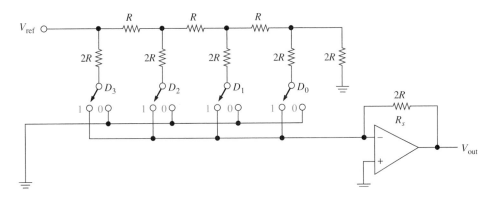

Figure 15–5 The *R/2R* ladder D/A converter.

In Figure 15–5, the 4-bit digital information to be converted to analog is entered on the D_0 to D_3 switches. (In an actual IC, those switches would be transistor switches.) The arrangement of the circuit is such that as the switches are moved to the 1 position, they cause a current to flow through the summing resistor, R_s, that is proportional to their binary equivalent value. (Each successive switch is worth double the previous one.)

This circuit, which is designed in the shape of a ladder, is an ingenious way to form a binary-weighted current-division circuit. Refer to Figure 15–6 to see how we arrive at the current levels and value of V_{out}. First, keep in mind the op-amp rules presented in Section 15–2. In particular, remember that the (−) input to the op amp is at virtual ground, and any current that reaches this point will continue to flow past it, up through the summing resistor (R_s). With this knowledge we can determine that the resistance of and the current through each rung of the ladder is unaffected by the position of any of the data switches (D_3 to D_0). This is because: (1) With a data switch in the 0 position, the bottom of the corresponding 20-kΩ resistor is connected to ground, and (2) when it is in the 1 position, it is connected to virtual ground, which acts the same as ground.

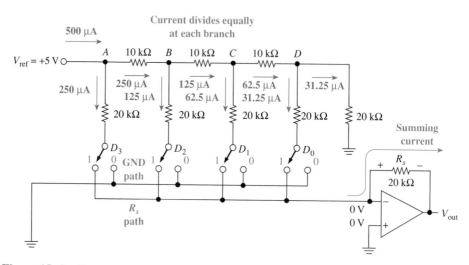

Figure 15–6 Current division in each rung of an *R/2R* ladder D/A converter.

To calculate the current contributed by each rung in Figure 15–6, we must first calculate the total current leaving V_{ref}. By collapsing the circuit from the right, we have 20 kΩ in parallel with 20 kΩ, which gives 10 kΩ. This 10 kΩ is in series with the 10-kΩ resistor between C and D, making 20 kΩ. This 20 kΩ is now in parallel with the 20-kΩ resistor above D_1. This procedure is repeated over and over until we arrive at the total resistance seen by V_{ref} being equal to 10 kΩ. Therefore, the total current leaving V_{ref} is 5 V divided by 10 kΩ, which equals 500 μA.

When that current reaches point A, it divides equally into the two branches, because each branch is 20 kΩ. The 250 μA that reaches point B also splits equally, sending 125 μA down the D_2 rung. This current splitting continues for each rung. Notice that each resulting current is one-half the current to its left, forming a binary-weighted ratio that is then available for the op-amp summing resistor.

If D_3 is in the 1 position, the 250 μA is routed through the 20-kΩ summing resistor (R_s), creating −5.0 V at V_{out}. (If it is in the 0 position, the current is routed directly to ground and does not contribute to V_{out}.) The portion of V_{out} contributed by the current through each data switch can be summarized as follows:

$$(D_3) \quad V_{out} = -250 \, \mu A \times 20 \, k\Omega \quad = -5.000 \text{ V}$$

$$(D_2) \quad V_{out} = -125 \, \mu A \times 20 \, k\Omega \quad = -2.500 \text{ V}$$

$$(D_1) \quad V_{out} = -62.5 \, \mu A \times 20 \, k\Omega \quad = -1.250 \text{ V}$$

$$(D_0) \quad V_{out} = -31.25 \, \mu A \times 20 \, k\Omega = \underline{-0.625 \text{ V}}$$

$$\text{Total } (D_3 \text{ to } D_0 = 1111) \qquad V_{out} = -9.375 \text{ V}$$

A 4-bit D/A converter such as this can have 16 different combinations of D_3 to D_0. The output voltage for any combination of binary input (B_{in}) of a 4-bit D/A converter can be determined by the following equation:

$$V_{out} = -\left(V_{ref} \times \frac{B_{in}}{8}\right) \qquad (15\text{–}1)$$

The analog output voltages for our 4-bit converter are given in Figure 15–7.

D_3	D_2	D_1	D_0	V_{out} (V)
0	0	0	0	0.000
0	0	0	1	−0.625
0	0	1	0	−1.250
0	0	1	1	−1.875
0	1	0	0	−2.500
0	1	0	1	−3.125
0	1	1	0	−3.750
0	1	1	1	−4.375
1	0	0	0	−5.000
1	0	0	1	−5.625
1	0	1	0	−6.250
1	0	1	1	−6.875
1	1	0	0	−7.500
1	1	0	1	−8.125
1	1	1	0	−8.750
1	1	1	1	−9.375

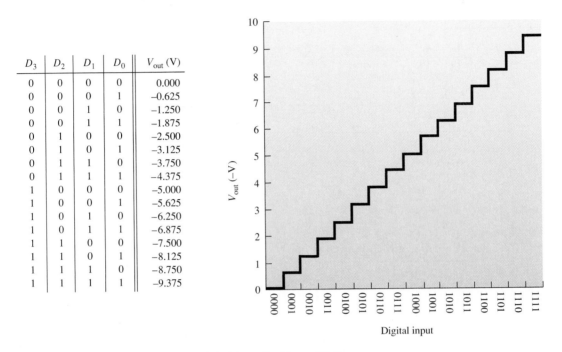

Figure 15–7   Analog output versus digital input for Figure 15–6.

Review Questions

15–4. If the first three resistors in a binary-weighted D/A converter are 30, 60, and 120 kΩ, the fourth resistor, used for the D_0 input, must be _____ ohms.

15–5. Why is it difficult to build an accurate *8-bit* binary-weighted D/A converter?

15–6. To build an 8-bit $R/2R$ ladder D/A converter, you would need at least eight different resistor sizes. True or false?

15–7. If V_{ref} is changed to 6 V in the $R/2R$ converter of Figure 15–6, the maximum value at V_{out} will be 11.25 V. True or false?

15–5   Integrated-Circuit Digital-to-Analog Converters

One very popular and inexpensive 8-bit D/A converter (DAC) is the DAC0808 and its equivalent, the MC1408. A block diagram, pin configuration, and typical application are shown in Figure 15–8. The circuit in Figure 15–8(c) is set up to accept an 8-bit digital input and provide a 0- to +10-V analog output. A reference current (I_{ref}) is required

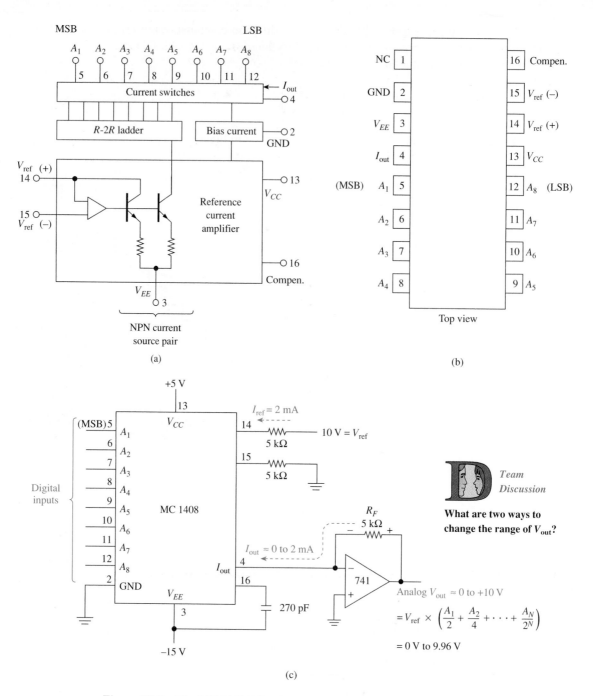

Figure 15–8 The MC 1408 D/A converter: (a) block diagram; (b) pin configuration; (c) typical application.

for the D/A and is provided by the 10-V, 5-kΩ combination shown. The negative reference (pin 15) is then tied to ground via an equal-size (5-kΩ) resistor.

The 2-mA reference current dictates the full-scale output current (I_{out}) to also be approximately 2 mA. To calculate the *actual* output current, use the formula

$$I_{out} = I_{ref} \times \left(\frac{A_1}{2} + \frac{A_2}{4} + \ldots + \frac{A_8}{256} \right) \tag{15–2}$$

[For example, with all inputs (A_1 to A_8) HIGH, $I_{out} = I_{ref} \times (0.996)$]. To convert an output current to an output voltage, a series resistor could be connected from pin 4 to ground and the output taken across the resistor. This method is simple, but it may cause inaccuracies as various-size loads are connected to it.

A more accurate method uses an op amp such as the 741 shown in Figure 15–8(c). The output current flows through R_F, which develops an output voltage equal to $I_{out} \times R_F$. The range of output voltage can be changed by changing R_F and is limited only by the specifications of the op amp used.

To test the circuit, an oscillator and an 8-bit counter can be used to drive the digital inputs, and the analog output can be observed on an oscilloscope, as shown in Figure 15–9. In Figure 15–9, as the counters count from 0000 0000 up to 1111 1111, the analog output will go from 0 V up to almost +10 V in 256 steps. The time per step will be equal to the reciprocal of the input clock frequency.

Team Discussion

How could the D/A converter be driven to create a triangle waveform? A square wave?

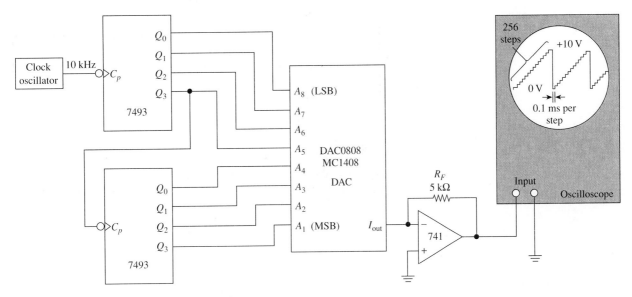

Figure 15–9   Using an 8-bit counter to test the 256-step output of a DAC.

EXAMPLE 15–3

Determine I_{out} and V_{out} in Figure 15–8(c) if the following binary strings are input at A_1 to A_8: (a) 1111 1111; (b) 1001 1011.

Solution:

(a) Using Equation 15–2,

$$I_{out} = I_{ref} \times \left(\frac{A_1}{2} + \frac{A_2}{4} + \ldots + \frac{A_8}{256} \right)$$

$$= 2 \text{ mA} \times \left(\frac{1}{2} + \frac{1}{4} + \frac{1}{8} + \frac{1}{16} + \frac{1}{32} + \frac{1}{64} + \frac{1}{128} + \frac{1}{256} \right)$$

$$= 2 \text{ mA} \times \left(\frac{255}{256} \right)$$

$$= 1.99 \text{ mA}$$

and $V_{out} = I_{out} \times R_f$

$$= 1.99 \text{ mA} \times 5 \text{ k}\Omega$$

$$= 9.96 \text{ V}$$

(b) Using Equation 15–2,

$$I_{out} = I_{ref} \times \left(\frac{A_1}{2} + \frac{A_2}{4} + \ldots + \frac{A_8}{256} \right)$$

$$= 2 \text{ mA} \times \left(\frac{1}{2} + \frac{1}{16} + \frac{1}{32} + \frac{1}{128} + \frac{1}{256} \right)$$

$$= 2 \text{ mA} \times \left(\frac{155}{256} \right)$$

$$= 1.21 \text{ mA}$$

and $V_{out} = I_{out} \times R_f$

$$= 1.21 \text{ mA} \times 5 \text{ k}\Omega$$

$$= 6.05 \text{ V}$$

(There are two ways to arrive at the fraction 155/256 in the above solution. One way is to find a common denominator for all the fractions and then add them together, as we did. The other way is to convert the binary input 1001 1011 into decimal, which equals 155, and then divide that by 256.)

15–6 IC Data Converter Specifications

Besides resolution, several other specifications are important in the selection of D/A and A/D converters (DAC and ADC). It is important that the specifications and their definitions given in the manufacturer's data book be studied and understood before selecting a particular DAC or ADC. Figure 15–10 lists some of the more important specifications as presented in the Signetics Linear LSI Data Manual.

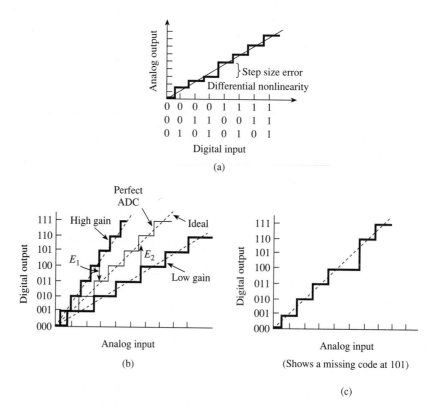

Figure 15–10   DAC and ADC specification definition: (a) differential **nonlinearity;** (b) gain error; (c) missing codes;

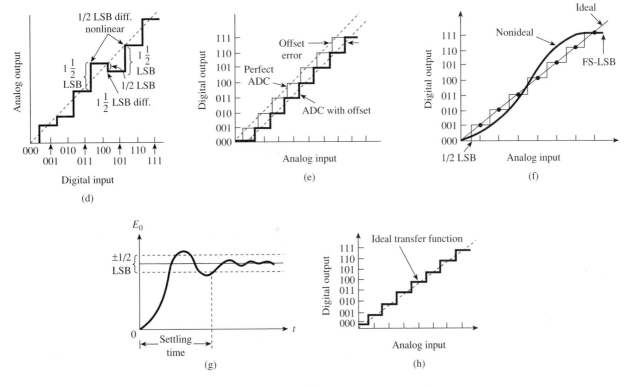

Figure 15–10 *(Continued)* (d) **nonmonotonic** (must be $>\pm\frac{1}{2}$ LSB nonlinear); (e) offset error; (f) relative accuracy; (g) setting time; (h) 3-bit ADC transfer characteristic. (Courtesy of Signetics Corporation)

Review Questions

15–8. The analog output of the MC1408 DAC IC is represented by current or voltage?

15–9. Which digital input to the MC1408 DAC IC has the most significant effect on the analog output: A_1 or A_8?

15–10. The *resolution* of a DAC or ADC specifies the _____.

15–11. A DAC is *nonmonotonic* if its analog output *drops* after a 1-bit increase in digital input. True or false?

15–12. Which error affects the rate of change, or slope, of the ideal transfer function of an ADC, the gain error or the offset error?

15–7 Parallel-Encoded Analog-to-Digital Converters

The process of taking an analog voltage and converting it to a digital signal can be done in several ways. One simple way that is easy to visualize is by means of parallel encoding (also known as *simultaneous, multiple comparator,* or *flash* converting). In this method, several comparators are set up, each at a different voltage reference level with their outputs driving a priority encoder as shown in Figure 15–11. The voltage-divider network in Figure 15–11 is designed to drop 1 V across each resistor. This sets up a voltage reference at each comparator input in 1-V steps.

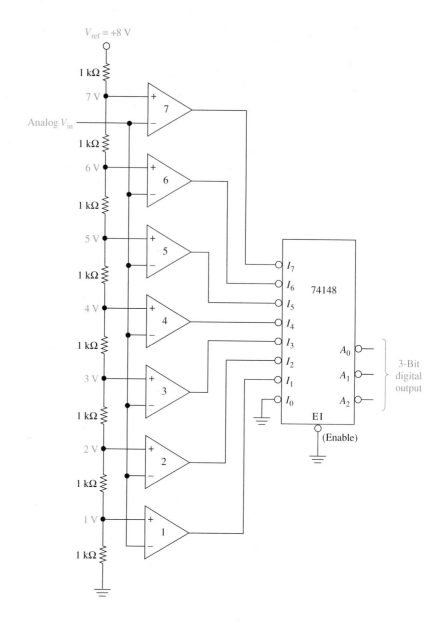

Figure 15–11 Three-bit parallel-encoded ADC.

When V_{in} is 0 V, the + input on all seven comparators will be higher than the − input, so they will all output a HIGH. In this case, $\overline{I_0}$ is the only active-LOW input that is enabled, so the 74148 will output an active-LOW binary 0 (111).

When V_{in} exceeds 1.0 V, comparator 1 will output a LOW. Now $\overline{I_0}$ and $\overline{I_1}$ are both enabled, but because it is a *priority* encoder, the output will be a binary 1 (110). As V_{in} increases further, each successive comparator outputs a LOW. The highest input that receives a LOW is encoded into its binary equivalent output.

The A/D converter in Figure 15–11 is set up to convert analog voltages in the range from 0 to 7 V. The range can be scaled higher or lower, depending on the input voltage levels that are expected. The resolution of this converter is only 3 bits, so it can only distinguish among eight different analog input levels. To expand to 4-bit resolution, eight more comparators are required to differentiate the 16 different voltage levels. To expand to 8-bit resolution, 256 comparators would be required! As you can see, circuit complexity becomes a real problem when using parallel encoding for high-resolution conversion. However, a big advantage of using parallel encoding is its high speed. The conversion speed is limited only by the propagation delays of the comparators and encoder (less than 20 ns total).

15–8 Counter-Ramp Analog-to-Digital Converters

Helpful Hint

This circuit is made completely of devices that have already been discussed, which makes it a good circuit to illustrate timing and interface between multiple ICs.

The counter-ramp method of A/D conversion (ADC) uses a counter in conjunction with a D/A converter (DAC) to determine a digital output that is equivalent to the unknown analog input voltage. In Figure 15–12, depressing the start conversion push button clears the counter outputs to 0, which sets the DAC output to 0 V. The (−) input to the comparator is now 0 V, which is less than the positive analog input voltage at the (+) input. Therefore, the comparator outputs a HIGH, which enables the AND gate, allowing the counter to start counting. As the counter's binary output increases, so does the DAC output voltage in the form of a staircase.

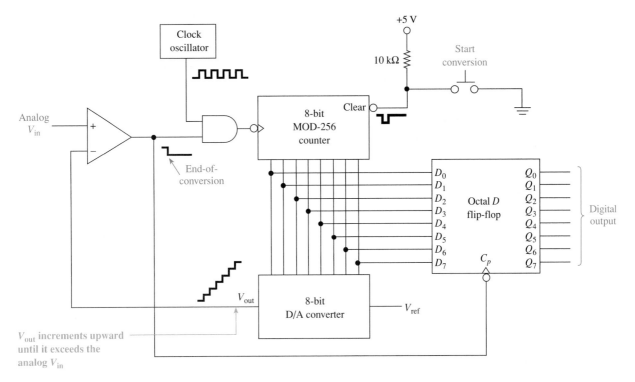

Figure 15–12 Counter-ramp A/D converter.

When the staircase voltage reaches and then exceeds the analog input voltage, the comparator output goes LOW, disabling the clock and stopping the counter. The counter output at that point is equal to the binary number that caused the DAC to output a voltage slightly greater than the analog input voltage. Thus we have the binary equivalent of the analog voltage!

The HIGH-to-LOW transition of the comparator is also used to trigger the D flip-flop to latch on to the binary number at that instant. To perform another conversion, the start push button is depressed again and the process repeats. The result from the previous conversion remains in the D flip-flop until the next end-of-conversion HIGH-to-LOW edge comes along.

To change the circuit to perform **continuous conversions,** the end-of-conversion line could be tied back to the clear input of the counter. A short delay needs to be inserted into this new line, however, to allow the D flip-flop to read the binary number before the counter is Reset. Two inverters placed end to end in the line will produce a sufficient delay.

The main *disadvantage* of the counter-ramp method of conversion is its slow conversion speed. The worst-case maximum **conversion time** will occur when the counter has to count all 255 steps before the DAC output voltage matches the analog input voltage.

15–13. In the parallel-encoded ADC of Figure 15–11, *only one* of the comparators will output a LOW for each analog input value. True or false?

15–14. One difficulty in building a high-resolution, 10-bit parallel-encoded ADC is that it would take _____ comparators to complete the design.

15–15. The counter-ramp ADC of Figure 15–12 signifies an "end of conversion" when the (−) input voltage to the comparator drops below the (+) input voltage. True or false?

15–16. The digital output of the octal *D* flip-flop in Figure 15–12 is continuously changing at the same rate as the MOD-16 counter. True or false?

15–9 Successive-Approximation Analog-to-Digital Conversion

Other methods of A/D conversion employ *up/down-counters* and *integrating slope converters* to track the analog input, but the method used in most modern integrated-circuit ADCs is called **successive approximation.** This converter circuit is similar to the counter-ramp ADC circuit except that the method of narrowing in on the unknown analog input voltage is much improved. Instead of counting up from 0 and comparing the DAC output each step of the way, a successive-approximation register (SAR) is used in place of the counter (see Figure 15–13).

Team Discussion

Why is the SAR converter much faster than the counter-ramp?

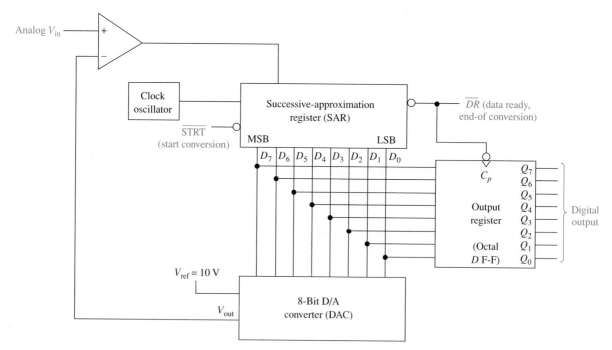

Figure 15–13 Simplified SAR A/D converter.

In Figure 15–13 the conversion is started by dropping the $\overline{STRT}$ line LOW. Then the SAR first tries a HIGH on the MSB (D_7) line to the DAC. (Remember, D_7 will cause the DAC to output half of its full-scale output.) If the DAC output is then *higher* than the unknown analog input voltage, the SAR returns the MSB LOW. If the DAC output was still *lower* than the unknown analog input voltage, the SAR leaves the MSB HIGH.

Now, the next lower bit (D_6) is tried. If a HIGH on D_6 causes the DAC output to be higher than the analog V_{in}, it is returned LOW. If not, it is left HIGH. The process continues until all 8 bits, down to the LSB, have been tried. At the end of this eight-step conversion process, the SAR contains a valid 8-bit binary output code that represents the unknown analog input. The $\overline{DR}$ output now goes LOW, indicating that the *conversion is complete* and the data are ready. The HIGH-to-LOW edge on $\overline{DR}$ clocks the D_0 to D_7 data into the octal D flip-flop to make the digital output results available at the Q_0 to Q_7 lines.

The main advantage of the SAR ADC method is its high speed. The ADC in Figure 5–13 takes only eight clock periods to complete a conversion, which is a vast improvement over the counter-ramp method.

EXAMPLE 15–4

Show the timing waveforms that would occur in the successive approximation ADC of Figure 15–13 when converting the analog voltage of 6.84 V to 8-bit binary, assuming that the full-scale input voltage to the DAC is 10 V ($V_{ref} = 10$ V).

Solution: Each successive bit, starting with the MSB, will cause the DAC part of the system to output a voltage to be compared. If the full-scale output is 10 V, D_7 will be worth 5 V, D_6 will be worth 2.5 V, D_5 will be worth 1.25 V, and so on, as shown in Table 15–1.

Now, when $\overline{STRT}$ goes LOW, successive bits starting with D_7 will be tried, creating the waveforms shown in Figure 15–14. The HIGH-to-LOW edge on $\overline{DR}$ clocks the final binary number 1010 1111 into the D flip-flop and Q_0 to Q_7 outputs.

Now the Q_0 to Q_7 lines contain the 8-bit binary representation of the analog number 6.8359375, which is an error of only 0.0594% from the target number of 6.84:

Team Discussion

What is the highest percent error that you could have with the 8-bit ADC in Example 15–4?

$$\% \text{ error} = \frac{\text{actual voltage} - \text{final DAC output}}{\text{actual voltage}} \times 100\% \quad (15\text{–}3)$$

Table 15–1   Voltage-Level Contributions by Each Successive Approximation Register Bit

DAC Input	DAC V_{out}
D_7	5.0000
D_6	2.5000
D_5	1.2500
D_4	0.6250
D_3	0.3125
D_2	0.15625
D_1	0.078125
D_0	0.0390625

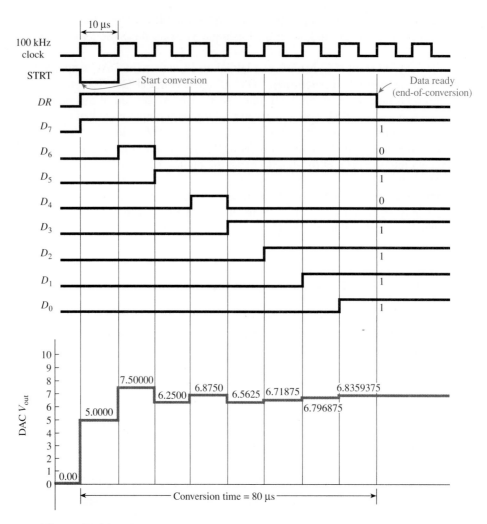

Figure 15–14 Timing waveforms for a successive approximation A/D conversion.

To watch the conversion in progress, an eight-channel oscilloscope or logic analyzer can be connected to the D_0 to D_7 outputs of the SAR.

For *continuous conversions* the $\overline{DR}$ line can be connected back to the $\overline{STRT}$ line. That way, as soon as the conversion is complete, the HIGH-to-LOW on $\overline{DR}$ will issue another start conversion ($\overline{STRT}$), which forces the data ready ($\overline{DR}$) line back HIGH for eight clock periods while the new conversion is being made. The latched Q_0 to Q_7 digital outputs will always display the results of the *previous conversion*.

15–10 Integrated-Circuit Analog-to-Digital Converters

Examples of two popular, commercially available ADCs are the NE5034 and the ADC0801 manufactured by Signetics Corporation.

The NE5034

The block diagram and pin configuration for the NE5034 are given in Figure 15–15. Operation of the NE5034 is almost identical to that of the SAR ADC presented in Section 15–9. One difference is that the NE5034 uses a three-state output buffer instead of

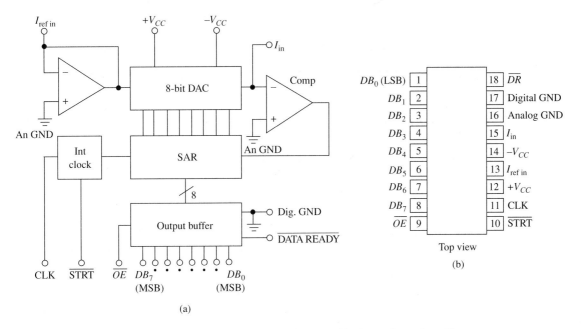

(a)

Top view

(b)

Figure 15–15 The NE5034 A/D converter: (a) block diagram; (b) pin configuration. (Courtesy of Signetics Corporation)

a *D* flip-flop. With three-state outputs, when $\overline{OE}$ (Output Enable) is LOW, the DB_7 to DB_0 outputs display the continuous status of the eight SAR lines, and when the $\overline{OE}$ line goes HIGH, the DB_7 to DB_0 outputs return to a float or high-impedance state. This way, if the ADC outputs go to a common *data* **bus** shared by other devices, when DB_7 to DB_0 float, one of the other devices can output information to the data bus without interference.

The NE5034 can provide conversion speeds as high as one per 17 *μ*s, and its three-state outputs make it compatible with bus-oriented microprocessor systems. It also has its own internal clock for providing timing pulses. The frequency is determined by an external capacitor placed between pins 11 and 17. Figure 15–16 shows the frequency and conversion time that can be achieved using the internal clock.

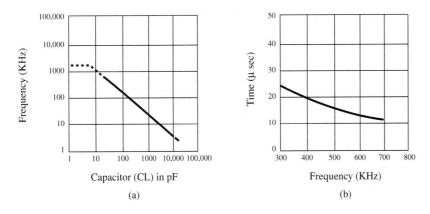

Figure 15–16 The NE5034 internal clock characteristics: (a) internal clock frequency versus external capacitor *(CL)*; (b) conversion time versus clock frequency. (Courtesy of Signetics Corporation)

The ADC0801

The pin configuration and block diagram for the ADC0801 are given in Figure 15–17. The ADC0801 uses the successive-approximation method to convert an analog input

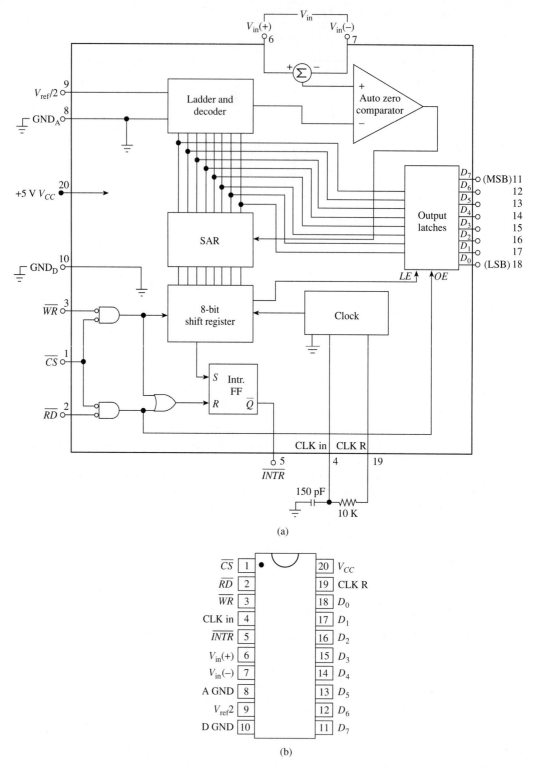

Figure 15–17 The ADC0801 converter: (a) block diagram; (b) pin configuration. (Courtesy of Signetics Corporation)

to an 8-bit binary code. Two analog inputs are provided to allow differential measurements [analog $V_{in} = V_{in(+)} - V_{in(-)}$]. It has an internal clock that generates its own timing pulses at a frequency equal to $f = 1/(1.1RC)$ (Figure 15–18 shows the connections

CHAPTER 15 I INTERFACING TO THE ANALOG WORLD

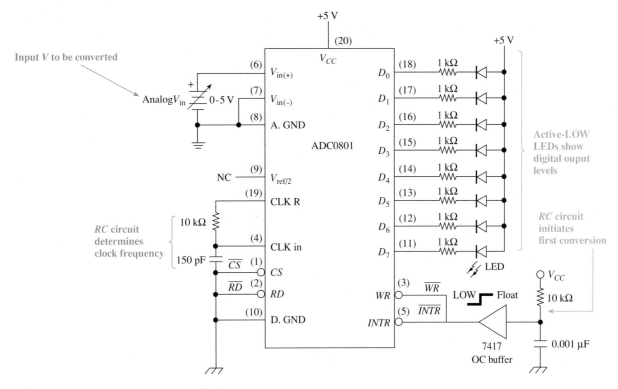

Figure 15–18 Connections for continuous conversions using the ADC0801.

for the external *R* and *C*). It uses output *D* latches that are three-stated to facilitate easy bus interfacing.

The convention for naming the ADC0801 pins follows that used by microprocessors to ease interfacing. Basically, the operation of the ADC0801 is similar to that of the NE5034. The ADC0801 pins are defined as follows:

$\overline{CS}$—active-LOW *Chip Select*

$\overline{RD}$—active-LOW *Output Enable*

$\overline{WR}$—active-LOW *Start Conversion*

CLK IN—external clock input or capacitor connection point for the internal clock

$\overline{INTR}$—active-LOW *End-of-Conversion* (Data Ready)

$V_{in(+)}$, $V_{in(-)}$—differential analog inputs (ground one pin for single-ended measurements)

A. GND—analog ground

$V_{ref}/2$—optional **reference voltage** (used to override the reference voltage assumed at V_{CC})

D. GND—digital ground

V_{CC} —5-V power supply and assumed reference voltage

CLK R—resistor connection for the internal clock

D_0 to D_7 —digital outputs

To set the ADC0801 up for continuous A/D conversions, the connections shown in Figure 15–18 should be made. The external *RC* will set up a clock frequency of

$$f = \frac{1}{1.1RC} = \frac{1}{1.1(10 \text{ k}\Omega)150 \text{ pF}} = 606 \text{ kHz}$$

The connection from $\overline{INTR}$ to $\overline{WR}$ will cause the ADC to start a new conversion each time the $\overline{INTR}$ (end-of-conversion) line goes LOW. The *RC* circuit with the 7417 open-collector buffer will issue a LOW-to-float pulse at power-up to ensure initial startup. An *open-collector* output gate is required instead of a totem-pole output because the $\overline{INTR}$ is forced LOW by the internal circuitry of the 0801 at the end of each conversion. This LOW would conflict with the HIGH output level if a totem-pole output were used. The $\overline{CS}$ is grounded to enable the ADC chip. $\overline{RD}$ is grounded to enable the D_0 to D_7 outputs. The analog input voltage is positive, 0 to 5 V, so it is connected to $V_{in(+)}$. If it were negative, $V_{in(+)}$ would be grounded and the input voltage would be connected to $V_{in(-)}$. **Differential measurements** (the difference between two analog voltages) can be made by using both $V_{in(+)}$ and $V_{in(-)}$. The LEDs connected to the digital output will monitor the operation of the ADC outputs. An LED ON indicates a LOW and an LED OFF indicates a HIGH. (In other words, they are displaying the *complement* of the binary output.) To test the circuit operation, you could watch the OFF LEDs count up in binary from 0 to 255 as the analog input voltage is slowly increased from 0 to +5 V.

The analog input voltage range can be changed to values other than 0 to 5 V by using the $V_{ref}/2$ input. This provides the means of encoding small analog voltages to the full 8 bits of resolution. The $V_{ref}/2$ pin is normally not connected, and it sits at 2.500 V ($V_{CC}/2$). By connecting 2.00 V to $V_{ref}/2$, the analog input voltage range is changed to 0 to 4 V; 1.5 V would change it to 0 to 3.0 V; and so on. However, the accuracy of the ADC suffers as the input voltage range is decreased.

Because the analog input voltage is directly proportional to the digital output, the following ratio can be used as an equation to solve for the digital output value:

$$\frac{A_{in}}{V_{ref}} = \frac{D_{out}}{256} \tag{15–4}$$

or

$$D_{out} = \frac{A_{in}}{V_{ref}} \times 256 \tag{15–5}$$

where   Ain = analog input voltage
   V_{ref} = reference voltage ($V_{pin\ 20}$ or $V_{pin\ 9} \times 2$)
   D_{out} = digital output (base 10)
   256 = total number of digital output steps (0 to 255, inclusive)

One final point on the ADC0801. An analog ground *and* a digital ground are both provided to enhance the accuracy of the system. The V_{CC}-to-digital ground lines are inherently noisy due to the switching transients of the digital signals. Using separate analog and digital grounds is not mandatory, but when used, they ensure that the analog voltage comparator will not switch falsely due to digital noise and jitter.

EXAMPLE 15–5

Change V_{CC} (pin 20) in Figure 15–18 to 5.12 V. Determine which LEDs will be on for the following analog input voltages:
(a) 5.100 V; **(b)** 2.26 V.

Solution:

(a)
$$D_{out} = \frac{A_{in}}{V_{ref}} \times 256$$
$$= \frac{5.100 \text{ V}}{5.120 \text{ V}} \times 256$$
$$= 255_{10}$$
$$= 1111\ 1111_2$$

Because the LEDs are active-LOW, none of them will be ON.

(b)
$$D_{out} = \frac{A_{in}}{V_{ref}} \times 256$$
$$= \frac{2.26 \text{ V}}{5.12 \text{ V}} \times 256$$
$$= 113_{10}$$
$$= 0111\ 0001_2$$

The following LEDs will be ON: $D7, D3, D2, D1$.

Review Questions

15–17. The SAR ADC in Figure 15–13 starts making a conversion when _____ goes LOW, and signifies that the conversion is complete when _____ goes LOW.

15–18. The SAR method of A/D conversion is faster than the counter/ramp method because the SAR clock oscillator operates at a higher speed. True or false?

15–19. Decide if the following pins on the ADC0801 are for input or output signals.

(a) $\overline{CS}$ (c) $\overline{WR}$

(b) $\overline{RD}$ (d) $\overline{INTR}$

15–20. List the order in which the signals listed in Question 15–19 become active to perform an A/D conversion.

15–11 Data Acquisition System Application

The computerized acquisition of analog quantities is becoming more important than ever in today's automated world. Computer systems are capable of scanning several analog inputs on a particular schedule and sequence to monitor critical quantities and acquire data for future recall. A typical eight-channel computerized **data acquisition system (DAS)** is shown in Figure 15–19.

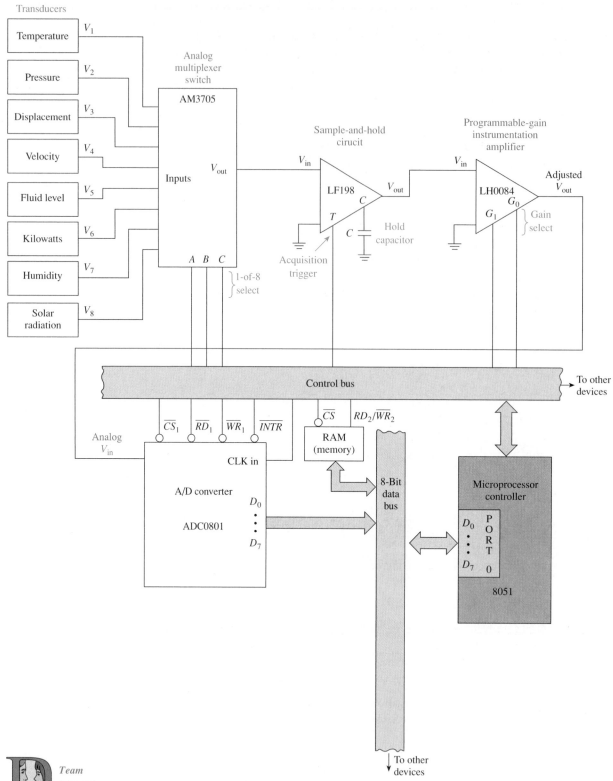

Figure 15–19 Data acquisition system.

Team Discussion

Discuss the flow of a signal as it travels from the temperature transducer through each of the ICs to the microcontroller.

The entire system in Figure 15–19 communicates via two common buses, the *data bus* and the *control bus*. The data bus is simply a common set of eight electrical conductors shared by as many devices as necessary to send and receive 8 bits of parallel data to and from anywhere in the system. In this case there are three devices on the data bus: the ADC, the microprocessor (or microcontroller), and memory. The control bus passes control signals to and from the various devices for such things as chip select $(\overline{CS})$, output enable $(\overline{RD})$, system clock, triggers, and selects.

Each of the eight transducers is set up to output a voltage that is proportional to the analog quantity being measured. The task of the microprocessor is to scan all the quantities at some precise interval and store the digital results in memory for future use. Since the analog values vary so slowly, scanning (reading) the variables at 1-s intervals is usually fast enough to get a very accurate picture of their levels. This can easily be accomplished by **microprocessors** with clock rates in the microsecond range.

To do this, the microprocessor must enable and send the proper control signals to each of the devices, in order, starting with the multiplexer and ending with the ADC. This is called **handshaking,** or *polling,* and is all done with software statements. If you are fortunate enough to take a course in microprocessor programming, you will learn how to perform some of these tasks.

All the hardware **interfacing** and handshaking that takes place between the microprocessor and the transducers can be explained by taking a closer look at each of the devices in the system.

Analog Multiplexer Switch (AM3705)

The multiplexer reduces circuit complexity and eliminates duplication of circuitry by allowing each of the eight transducer outputs to take turns traveling through the other devices. The microprocessor selects each of the transducers at the appropriate time by setting up the appropriate binary select code on the *A, B, C* inputs via the control bus. This allows the selected transducer signal to pass through to the next device.

Sample-and-Hold Circuit (LF198)

Because analog quantities can be constantly varying, it is important to be able to select a precise time to take the measurement. The **sample-and-hold** circuit, with its external *Hold capacitor,* allows the system to take (Sample) *and Hold* an analog value at the precise instant that the microprocessor issues the *acquisition trigger.*

Programmable-Gain Instrumentation Amplifier (LH0084)

Each of the eight transducers has different full-scale output ratings. For instance, the temperature transducer may output in the range from 0 to 5 V, while the pressure transducer may only output 0 to 500 mV. The LH0084 is a **programmable-gain amplifier** capable of being programmed, via the gain select inputs, for gains of 1, 2, 5, or 10. When it is time to read the pressure transducer, the microprocessor will program the gain for 10 so that the range will be 0 to 5 V, to match that of the other transducers. This way the ADC can always operate in its most accurate range, 0 to 5 V.

Analog-to-Digital Converter (ADC0801)

The ADC receives the adjusted analog voltage and converts it to an equivalent 8-bit binary string. To do this, the microprocessor issues chip select $(\overline{CS_1})$ and start conversion $(\overline{WR_1})$ pulses. When the end-of-conversion $(\overline{INTR})$ line goes LOW, the microprocessor issues an output enable $(\overline{RD_1})$ to read the data (D_0 and D_7) that pass, via the data bus, into the microprocessor and then into the random-access memory (RAM) chip (more on memory in Chapter 16).

This cycle repeats for all eight transducers whenever the microprocessor determines that it is time for the next scan. Other software routines executed by the microprocessor will act on the data that have been gathered. Some possible responses to the measured results might be to sound an alarm, speed up a fan, reduce energy consumption, increase a fluid level, or simply produce a tabular report of the measured quantities.

15–12 Transducers and Signal Conditioning

Hundreds of *transducers* are available today that convert physical quantities such as heat, light, or force into electrical quantities (or vice versa). The *electrical quantities* (or signal levels) must then be *conditioned* (or modified) before they can be interpreted by a digital computer.

Signal conditioning is required because transducers each output different ranges and types of electrical signals. For example, transducers can produce output voltages or output currents or act like variable resistances. A transducer may have a nonlinear response to input quantities, may be inversely proportional, and may output signals in the microvolt range.

A transducer's response specifications are given by the manufacturer and must be studied carefully to determine the appropriate analog signal-conditioning circuitry required to interface it to an analog-to-digital converter. After the information is read into a digital computer, software instructions convert the binary input into a meaningful output that can be used for further processing. Let's take a closer look at three commonly used transducers: a thermistor, an IC temperature sensor, and a strain gage.

Thermistors

A thermistor is an electronic component whose resistance is highly dependent on temperature. Its resistance changes by several percent with each degree change in temperature. It is a very sensitive temperature-measuring device. One problem, however, is that its response is nonlinear, meaning that 1° step changes in temperature will not create equal step changes in resistance. This fact is illustrated in the characteristic curve of the 10-kΩ (at 25°C) thermistor shown in Figure 15–20.

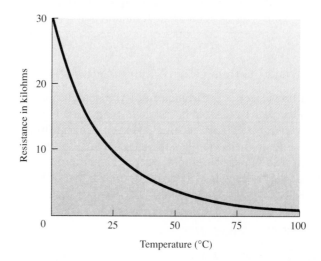

Figure 15–20   Thermistor characteristic curve of resistance versus temperature.

From the characteristic curve you can see that not only is the thermistor nonlinear, but it also has a negative temperature coefficient (that is, its resistance decreases with increasing temperatures).

To use a thermistor with an analog-to-digital converter like the ADC0801, we need to convert the thermistor resistance to a voltage in the range of 0 to 5 V. One way to accomplish this task is with the circuit shown in Figure 15–21.

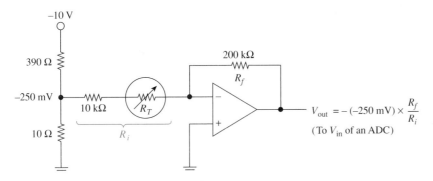

Figure 15–21 Circuit used to convert thermistor ohms to a dc voltage.

This circuit operates similarly to the op-amp circuit explained in Section 15–2. The output of the circuit is found using the formula

$$V_{out} = -V_{in} \times \frac{R_f}{R_i}$$

where V_{in} is a fixed reference voltage of –250 mV and R_i is the sum of the thermistor's resistance, R_T, plus 10 kΩ. Using specific values for R_T found in the manufacturer's data manual, we can create Table 15–2, which shows V_{out} as a function of temperature.

Table 15–2 Tabulation of Output Voltage Levels for a Temperature Range of 0° to 100°C in Figure 15–21

Temperature (in °C)	R_T (in kΩ)	R_i (in kΩ)	V_{out} (in V)
0	29.490	39.490	1.27
25	10.000	20.000	2.50
50	3.893	13.893	3.60
75	1.700	11.700	4.27
100	0.817	10.817	4.62

The output voltage, V_{out}, is fed into an ADC that converts it into an 8-bit binary number. The binary number is then read by a microprocessor that converts it into the corresponding degrees Celsius using software program instructions.

Linear IC Temperature Sensors

The computer software required to convert the output voltages of the previous thermistor circuit is fairly complicated because of the nonlinear characteristics of the device. *Linear temperature sensors* were developed to simplify the procedure. One such device is the LM35 integrated-circuit temperature sensor. It is fabricated in a three-terminal transistor package and is designed to output 10 mV for each degree Celsius above zero. (Another temperature sensor, the LM34, is calibrated in degrees Fahrenheit.) For example, at 25°C the sensor outputs 250 mV, at 50°C it outputs 500 mV, and so on, in linear steps for its entire range. Figure 15–22 shows how we can interface the LM35 to an ADC and microprocessor.

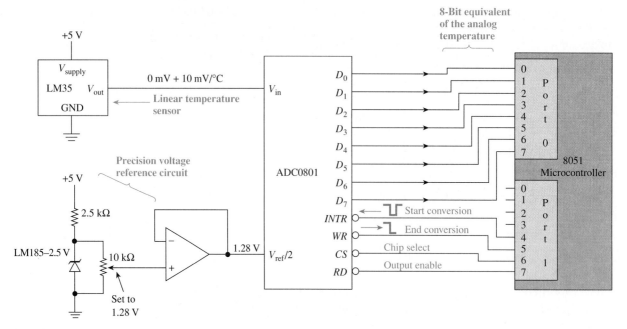

Team Discussion

Discuss the order of the control signals between the microcontroller and the ADC.

Figure 15–22 Interfacing the LM35 linear temperature sensor to an ADC and a microcontroller.

The 1.28-V reference level used in Figure 15–22 is the key to keeping the conversion software programming simple. With 1.28 V at $V_{ref}/2$, the maximum full-scale analog V_{in} is defined as 2.56 V (2560 mV). This value corresponds one-for-one with the 256 binary output steps provided by an 8-bit ADC. A 1° rise in temperature increases V_{in} by 10 mV, which increases the binary output by 1. Therefore, if V_{in} equals 0 V, D_7 to D_0 equals 0000 0000; if V_{in} equals 2.55 V, D_7 to D_0 equals 1111 1111; and if V_{in} equals 1.00 V, D_7 to D_0 equals the binary equivalent of 100, which is 0110 0100. Table 15–3 lists some representative values of temperature versus binary output.

Table 15–3 Tabulation of Temperature Versus Binary Output for a Linear Temperature Sensor and an ADC Set Up for 2560 mV Full Scale

Temperature (in °C)	V_{in} (in mV)	Binary Output (D_7 to D_0)
0	0	0000 0000
1	10	0000 0001
2	20	0000 0010
25	250	0001 1001
50	500	0011 0010
75	750	0100 1011
100	1000	0110 0100

In Figure 15–22, the LM185 is a 2.5-V precision voltage reference diode. This diode maintains a steady 2.5 V across the 10-kΩ potentiometer even if the 5-V power supply line fluctuates. The 10-kΩ potentiometer must be set to output exactly 1.280 V. The op amp is used as a unity-gain buffer between the potentiometer and ADC and will maintain a steady 1.280 V for the $V_{ref}/2$ pin.

The Strain Gage

The strain gage is a device whose resistance changes when it is stretched. The gage is stretched, or elongated, when it is "strained" by a physical force. This property makes it useful for measuring weight, pressure, flow, and acceleration.

Several types of strain gages exist, the most common being the foil type illustrated in Figure 15–23. The gage is simply a thin electrical conductor that is looped back and forth and bonded securely to the piece of material to be strained (see Figure 15–24). Applying a force to the metal beam bends the beam slightly, which stretches the strain gage in the direction of its sensitivity axis.

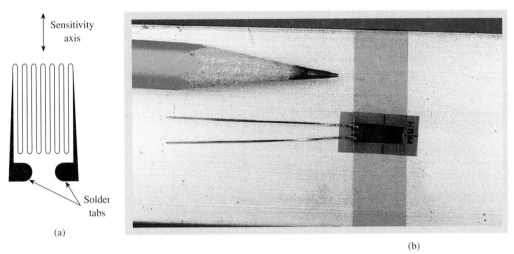

(a)

(b)

Figure 15–23   A foil-type strain gage: (a) sketch; (b) photograph.

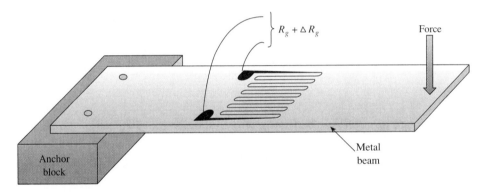

Figure 15–24   Using a strain gage to measure force.

As the strain gage conductor is stretched, its cross-sectional area decreases and its length increases, thus increasing the resistance measured at the solder tabs. The change in resistance is linear with respect to changes in the length of the strain gage. However, the change in resistance is very slight, usually milliohms, and it must be converted to a voltage and amplified before it is input to an ADC. Figure 15–25 shows the signal conditioning circuitry for a 120-Ω strain gage [R_g (unstrained) = 120 Ω]. The instrumentation amplifier is a special-purpose IC used to amplify small-signal differential voltages such as those at points A and B in the bridge circuit in Figure 15–25.

With the metal beam unstrained (no force applied), the 120-Ω potentiometer is adjusted so that V_{out} equals 0 V. Next, force is applied to the metal beam, elongating the strain gage and increasing its resistance slightly from its unstrained value of 120 Ω. This increase causes the bridge circuit to become unbalanced, thus creating a voltage at points A and B. From basic circuit theory, the voltage is calculated by the formula

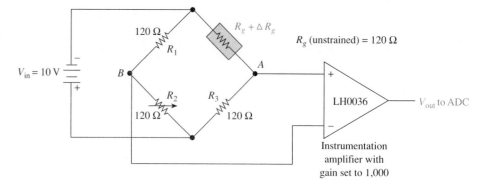

Figure 15–25 Signal conditioning for a strain gage.

$$V_{AB} = V_{in}\left[\frac{R_3}{R_3 + (R_g + \Delta R_g)} - \frac{R_2}{R_1 + R_2}\right] \qquad (15\text{–}6)$$

For example, if R_2 is set at 120 Ω and ΔR_g is 150 mΩ, the voltage is

$$V_{AB} = -10\left[\frac{120}{120 + (120.150)} - \frac{120}{120 + 120}\right] = 3.12 \text{ mV}$$

Because the instrumentation amplifier gain is set at 1000, the output voltage sent to the ADC will be 3.12 V.

Through experimentation with several known weights, the relationship of force versus V_{out} can be established and can be programmed into the microprocessor reading the ADC output.

Review Questions

15–21. The AM7305 multiplexer in Figure 15–19 is used to apply the appropriate voltage gain to each of the transducers connected to it. True or false?

15–22. The 8-bit data out from the ADC in Figure 15–19 passes to the microprocessor via the data bus. True or false?

15–23. Signal conditioning is required to make transducer output levels compatible with ADC input requirements. True or false?

15–24. What is the advantage of using the LM35 linear temperature sensor over a thermistor for measuring temperature?

Summary

In this chapter we have learned that

1. Any analog quantity can be represented by a binary number. Longer binary numbers provide higher resolution, which gives a more accurate representation of the analog quantity.

2. Operational amplifiers are important building blocks in analog-to-digital (A/D) and digital-to-analog (D/A) converters. They provide a means for summing currents at the input and converting a current to a voltage at the output of converter circuits.

3. The binary-weighted D/A converter is the simplest to construct, but it has practical limitations in resolution (number of input bits).

4. The $R/2R$ ladder D/A converter uses only two different resistor values, no matter how many binary input bits are included. This allows for very high resolution and ease of fabrication in integrated-circuit form.

5. The DAC0808 (or MC1408) IC is an 8-bit D/A converter that uses the $R/2R$ ladder method of conversion. It accepts 8 binary input bits and outputs an equivalent analog current. Having 8 input bits means that it can resolve up to 256 unique binary values into equivalent analog values.

6. Applying an 8-bit counter to the input of an 8-bit D/A converter will produce a 256-step sawtooth waveform at its output.

7. The simplest way to build an analog-to-digital (A/D) converter is to use the parallel encoding method. The disadvantage is that it is practical only for low-resolution applications.

8. The counter-ramp A/D converter employs a counter, a D/A converter, and a comparator to make its conversion. The counter counts from zero up to a value that causes the D/A output to exceed the analog input value slightly. That binary count is then output as the equivalent to the analog input.

9. The method of A/D conversion used most often is called successive approximation. In this method, successive bits are tested to see if they contribute an equivalent analog value that is greater than the analog input to be converted. If they do, they are returned to zero. After all bits are tested, the ones that are left ON are used as the final digital equivalent to the analog input.

10. The NE5034 and the ADC0801 are examples of A/D converter ICs. To make a conversion, the *start-conversion* pin is made LOW. When the conversion is completed the *end-of-conversion* pin goes LOW. Then to read the digital output, the *output enable* pin is made LOW.

11. Data acquisition systems are used to read several different analog inputs, respond to the values read, store the results, and generate reports on the information gathered.

12. Transducers are devices that convert physical quantities such as heat, light, or force into electrical quantities. Those electrical quantities must then be conditioned (or modified) before they can be interpreted by a digital computer.

Glossary

A/D Converter (ADC): Analog-to-digital converter.

Binary Weighting: Each binary position in a string is worth double the amount of the bit to its right. By choosing resistors in that same proportion, binary-weighted current levels will flow.

Bus: A common set of electrical conductors shared by several devices and ICs.

Continuous Converter: An ADC that is connected to repeatedly perform analog-to-digital conversions by using the end-of-conversion signal to trigger the start-conversion input.

Conversion Time: The length of time between the start of conversion and end of conversion of an ADC.

D/A Converter (DAC): Digital-to-analog converter.

Data Acquisition: A term generally used to refer to computer-controlled acquisition and conversion of analog values.

Differential Measurement: The measurement of the difference between two values.

Handshaking: Devices and ICs that are interfaced together must follow a specific protocol, or sequence of control operations, in order to be understood by each other.

Interfacing: The device control and interconnection schemes required for electronic devices and ICs to communicate with each other.

Memory: A storage device capable of holding data that can be read by some other device.

Microprocessor: A large-scale IC capable of performing several functions, including the interpretation and execution of programmed software instructions.

Nonlinearity: Nonlinearity error describes how far the actual transfer function of an ADC or DAC varies from the ideal straight line drawn from zero up to the full-scale values.

Nonmonotonic: A nonmonotonic DAC is one in which, for every increase in the input digital code, the output level does not either remain the same or increase.

Op-Amp: An amplifier that exhibits almost ideal features (that is, infinite input impedance, infinite gain, and zero output impedance).

Programmable-Gain Amplifier: An amplifier that has a variable voltage gain that is set by inputting the appropriate digital levels at the gain select inputs.

Reference Voltage: In DAC and ADC circuits, a reference voltage or current is provided to the circuit to set the relative scale of the input and output values.

Resolution: The number of bits in an ADC or DAC. The higher the number, the closer the final representation can be to the actual input quantity.

Sample and Hold: A procedure of taking a reading of a varying analog value at a precise instant and holding that reading.

Successive Approximation: A method of arriving at a digital equivalent of an analog value by successively trying each of the individual digital bits, starting with the MSB.

Thermistor: An electronic component whose resistance changes with a change in temperature.

Transducer: A device that converts a physical quantity such as heat or light into an electrical quantity such as amperes or volts.

Virtual Ground: In certain op-amp circuit configurations, with one input at actual ground potential the other input will be held at a 0-V potential but will not be able to sink or source current.

Problems

Sections 15–1 and 15–2

15–1. Describe the function of a transducer.

15–2. How many different digital representations are allowed with:

(a) A 4-bit converter?   (c) An 8-bit converter?

(b) A 6-bit converter?   (d) A 12-bit converter?

15–3. List three characteristics of op amps that make them almost ideal amplifiers.

15–4. Determine V_{out} for the op-amp circuits of Figure P15–4.

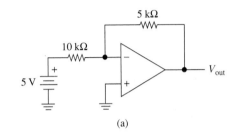

(a)

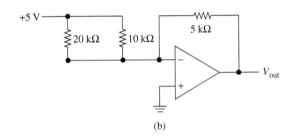

(b)

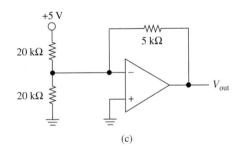

(c)

Figure P15–4

15–5. The virtual ground concept simplifies the analysis of op-amp circuits by allowing us to assume what?

Sections 15–3 and 15–4

15–6. Calculate the current through each switch and the resultant V_{out} in Figure 15–4 if the number 12_{10} is input.

D   **15–7.**

(a) Change the resistor that is connected to the D_3 switch in Figure 15–4 to

10 kΩ. What values must be used for the other three resistors to ensure the correct binary weighting factors?

(b) Reconstruct the data table in Figure 15–4 with new values for V_{out} using the resistor values found in part (a).

15–8. What effect would doubling the 20-kΩ resistor have on the values for V_{out} in Figure 15–4?

15–9. What effect would changing the reference voltage (V_{ref}) in Figure 15–6 from +5 V to –5 V have on V_{out}?

15–10. Change V_{ref} in Figure 15–6 to +2 V and calculate V_{out} for $D_0 = 0$, $D_1 = 0, D_2 = 0, D_3 = 1$.

15–11. Reconstruct the data table in Figure 15–7 for a V_{ref} of +2 V instead of +5 V.

Sections 15–5 and 15–6

15–12. Does the MC1408 DAC use a binary-weighted or an $R/2R$ method of conversion?

15–13. What is the purpose of the op amp in the DAC application circuit of Figure 15–8(c)?

15–14. What is the resolution of the DAC0808/MC1408 DAC shown in Figure 15–8?

C

15–15. Calculate V_{out} in Figure 15–8(c) for the following input values:

(a) $0100\ 0000_2$ **(c)** 32_{10}

(b) $0011\ 0110_2$ **(d)** 30_{10}

15–16. Sketch a partial transfer function of analog output versus digital input in Figure 15–8(c) for digital input values of 0000 0000 through 0000 0111.

D

15–17. How could the reference current (I_{ref}) in Figure 15–8 be changed to 1.5 mA? What effect would that have on the range of I_{out} and V_{out}?

15–18. In Figure 15–8(c), if V_{ref} is changed to 5 V, find V_{out} full-scale (A_1 to A_8 = HIGH).

Sections 15–7 and 15–8

15–19. Draw a graph of the transfer function (digital output versus analog input) for the parallel-encoded ADC of Figure 15–11.

15–20. What is one advantage and one disadvantage of using the multiple-comparator parallel encoding method of A/D conversion?

15–21. Refer to the counter-ramp ADC of Figure 15–12.

(a) What is the level at the DAC output the *instant after* the start conversion push button is pressed?

(b) What is the relationship between the $V(+)$ and $V(-)$ comparator inputs the *instant before* the HIGH-to-LOW edge of end of conversion?

15–22. In Figure 15–12, what is the worst-case (longest) conversion time that might be encountered if the clock frequency is 100 kHz?

Sections 15–9 and 15–10

15–23. Determine the conversion time for an 8-bit ADC that uses a successive-approximation circuit similar to Figure 15–13 if its clock frequency is 50 kHz.

15–24. What connections could be made in the ADC shown in Figure 15–13 to enable it to make continuous conversions?

C   **15–25.** Use the SAR ADC of Figure 15–13 to convert the analog voltage of 7.28 to 8-bit binary. If $V_{ref} = 10$ V, determine the final binary answer and the percent error.

15–26. Why is the three-state buffer at the output of the NE5034 ADC an important feature?

15–27. Referring to the block diagram of the ADC0801 (Figure 15–17), which inputs are used to enable the three-state output latches? Are they active-LOW or active-HIGH inputs?

15–28. What type of application might require the use of the differential inputs $[V_{in(+)}, V_{in(-)}]$ on the ADC0801?

15–29. Refer to Figures 15–17 and 15–18.

   (a) How would the operation change if $\overline{RD}$ were connected to +5 V instead of ground?

   (b) How would the operation change if $\overline{CS}$ were connected to +5 V instead of ground?

   (c) What is the purpose of the 10-kΩ–0.001-μF *RC* circuit?

   (d) What is the maximum range of the analog V_{in} if $V_{ref}/2$ is changed to 0.5 V?

C   **15–30.** Change V_{cc} to 5.12 V in Figure 15–18 and determine which LEDs will be ON for the following values of V_{in}:

   (a) 3.6 V   **(b)** 1.86 V

Sections 15–11 and 15–12

15–31. Briefly describe the flow of the signal from the temperature transducer as it travels through the circuit to the RAM memory in the data acquisition system of Figure 15–19.

C D   **15–32.** Table 15–2 gives the values for the output of the thermistor circuit of Figure 15–21 for various temperatures. Redesign the circuit by changing R_f so that V_{out} equals 5 V at 25°C.

C   **15–33.** The output voltage of Figure 15–21 is exactly 2.5 V at 25°C. Calculate the range of output voltage if the 10 kΩ thermistor has a tolerance of 5%.

15–34. What binary value will the 8051 microcontroller read in Figure 15–22 if the temperature is 60°C?

C   **15–35.** Through experimentation it is determined that the strain gage in

Figures 15–24 and 15–25 changes in resistance by 20 mΩ for each kilogram that is added to the metal beam. Determine how many kilograms of force are on the metal beam if V_{out} is 4 V in Figure 15–25.

Schematic Interpretation Problems

See Appendix G for the schematic diagrams.

S C D   **15–36.** Design a circuit interface that will provide analog output capability to the 4096/4196 control card. Assume that software will be written by a programmer to output the appropriate digital strings to port 1 (P1.7–P1.0) of the 8031 microcontroller. Devise an analog output circuit using an MC1408 DAC with a 741 op amp to output analog voltages in the range of 0 V to 5 V.

S C D   **15–37.** Design a circuit interface that will provide analog input capability to the 4096/4196 control card. The design must be capable of inputting the 8-bit digital results from *two* ADC0801 converters into port 1 (P1.7–P1.0) of the 8031 microcontroller. Assume that a single-bit control signal will be output on port 2, bit 0 (P2.0) to tell which ADC results are to be transmitted. (Assume 1 = ADC 1 and 0 = ADC 2). Set up the ADCs for continuous conversions and 0-V to 5-V analog input level.

Answers to Review Questions

15–1. True

15–2. 256

15–3. Infinite, 0

15–4. 240 kΩ

15–5. Because finding accurate resistances over such a large range of values would be very difficult

15–6. False

15–7. True

15–8. Current

15–9. A_1

15–10. Number of bits at the input or output

15–11. True

15–12. Gain error

15–13. False

15–14. 1023

15–15. False

15–16. False

15–17. $\overline{STRT}$, $\overline{DR}$

15–18. False

15–19. (a) Input (b) input (c) input (d) output

15–20. $\overline{CS}$, $\overline{WR}$, $\overline{INTR}$, $\overline{RD}$

15–21. False

15–22. True

15–23. True

15–24. It is a linear device, whereas the thermistor is a nonlinear device.

16

Semiconductor Memory and Programmable Arrays

OUTLINE

16–1   Memory Concepts
16–2   Static RAMs
16–3   Dynamic RAMs
16–4   Read-Only Memories
16–5   Memory Expansion and Address Decoding Applications
16–6   Programmable Logic Devices

Objectives

Upon completion of this chapter, you should be able to

- Explain the basic concepts involved in memory addressing and data storage.

- Interpret the specific timing requirements given in a manufacturer's data manual for reading or writing to a memory IC.

- Discuss the operation and application for the various types of semiconductor memory ICs.

- Design circuitry to facilitate memory expansion.

- Explain the refresh procedure for dynamic RAMs.

- Explain the programming procedure and applications for programmable array ICs.

Introduction

In digital systems, memory circuits provide the means of storing information (data) on a temporary or permanent basis for future recall. The storage medium can be either a semiconductor integrated circuit or a magnetic device such as magnetic tape or disk. **Magnetic memory** generally is capable of storing larger quantities of data than semiconductor memories, but the access time (time it takes to locate and then read or write data) is usually much more for magnetic devices. With magnetic tape or disk it takes time to physically move the read/write mechanism to the exact location to be written to or read from.

With **semiconductor memory** ICs, electrical signals are used to identify a particular memory location within the integrated circuit, and data can be stored in or read from that location in a matter of nanoseconds.

The technology used in the fabrication of memory ICs can be based on either bipolar or MOS transistors. In general, bipolar memories are faster than MOS memories, but MOS can be integrated more densely, providing much more memory locations in the same amount of area.

16–1 Memory Concepts

Let's say that you have an application where you must store the digital states of eight binary switches once every hour for 16 hours. This would require 16 *memory locations,* each having a *unique 4-bit* **memory address** (0000 to 1111) and each location being capable of containing 8 bits of data as the **memory contents.** A group of 8 bits is also known as 1 **byte,** so what we would have is a 16-byte memory, as shown in Figure 16–1.

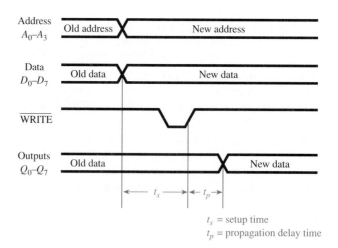

Figure 16–1   Layout for sixteen 8-bit memory locations.

To set up this memory system using actual ICs, we could use sixteen 8-bit flip-flop registers to contain the 16 bytes of data. To identify the correct address, a 4-line-to-16-line decoder can be used to decode the 4-bit address location into an active-LOW chip select to select the appropriate (1-of-16) data register for input/ output. Figure 16–3 shows the circuit used to implement this memory application.

Team Discussion

What is the order of operation between the data bus, the address bus, and the write pulse?

Figure 16–2   Timing requirements for writing data to the 16-byte memory circuit of Figure 16–3.

CHAPTER 16 | SEMICONDUCTOR MEMORY AND PROGRAMMABLE ARRAYS

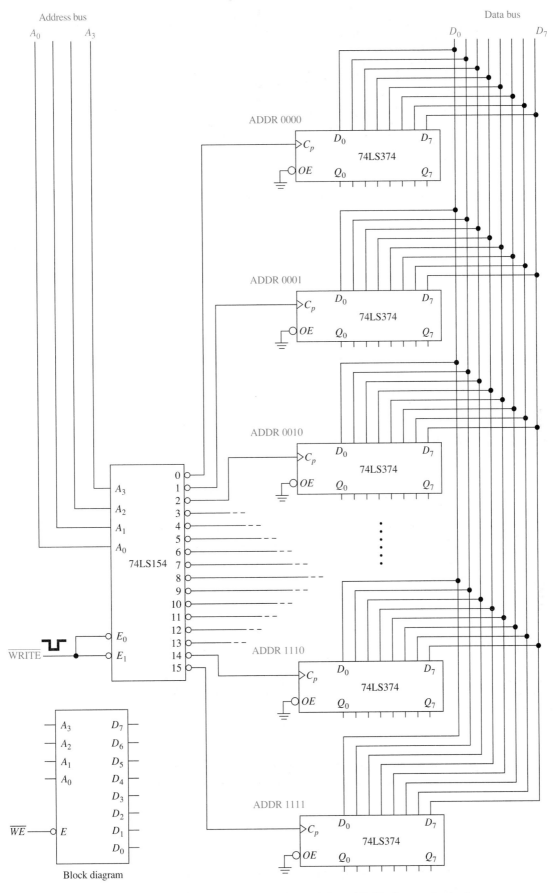

Figure 16–3 Writing to a 16-byte memory constructed from 16 octal *D* flip-flops and a 1-of-16 decoder. (Data travel down the data bus to the addressed *D* flip-flop.)

The 74LS374s are octal (eight) D flip-flops with three-state outputs. To store data in them, 8 bits of data are put on the D_0 to D_7 data inputs via the data bus. Then a LOW-to-HIGH edge on the C_p clock input will cause the data at D_0 to D_7 to be latched into each flip-flop. The value stored in the D flip-flops is observed at the Q_0 to Q_7 outputs by making the Output Enable $(\overline{OE})$ pin LOW.

To select the appropriate (1-of-16) memory location, a 4-bit address is input to the 74LS154 (4-line-to-16-line decoder), which outputs a LOW pulse on one of the output lines when the $\overline{WRITE}$ enable input is pulsed LOW.

As you can see, the timing for setting up the address bus and data bus, and pulsing the $\overline{WRITE}$ line is critical. Timing diagrams are necessary for understanding the operation of memory ICs, especially when you are using larger-scale memory ICs. The timing diagram for our 16-byte memory design of Figure 16–3 is given in Figure 16–2.

Figure 16–2 begins to show us some of the standard ways that manufacturers illustrate timing parameters for bus-driven devices. Rather than showing all four address lines and all eight data lines, they group them together and use an X (crossover) to show where any or all of the lines are allowed to change digital levels.

In Figure 16–2 the address and data lines must be set up some time (t_s) before the LOW-to-HIGH edge of $\overline{WRITE}$. In other words, the address and data lines must *be valid* (be at the appropriate levels) some period of time (t_s) *before* the LOW-to-HIGH edge of $\overline{WRITE}$ in order for the 74LS374 D flip-flop to interpret the input correctly.

When the $\overline{WRITE}$ line is pulsed, the 74LS154 decoder outputs a LOW pulse on one of its 16 outputs, which clocks the appropriate memory location to receive data from the data bus. After the propagation delay (t_p), the data output at Q_0 to Q_7 will be the new data just entered into the D flip-flop. t_p will include the propagation delay of the decoder *and* the C_p-to-Q of the D flip-flop.

In Figure 16–3, all the three-state outputs are continuously enabled so that their Q outputs are always active. To connect the Q_0 to Q_7 outputs of all 16 memory locations back to the data bus, the $\overline{OE}$ enables would have to be individually selected at the appropriate time to avoid a conflict on the data bus, called **bus contention.** Bus contention occurs when two or more devices are trying to send their own digital levels to the shared

Team Discussion

Discuss the entire sequence of events and timing required to store 1 byte of data at each memory location in Figure 16–2. Repeat for the retrieval of each byte of data using the decoder in Figure 16–4.

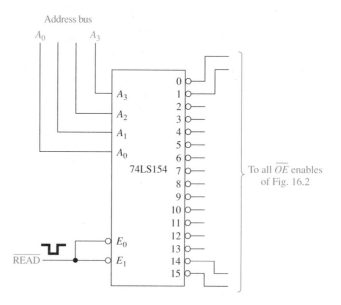

Figure 16–4 Using another decoder to individually select memory locations for Read operations.

CHAPTER 16 | SEMICONDUCTOR MEMORY AND PROGRAMMABLE ARRAYS

data bus at the same time. To individually select each group of Q outputs in Figure 16–3, the grounds on the $\overline{OE}$ enables would be removed and instead be connected to the output of another 74LS154 1-of-16 decoder, as shown in Figure 16–4.

In Figures 16–2, 16–3, and 16–4, we have designed a small 16-byte (16 × 8) random-access memory (**RAM**). Commercially available RAM ICs combine all the decoding and storage elements in a single package, as seen in the next section.

Review Questions

16–1. The _____ bus is used to specify the location of the data stored in a memory circuit.

16–2. Once a memory location is selected, data travel via the _____ bus.

16–3. In Figure 16–3, how is the correct octal D flip-flop chosen to receive data?

16–4. What is the significance of the X (crossover) on the address and data waveforms in Figure 16–2?

16–5. Why would a second 74LS154 decoder be required in Figure 16–3 to read the data from the D flip-flops if all Q outputs were connected back to the data bus? (*Hint:* See Figure 16–4.)

16–2 Static RAMs

Large-scale *random-access memory* (RAM), also known as *read/write memory,* is used for temporary storage of data and program instructions in microprocessor-based systems. The term *random access* means that the user can access (read or write) data at any location within the entire memory device randomly without having to sequentially read through several data values until positioned at the desired memory location. [An example of a sequential (nonrandom) memory device is magnetic tape.]

A better term for RAM is *read/write memory* (RWM), because all semiconductor and disk memories have random access. RWM is more specific because it tells us that data can be *read or written* to any memory location.

RAM is classified as either static or dynamic. *Static* RAMs (SRAMs) use flip-flops as basic storage elements, whereas *dynamic* RAMs (DRAMs) use internal capacitors as basic storage elements. Additional *refresh* circuitry is needed to maintain the charge on the internal capacitors of a dynamic RAM, which makes it more difficult to use. Dynamic RAMs can be packed very densely, however, yielding much more storage capacity per unit area than a static RAM. The cost per bit of dynamic RAM is also much less than that of the static RAM.

The 2147H Static MOS RAM

The 2147H is a very popular static RAM that uses MOS technology. The 2147H is set up with 4096 (abbreviated 4K, where 1K = 1024) memory locations, with each location containing 1 bit of data. This configuration is called 4096 × 1.

To develop a unique address for each of the 4096 locations, 12 address lines must be input ($2^{12} = 4096$). The storage locations are set up as a 64 × 64 array with A_0 to A_5 identifying the row and A_6 to A_{11} identifying the column to pinpoint the specific location to be used. The data sheet for the 2147H is given in Figure 16–5. This figure shows the

intel®

2147H
HIGH SPEED 4096 × 1 BIT STATIC RAM

	2147H-1	2147H-2	2147H-3	2147HL-3	2147H	2147HL
Max. Access Time (ns)	35	45	55	55	70	70
Max. Active Current (mA)	180	180	180	125	160	140
Max. Standby Current (mA)	30	30	30	15	20	10

- Pinout, Function, and Power Compatible to Industry Standard 2147
- HMOS II Technology
- Completely Static Memory—No Clock or Timing Strobe Required
- Equal Access and Cycle Times
- Single +5V Supply
- 0.8–2.0V Output Timing Reference Levels
- Direct Performance Upgrade for 2147
- Automatic Power-Down
- High Density 18-Pin Package
- Directly TTL Compatible—All Inputs and Output
- Separate Data Input and Output
- Three-State Output

The Intel® 2147H is a 4096-bit static Random Access Memory organized as 4096 words by 1-bit using HMOS-II, Intel's next generation high-performance MOS technology. It uses a uniquely innovative design approach which provides the ease-of-use features associated with non-clocked static memories and the reduced standby power dissipation associated with clocked static memories. To the user this means low standby power dissipation without the need for clocks, address setup and hold times, nor reduced data rates due to cycle times that are longer than access times.

$\overline{CS}$ controls the power-down feature. In less than a cycle time after $\overline{CS}$ goes high—deselecting the 2147H —the part automatically reduces its power requirements and remains in this low power standby mode as long as $\overline{CS}$ remains high. This device feature results in system power savings as great as 85% in larger systems, where the majority of devices are deselected.

The 2147H is placed in an 18-pin package configured with the industry standard 2147 pinout. It is directly TTL compatible in all respects: inputs, output, and a single +5V supply. The data is read out nondestructively and has the same polarity as the input data. A data input and a separate three-state output are used.

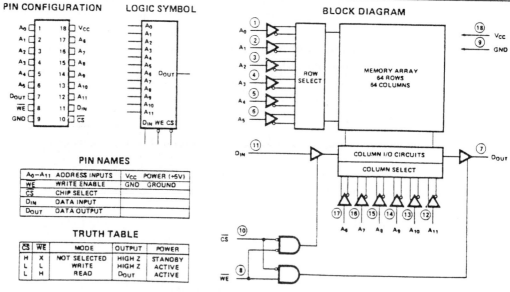

PIN CONFIGURATION   LOGIC SYMBOL   BLOCK DIAGRAM

PIN NAMES

A₀–A₁₁	ADDRESS INPUTS	Vcc	POWER (+5V)
$\overline{WE}$	WRITE ENABLE	GND	GROUND
$\overline{CS}$	CHIP SELECT		
D_IN	DATA INPUT		
D_OUT	DATA OUTPUT		

TRUTH TABLE

$\overline{CS}$	$\overline{WE}$	MODE	OUTPUT	POWER
H	X	NOT SELECTED	HIGH Z	STANDBY
L	L	WRITE	HIGH Z	ACTIVE
L	H	READ	D_OUT	ACTIVE

Team Discussion

You should be able to follow the logic of $\overline{CS}$ and $\overline{WE}$ in the block diagram for writing data and reading data.

Figure 16–5   The 2147H 4K × 1 static RAM. (Courtesy of Intel Corporation)

row and column circuitry used to pinpoint the **memory cell** within the 64 × 64 array. The box labeled "Row Select" is actually a 6-to-64 decoder for identifying the appropriate 1-of-64 row. The box labeled "Column Select" is also a 6-to-64 decoder for identifying the appropriate 1-of-64 column. Once the location is selected, the AND gates at the bottom of the block diagram allow the data bit to either pass into (D_{in}) or come out of (D_{out}) the memory location selected. Each memory location, or cell, is actually a configuration of transistors that functions like a flip-flop that can be Set (1) or Reset (0).

During *write operations,* in order for D_{in} to pass through its three-state buffer, the Chip Select *($\overline{CS}$)* must be LOW *and* the Write Enable *($\overline{WE}$)* must also be LOW. During *read operations,* in order for D_{out} to receive data from its three-state buffer, the Chip Select *($\overline{CS}$)* must be LOW *and* the Write Enable *($\overline{WE}$)* must be HIGH, signifying a *read operation.* The timing waveforms for the Read and Write cycles are given in Figure 16–6(a) and (b).

Read Operation: The circuit connections and waveforms for reading data from a location in a 2147H are given in Figure 16–6. The 12 address lines are brought in from the address bus for address selection. The $\overline{WE}$ input is held HIGH to enable the Read operation.

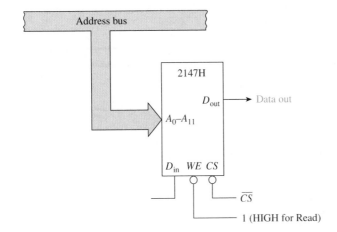

Team Discussion

Discuss the differences in the timing waveforms of D_{out} and D_{in} in Figure 16–6(a) and (b).

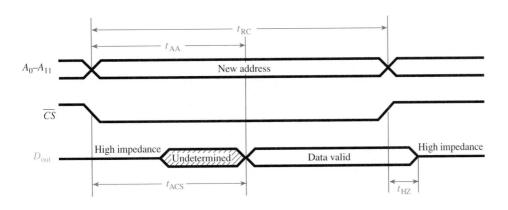

Symbol	Parameter	Min.	Max.	Unit
t_{RC}	Read cycle time	35		ns
t_{AA}	Address access time		35	ns
t_{ACS}	Chip select access time		35	ns
t_{HZ}	Chip deselection to high-Z out	0	30	ns

(a)

Figure 16–6   The 2147H static RAM timing waveforms: (a) Read cycle;

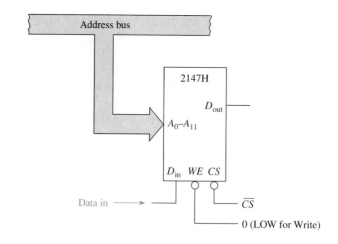

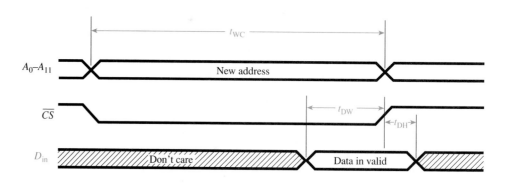

Symbol	Parameter	Min.	Max.	Unit
t_{WC}	Write cycle time	35		ns
t_{DW}	Data valid to end of write (Setup)	20		ns
t_{DH}	Data hold	0		ns

(b)

Figure 16–6   *(Continued)* (b) Write cycle.

Referring to the timing diagram, when the new address is entered in the A_0 to A_{11} inputs and the $\overline{CS}$ lines goes LOW, it takes a short period of time, called the *access time,* before the data output is valid. The access time is the length of time from the beginning of the read cycle to the end of t_{ACS}, or t_{AA}, whichever ends last. Before the $\overline{CS}$ is brought LOW, D_{out} is in a high-impedance (float) state. The $\overline{CS}$ *and* A_0 to A_{11} inputs must both be held stable for a minimum length of time, t_{RC}, before another Read cycle can be initiated.

After $\overline{CS}$ goes back HIGH, the data out is still valid for a short period of time, t_{HZ}, before returning to its high-impedance state.

Write Operation: A similar set of waveforms is used for the write operation [Figure 16–6(b)]. In this case the D_{in} is written into memory while the $\overline{CS}$ and $\overline{WE}$ are both LOW. The D_{in} must be set up for a length of time *before* either $\overline{CS}$ or $\overline{WE}$ go back HIGH (t_{DW}), and it must also be held for a length of time *after* either $\overline{CS}$ or $\overline{WE}$ go back HIGH (t_{DH}).

Memory Expansion: Because the contents of each memory location in the 2147H is only 1 bit, to be used in an 8-bit computer system, eight 2147Hs must be set up in such a way that, when an address is specified, 8 bits of data will be read or written.

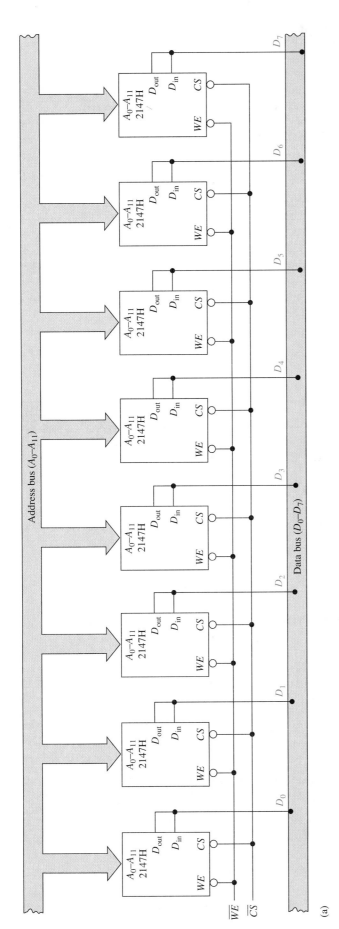

Figure 16–7   RAM memory expansion: (a) eight 2147Hs configured as a 4K × 8; (b) thirty-two 2114s configured as two banks of 8K × 8.

(b)

571

With eight 2147s we have a 4096 × 8 (4K × 8) memory system, as shown in Figure 16–7(a).

The address selection for each 2147H in Figure 16–7(a) is identical, because they are all connected to the same address bus lines. This way, when reading or writing from a specific address, 8 bits, each at the same address, will be sent to or received from the data bus simultaneously. The $\overline{WE}$ input determines which internal three-state buffer is enabled, connecting *either* D_{in} or D_{out} to the data bus. The $\overline{WE}$ input is sometimes labeled READ/$\overline{WRITE}$, meaning that it is HIGH for a Read operation, which puts data out to the data bus via D_{out}, and it is LOW for a Write operation, which writes data into the memory via D_{in}.

Several other configurations of RAM memory are available. For example, the 2114 is configured as a 1024 × 4-bit (1K × 4) RAM, instead of the 4096 × 1 used by the 2147H. A 1024 × 4-bit RAM will input/output 4 bits at a time for each address specified. This way, interfacing to an 8-bit data bus is simplified by having to use only two 2114s, one for the LOW-order data bits (D_0 to D_3) and the other for the HIGH-order data bits (D_4 to D_7). For example, the photograph in Figure 16–7(b) shows a memory expansion printed circuit board with thirty-two 2114 RAM ICs. They are set up as two memory banks, each configured as 8K × 8.

Table 16–1 provides a sample of the variety of RAM ICs that are available to the digital design engineer.

Table 16–1   Sample RAM ICs

Part No.	Organization	Description
2147H	4K × 1	High-speed MOS static RAM
2148H	1K × 4	High-speed MOS static RAM
2114A	1K × 4	Low-power MOS static RAM
6164	8K × 8	Low-power, high-speed MOS static RAM
6206	32K × 8	High-speed, high-density MOS static RAM
6207	256K × 1	High-speed, high-density MOS static RAM

Review Questions

16–6. The 2114 memory IC is a 1K × 4 static RAM, which means that it has _____ memory locations with _____ data bits at each location.

16–7. To perform a *read operation* with a 2147H RAM, $\overline{CS}$ must be _____ (LOW, HIGH) and $\overline{WE}$ must be _____ (LOW, HIGH).

16–8. In the 2147H memory array, address lines _____ through _____ are used to select the *row* and _____ through _____ are used to select the *column*.

16–9. According to the Read cycle timing waveforms for the 2147H, how long must you wait after a chip select ($\overline{CS}$) before valid data are available at D_{out}?

2118 FAMILY
16,384 x 1 BIT DYNAMIC RAM

	2118-10	2118-12	2118-15
Maximum Access Time (ns)	100	120	150
Read, Write Cycle (ns)	235	270	320
Read–Modify–Write Cycle (ns)	285	320	410

- Single +5V Supply, ±10% Tolerance

- HMOS Technology

- Low Power: 150 mW Max. Operating
  11 mW Max. Standby

- Low V_{DD} Current Transients

- All Inputs, Including Clocks,
  TTL Compatible

- $\overline{CAS}$ Controlled Output is
  Three-State, TTL Compatible

- $\overline{RAS}$ Only Refresh

- 128 Refresh Cycles Required
  Every 2ms

- Page Mode and Hidden
  Refresh Capability

- Allows Negative Overshoot
  V_{IL} min = -2V

The Intel® 2118 is a 16,384 word by 1-bit Dynamic MOS RAM designed to operate from a single +5V power supply. The 2118 is fabricated using HMOS — a production proven process for high performance, high reliability, and high storage density.

The 2118 uses a single transistor dynamic storage cell and advanced dynamic circuitry to achieve high speed with low power dissipation. The circuit design minimizes the current transients typical of dynamic RAM operation. These low current transients contribute to the high noise immunity of the 2118 in a system environment.

Multiplexing the 14 address bits into the 7 address input pins allows the 2118 to be packaged in the industry standard 16-pin DIP. The two 7-bit address words are latched into the 2118 by the two TTL clocks, Row Address Strobe ($\overline{RAS}$) and Column Address Strobe ($\overline{CAS}$). Non-critical timing requirements for $\overline{RAS}$ and $\overline{CAS}$ allow use of the address multiplexing technique while maintaining high performance.

The 2118 three-state output is controlled by $\overline{CAS}$, independent of $\overline{RAS}$. After a valid read or read-modify-write cycle, data is latched on the output by holding $\overline{CAS}$ low. The data out pin is returned to the high impedance state by returning $\overline{CAS}$ to a high state. The 2118 hidden refresh feature allows $\overline{CAS}$ to be held low to maintain latched data while $\overline{RAS}$ is used to execute $\overline{RAS}$-only refresh cycles.

The single transistor storage cell requires refreshing for data retention. Refreshing is accomplished by performing $\overline{RAS}$-only refresh cycles, hidden refresh cycles, or normal read or write cycles on the 128 address combinations of A_0 through A_6 during a 2ms period. A write cycle will refresh stored data on all bits of the selected row except the bit which is addressed.

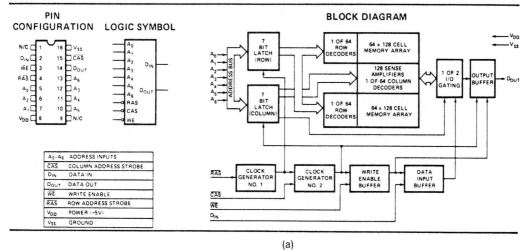

(a)

Figure 16–8 (a) The 2118 16K × 1 dynamic RAM;

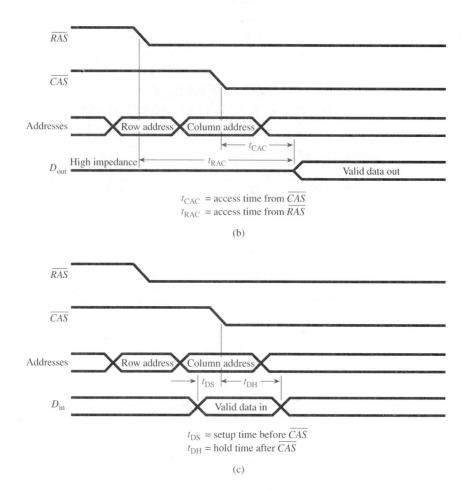

t_{CAC} = access time from $\overline{CAS}$
t_{RAC} = access time from $\overline{RAS}$

(b)

t_{DS} = setup time before $\overline{CAS}$
t_{DH} = hold time after $\overline{CAS}$

(c)

Figure 16–8 *(Continued)* (b) dynamic RAM Read cycle timing (WE = HIGH); (c) dynamic RAM Write cycle timing (WE = LOW). [(a) Courtesy of Intel Corporation]

16–3 Dynamic RAMs

Although **dynamic** RAMs require more support circuitry and are more difficult to use than static RAMs, they are less expensive per bit and have a much higher density, minimizing circuit-board area. Most applications requiring large amounts of read/write memory will use dynamic RAMs instead of static.

Dynamic MOS RAMs store information on a small internal capacitor instead of a flip-flop. All the internal capacitors require recharging, or refreshing, every 2 ms or less to maintain the stored information. An example of a 16K × 1-bit dynamic RAM is the Intel 2118, whose data sheet is shown in Figure 16–8(a) on page 573.

To uniquely address 16,384 locations, 14 address lines are required (2^{14} = 16,384). However, Figure 16–8(a) shows only seven address lines (A_0 to A_6). This is because with larger memories such as this, in order to keep the IC pin count to a minimum, the address lines are *multiplexed* into two groups of seven. An external 14-line-to-7-line multiplexer is required in conjunction with the control signals, $\overline{RAS}$ and $\overline{CAS}$, in order to access a complete 14-line address.

The controlling device must put the valid 7-bit address of the desired memory array *row* on the A_0 to A_6 inputs and then send the Row Address Strobe ($\overline{RAS}$) LOW. Next, the controlling device must put the valid 7-bit address of the desired memory array *column* on the *same* A_0 to A_6 inputs and then send the Column Address Strobe ($\overline{CAS}$) LOW. Each of these 7-bit addresses is latched and will pinpoint the desired 1-bit memory location by its row–column coordinates.

Once the memory location is identified, the $\overline{WE}$ input is used to direct either a Read or Write cycle similar to the static RAM operation covered in Section 16–2. When $\overline{WE}$ is LOW, data are written to the RAM via D_{in}; when $\overline{WE}$ is HIGH, data are read from the RAM via D_{out}.

Read Cycle Timing [Figure 16–8(b)]

1. $\overline{WE}$ is HIGH.

2. A_0 to A_6 are set up with the row address and $\overline{RAS}$ is sent LOW.

3. A_0 to A_6 are set up with the column address and $\overline{CAS}$ is sent LOW.

4. After the access time from $\overline{RAS}$ or $\overline{CAS}$ (whichever is longer), the D_{out} line will contain valid data.

Write Cycle Timing [Figure 16–8(c)]

1. $\overline{WE}$ is LOW.

2. A_0 to A_6 are set up with the row address and $\overline{RAS}$ is sent LOW.

3. A_0 to A_6 are set up with the column address and $\overline{CAS}$ is sent low.

4. At the HIGH-to-LOW edge of $\overline{CAS}$, the level at D_{in} is stored at the specified row–column memory address. D_{in} must be set up prior to and held after the HIGH-to-LOW edge of $\overline{CAS}$ to be interpreted correctly. (There are other setup, hold, and delay times that are not shown. Refer to a memory data book for more complete specifications.)

Refresh Cycle Timing

Each of the 128 rows of the 2118 must be *refreshed* every 2 ms or sooner to replenish the charge on the internal capacitors. There are three ways to refresh the memory cells:

1. Read cycle

2. Write cycle

3. $\overline{RAS}$-only cycle

Unless you are reading or writing from all 128 rows every 2 ms, the $\overline{RAS}$-only cycle is the preferred technique to provide data retention. To perform a $\overline{RAS}$-only cycle, the following procedure is used:

1. $\overline{CAS}$ is HIGH.

2. A_0 to A_6 are set up with the row address 000 0000.

3. $\overline{RAS}$ is pulsed LOW.

4. Increment the A_0 to A_6 row address by 1.

5. Repeat steps 3 and 4 until all 128 rows have been accessed.

Dynamic RAM Controllers

It seems like a lot of work demultiplexing the addresses and refreshing the memory cells, doesn't it? Well, most manufacturers of dynamic RAMs (DRAMs) have devel-

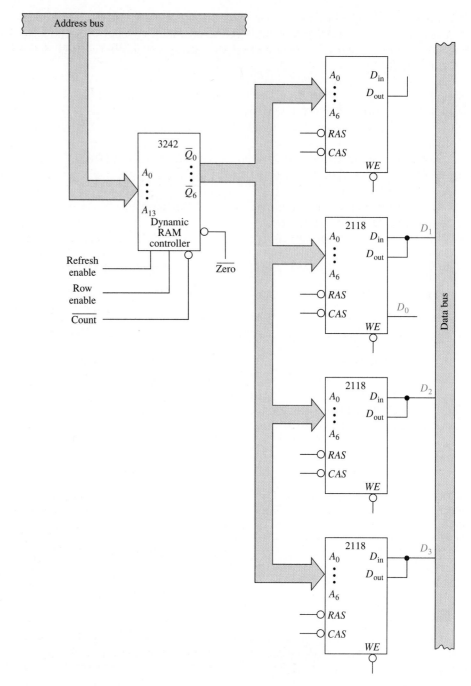

Figure 16–9 Using a 3242 address multiplexer and refresh counter in a 16K × 4 dynamic RAM memory system.

oped controller ICs to simplify the task. Some of the newer dynamic RAMs have refresh and error detection/correction circuitry built right in, which makes the DRAM look **static** to the user.

A popular controller IC is the Intel 3242 address multiplexer and refresh counter for 16K dynamic RAMs. Figure 16–9 shows how this controller IC is used in conjunction with four 2118 DRAMs.

The 3242 in Figure 16–9 is used to multiplex the 14 input addresses A_0 to A_{13} to seven active-LOW output addresses $\overline{Q}_0$ to $\overline{Q}_6$. When the Row Enable input is HIGH,

A_0 to A_6 are output inverted to $\overline{Q}_0$ to $\overline{Q}_6$ as the row addresses. When the Row Enable input is LOW, A_7 to A_{13} are output inverted to $\overline{Q}_0$ to $\overline{Q}_6$ as the column address. Of course, the timing of the $\overline{RAS}$ and $\overline{CAS}$ on the 2118s must be synchronized with the Row Enable signal.

To provide a "burst" refresh to all 128 rows of the 2118s, the Refresh Enable input in Figure 16–9 is made HIGH. This causes the $\overline{Q}_0$ to $\overline{Q}_6$ outputs to count from 0 to 127 at a rate determined by the $\overline{\text{count}}$ input clock signal. When the first 6 significant bits of the counter sequence to all zeros, the $\overline{\text{zero}}$ output goes LOW, signifying the completion of the first 64 refresh cycles.

One commonly used method of setting up the timing for $\overline{RAS}$, $\overline{CAS}$, and Row Enable is with a multitap **delay line,** as shown in Figure 16–10. Basically, the four-tap delay line IC of Figure 16–10(a) is made up of four inverters with precision RCs to develop a 50-ns delay between each inverter. The pulses out of each tap have the same width, but each successive tap is inverted and delayed by 50 ns. [In Figure 16–10(b), every other tap was used to arrive at noninverted, 100-ns delay pulses.] Delay lines are very useful for circuits requiring sequencing, as dynamic RAM memory systems do.

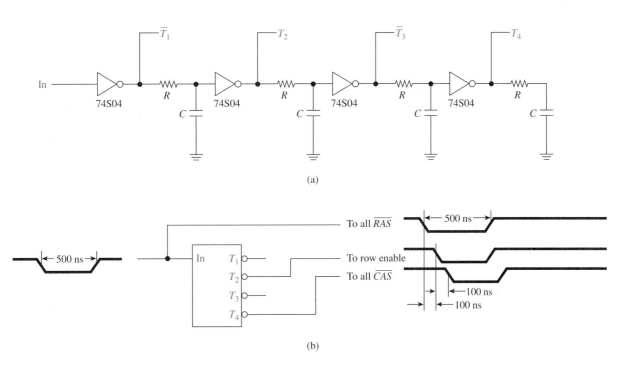

(a)

(b)

Figure 16–10 Four-tap, 50-ns delay line used for dynamic RAM timing: (a) logic diagram; (b) logic symbol and timing.

The waveforms produced by the delay line of Figure 16–10(b) can be used to drive the control inputs to the 16K × 4 dynamic RAM memory system of Figure 16–9 (a LOW $\overline{RAS}$ pulse, then a LOW Row Enable pulse, then a LOW $\overline{CAS}$ pulse). Careful inspection of the data sheets for the 3242 and 2118 is required to determine the maximum and minimum allowable values for pulse widths and delay times. To design the absolute fastest possible memory circuit, all the times would be kept at their minimum value. But it is a good practice to design in a 10% to 20% margin to be safe.

Because of their extremely high circuit density, dynamic RAMs are usually used in computer systems that typically require up to 1 megabyte of RAM memory area. Table 16–2 lists several popular high-density dynamic RAM ICs.

Table 16–2 Common Dynamic RAM ICs

Part No.	Organization	Description
2118	$16K \times 1$	16K-bit MOS dynamic RAM
4116	$16K \times 1$	16K-bit MOS dynamic RAM
4164	$64K \times 1$	64K-bit MOS dynamic RAM
6256	$256K \times 1$	256K-bit MOS dynamic RAM
511000	$1M \times 1$	1Meg-bit MOS dynamic RAM

Review Questions

16–10. Why are address lines on larger memory ICs multiplexed?

16–11. What is meant by *refreshing* a dynamic RAM?

16–4 Read-Only Memories

ROMs are memory ICs used to store data on a permanent basis. They are capable of random access and are *nonvolatile,* meaning that they do not lose their memory contents when power is removed. This makes them very useful for the storage of computer operating systems, software language compilers, table look-ups, specialized code conversion routines, and programs for dedicated microprocessor applications.

ROMs are generally used for read-only operations and are not written to after they are initially programmed. However, there is an erasable variety of ROM called an **EPROM** (erasable-programmable-read-only memory) that is very useful because it can be erased and then reprogrammed if desired.

To use a ROM, the user simply specifies the correct address to be read and then enables the chip select $(\overline{CS})$. The data contents at that address (usually 8 bits) will then appear at the outputs of the ROM (some ROM outputs will be three stated, so you will have to enable the output with a LOW on $\overline{OE}$).

Mask ROMs

Manufacturers will make a custom **mask** ROM for users who are absolutely sure of the desired contents of the ROM and have a need for at least 1000 or more chips. To fabricate a custom IC like the mask ROM, the manufacturer charges a one-time fee of more than $1000 for the design of a unique mask that is required in the fabrication of the integrated circuit. After that, each identical ROM that is produced is very inexpensive. In basic terms, a mask is a cover placed over the silicon chip during fabrication that determines the permanent logic state to be formed at each memory location. Of course, before the mass production of a quantity of mask ROMs, the user should have thoroughly tested the program or data that will be used as the model for the mask. Most desktop computers use mask ROMs to contain their operating system and for executing procedures that do not change, such as decoding the keyboard and the generation of characters for the CRT.

Fusible-Link PROMs

To avoid the high one-time cost of producing a custom mask, IC manufacturers provide user-programmable ROMs (**PROMs**). They are available in standard configurations such as $4K \times 4$, $4K \times 8$, $8K \times 4$, and so on.

Initially, every memory cell has a fusible link, keeping its output at 0. A 0 is changed to a 1 by sending a high-enough current through the fuse to permanently open it, making the output of that cell a 1. The programming procedure involves addressing each memory location, in turn, and placing the 4-bit or 8-bit data to be programmed at the PROM outputs and then applying a programming pulse (either a HIGH voltage or a constant current to the programming pin). Details for programming are given in the next section.

Once the fusible link is burned open, the data are permanently stored in the PROM and can be read over and over again just by accessing the correct memory address. The process of programming such a large number of locations is best done by a PROM programmer or microprocessor development system (MDS). These systems can copy a good PROM or the data can be input via a computer keyboard or from a magnetic disk.

EPROMs and EEPROMs

When using mask ROMs or PROMs, if you need to make a change in the memory contents or if you make a mistake in the initial programming, you are out of luck! One solution to that problem is to use an erasable PROM (EPROM). These PROMs are erased by exposing an open "window" (see Figure 16–11) in the IC to an ultraviolet (UV) light source for a specified length of time. Another type of EPROM is also available, called an electrically erasable PROM (**EEPROM** or E²PROM) or electrically alterable PROM (EAROM). By applying a high voltage (about 21 V), a single byte, or the entire chip, can be erased in 10 ms. This is a lot faster than UV erasing and can be done easily while the chip is still in the circuit. One application of the EEPROM is in the tuner of a modern TV set. The EEPROM remembers (1) the

Figure 16–11 A 2716 EPROM IC showing the window that allows UV light to strike the memory cells for erasure.

channel you were watching when you turned off the set and (2) the volume setting of the audio amplifier.

The 2716 EPROM: The data sheet for the 2716 EPROM is given in Appendix B. Referring to the data sheet, notice that the 2716 has 16K bits of memory, organized as $2K \times 8$. 2K locations require 11 address inputs ($2^{11} = 2048$), which are labeled A_0 to A_{10}.

To read a byte (8 bits) of data from the chip, the 11 address lines are set up, and then $\overline{CE}$ and $\overline{OE}$ are brought LOW to enable the chip and to enable the output. The ac waveforms for the chip show that the data outputs (O_0 to O_7) become valid after a time delay for setting up the addresses (t_{ACC}), or enabling the chip (t_{CE}), or enabling the output (t_{OE}), whichever is completed last. Figure 16–12(a) shows the circuit connections and waveforms for reading the 2716 EPROM.

In Figure 16–12(a), the X in the address waveform signifies the point where the address lines must change (1 to 0 or 0 to 1), if they are going to change. The $\overline{CE}/PGM$ line is LOW for Chip Enable and HIGH for programming mode. Outputs O_0 to O_7 are in the high-impedance state (float) until $\overline{OE}$ goes LOW. The outputs are then undetermined until the delay time t_{OE} has expired, at which time they become the valid levels from the addressed memory contents.

Programming the 2716: Initially, and after an erasure, all bits in the 2716 are 1's. To program the 2716, the following procedure is used:

1. Set V_{pp} to 25 V and $\overline{OE}$ = HIGH (5 V).

2. Set up the address of the byte location to be programmed.

3. Set up the 8-bit data to be programmed on the O_0 to O_7 outputs.

4. Apply a 50-ms positive TTL pulse to the $\overline{CE}/PGM$ input.

5. Repeat steps 2, 3, and 4 until all the desired locations have been programmed.

Figure 16–12(b) shows the circuit connections and waveforms for programming a 2716.

Table 16–3 shows a partial listing of the wide variety of EPROMs and EEPROMs available to the digital designer. You will notice that the bit size of EEPROMs is somewhat limited, but with the advent of the newer "Flash" technology, the bit density of electrically erasable EEPROMs is approaching that of the UV-erasable EPROM.

Table 16–3 Representative EPROMs and EEPROMs

Part No.	Organization	Description
2716	$2K \times 8$	UV-erasable MOS EPROM
2764	$8K \times 8$	UV-erasable MOS EPROM
27256	$32K \times 8$	UV-erasable MOS EPROM
27512	$64K \times 8$	UV-erasable MOS EPROM
271024	$64K \times 16$	UV-erasable MOS EPROM
272048	$128K \times 16$	UV-erasable MOS EPROM
2816	$2K \times 8$	Electrically erasable MOS EEPROM
2864	$8K \times 8$	Electrically erasable MOS EEPROM
27F64	$8K \times 8$	Flash electrically erasable MOS EEPROM
27F256	$32K \times 8$	Flash electrically erasable MOS EEPROM

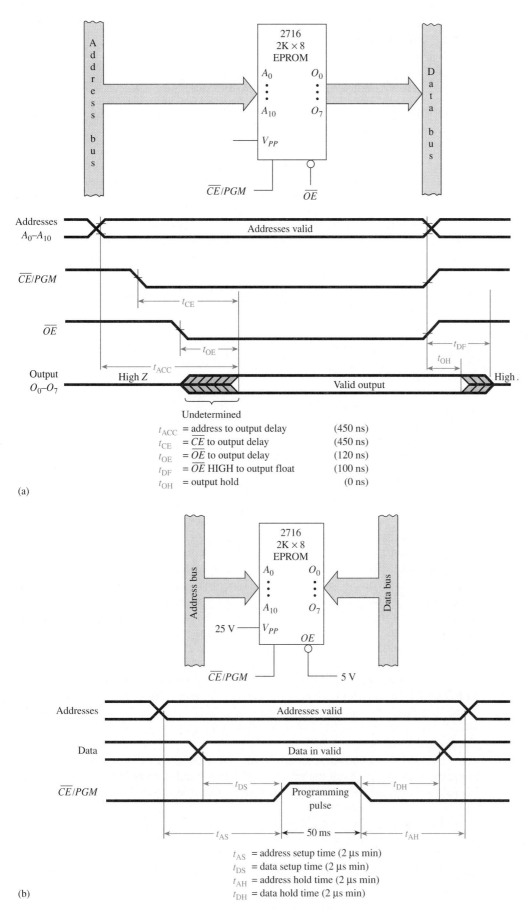

t_{ACC} = address to output delay   (450 ns)
t_{CE} = $\overline{CE}$ to output delay   (450 ns)
t_{OE} = $\overline{OE}$ to output delay   (120 ns)
t_{DF} = $\overline{OE}$ HIGH to output float   (100 ns)
t_{OH} = output hold   (0 ns)

(a)

t_{AS} = address setup time (2 μs min)
t_{DS} = data setup time (2 μs min)
t_{AH} = address hold time (2 μs min)
t_{DH} = data hold time (2 μs min)

(b)

Figure 16–12 The 2716 EPROM: (a) Read cycle; (b) Program cycle.

581

16–12. What is meant by the term *volatile*?

16–13. Describe a situation where you would want to convert your EPROM memory design over to the Mask-ROMs.

16–14. According to the 2716 EPROM Read cycle timing waveforms, you must wait _____ nanoseconds after $\overline{OE}$ goes LOW before the data at O_0 to O_7 are valid (assuming $\overline{CE}$ has already been LOW for at least _____ nanosecond).

16–15. The time for the outputs to return to a float state for the 2716 is $t_{DF} =$ 100 ns. Under what circumstances would that time be important to know?

16–5 Memory Expansion and Address Decoding Applications

When more than one memory IC is used in a circuit, a decoding technique (called **address decoding**) must be used to identify *which IC* is to be read or written to. Most 8-bit microprocessors use 16 separate address lines to identify unique addresses within the computer system. Some of those 16 lines will be used to identify the chip to be accessed, whereas the others pinpoint the exact memory location. For instance, the 2732 is a 4K × 8 EPROM that requires 12 of these address lines (A_0 to A_{11}) just to locate specific contents within its memory. This leaves four address lines (A_{12} to A_{15}) free for chip address decoding. A_{12} to A_{15} can be used to identify which IC within the system is to be accessed.

With 16 total address lines, there will be 64K, or 65,536 (2^{16} = 65,536), unique address locations. One 2732 will use up 4K of those. To design a large EPROM memory system, let's say, 16K bytes, four 2732s would be required. The address decoding scheme shown in Figure 16–13 could be used to set up the four EPROMS consecutively in the first 16K addresses of a computer system.

The four EPROMS in Figure 16–13 are set up in consecutive memory locations between 0 to 16K and are individually enabled by the 74LS138 address decoder. The 4K × 8 EPROMS each require 12 address lines for internal memory selection, leaving the four HIGH-order address lines (A_{12} to A_{15}) free for chip selection by the 74LS138.

To read from the EPROMS, the microprocessor first sets up on the address bus the unique 16-bit address that it wants to read from. Then it issues a LOW level on its $\overline{RD}$ output. This satisfies the three enable inputs for the 74LS138, which then uses A_{12}, A_{13}, and A_{14} to determine which of its outputs is to go LOW, selecting one of the four EPROMs. Once an EPROM has been selected, it outputs its addressed 8-bit contents to the data bus. The outputs of the other EPROMs will float because their $\overline{CE}$s are HIGH. The microprocessor gives all the chips time to respond and then reads the data that it requested from the data bus.

The address decoding scheme shown in Figure 16–13 is a very common technique used for mapping out the memory allocations in microprocessor-based systems (called *memory mapping*). RAM (or RWM) is added to the memory system the same way.

For example, if we wanted to add four 4K × 8 RAMs, their chip enables would be connected to the 4–5–6–7 outputs of the 74LS138, and they would occupy locations 4XXX, 5XXX, 6XXX, and 7XXX. Then, when the microprocessor issues a Read or Write command for, let's say, address 4007H (4007 Hex), the first RAM would be accessed.

Team Discussion

This is a very popular scheme for address decoding (Figure 16–13). To be sure that you thoroughly understand it, determine the new addresses if A_{15} is moved to the EN3 input.

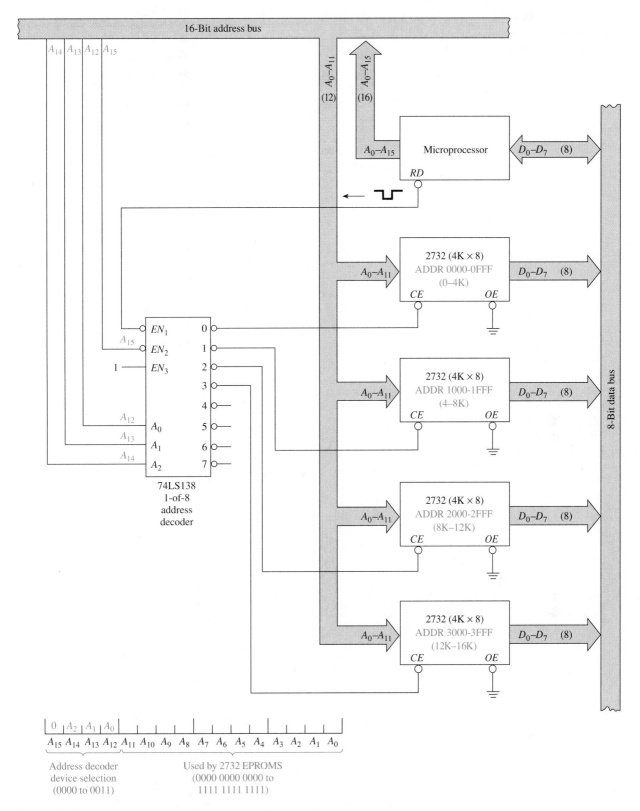

Figure 16–13 Address decoding scheme for a 16K-byte EPROM memory system.

EXAMPLE 16–1

Determine which EPROM and which EPROM address are accessed when the microprocessor of Figure 16–13 issues a Read command for the following hex addresses: (a) READ 0007H; (b) READ 26C4H; (c) READ 3FFFH; (d) READ 5007H.

Solution:

(a) The HIGH-order hex digit (0) will select the first EPROM. Address 007 (0000 0000 0111) in the first EPROM will be accessed. (Address 007H is actually the *eighth* location in that EPROM.)

(b) The HIGH-order hex digit (2) will select the third EPROM (A_{15} = 0, A_{14} = 0, A_{13} = 1, A_{12} = 0). Address 6C4H in the third EPROM will be accessed.

(c) The HIGH-order hex digit (3) will select the fourth EPROM (A_{15} = 0, A_{14} = 0, A_{13} = 1, A_{12} = 1). Address FFFH (the last location) in the fourth EPROM will be accessed.

(d) The HIGH-order hex digit (5) will cause the output 5 of the 74LS138 to go LOW. Because no EPROM is connected to it, nothing will be read.

Helpful Hint

This decoding scheme can also be used to access RAM ICs. In that case the $\overline{RD}$ line would need to be ORed with $\overline{WR}$.

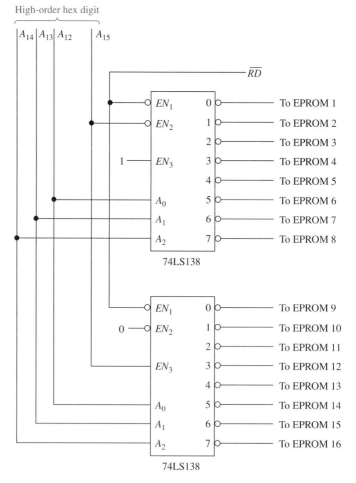

Figure 16–14   Expanding the memory of Figure 16–13 to 64K bytes.

CHAPTER 16 I SEMICONDUCTOR MEMORY AND PROGRAMMABLE ARRAYS

Expansion to 64K

The memory system of Figure 16–13 can be expanded to 64K bytes by utilizing two 74LS138 decoders, as shown in Figure 16–14. Address lines A_0 to A_{12} are not shown in Figure 16–14, but they would go to each 2732 EPROM, just as they did in Figure 16–13. The HIGH-order addresses (A_{12} to A_{15}) are used to select the individual EPROMS. When A_{15} is LOW, the upper decoder in Figure 16–14 is enabled and EPROMs 1 to 8 can be selected. When A_{15} is HIGH, the lower decoder is enabled and EPROMs 9 to 16 can be selected. Using the circuit in Figure 16–14 will allow us to map in sixteen $4K \times 8$ EPROMs, which will *completely* fill the memory map in a 16-bit address system. Actually, this would not be practical because some room must be set aside for RAM and input/output devices.

One final point on memory and bus operation: microprocessors and MOS memory ICs are generally designed to drive only a single TTL load. Therefore, when several inputs are being fed from the same bus, an MOS device driving the bus must be buffered. An octal **buffer** IC such as the 74241 connected between an MOS IC output and the data bus will provide the current capability to drive a heavily loaded data bus. Bidirectional bus drivers (or transceivers) such as the 74LS640 provide buffering in both directions for use by read/write memories (RAM or RWM).

APPLICATION 16–1

A PROM Look-Up Table

Besides being used strictly for memory, ROMs, PROMs, and EPROMS can also be programmed to provide special-purpose functions. One common use is as a **look-up table.** A simple example is to use a PROM as a 4-bit binary-to-Gray code converter, as shown in Figure 16–15.

Team Discussion

How could a PROM or EPROM be used as an ASCII look-up table?

Binary				Gray code			
0	0	0	0	0	0	0	0
0	0	0	1	0	0	0	1
0	0	1	0	0	0	1	1
0	0	1	1	0	0	1	0
0	1	0	0	0	1	1	0
0	1	0	1	0	1	1	1
0	1	1	0	0	1	0	1
0	1	1	1	0	1	0	0
1	0	0	0	1	1	0	0
1	0	0	1	1	1	0	1
1	0	1	0	1	1	1	1
1	0	1	1	1	1	1	0
1	1	0	0	1	0	1	0
1	1	0	1	1	0	1	1
1	1	1	0	1	0	0	1
1	1	1	1	1	0	0	0

Used as PROM addresses

Data at the specified address

Figure 16–15 Using a PROM look-up table to convert binary to Gray code.

The PROM chosen for Figure 16–15 must have 16 memory locations, each location containing a 4-bit Gray code. The 4-bit binary string to be converted is used as the address inputs to the PROM. The PROM must be programmed such that each memory contains the equivalent Gray code to be output. For example, address location 0010 will contain 0011, 0100 will contain 0110, and so on, for the complete binary-to-Gray code data table.

A more practical application would be to use a PROM to convert 7-bit binary to two BCD digits, which is a very complicated procedure using ordinary logic gates.

APPLICATION 16–2
A Digital LCD Thermometer

Another application, one that covers several topics from within this text, is a digital Celsius thermometer. In this application, using a PROM look-up table simplifies the task of converting meaningless digital strings into decimal digits. Figure 16–16 shows a block diagram of a two-digit Celsius thermometer.

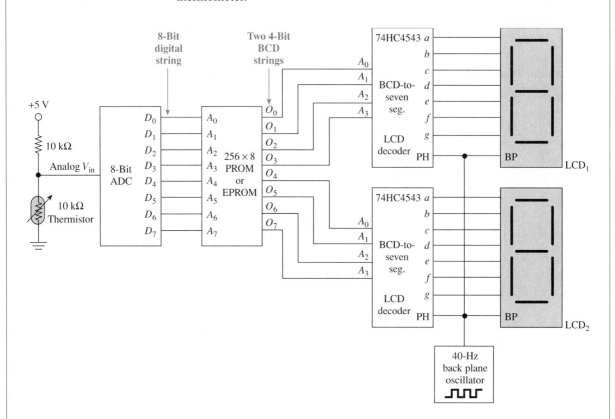

Figure 16–16   Using a PROM as a look-up table for binary-to-BCD conversion for an LCD thermometer.

For the circuit of Figure 16–16 to work, a binary-to-two-digit BCD look-up table has to be programmed into the PROM. Because a standard thermistor is a nonlinear device, as the temperature varies the binary output of the ADC will not change in proportional steps. Programming the PROM with the appropriate codes can compensate for that and can also ensure that the output being fed to the two decoders is in the form of two BCD codes, each within the range 0 to 9. The appropriate codes for the PROM contents are best determined through experimentation. For example, if at 30°C the output of the ADC is 0100 1100 ($4C_{16}$), then location $4C_{16}$ of the PROM should be programmed with 0011 0000 (30_{BCD}).

CHAPTER 16 | SEMICONDUCTOR MEMORY AND PROGRAMMABLE ARRAYS

The 74HC4543 will convert its BCD input into a seven-segment code for the liquid-crystal displays (**LCDs**). Liquid-crystal displays consume significantly less power than LED displays but require a separate square-wave oscillator to drive their backplane. As shown in Figure 16–16, a 40-Hz oscillator is connected to the phase input (PH) of each decoder and the backplane (BP) of each LCD.

Review Questions

16–16. Determine which EPROM and which EPROM address are accessed when the microprocessor of Figure 16–13 issues a Read command for the following hexadecimal addresses:

(a) READ 2002H   (b) READ 0AF7H

16–17. In Figure 16–13, connect A_{15} to EN_3 and ground EN_2. What is the new range of addresses that will access the second EPROM that was formerly accessed at 1000H through 1FFFH?

16–18. In Figure 16–16, assume that the thermistor resistance is 10 kΩ at 25°C, thus making D_7 to D_0 equal to 1000 0000 (one-half of full scale). Determine what data value should be programmed into the EPROM at address 1000 0000.

16–6   Programmable Logic Devices

In Chapter 5 we saw that complex combinational logic circuits could be reduced to their simplest sum-of-products (SOP) form using De Morgan's theorem and Karnaugh mapping. An example of an SOP expression is

$$X = A\overline{B}\,\overline{C} + \overline{A}\,\overline{B}C + \overline{A}B\overline{C}$$

To implement this expression using conventional logic would require three different ICs: a hex inverter, a triple 3-input AND gate, and a 3-input OR gate. As SOP logic complexity increases, the number of SSI or MSI ICs becomes excessive.

An increasingly popular solution being used today is to implement the logic function using programmable logic devices (PLDs). PLD ICs can be selected from the TTL, CMOS, or ECL families, depending on your requirements for high speed, low power, and logic function availability. The three basic forms of PLDs are Programmable Read-Only Memory (PROM), **Programmable Array** Logic (**PAL®**),[*] and Programmable Logic Array (PLA). PROM, which was discussed earlier in this chapter, is most commonly used as a memory device and is not well suited for implementing complex logic equations.

Standard PAL has several multiinput AND gates connected to the input of an OR gate and inverter. The inputs are set up as a fusible-link programmable multivariable array. By burning specific fuses in an array, the user can program an IC to solve a mul-

---

[*]PAL® is a registered trademark of Monolithic Memories, Inc., a wholly owned subsidiary of Advanced Micro Devices, Inc.

titude of various combinational logic problems. Hard array logic **(HAL),** which requires custom mask design by the manufacturer, is also available (a HAL is to a PAL as a ROM is to a PROM).

Standard PLA (or Field-Programmable Logic Array, **FPLA**) has multiinput programmable AND gates as well, but it has the additional flexibility of having its AND gates connected to several *programmable OR gates.* Because of this additional flexibility, PLA is slightly more difficult to program than PAL, and the additional level of logic gates increases the overall propagation delay times.

To keep logic diagrams for PLDs easy to read, a one-line convention has been adopted, as illustrated in Figure 16–17.

Common Misconception

Initially, you may have a hard time understanding how you can have multiple input variables to an AND gate having only one input. It is hoped that this illustration will eliminate the confusion.

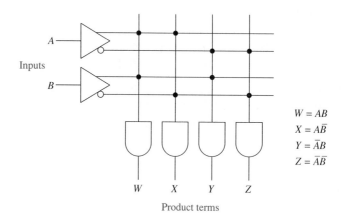

Figure 16–17 One-line convention for PLDs.

Each intersection of the straight lines has a **fusible link.** The programmer selects which fuses are to be left intact and which fuses will be blown. A dot signifies the fuses that are left intact. All other fuses are blown, thus breaking their connection. Figure 16–17 shows four *product terms*: W, X, Y, and Z.

To form an SOP expression, we need to feed the product terms into an OR gate. This difference distinguishes PALs from PLAs. PALs have *fixed* (hard-wired) OR gates, as shown in Figure 16–18, whereas PLAs have *programmable* OR gates, as shown in Figure 16–19.

Improvements in standard PALs and PLAs have led to the introduction of *sequential logic* (*D* and *S-R* flip-flops) within the IC. Field-Programmable Logic Sequencers **(FPLSs)** and sequential PALs provide the additional capability to allow the user to program *sequential operations* such as those required for counter and shift register operations. Electrically Erasable PLDs (EEPLDs) are also available. They allow the user to erase a previous design and program a new one, thus saving the cost of buying a new PLD.

Programmable arrays are sometimes called *semicustom logic.* They are provided as a standard IC part with programmable links that allow users to develop their own custom parts. These programmable arrays bridge the gap between random combinational logic gates and expensive manufacturer-designed, mask-type custom ICs. Programmable arrays reduce the chip count on a typical PC board and allow a design engineer to create a desired logic circuit from a family of blank programmable ICs, instead of stocking a full line of TTL and CMOS chips.

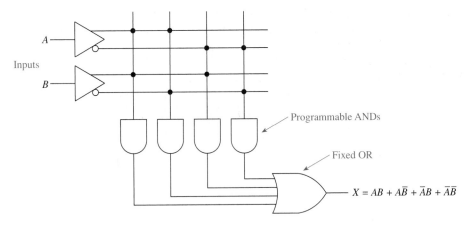

Figure 16–18 PAL architecture.

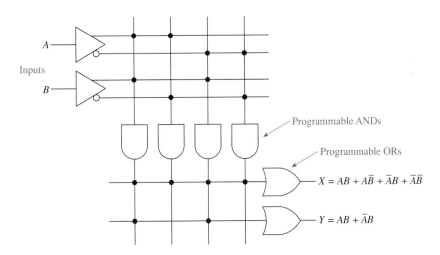

Figure 16–19 PLA (FPLA) architecture.

FPLA Operation

Now let's look at an actual FPLA, the PLS100 manufactured by Signetics Corporation. The PLS100 is configured as a $16 \times 48 \times 8$ array and implements SOP expressions that have a maximum of 16 inputs, 48 product terms, and eight sum outputs. The logic diagram for the PLS100 is given in Figure 16–20. In the figure the variables to be input to the circuit are connected to I_0 to I_{15}, which provide both true and complemented levels. Each intersection in the matrix is actually a fused connection that is left intact or blown by the user. There are 48 AND gates, providing for 48 product terms in an SOP expression. Each AND gate actually has 32 inputs intersecting its input line. The actual number of inputs to an AND gate is determined by the number of fuses that are left intact at each intersection.

The AND-OR matrix in the lower part of Figure 16–20 provides each of the eight OR gates with 48 product-term inputs. The OR gate inputs are programmable to determine which product terms (AND gates) are to be ORed together. An active-HIGH, active-LOW output function is provided by the programmable X_0 to X_7 fusible-

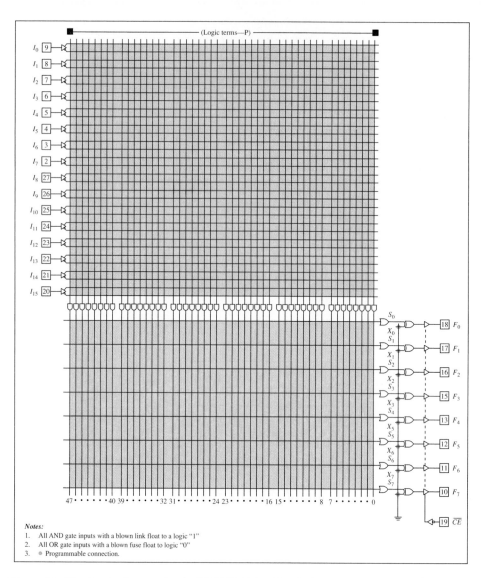

Figure 16–20 Logic diagram for the PLS100 ($16 \times 48 \times 8$) FPLA. (Courtesy of Signetics Corporation)

link connections to the exclusive-OR gates. If the X_0 to X_7 fuses are left intact, the input to the Ex-OR is 0, providing true output. Any fuse blown at X_0 to X_7 places a 1 at the input to that Ex-OR, providing the complement output.

To summarize, all matrix intersections are initially intact. The AND input matrix provides 32 true and complemented inputs to 48 product-term AND gates. The AND-OR matrix allows the user to program the connections of any or all of the 48 AND gates to any of the eight output OR gates. The eight outputs can be programmed as either true or complement. The PLS100 also has an active-LOW Chip Enable $(\overline{CE})$ to provide three-state output capability.

The chip is programmed by blowing all the intersecting fusible links except those that must be left intact to complete the required functions. A special programming table is filled out first, and the data from it are entered into a special PROM programming system that interprets the data and blows the appropriate fuses. Free software is provided by the IC manufacturer to aid in the construction of the data table and to execute the data translation (interpretation) and blowing of fuses.

CHAPTER 16 I SEMICONDUCTOR MEMORY AND PROGRAMMABLE ARRAYS

FPLA Programming*

A data programming table for a PLS100 FPLA set up to implement two SOP equations, $X_0 = AB + \overline{C}D + B\overline{D}$ and $\overline{X}_1 = \overline{A}B + \overline{C}D + EFG$, is shown in Figure 16–21.

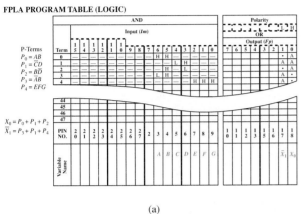

FPLA PROGRAM TABLE (LOGIC)

Step 1
Select which input pins I_0–I_{15} will correspond to the input variables. In this case A-G are the input variable names I_4 through I_0 were selected to accept inputs A-G respectively.

Step 2
Transfer the Boolean Terms to the FLPA Program Table. This is done simply by defining each term and entering it on the Program Table.

e.g., $P_0 = AB$

The P-term translates to the Program Table by selecting A = I_6 = H and B = I_5 = H and entering the information in the appropriate column.

$P_1 = \overline{C}D$

This term is defined by selecting C = I_4 = L and D = I_3 = H, and entering the data into the Program Table. Continue this operation until all P-terms are entered into the Program Table.

Step 3
Select which ouput pins correspond to each output function. In this case F_0 = Pin 18 = X_0 and F_1 = Pin 17 = X_1.

Step 4
Select the Output Active Level desired for each Output Function. For X_0 the active level is high for a positive logic expression of this equation. Therefore it is only necessary to place an (H) in the Active Level box above Output Function. 0, (F_0). Conversely, X_1 can be expressed as $\overline{X}_1$ by placing an (L) in the Active Level box above Output Function 1. (F_1).

Step 5
Step the P-terms you wish to make active for each Output Function. In this case $X_0 = P_0 + P_1 + P_2$, so an A has been placed in the intersection box for P_0 and X_0, P_1 and X_0 and P_2 and X_0.
Terms which are not active for a given output are made inactive by placing a (•) in the box under that P-term. Leave all unused P-terms unprogrammed.
Continue this operation until all outputs have been defined in the Program Table.

Step 6
Enter the data into a Signetics approved programmer. The input format is identical to the Signetics Program Table. You specify the P-Terms, Output Active Level, and which P-Terms are active for each output exactly the way it appears on the Program Table.

(a)   (b)

Figure 16–21 (a) Completed PLS100 FPLA program table; (b) steps for completing the table. (Courtesy of Signetics Corporation)

APPLICATION 16–3
FPLA Design and Programming

Use a PLS100 to function as an active-LOW output BCD-to-seven-segment decoder. Show the connection points that are to be left intact by placing a dot at the correct intersection points in the FPLA logic diagram of Figure 16–22(a). Also, fill in the program table in Figure 16–22(b) with the data to be entered in the FPLA programmer computer.

Solution: First, complete the table in Figure 16–23, showing the input/output characteristics of a BCD-to-seven-segment decoder. Figure 16–23 illustrates the input conditions that make each segment LOW. For example, you can see that the $\overline{a}$ segment is LOW for the numbers 0, 2, 3, 5, 7, 8, 9 ($\overline{A}_0\overline{A}_1\overline{A}_2\overline{A}_3$, $\overline{A}_0A_1\overline{A}_2\overline{A}_3$, $A_0A_1\overline{A}_2\overline{A}_3$, and so on). Figure 16–22(a) shows a dot at each array intersection that is to be *left intact*. In Figure 16–22(a), a four-variable product term is set up for each of the 10 decimal digits and is shown by the connection dots in the upper matrix. The SOP expression for each of the seven segments is then set up by connecting the appropriate product terms to each OR gate. For example, the S_0 OR gate ($\overline{a}$ segment) is connected to seven product term AND gates (0, 2, 3, 5, 7, 8, 9). The S_4 OR gate ($\overline{e}$ segment) is connected to four product term AND gates (0, 2, 6, 8), making the equation for a LOW at $\overline{e}$ equal to $\overline{A}_0\overline{A}_1\overline{A}_2\overline{A}_3 + \overline{A}_0A_1\overline{A}_2\overline{A}_3 + \overline{A}_0A_1A_2\overline{A}_3 + \overline{A}_0\overline{A}_1A_2A_3$.

---

*The text and application that follow discuss the step-by-step procedure for completing a logic diagram and programming table. Although this procedure is important to understand, it is also important to understand how to use the software provided by PLD manufacturers to complete a practical PLD design. Appendix E (PLD Development Software) explains several applications involving software development and programming of PLD systems using manufacturer's software.

The outputs are all made active-LOW by blowing the fuses at X_0 to X_6 (no connection dot shown). A blown link at the input to the Ex-OR gates floats to a logic 1, causing its other input to be complemented.

The final solution to this application is to fill in data in Figure 16–22(b) as shown. [Notice the similarities between Figures 16–22(b) and 16–23.] The entries in the data table in Figure 16–22(b) are then used as input records to the software programs supplied by the IC manufacturer to execute the programming of the FPLA.

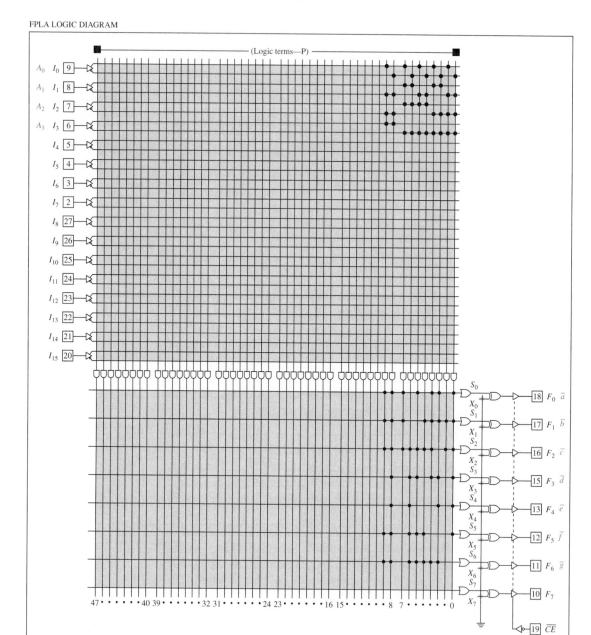

FPLA LOGIC DIAGRAM

Notes:
1. All AND gate inputs with a blown link float to a logic "1"
2. All OR gate inputs with a blown fuse float to logic "0"
3. ● Programmable connection.

(a)

Figure 16–22   (a) The PLS100 FPLA logic diagram showing intersections left intact to form a BCD-to-seven-segment decoder;

CHAPTER 16 I SEMICONDUCTOR MEMORY AND PROGRAMMABLE ARRAYS

FPLA PROGRAM TABLE

PROGRAM TABLE ENTRIES (legend)

OUTPUT ACTIVE LEVEL	
Active High	H
Active Low	L

NOTES — 1 Polarity programmed once only · 2 Enter (H) for all unused outputs

OUTPUT FUNCTION	
Prod. Term Present in Fp	A
Prod. Term Not Present in Fp	. (period)

NOTES — 1 Entries independent of output polarity · 2 Enter (A) for unused outputs of used P-terms

INPUT VARIABLE	
Im H	H
Im L	L
Don't care	— (dash)

NOTE — Enter (—) for unused inputs of used P terms

Customer fields: CUSTOMER NAME ___ · PURCHASE ORDER # ___ · SIGNETICS DEVICE # ___ · CUSTOMER SYMBOLIZED PART # ___ · TOTAL NUMBER OF PARTS ___ · PROGRAM TABLE # ___ · CF (XXXX) ___ · REV ___ · DATE ___

Main program table (AND / INPUT (Im) columns 15–0, POLARITY / OR / OUTPUT (Fp) columns 7–0). Polarity row: L L L L L L L L

TERM	I15	I14	I13	I12	I11	I10	I9	I8	I7	I6	I5	I4	I3	I2	I1	I0	O7	O6	O5	O4	O3	O2	O1	O0
0	—	—	—	—	—	—	—	—	—	—	—	—	L	L	L	L	A	•	A	A	A	A	A	A
1	—	—	—	—	—	—	—	—	—	—	—	—	L	L	L	H	A	•	•	•	•	A	A	•
2	—	—	—	—	—	—	—	—	—	—	—	—	L	L	H	L	A	A	•	A	A	•	A	A
3	—	—	—	—	—	—	—	—	—	—	—	—	L	L	H	H	A	A	•	•	A	A	A	A
4	—	—	—	—	—	—	—	—	—	—	—	—	L	H	L	L	A	A	A	•	•	A	A	•
5	—	—	—	—	—	—	—	—	—	—	—	—	L	H	L	H	A	A	A	•	A	A	•	A
6	—	—	—	—	—	—	—	—	—	—	—	—	L	H	H	L	A	A	A	A	A	A	•	•
7	—	—	—	—	—	—	—	—	—	—	—	—	L	H	H	H	A	•	•	•	•	A	A	A
8	—	—	—	—	—	—	—	—	—	—	—	—	H	L	L	L	A	A	A	A	A	A	A	A
9	—	—	—	—	—	—	—	—	—	—	—	—	H	L	L	H	A	A	A	•	•	A	A	A
10																								
11																								
12																								
13																								
14																								
15																								
16																								
17																								
18																								
19																								
20																								
21																								
22																								
23																								
24																								
25																								
26																								
27																								
28																								
29																								
30																								
31																								
32																								
33																								
34																								
35																								
36																								
37																								
38																								
39																								
40																								
41																								
42																								
43																								
44																								
45																								
46																								
47																								
PIN NO.	20	21	22	23	24	25	26	27	2	3	4	5	6	7	8	9	10	11	12	13	15	16	17	18
VARIABLE NAME													A_3	A_2	A_1	A_0		$\bar g$	$\bar f$	$\bar e$	$\bar d$	$\bar c$	$\bar b$	$\bar a$

(b)

Figure 16–22 *(Continued)* (b) the PLS100 FPLA program table solution for Application 16–3.

Helpful Hint

It is seldom necessary for designers to fill in a programming table like this now that the software to solve PLD designs is so widespread; however, it is an excellent illustration of the process performed by the software.

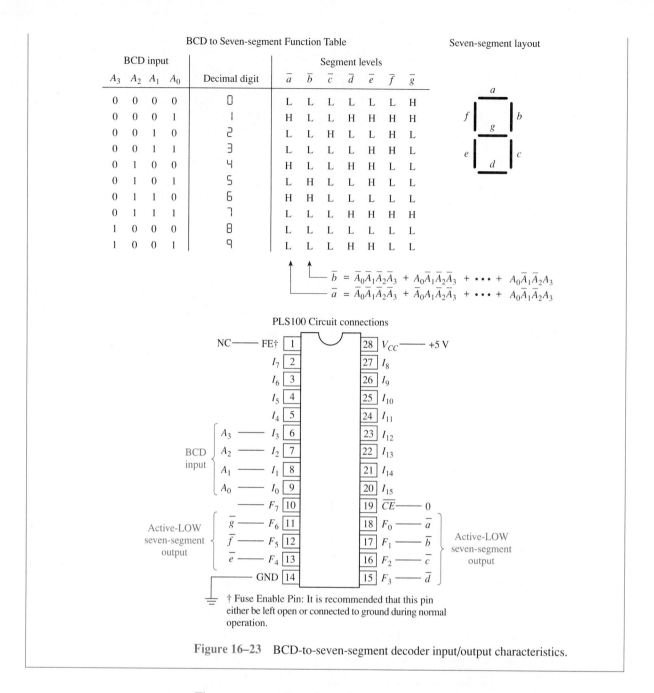

Figure 16–23 BCD-to-seven-segment decoder input/output characteristics.

There are several configurations of programmable logic devices on the market today and the list is continuously expanding. PLD applications are growing rapidly and will be used to solve many combinational logic applications in the future. Table 16–4 gives a small sample of the PLD devices available today.

Table 16–4 Sample of Programmable Logic Devices

Part No.	Organization	Description
PLS100	$16 \times 48 \times 8$	Programmable logic array: 16 input variables, 48 product terms, 8 programmable-OR outputs
PLS105	$16 \times 48 \times 8$	PLA-type logic sequencer: field-programmable replacement for sequential control logic

(continued)

PAL16L8[a]	16 × 8	Programmable array logic: 16 inputs, 8 product-term active-LOW outputs
PAL16R8[a]	16 × 8	Programmable array logic: 16 inputs, 8 registered (D flip-flop) outputs
PAL18V8	18 × 8	V (for versatile) PAL [also known as Generic Array Logic (**GAL®**)[b]]: PAL-type replacement for 10 × 8 up to 18 × 8 PALs; UV-erasable with registered outputs
PALCE22V10	22 × 10	GAL-type device like the 18V8 with higher complexity and electrically erasable capability

[a]The PAL16L8 and PAL16R8 are utilized in Appendix E to demonstrate the use of PLD programming software.
[b]GAL® is a registered trademark of Lattice Semiconductor Corp.

Review Questions

16–19. Using the one-line convention for PLDs, a dot signifies fuses that are _____ (blown, left intact).

16–20. A single PLD can be used to implement several _____ (POS, SOP) expressions.

16–21. An advantage that basic PLA architecture has over basic PAL architecture is that PLA has _____ (fixed, programmable) OR gates at its outputs.

16–22. The PLS100 PLA is capable of implementing up to _____ unique SOP expressions, each expression having up to _____ product terms. The number of input variables in each product term can be as high as _____.

Summary

In this chapter we have learned that

1. A simple 16-byte memory circuit can be constructed from 16 octal D flip-flops and a decoder. This circuit would have 16 memory locations (addresses) selectable by the decoder, with 1 byte (8 bits) of data at each location.

2. Static RAM (random-access memory) ICs are also called read/write memory. They are used for the temporary storage of data and program instructions in microprocessor-based systems.

3. A typical RAM IC is the 2114A. It is organized as 1K × 4, which means that it has 1K locations, with 4 bits of data at each location. (1K is actually an abbreviation for 1024.) An example of a higher-density RAM IC is the 6206, which is organized as 32K × 8.

4. Dynamic RAMs are less expensive per bit and have a much higher density than static RAMs. Their basic storage element is an internal capacitor at each memory cell. External circuitry is required to refresh the charge on all capacitors every 2 ms or less.

5. Dynamic RAMs generally multiplex their address bus. This means that the high-order address bits share the same pins as the low-order address bits. They are demultiplexed by the RAS and CAS (Row Address Strobe and Column Address Strobe) control signals.

6. Read-only memory (ROM) is used to store data on a permanent basis. It is nonvolatile, which means that it does not lose its memory contents when power is removed.

7. The three most common ROMs are (1) the mask ROM, which is programmed once by a masking process by the manufacturer; (2) the fusible-link programmable ROM (PROM), which is programmed once by the user; and (3) the erasable-programmable ROM (EPROM), which is programmable and erasable by the user.

8. Memory expansion in microprocessor systems is accomplished by using octal or hexadecimal decoders as address decoders to select the appropriate memory IC.

9. Complex combinational and sequential logic circuits can be implemented in a single IC called a Programmable Logic Device (PLD).

Glossary

Address Decoding: A scheme used to locate and enable the correct IC in a system with several addressable ICs.

Buffer: An IC placed between two other ICs to boost the load-handling capability of the source IC and to provide electrical isolation.

Bus Contention: Bus contention arises when two or more devices are outputting to a common bus at the same time.

Byte: A group of 8 bits.

CAS: Column address strobe. An active-LOW signal provided when the address lines contain a valid column address.

Delay Line: An integrated circuit that has a single pulse input and provides a sequence of true and complemented output pulses, with each output delayed from the preceding one by some predetermined time period.

Dynamic: A term used to describe a class of semiconductor memory that uses the charge on an internal capacitor as its basic storage element.

EEPROM: Electrically erasable, programmable read-only memory.

EPROM: Erasable, programmable read-only memory.

FPLA (or PLA): Field-programmable logic array. Programmable arrays of AOI logic.

FPLS: Field-programmable logic sequencer. Programmable arrays of AOI and sequential logic.

Fusible Link: Used in programmable ICs to determine the logic level at that particular location. Initially, all fuses are intact. Programming the IC either blows the fuse to change the logic state or leaves it intact.

GAL: Generic array logic. A versatile PLD capable of substituting for several different PAL-type devices.

HAL: Hard array logic. Mask programmable arrays of AOI and sequential logic.

LCD: Liquid-crystal display. A multisegmented display similar to LED displays except that it uses liquid-crystal technology instead of light-emitting diodes.

Look-Up Table: A table of values that is sometimes programmed into an IC to provide a translation between two quantities.

Magnetic Memory: A storage medium such as tape or disk that holds a magnetic image of large amounts of binary data.

Mask: A material covering the silicon of a masked ROM during the fabrication process. It determines the permanent logic state to be formed at each memory location.

Memory Address: The location of the stored data to be accessed.

Memory Cell: The smallest division of a memory circuit or IC. It contains a single bit of data (1 or 0).

Memory Contents: The binary data quantity stored at a particular memory address.

PAL: Programmable array logic. Programmable arrays of AOI and sequential logic.

Programmable Array: A user-programmable array of AND-OR-INVERT logic on a single IC, capable of replacing the combinational logic requirements of several SSI and MSI ICs.

PROM: Programmable read-only memory.

RAM: Random-access memory (read/write memory).

RAS: Row address strobe. An active-LOW signal provided when the address lines contain a valid row address.

ROM: Read-only memory.

Semiconductor Memory: Digital integrated circuits used for the storage of large amounts of binary data. The binary data at each memory cell are stored as the state of a flip-flop or charge on a capacitor.

Static: A term used to describe a class of semiconductor memory that uses the state on an internal flip-flop as its basic storage element.

Problems

Section 16–1

16–1.

(a) In general, which type of memory technology is faster, bipolar or MOS?

(b) Which is more dense, bipolar or MOS?

16–2. Describe the difference between the columns labeled "address" and "data" in Figure 16–1.

16–3. In Figure 16–2, all $\overline{OE}$s are grounded. Why isn't there a problem with bus contention?

T   **16–4.** Assume that the memory system in Figure 16–3 is loaded with the output of a hex counter from 00H to 0FH (H = Hex). Connecting the circuit of Figure 16–4 to Figure 16–3 allows you to test the memory system. When you read the data from the memory, you find that addresses 0000 to 0111 have the hex numbers 00H to 07H as they are supposed to, but addresses 1000 to 1111 have the same data (00H to 07H). What do you suppose is wrong, and how would you troubleshoot the system?

D   **16–5.** Design and sketch an 8-byte memory system similar to Figure 16–2, using eight 74LS374s and one 74LS138.

Section 16–2

16–6. Briefly describe the difference between static and dynamic RAMs. What are the advantages and disadvantages of each?

16–7. Use the block diagram for the 2147H in Figure 16–5 to determine what state $\overline{CS}$ and $\overline{WE}$ must be in to enable the three-state buffer connected to D_{in}.

16–8. What is the level of D_{out} on a 2147H when $\overline{CE}$ and $\overline{WE}$ are both LOW?

16–9. How many address lines are required to select a specific memory location within a RAM having:

(a) 1024 locations?   **(c)** 8192 locations?

(b) 4096 locations?

16–10. How many memory *locations* do the following RAM configurations have?

(a) 2048 × 1   **(d)** 1024 × 4

(b) 2K × 4   **(e)** 4K × 8

(c) 8192 × 8   **(f)** 16K × 1

16–11. What is the total number of *bits* that can be stored in the following RAM configurations?

(a) 1K × 8   **(c)** 8K × 8

(b) 4K × 4   **(d)** 16K × 1

D   **16–12.** Design and sketch a 1K × 8 RAM memory system using two 2148Hs.

T   **16–13.** When troubleshooting the memory system in Figure 16–7, you keep reading incorrect values at D_0 to D_7. Using a logic analyzer, you observe the waveforms of Figure P16–13 at A_0 to A_{11} and $\overline{CE}$. What is wrong and how would you correct it?

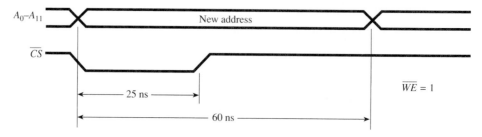

Figure P16–13

CHAPTER 16 I SEMICONDUCTOR MEMORY AND PROGRAMMABLE ARRAYS

D   **16–14.** Use Table 16–1 to determine how many 6207 RAM ICs it would take to design a 1-megabyte memory system for a computer that uses 8-bit data storage.

Section 16–3

16–15. Which lines are multiplexed on dynamic RAMs, and why?

16–16. What is the purpose of $\overline{RAS}$ and $\overline{CAS}$ on dynamic RAMs?

C   **16–17.**

(a) Draw the timing diagrams for a Read cycle and a Write cycle of a dynamic RAM similar to Figure 16–8(b) and (c). Assume that $\overline{CAS}$ is delayed from $\overline{RAS}$ by 100 ns. Also assume that $t_{CAC} = 120$ ns (max), $t_{RAC} = 180$ ns (max), $t_{DS} = 40$ ns (min), and $t_{DH} = 30$ ns (min).

(b) How long after the falling edge of $\overline{RAS}$ will the *data out* be valid?

(c) How soon after the falling edge of $\overline{RAS}$ must the *data in* be set up?

16–18. How often does a 2118 dynamic RAM have to be *refreshed,* and why?

16–19. What functions does the 3242 dynamic RAM controller take care of?

Section 16–4

16–20. Are the following memory ICs volatile or nonvolatile?

(a) Mask ROM   **(c)** Dynamic RAM

(b) Static RAM   **(d)** EPROM

16–21. Which EPROM is electrically erasable, the 2716 or the 2816?

Section 16–5

16–22. Which EPROM and which EPROM address are accessed when the microprocessor of Figure 16–13 issues a read command for the following addresses: (a) READ 1020H; (b) READ 0ABCH; (c) READ 7001H; (d) READ 3FFFH?

D   **16–23.** Redesign the connections to the 74LS138 in Figure 16–13 so that the four EPROMs are accessed at addresses 8000H to BFFFH.

T   **16–24.** When testing the EPROM memory system of Figure 16–13, the microprocessor reads valid data from all EPROMs except EPROM2(1000H–1FFFH). What are two probable causes?

T   **16–25.** When the microprocessor of Figure 16–13 reads data from addresses 8000H to 8FFFH, it finds the same data as that at 0000H to 0FFFH. What is the problem?

D   **16–26.** In Figure 16–13, should the microprocessor software be designed to read data from the EPROMs when it issues the HIGH-to-LOW or LOW-to-HIGH edge on $\overline{RD}$? Why?

C D   **16–27.** Design and sketch an address decoding scheme similar to Figure 16–13 for an 8K × 8 EPROM memory system using 2716 EPROMs. (The 2716 is a 2K × 8 EPROM.)

16–28. What single decoder chip could be used in Figure 16–14 in place of the two 74LS138s?

C D **16–29.** Design a PROM IC to act like a 3-bit *controlled inverter*. Use a 16×4 PROM similar to that in Figure 16–15. When A_3 is HIGH, the input at A_0 to A_2 is to be inverted; otherwise, it is not. Build a truth table showing the 16 possible inputs and the resultant output at Q_0 to Q_2 (Q_3 is not used).

16–30. If at 65°C the output of the ADC in Figure 16–16 is 0111 1010 (7AH), determine what location 7AH in the EPROM should contain.

Section 16–6

16–31. Describe how a PAL or FPLA is different from a AOI gate such as the 74LS54.

16–32. Draw a 2×4 PAL structure similar to Figure 16–18. Place dots at the appropriate intersections to implement the following SOP expression: $X = \overline{A}\,\overline{B} + AB + \overline{A}B$.

16–33. Which of the following Boolean equations could *not* be implemented with a PLS100?

(a) $X = AB + \overline{B}\,\overline{C}D + CD + A\overline{B}\,\overline{C}\,\overline{D}$

(b) $Y = ABCDE + \overline{A} + CDE + \overline{B} + D\overline{E} + AC\overline{D}$

(c) $Z = BCDEFGHJKLM + \overline{A}\,\overline{B}\,\overline{C}$

C **16–34.** Write the Boolean equation for the g segment (F_6 output) in the PLS100 application of Figure 16–22(a) by determining the program connections indicated by the dots.

Schematic Interpretation Problems

See Appendix G for the schematic diagrams.

S **16–35.** The 62256 (U10) IC in the 4096/4196 schematic is a MOS static RAM. By looking at the number of address lines and data lines, determine the size and configuration of the RAM.

S **16–36.** Repeat Problem 16–35 for U6 of the HC11D0 schematic. Looking at the connections to the address lines, determine how much of the RAM is actually accessible.

S C **16–37.** The HC11D0 schematic uses two 27C64 EPROMS.

(a) What are their size and configuration?

(b) What are the labels of the control signals used to determine which EPROM is selected?

(c) Place a jumper from pin 2 to pin 3 of jumper J1 (grid location D-6). Determine the range of addresses that make SMN_SL active (active-LOW).

(d) Determine the range of addresses that make MON_SL active (active-LOW).

Answers to Review Questions

16–1. Address

16–2. Data

16–3. By inputting the 4-bit address to the 74LS154, which outputs a LOW pulse on one of the output lines when $\overline{\text{WRITE}}$ is pulsed LOW

16–4. It shows when any or all of the lines are allowed to change digital levels.

16–5. To avoid a bus conflict

16–6. 1024, 4

16–7. LOW, HIGH

16–8. $A_0\,A_5$, $A_6\,A_{11}$

16–9. 35 ns max

16–10. To keep the IC pin count to a minimum. They are demultiplexed by using control signals $\overline{\text{RAS}}$ and $\overline{\text{CAS}}$.

16–11. The charge on the internal capacitors in the RAM is replenished.

16–12. It means that when the power is removed, the memory contents are lost.

16–13. After the EPROM program has been thoroughly tested

16–14. 120, 330 (450 total)

16–15. When another device has to use the bus

16–16. (a) third EPROM, address 002H (b) first EPROM, address AF7H.

16–17. 9000H–9FFFH

16–18. 0010 0101

16–19. Left intact

16–20. SOP

16–21. Programmable

16–22. 8, 48, 16

17 Microprocessor Fundamentals

OUTLINE

17–1 Introduction to System Components and Buses
17–2 Software Control of Microprocessor Systems
17–3 Internal Architecture of a Microprocessor
17–4 Instruction Execution Within a Microprocessor
17–5 Hardware Requirements for Basic I/O Programming
17–6 Writing Assembly Language and Machine Language Programs
17–7 Survey of Microprocessors and Manufacturers

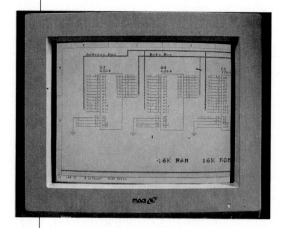

Objectives

Upon completion of this chapter you should be able to

- Describe the benefits that microprocessor design has over hard-wired IC logic design.

- Discuss the functional blocks of a microprocessor-based system having basic input/output capability.

- Describe the function of the address, data, and control buses.

- Discuss the timing sequence on the three buses required to perform a simple input/output operation.

- Explain the role of software program instructions in a microprocessor-based system.

- Understand the software program used to read data from an input port and write it to an output port.

- Discuss the basic function of each of the internal blocks of the 8085A microprocessor.

- Follow the flow of data as they pass through the internal parts of the 8085A microprocessor.

- Make comparisons between assembly language, machine language, and high-level languages.

- Discuss the fundamental circuitry and timing sequence for external microprocessor I/O.

Introduction

The design applications studied in the previous chapters have all been based on combinational logic gates and sequential logic ICs. One example is a traffic light controller that goes through the sequence green–yellow–red. To implement the circuit using combinational and sequential logic, we would use some counter ICs for the timing, a shift register for sequencing the lights, and a *D* flip-flop if we want to interrupt the sequence with a pedestrian crosswalk push button. A complete design solution is easily within the realm of SSI and MSI ICs.

On the other hand, think about the complexity of electronic control of a modern automobile. There are several analog quantities to monitor, such as engine speed, manifold pressure, and coolant temperature; and there are several digital control functions to perform, such as spark plug timing, fuel mixture control, and radiator circulation control. The operation is further complicated by the calculations and decisions that have to be made on a continuing basis. This is definitely an application for a *microprocessor-based system.*

A system designer should consider a microprocessor-based solution whenever an application involves making calculations, making decisions based on external stimulus, and maintaining memory of past events. A microprocessor offers several advantages over the hard-wired SSI/MSI IC approach. First, the microprocessor itself is a general-purpose device. It takes on a unique personality by the software program instructions given by the designer. If you want it to count, you tell it to do so, with software. If you want to shift its output level left, there's an instruction for that. And if you want to add a new quantity to a previous one, there's another instruction for that. Its capacity to perform arithmetic, make comparisons, and update memory makes it a very powerful digital problem solver. Making changes to an application can usually be done by changing a few program instructions, unlike the hard-wired system that may have to be totally redesigned and reconstructed.

New microprocessors are introduced every year to fill the needs of the design engineer. However, the theory behind microprocessor technology remains basically the same. It is a general-purpose digital device that is driven by software instructions and communicates with several external *support chips* to perform the necessary input/output of a specific task. Once you have a general understanding of one of the earlier microprocessors that came on the market, such as the Intel 8080/8085, the Motorola 6800, or the Zilog Z80, it is an easy task to teach yourself the necessary information to upgrade to the new microprocessors as they are introduced. Typically, when a new microprocessor is introduced, it will have a few new software instructions available and will have some of the I/O features, previously handled by external support chips, integrated right into the microprocessor chip. Learning the basics on these new microprocessor upgrades is more difficult, however, because some of their advanced features tend to hide the actual operation of the microprocessor and may hinder your complete understanding of the system.

17–1 Introduction to System Components and Buses

Figure 17–1 shows a **microprocessor** with the necessary **support circuitry** to perform basic input and output functions. We will use this figure to illustrate how the microprocessor acts like a general-purpose device, driven by software, to perform a specific task related to the input data switches and output data LEDs. First, let's discuss the components of the system.

Microprocessor

The heart of the system is an 8-bit microprocessor. It could be any of the popular 8-bit microprocessors such as the Intel 8085, the Motorola 6800, or the Zilog Z80. They are called 8-bit microprocessors because external and internal data movement is performed on 8 bits at a time. It will read *program instructions* from memory and execute those instructions that drive the three *external buses* with the proper levels and timing to make the connected devices perform specific operations. The buses are simply groups of conductors that are routed throughout the system and tapped into by various devices (or ICs) that need to share the information that is traveling on them.

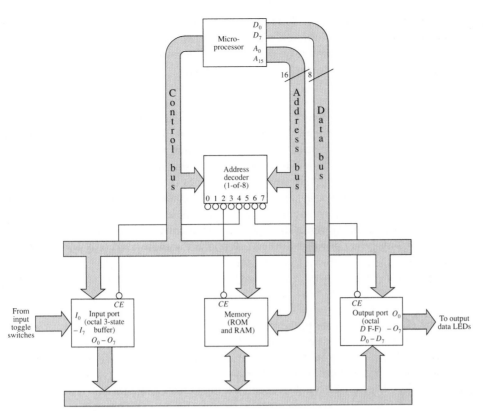

Team Discussion

Discuss the flow of data as the microprocessor reads program instructions from memory, which then tell it to read the input data switches and transfer what it has read out to the LEDs.

Team Discussion

Which port could be connected to a D/A converter? To an A/D converter?

Figure 17–1 Example of a microprocessor-based system used for simple input/output operations.

Address Bus

The **address bus** is 16 bits wide and is generated by the microprocessor to select a particular location or IC to be active. In the case of a selected memory IC, the low-order bits on the address bus select a particular location within the IC (see Section 16–5). Because the address bus is 16 bits wide, it can actually specify 65,536 (2^{16}) different addresses. The input port is one address, the output port is one address, and the memory in a system of this size may be 4K (4096) addresses. This leaves about 60K addresses available for future expansion.

Data Bus

Once the address bus is set up with the particular address that the microprocessor wants to access, the microprocessor then sends or receives 8 bits of data to or from that address via the **bidirectional** (two-way) **data bus.**

Control Bus

The **control bus** is of varying width, depending on the microprocessor being used. It carries control signals that are tapped into by the other ICs to tell what type of operation is being performed. From these signals, the ICs can tell if the operation is a read, a write, an I/O, a memory access, or some other operation.

Address Decoder

The address decoder is usually an octal decoder like the 74LS138 studied in Chapters 8 and 16. Its function is to provide active-LOW Chip Enables $(\overline{CE})$ to the external ICs based on information it receives from the microprocessor via the control and address buses. Because there are multiple ICs on the data bus, the address decoder ensures that only one IC is active at a time to avoid a bus conflict caused by two ICs writing different data to the same bus.

Memory

There will be at least two memory ICs: ROM or EPROM and a RAM. The ROM will contain the *initialization* instructions, telling the microprocessor what to do when power is first turned on. This includes tasks like reading the keyboard and driving the CRT display. It will also contain several subroutines that can be called by the microprocessor to perform such tasks as time delays or input/output data translation. These instructions, which are permanently stored in ROM, are referred to as the **monitor program** or **operating system.** The RAM part of memory is volatile, meaning that it loses its contents when power is turned off, and is therefore used only for temporary data storage.

Input Port

The input port provides data to the microprocessor via the data bus. In this case, it is an octal buffer with three-stated outputs. The input to the buffer will be provided by some input device like a keyboard or, as in this case, from eight HIGH–LOW toggle switches. The input port will dump its information to the data bus when it receives a Chip Enable $(\overline{CE})$ from the address decoder and a Read command $(\overline{RD})$ from the control bus.

Output Port

The output port provides a way for the microprocessor to talk to the outside world. It could be sending data to an output device like a printer or, as in this case, it could send data to eight LEDs. An octal D flip-flop is used as the interface because, after the microprocessor sends data to it, the flip-flop will latch on to the data, allowing the microprocessor to continue with its other tasks.

To load the D flip-flop, the microprocessor must first set up the data bus with the data to be output. Then it sets up the address of the output port so that the address decoder will issue a LOW $\overline{CE}$ to it. Finally, it issues a pulse on its $\overline{WR}$ (write) line that travels the control bus to the clock input of the D flip-flop. When the D flip-flop receives the clock trigger pulse, it latches onto the data that are on the data bus at that time and drives the LEDs.

Review Questions

17–1. What are the names of the three buses associated with microprocessors?

17–2. How much of the circuitry shown in Figure 17–1 is contained inside a microcontroller IC?

17–3. Why must the data bus be bidirectional?

17–4. The purpose of the address decoder IC in Figure 17–1 is to enable two or more of the external ICs to be active at the same time to speed up processing. True or false?

17–5. The input port in Figure 17–1 must have *three-state* outputs so that its outputs are floating whenever any other IC is writing data to the data bus. True or false?

17–6. What would be the consequence of using an octal buffer for the output port in Figure 17–1 instead of an octal *D* flip-flop?

17–2 Software Control of Microprocessor Systems

The nice thing about microprocessor-based systems is that once you have a working prototype you can put away the soldering iron because all operational changes can then be made with software. The student of electronics has a big advantage when writing microprocessor software because he or she understands the hardware at work as well as the implications that software will have on the hardware. Areas such as address decoding, chip enables, instruction timing, and hardware interfacing become important when programming microprocessors.

As a brief introduction to microprocessor software, let's refer back to Figure 17–1 and learn the statements required to perform some basic input/output operations. To route the data from the input switches to the output LEDs, the data from the input port must first be read into the microprocessor before they can be sent to the output port. The microprocessor has an 8-bit internal register called the **accumulator** that can be used for this purpose.

The software used to drive microprocessor-based systems is called **assembly language.** The Intel 8080/8085 assembly language statement to load the contents of the input port into the accumulator is LDA *addr*. LDA is called a **mnemonic,** an abbreviation of the operation being performed, which in this case is "Load Accumulator." The suffix *addr* will be replaced with a 16-bit address (4 hex digits) specifying the address of the input port.

After the execution of LDA *addr,* the accumulator will contain the digital value that was on the input switches. Now, to write these data to the output port, we use the command STA *addr*. STA is the mnemonic for "Store Accumulator" and *addr* is the 16-bit address where you want the data stored.

Execution of those two statements is all that is necessary to load the value of the switches into the accumulator and then transfer these data to the output LEDs. The microprocessor takes care of the timing on the three buses, and the address decoder takes care of providing chip enables to the appropriate ICs.

If the system is based on Motorola or Zilog technology, the software in this case will be almost the same. Table 17–1 makes a comparison of the three assembly languages.

Table 17–1 Comparison of Input/Output Software on Three Different Microprocessors

Operation	Intel 8080/8085	Motorola 6800	Zilog Z80
Load accumulator with contents of location *addr*	LDA *addr*	LDAA *addr*	LD A, *(addr)*
Store accumulator to location *addr*	STA *addr*	STAA *addr*	LD *(addr)*, A

Helpful Hint

It is beyond the scope of this text to cover an entire software instruction set, but it might be instructive for you to locate a programmer's manual for one of the more popular microprocessors to see an entire list of instructions so that you can get a feel for its power.

17–3 Internal Architecture of a Microprocessor

The design for the Intel 8085A microprocessor was derived from its predecessor, the 8080A. The 8085A is **software** *compatible* with the 8080A, meaning that software programs written for the 8080A can run on the 8085A without modification. The 8085A has a few additional features not available on the 8080A. The 8085A also has a higher level of hardware integration, allowing the designer to develop complete microprocessor-based systems with fewer external support ICs than were required by the 8080A. Studying the internal **architecture** of the 8085A in Figure 17–2 and its pin configuration, Figure 17–3, will give us a better understanding of its operation.

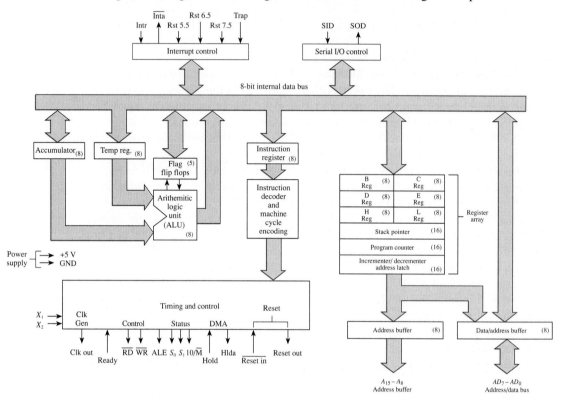

Figure 17–2   The 8085A CPU functional block diagram. (Courtesy of Intel Corporation)

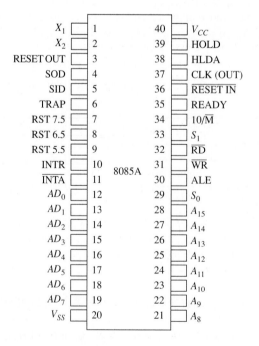

Figure 17–3   The 8085A pin configuration. (Courtesy of Intel Corporation)

CHAPTER 17 | MICROPROCESSOR FUNDAMENTALS

The 8085A is an 8-bit parallel **central processing unit (CPU).** The accumulator discussed in the previous section is connected to an 8-bit *internal data bus.* Six other *general-purpose registers* labeled B, C, D, E, H, and L are also connected to the same bus.

All arithmetic operations take place in the **arithmetic logic unit (ALU).** The accumulator, along with a temporary register, is used as input to all arithmetic operations. The output of the operations is sent to the internal data bus and to five *flag flip-flops* that record the status of the arithmetic operation.

The **instruction register** and **decoder** receive the software instructions from external memory, interpret what is to be done, and then create the necessary timing and control signals required to execute the instruction.

The block diagram also shows **interrupt** *control,* which provides a way for an external digital signal to interrupt a software program while it is executing. This is accomplished by applying the proper digital signal on one of the interrupt inputs: INTR, RSTx.x, or TRAP. *Serial communication* capabilities are provided via the SID and SOD I/O pins (Serial Input Data, Serial Output Data).

The *register array* contains the six general-purpose 8-bit registers and three 16-bit registers. Sixteen-bit registers are required whenever you need to store addresses. The **stack pointer** stores the address of the last entry on the stack. The stack is a data storage area in RAM used by certain microprocessor operations. The **program counter** contains the 16-bit address of the next software instruction to be executed. The third 16-bit register is the *address latch,* which contains the current 16-bit address that is being sent to the address bus.

The six general-purpose 8-bit registers can also be used in pairs (*B–C, D–E, H–L*) to store addresses or 16-bit data.

Review Questions

17–7. The suffix *addr* in the LDA and STA mnemonics is used to specify the address of the accumulator. True or false?

17–8. The LDA command is used by the microprocessor to _____ (read, write) data, and the STA command is used to _____ (read, write) data.

17–9. The 8085A microprocessor has an *internal* accumulator and six *external* registers called B, C, D, E, H, and L. True or false?

17–10. The ALU block inside the 8085A determines the timing and control signals required to execute an instruction. True or false?

17–4   Instruction Execution Within a Microprocessor

Now, referring back to the basic I/O system diagram of Figure 17–1, let's follow the flow of the LDA and STA instructions as they execute in the block diagram of the 8085A. Figure 17–4 shows the 8085A block diagram with numbers indicating the succession of events that occurs when executing the LDA instruction.

Remember, LDA *addr* and STA *addr* are assembly language instructions, stored in an external memory IC, that tell the 8085A CPU what to do. LDA *addr* tells the CPU to load its accumulator with the data value that is at address *addr.* STA *addr* tells the CPU to store (or send) the 8-bit value that is in the accumulator to the output port at address *addr.*

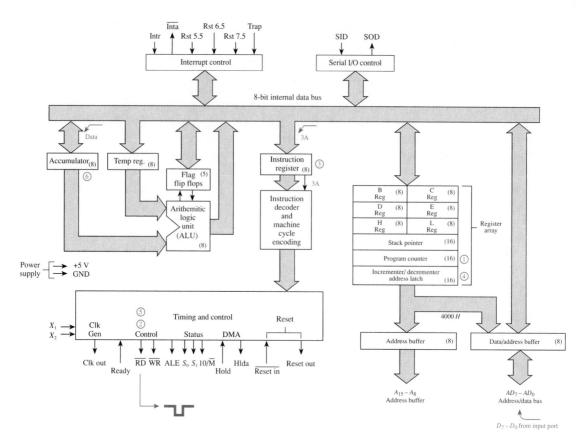

Figure 17–4 Execution of the LDA instruction within the 8085A.

The mnemonics LDA and STA cannot be understood by the CPU as they are; they have to be *assembled,* or converted, into a binary string called **machine code.** Binary or hexadecimal machine code is what is actually read by the CPU and passed to the instruction register and decoder to be executed. The Intel 8085A Users Manual gives the machine code translation for LDA as $3A_{16}$ (or 3AH) and STA as 32H.

Before studying the flow of execution in Figure 17–4, we need to make a few assumptions. Let's assume that the input port is at address 4000H and the output port is at address 6000H. Let's also assume that the machine code program LDA 4000H, STA 6000H is stored in RAM starting at address 2000H.

Load Accumulator

The sequence of execution of LDA 4000H in Figure 17–4 will be as follows:

1. The program counter will put the address 2000H on the address bus.

2. The timing and control unit will issue a LOW pulse on the $\overline{RD}$ line. This pulse will travel the control bus to the RAM in Figure 17–1 and will cause the contents at location 2000H to be put onto the external data bus. RAM (2000H) has the machine code 3AH, which will travel across the internal data bus to the instruction register.

3. The instruction register passes the 3AH to the instruction decoder, which determines that 3AH is the code for LDA and that a 16-bit (2-byte) address must follow. Because the entire instruction is 3 bytes (one for the 3AH and two for the address 4000H), the instruction decoder increments the program counter two more times so that the address latch register can read and store bytes 2 and 3 of the instruction.

4. The address latch and address bus now have 4000H on them, which provides the LOW $\overline{CE}$ for the input port in Figure 17–1.

5. The timing and control unit again issues a LOW pulse on the $\overline{RD}$ line. This pulse will travel the control bus to the input port, causing the data at the input port (4000H) to be put onto the external data bus.

6. That data will travel across the external data bus in Figure 17–1, to the internal data bus in Figure 17–4, to the accumulator, where they are now stored. The instruction is complete.

Store Accumulator

Figure 17–5 shows the flow of execution of the STA 6000H instruction.

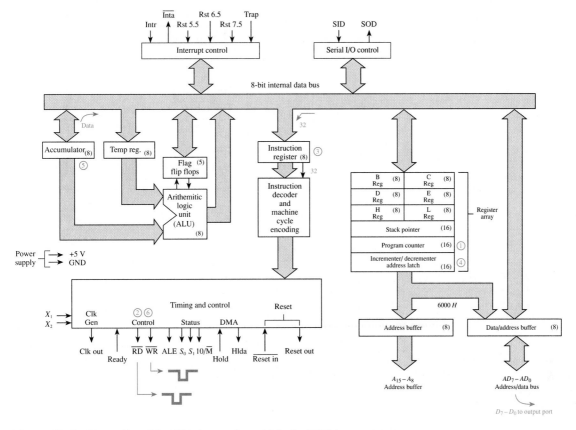

Figure 17–5 Execution of the STA instruction within the 8085A.

1. After the execution of the 3-byte LDA 4000H instruction, the program counter will have 2003H in it. (Instruction LDA 4000H resided in locations 2000H, 2001H, 2002H.)

2. The timing and control unit will issue a LOW pulse on the $\overline{RD}$ line. This will cause the contents of RAM location 2003H to be put onto the external data bus. RAM (2003H) has the machine code 32H, which will travel up the internal data bus to the instruction register.

3. The instruction register passes the 32H to the instruction decoder, which determines that 32H is the code for STA and that a 2-byte address must follow. The program counter gets incremented two more times, reading and storing bytes 2 and 3 of the instruction into the address latch.

4. The address latch and address bus now have 6000H on them, which is the address of the output port in Figure 17–1.

5. The instruction decoder now issues the command to place the contents of the accumulator onto the data bus.

6. The timing and control unit issues a LOW pulse on the $\overline{WR}$ line. Since the $\overline{WR}$ line is used as a clock input to the D flip-flop of Figure 17–1, the data from the data bus will be stored and displayed on the LEDs. (The $\overline{WR}$ line from the microprocessor is part of the control bus in Figure 17–1.)

The complete assembly language and machine code program for the preceding input/output example is given in Table 17–2.

Table 17–2   Assembly Language and Machine Code Listing for the LDA-STA Program

Memory Location	Assembly Language	Machine Code	
2000H	LDA 4000H	3A	Three-byte instruction to load accumulator with contents from address 4000H
2001H		00	
2002H		40	
2003H	STA 6000H	32	Three-byte instruction to store accumulator out to address 6000H
2004H		00	
2005H		60	

Review Questions

17–11. What is the difference between assembly language and machine code?

17–12. Why is the instruction LDA 4000H called a 3-byte instruction?

17–13. When executing the instruction LDA 4000H, the microprocessor fetches the machine code from RAM location 4000H. True or false?

17–14. Which instruction, STA or LDA, issues a pulse on the $\overline{WR}$ line? Why?

17–15. How many bytes of RAM does the program in Table 17–2 occupy?

17–5   Hardware Requirements for Basic I/O Programming

A good way to start out in microprocessor programming is to illustrate program execution by communicating to the outside world. In Section 17–4 we read input switches at memory location 4000H using the LDA instruction and wrote their value to output LEDs at location 6000H using the STA instruction. This was an example of **memory-mapped I/O.** Using this method, the input and output devices were accessed *as if they were memory locations* by specifying their unique 16-bit address (4000H or 6000H).

The other technique used by the 8085A microprocessor for I/O mapping is called *standard I/O* or **I/O-mapped I/O.** *I/O-mapped systems* identify their input and output devices by giving them an 8-bit **port number.** The microprocessor then accesses the I/O ports by using the instructions OUT *port* and IN *port,* where port is 00H to FFH.

Special **hardware** external to the 8085A is required to provide the source for the IN instruction and the destination for the OUT instruction. Figure 17–6 shows a basic hardware configuration, using standard SSI and MSI ICs, that could be built to input data from eight switches and to output data to eight LEDs using I/O-mapped I/O.

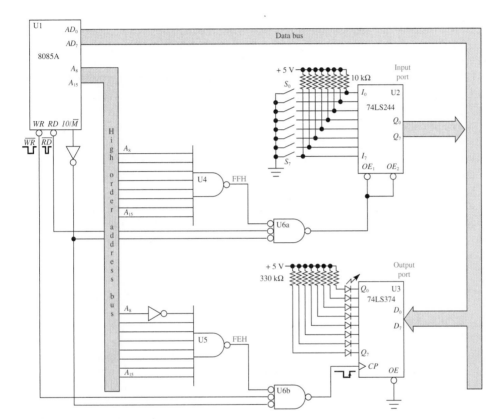

Helpful Hint

Review the operation of the 74LS244 octal buffer and the 74LS374 octal *D* flip-flop before studying this figure.

Figure 17–6 Hardware requirements for the IN FFH and OUT FEH instructions.

Figure 17–6 is set up to decode the input switches as port FFH and the output LEDs as port FEH. The *IO/$\overline{M}$* line from the microprocessor goes HIGH whenever an IN or OUT instruction is being executed (I/O-mapped I/O). All instructions that access memory and memory-mapped devices will cause the *IO/$\overline{M}$* line to go LOW. The $\overline{RD}$ line from the microprocessor will be pulsed LOW when executing the IN instruction, and the $\overline{WR}$ line will be pulsed LOW when executing the OUT instruction.

IN FFH

The 74LS244 is an octal three-state buffer that is set up to pass the binary value of the input switches over to the data bus as soon as $\overline{OE}_1$ and $\overline{OE}_2$ are brought LOW. To get that LOW, U6a, the inverted-input NAND gate (OR gate), must receive three LOWs at its input. We know that the IN instruction will cause the *inverted IO/$\overline{M}$* line to go LOW and the $\overline{RD}$ line to go LOW. The other input is dependent on the output from the right-input NAND gate (U4). Gate U4 *will* output a LOW because the binary value of the port number (1111 1111) used in the IN instruction is put onto the high-order address bus during the execution of the IN FFH instruction.

All conditions are now met; U6a will output a LOW pulse (the same width as the LOW $\overline{RD}$ pulse), which will enable the outputs of U2 to pass to the data bus. After the microprocessor drops the $\overline{RD}$ line LOW, it waits a short time for external devices (U2 in this case) to respond; then it reads the data bus and raises the $\overline{RD}$ line back HIGH. The data from the input switches are now stored in the accumulator.

OUT FEH

The 74LS374 is an octal D flip-flop set up to sink current to illuminate individual LEDs based on the binary value it receives from the data bus. The outputs at Q_0 to Q_7 will latch onto the binary values at D_0 to D_7 at the LOW-to-HIGH edge of C_p. U5 and U6b are set up similarly to U4 and U6a, except U5's output goes LOW when FEH (1111 1110) is input. Therefore, during the execution of OUT FEH, U6b will output a LOW pulse, the same width as the $\overline{WR}$ pulse issued by the microprocessor.

The setup time of the 74LS374 latch is accounted for by the microprocessor timing specifications. The microprocessor issues a HIGH-to-LOW edge at $\overline{WR}$ that makes its way to C_p. At the same time, the microprocessor also sends the value of the accumulator to the data bus. After a time period greater than the setup time for U3, $\overline{WR}$ goes back HIGH, which applies the LOW-to-HIGH trigger edge for U3, latching the data at Q_0 to Q_7.

To summarize, the instruction IN FFH reads the binary value at port FFH into the accumulator. The instruction OUT FEH writes the binary value in the accumulator out to port FEH. Port selection is taken care of by eight-input NAND gates attached to the high-order address bus and by use of the $\overline{RD}$, $\overline{WR}$, and $IO/\overline{M}$ lines.

17–6   Writing Assembly Language and Machine Language Programs

The microprocessor is driven by software instructions to perform specific tasks. The instructions are first written in assembly language using mnemonic abbreviations and then converted to machine language so that they can be interpreted by the microprocessor. The conversion from assembly language to machine language involves translating each mnemonic into the appropriate hexadecimal machine code and storing the codes in specific memory addresses. This can be done by a software package called an **assembler,** provided by the microprocessor manufacturer, or it can be done by the programmer by looking up the codes and memory addresses (called **hand assembly**).

Assembly language is classified as a low-level language because the programmer has to take care of all the most minute details. High-level languages such as Pascal, FORTRAN, and BASIC are much easier to write but are not as memory efficient or as fast as assembly language. All languages, whether Pascal, BASIC, or FORTRAN, get reduced to machine language code before they can be executed by the microprocessor. The conversion from high-level languages to machine code is done by a **compiler.** The compiler makes memory assignments and converts the English-language-type instructions into executable machine code.

On the other hand, assembly language translates *directly* into machine code. This allows the programmer to write the most streamlined, memory-efficient, and fastest programs possible on the specific hardware configuration that is being used.

Assembly language and its corresponding machine code differ from processor to processor. The fundamentals of the different assembly languages are the same, however, and once you have become proficient on one microprocessor, it is easy to pick it up on another.

Let's start off our software training by studying a completed assembly language program and comparing it to the same program written in the **BASIC** computer **language.** BASIC is a high-level language that uses English-language-type commands that are fairly easy to figure out, even by the inexperienced programmer.

Program Definition

Write a program that will function as a down-counter, counting 9 to 0 repeatedly. First draw a **flowchart;** then write the program statements in the BASIC language, assembly language, and machine language.

Solution

The flowchart in Figure 17–7 is used to show the sequence of program execution, including the branching and looping that takes place.

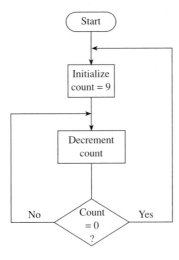

Figure 17–7 Flowchart for Table 17–3.

According to the flowchart, the counter is decremented repeatedly until zero is reached, at which time the counter is reinitialized to 9 and the cycle repeats. The instructions used to implement the program are given in Table 17–3.

Table 17–3 Down-Counter Program in Three Languages

BASIC Language		8085A Assembly Language		8085A Machine Language	
Line	Instruction	Label	Instruction	Address	Contents
10	COUNT = 9	START:	MVI A,09H	2000	3E (opcode)
				2001	09 (data)
20	COUNT = COUNT-1	LOOP:	DCR A	2002	3D (opcode)
30	IF COUNT = 0		JZ START	2003	CA (opcode)
	THEN GO TO 10			2004	00 ⎫ (address)
				2005	20 ⎭
40	GO TO 20		JMP LOOP	2006	C3 (opcode)
				2007	02 ⎫ (address)
				2008	20 ⎭

BASIC

BASIC uses the variable COUNT to hold the counter value. Line 30 checks the count. If COUNT is equal to zero, then the program goes back to the beginning. Otherwise, it goes back to subtract 1 from COUNT and checks COUNT again.

The 8085A version of the program is first written in assembly language and then it is either hand assembled into machine language or it could be computer assembled using a personal computer with an assembler software package.

Assembly Language

Assembly language is written using *mnemonics:* MVI, DCR, JZ, and the like. The term mnemonics is defined as "abbreviations used to assist the memory." The first mnemonic, MVI, stands for "Move Immediate." The instruction MVI A,09H will move the data value 09H into register *A* (register *A* and the accumulator are the same). The next instruction, DCR A, decrements register *A* by 1.

The third instruction, JZ START, is called a *conditional jump.* The condition that it is checking for is the *zero* condition. As the *A* register is decremented, if *A* reaches 0, then a flag bit, called the **zero flag,** gets set (a *set* flag is equal to 1). The instruction JZ START is interpreted as "jump to **statement label** START if the zero flag is set."

If the condition is not met (zero flag not set), then control passes to the next instruction, JMP LOOP, which is an *unconditional jump.* This instruction is interpreted as "jump to label LOOP regardless of any condition flags."

At this point you should see how the assembly language program functions exactly like the BASIC language program.

Machine Language

Machine language is the final step in creating an executable program for the microprocessor. In this step we must determine the actual hexadecimal codes that will be stored in memory to be read by the microprocessor. First, we have to determine what memory locations will be used for our program. This depends on the memory-map assignments made in the system hardware design. We have 64K of addressable memory locations (0000H to FFFFH). We'll make an assumption that the user program area was set up in the hardware design to start at location 2000H. The length of the program memory area depends on the size of the ROM or RAM memory IC being used. A 256×8 RAM memory is usually sufficient for introductory programming assignments and is commonly used on educational microprocessor trainers. The machine language program listed in Table 17–3 fills up 9 bytes of memory (2000H to 2008H).

The first step in the hand assembly is to determine the code for MVI A. This is known as the **opcode** (operation code) and is found in an 8085A Assembly Language Reference Chart. The opcode for MVI A is 3E. The programmer will store the binary equivalent for 3E (0011 1110) into memory location 2000H. Instructions for storing your program into memory are given by the manufacturer of the microprocessor trainer that you are using. If you are using an assembler software package, then the machine code that is generated will usually be saved on a computer disk or used to program an EPROM to be placed in a custom microprocessor hardware design.

The machine language instruction MVI A,09H in Table 17–3 requires 2 bytes to complete. The first byte is the opcode, 3E, which identifies the instruction for the microprocessor. The second byte (called the **operand**) is the data value, 09H, which is to be moved into register *A*.

The second instruction, DCR A, is a 1-byte instruction. It requires just its opcode, 3D, which is found in the reference chart.

The opcode for the JZ instruction is CA. It must be followed by the 16-bit (2-byte) address to jump to if the condition (zero) is met. This makes it a 3-byte instruction. Byte 2 of the instruction (location 2004H) is the low-order byte of the address, and byte 3 is the high-order byte of the address to jump to. (Be careful to always enter addresses as low-order first, then high-order.)

The opcode for JMP is C3 and must also be followed by a 16-bit (2-byte) address specifying the location to jump to. Therefore, this is also a 3-byte instruction where byte 2–byte 3 gives a jump address of 2002H.

17–7   Survey of Microprocessors and Manufacturers

Since its introduction in the early 1970s, the microprocessor has had a huge impact on the electronics industry. The first microprocessors had a 4-bit internal data width. In 1974, Intel introduced the first 8-bit microprocessor, the 8008. Within a year it offered an upgrade, the 8080, which served as a point of comparison for all other manufacturers.

The first challenger to the 8080 was the Motorola 6800. Other IC manufacturers (National Semiconductor, Texas Instruments, Zilog, RCA, and Fairchild) soon introduced their own versions of the microprocessor. The race had begun. Since then, 16- and 32-bit architectures have been developed and are finding their way into most new high-end applications.

Along the way, manufacturers started integrating whole multichip systems with RAM, ROM, and I/O into a single package called a *microcontroller*. Today the microcontroller is the most popular choice for embedded control applications such as those found in automobiles, home entertainment systems, and data acquisition and control systems.

Each microprocessor and microcontroller has its own special niche, but throughout the years the two most important players have been Intel and Motorola. Table 17–4 lists the most popular processors manufactured by these two companies. You can identify the manufacturer by the first two numbers in the part number (68 for Motorola and 80 for Intel). Table 17–5 lists the addresses of several microprocessor manufacturers that will be more than happy to send your school data on their microprocessor lines.

Table 17–4   Popular Intel and Motorola Microprocessors and Microcontrollers

Part No.	Data Bits	Address Bits	Comments
8085	8	16	Upgrade of the 8080
8051	8	16	Microcontroller with on-chip ROM, RAM, and I/O
6809	8	16	Upgrade of the original 6800
68HC11	8	16	Microcontroller with on-chip ROM, RAM, I/O, and A/D Converter
8088	8	20	8-bit downgrade of the 8086
8086	16	20	Made popular by its use in IBM PC-compatible computers
80186	16	20	8086 with on-chip support functions
80286	16	24	8086 upgrade with extended addressing capability; used in IBM AT-compatible computers
8096	16	16	Microcontroller with on-chip ROM, RAM, I/O, and A/D converter
68000	16	23	Made popular by its use in Apple Macintosh II and Unix-based workstations
68010	16	24	Upgrade of the 68000
80386	32	32	32-bit upgrade of the 80286
80486	32	32	Upgrade of the 80386
68020	32	32	32-bit upgrade of the 68010
68030	32	32	Upgrade of the 68020

Table 17–5 Listing of Several Microprocessor Manufacturers

Advanced Micro Devices
Austin, TX
(512) 462-4360

Cirrus Logic
Fremont, CA
(510) 226-2262

Cyrix Corp.
Richardson, TX
(214) 234-8388

Dallas Semiconductor
Dallas, TX
(214) 450-0448

Digital Equipment Corp.
Semiconductor Operations
Hudson, MA
(508) 568-6868

Fujitsu Microelectronics Inc.
IC Division
San Jose, CA
(800) 642-7616

GEC-Plessey Semiconductor
Farmingdale, NY
(516) 293-8686

Harris Semiconductor Corp.
Melbourne, FL
(407) 724-7000

Hewlett-Packard Co.
Cupertino, CA
(800) 752-0900

Hitachi America Ltd.
Semiconductor and IC Division
Brisbane, CA
(800) 285-1601

IBM Microelectronics Inc.
Fishkill, NY
(800) 426-0181

Integrated Device Technology
Santa Clara, CA
(800) 345-7015

Intel Literature Center
Mount Prospect, IL
(800) 468-8118

LSI Logic Corp.
Milpitas, CA
(408) 433-4008

Mitsubishi Electronics
America Inc.
Sunnyvale, CA
(408) 730-5900

Motorola Inc.
Austin, TX
(512) 891-2000

National Semiconductor Corp.
Santa Clara, CA
(800) 272-9959

NEC Electronics Inc.
Mountain View, CA
(800) 366-9782

Oki Semiconductor Inc.
Sunnyvale, CA
(800) 654-6388

Philips Semiconductors
Sunnyvale, CA
(408) 991-3445

Rockwell International Corp.
Digital Communications Division
Newport Beach, CA
(800) 854-8099;
in CA, (800) 422-4230

SGS-Thomson Microelectronics
Phoenix, AZ
(602) 867-6200

Sharp Microelectronics Corp.
Mahwah, NJ
(201) 529-8200

Siemens Components
Cupertino, CA
(408) 777-4518

Sun Microsystems Computer
Corp.
Mountain View, CA
(408) 779-8119

Texas Instruments Inc.
Literature Response Center
Denver, CO
(800) 477-8924, ext. 4500

Toshiba America Electronic
Components Inc.
Irvine, CA
(714) 455-2000

VLSI Technology Inc.
San Jose, CA
(408) 434-7877

Zilog Inc.
Campbell, CA
(408) 370-8000

Review Questions

17–16. Programs written in assembly language must be converted to machine code before being executed by a microprocessor. True or false?

17–17. Programs written in a high-level language are more memory efficient than those written in assembly language. True or false?

17–18. What is the *port number* of the input switches and of the output LEDs in Figure 17–6?

17–19. The OUT FEH statement is used to output the switch data to the microprocessor, and the IN FFH statement is used to input data to the LEDs. True or false?

17–20. In Table 17–3, addresses 2000 through 2008 are where the _____ is stored.

17–21. How does the JZ instruction differ from the JMP instruction?

Summary of Instructions

LDA Addr: (Load Accumulator Direct) Load the accumulator with the contents of memory whose address *(addr)* is specified in byte-2–byte 3 of the instruction.

STA Addr: (Store Accumulator Direct) Store the contents of the accumulator to memory whose address *(addr)* is specified in byte 2–byte 3 of the instruction.

IN Port: (Input) Load the accumulator with the contents of the specified port.

OUT Port: (Output) Move the contents of the accumulator to the specified port.

MVI r,data: (Move Immediate) Move into register *r* the data specified in byte 2 of the instruction.

DCR r: (Decrement Register) Decrement the value in register *r* by 1.

JMP Addr: (Jump) Transfer control to address *addr* specified in byte 2–byte 3 of the instruction.

JZ Addr: (Jump If Zero) Transfer control to address *addr* if the zero flag is set.

Summary

In this chapter we have learned that

1. A system designer should consider using a microprocessor instead of logic circuitry whenever an application involves making calculations, making decisions based on external stimuli, and maintaining memory of past events.

2. A microprocessor is the heart of a computer system. It reads and acts on program instructions given to it by a programmer.

3. A microprocessor system has three buses: address, data, and control.

4. Microprocessors operate on instructions given to them in the form of machine code (1's and 0's). The machine code is generated by a higher-level language like C or assembly language.

5. The Intel 8085A is an 8-bit microprocessor. It has 7 internal registers, an 8-bit data bus, an arithmetic/logic unit, and several input/output functions.

6. Program instructions are executed inside the microprocessor by the instruction decoder, which issues the machine cycle timing and initiates input/output operations.

7. The microprocessor provides the appropriate logic levels on the data and address buses and takes care of the timing of all control signals output to the connected interface circuitry.

8. Assembly language instructions are written using mnemonic abbreviations and then converted into machine language so that they can be interpreted by the microprocessor.

9. Higher-level languages like C or Pascal are easier to write than assembly language, but they are not as memory efficient or as fast. All languages must be converted into a machine language matching that of the microprocessor before they can be executed.

Glossary

Accumulator: The parallel register in a microprocessor that is the focal point for all arithmetic and logic operations.

Address Bus: A group of conductors that are routed throughout a computer system and are used to select a unique location based on their binary value.

Architecture: The layout and design of a system.

Arithmetic Logic Unit (ALU): The part of a microprocessor that performs all the arithmetic and digital logic functions.

Assembler: A software package that is used to convert assembly language into machine language.

Assembly Language: A low-level programming language unique to each microprocessor. It is converted, or assembled, into machine code before it can be executed.

BASIC Language: A high-level computer programming language that uses English-language-type instructions that are converted to executable machine code.

Bidirectional: Systems capable of transferring digital information in two directions.

Central Processing Unit (CPU): The "brains" of a computer system. The term is used interchangeably with "microprocessor."

Compiler: A software package that converts a high-level language program into machine language code.

Control Bus: A group of conductors that is routed throughout a computer system and used to signify special control functions, such as Read, Write, I/O, Memory, and Ready.

Data Bus: A group of conductors that is routed throughout a computer system and contains the binary data used for all arithmetic and I/O operations.

Flowchart: A diagram used by the programmer to map out the looping and conditional branching that a program must make. It becomes the blueprint for the program.

Hand Assembly: The act of converting assembly language instructions into machine language codes by hand, using a reference chart.

Hardware: The integrated circuits and electronic devices that make up a computer system.

Instruction Decoder: The circuitry inside a microprocessor that interprets the machine code and produces the internal control signals required to execute the instruction.

Instruction Register: A parallel register in a microprocessor that receives the machine code and produces the internal control signals required to execute the instruction.

Interrupt: A digital control signal that is input to a microprocessor IC pin that suspends current software execution and performs another predefined task.

I/O-Mapped I/O: A method of input/output that addresses each I/O device as a port selected by an 8-bit port number.

Machine Code: The binary codes that make up a microprocessor's program instructions.

Memory-Mapped I/O: A method of input/output that addresses each I/O device as a memory location selected by a 16-bit address.

Microprocessor: An LSI or VLSI integrated circuit that is the fundamental building block of a digital computer. It is controlled by software programs that allow it to do all digital arithmetic, logic, and I/O operations.

Mnemonic: The abbreviated spellings of instructions used in assembly language.

Monitor Program: The computer software program initiated at power-up that supervises system operating tasks, such as reading the keyboard and driving the CRT.

Opcode: Operation code. It is the unique 1-byte code given to identify each instruction to the microprocessor.

Operand: The parameters that follow the assembly language mnemonic to complete the specification of the instruction.

Operating System: *See* Monitor program.

Port Number: An 8-bit number used to select a particular I/O port.

Program Counter: A 16-bit internal register that contains the address of the next program instruction to be executed.

Software: Computer program statements that give step-by-step instructions to a computer to solve a problem.

Stack Pointer: A 16-bit internal register that contains the address of the last entry on the RAM stack.

Statement Label: A meaningful name given to certain assembly language program lines so that they can be referred to from different parts of the program, using statements like JUMP or CALL.

Support Circuitry: The integrated circuits and electronic devices that assist the microprocessor in performing I/O and other external tasks.

Zero Flag: A bit internal to the microprocessor that, when set (1), signifies that the last arithmetic or logic operation had a result of zero.

Problems

Section 17–1

17–1. Describe the circumstances that would prompt you to use a microprocessor-based design solution instead of a hard-wired IC logic design.

17–2. In an 8-bit microprocessor system, how many lines are in the data bus? The address bus?

17–3. What is the function of the address bus?

D   **17–4.** Use a TTL data manual to find an IC that you could use for the *output port* in Figure 17–1. Draw its logic diagram and external connections.

D   **17–5.** Repeat Problem 17–4 for the *input port*.

C D   **17–6.** Repeat Problem 17–4 for the *address decoder*. Assume that the input port is at address 4000H, the output port is at address 6000H, and memory is at address 2000H. (*Hint:* Use an address decoding scheme similar to that found in Section 16–5.)

17–7. Why does the input port in Figure 17–1 have to have three-stated outputs?

17–8. What two control signals are applied to the input port in Figure 17–1 to cause it to transfer the switch data to the data bus?

17–9. How many different addresses can be accessed using a 16-bit address bus?

Sections 17–2 and 17–3

17–10. In the assembly language instruction LDA 4000H, what does the LDA signify and what does the 4000H signify?

17–11. Describe what the statement STA 6000H does.

17–12. What are the names of the six internal 8085A general-purpose registers?

17–13. What is the function of the 8085A's instruction register and instruction decoder?

17–14. Why is the program counter register 16 bits instead of 8?

Section 17–4

C   **17–15.** During the execution of the LDA 4000 instruction in Figure 17–4, the $\overline{RD}$ line goes LOW four times. Describe the activity initiated by each LOW pulse.

17–16. What action does the LOW $\overline{WR}$ pulse initiate during the STA 6000 instruction in Figure 17–5?

Section 17–5

17–17. Describe one advantage and one disadvantage of writing programs in a high-level language instead of assembly language.

17–18. Are the following instructions used for memory-mapped I/O or for I/O-mapped I/O?

(a) LDA *addr*   (c) IN *port*

(b) STA *addr*   (d) OUT *port*

17–19. What is the digital level on the microprocessor's $IO/\overline{M}$ line for each of the following instructions?

(a) LDA *addr*   (c) IN *port*

(b) STA *addr*   (d) OUT *port*

D   **17–20.** List the new IN and OUT instructions that would be used to I/O to the switches and LEDs if the following changes to U4 and U5 were made in Figure 17–6.

(a) Add inverters to inputs A_8 and A_9 of U4 and to A_9 and A_{10} of U5.

(b) Add inverters to inputs A_{14} and A_{15} of U4 and to A_{14} and A_{15} of U5.

17–21. U6a and U6b in Figure 17–6 are OR gates. Why are they drawn as inverted-input NAND gates?

17–22. Are the LEDs in Figure 17–6 active-HIGH or active-LOW?

17–23. Is the $\overline{RD}$ line or the $\overline{WR}$ line pulsed LOW by the microprocessor during the:

(a) IN instruction?   (b) OUT instruction?

17–24. What three conditions must be met to satisfy the output enables of U2 in Figure 17–6?

17–25. What three conditions must be met to provide a pulse to the C_p input of U3 in Figure 17–6?

17–26. Which internal data register is used for the IN and OUT instructions?

Section 17–6

17–27. Write the assembly language instruction that would initialize the accumulator to 4FH.

17–28. Describe in words what the instruction JZ LOOP does.

C   **17–29.** Write the machine language code for the following assembly language program. (Start the machine code at address 2010H.)

```
INIT:   MVI A,04H
  X1:   DCR A
        JZ INIT
        JMP X1
```

Schematic Interpretation Problems

See Appendix G for the schematic diagrams.

S   **17–30.** Locate the 68HC11 microcontroller in the HC11D0 schematic. (A microcontroller is a microprocessor with built in RAM, ROM, and I/O ports). Pins 31–38 are the low-order address bus (A0–A7) multiplexed (shared) with the data bus (D0–D7). Pins 9–16 are the high-order address bus (A8–A15). The low-order address bus is demultiplexed (selected and latched) from the shared address/data lines by U2 and the AS (Address Strobe) line.

(a) Which ICs are connected to the data bus (DB0–DB7)?

(b) Which ICs are connected to the address bus (AD0–AD15)?

S C   **17–31.** U9 and U5 in the HC11D0 schematic are used for address decoding. Determine the levels on AD11–AD15 and AD3–AD5 to select (a) the LCD (LCD_SL) and (b) the keyboard (KEY_SL).

S   **17–32.** Locate the microcontroller in the 4096/4196 schematic.

(a) What is its grid location and part number?

(b) Its low-order address is multiplexed like the 68HC11 in Problem 17–30. What IC and control signal are used to demultiplex the address/data bus (AD0–AD7) into the low-order address bus (A0–A7)?

(c) What IC and control signal are used to demultiplex the address/data bus (AD0–AD7) into the data bus (D0–D7)?

Answers to Review Questions

17–1. Address, data, control

17–2. All of it

17–3. To use the same path for both input and output data

17–4. False

17–5. True

17–6. The output to the LEDs would float when $\overline{WR}$ returns HIGH.

17–7. False

17–8. Read, write

17–9. False

17–10. False

17–11. Assembly language, which is written in short mnemonics, needs to be assembled or converted into a binary string called a machine code, which is read by the CPU.

17–12. 1 byte is used to store 3AH, 2 bytes are used to store 4000H.

17–13. False

17–14. STA, the $\overline{WR}$ is used as the clock input to the D flip-flop.

17–15. 6 bytes

17–16. True

17–17. False

17–18. Input is FFH, output is FEH

17–19. False

17–20. Machine code

17–21. The JZ instruction looks for a zero flag and jumps to the label START if the flag is set. The JMP instruction jumps to the label LOOP, regardless of any flags.

Bibliography

ABT Advanced BiCMOS Logic Databook. Sunnyvale, Calif.: Philips Semiconductors, 1994.

Analog Data Manual. Sunnyvale, Calif.: Philips Semiconductors, 1987.

CMOS Databook. Santa Clara, Calif.: National Semiconductor Corporation, 1987.

CMOS HE4000B I.C. Family. Sunnyvale, Calif.: Signetics Corporation, 1990.

Component Data Catalog. Santa Clara, Calif.: Intel Corporation, 1981.

Data Conversion/Acquisition Databook. Santa Clara, Calif.: National Semiconductor Corporation, 1993.

Embedded Controller Handbook. Santa Clara, Calif.: Intel Corporation, 1988.

FAST Logic Databook. Sunnyvale, Calif.: Philips Semiconductors, 1994.

High-Speed CMOS Data Manual. Sunnyvale, Calif.: Philips Semiconductors, 1990.

High-Speed CMOS Logic Data Book. Dallas: Texas Instruments, Inc., 1984.

Linear LSI Data and Applications Manual. Sunnyvale, Calif.: Philips Semiconductors, 1987.

Low-Voltage Logic. Sunnyvale, Calif.: Philips Semiconductors, 1994.

Low-Voltage Logic. Dallas: Texas Instruments, Inc., 1993.

MCS-80/85 Family User's Manual. Santa Clara, Calif.: Intel Corporation, 1983.

Memory Data. Phoenix, Ariz.: Motorola Inc., 1988.

Memory Databook. Santa Clara, Calif.: National Semiconductor Corporation, 1993.

Memory Data Manual. Sunnyvale, Calif.: Philips Semiconductors, 1994.

PALASM User's Manual. Sunnyvale, Calif.: Advanced Micro Devices, Inc., 1990.

Programmable Logic Devices. Sunnyvale, Calif.: Philips Semiconductors, 1993.

TTL Data Manual. Sunnyvale, Calif.: Philips Semiconductors, 1990.

TTL Logic Data Book. Dallas: Texas Instruments, Inc., 1988.

B

Manufacturers' Data Sheets*

IC Numbers:

74HC00

74LV00

74ABT244

7400

7414

74121

2716

ADC0801

MC1508/1408

μA741

NE555

———

*Courtesy of Signetics (Philips Semiconductors) Corporation, Intel Corporation, and Texas Instruments Inc.

DESCRIPTION

The 54/74HC00 and 54/74HCT00 are high-speed Si-gate CMOS devices and are pin compatible with low power Schottky TTL (LSTTL). They are specified in compliance with JEDEC standard no. 7.

The 54/74HC00 and 54/74HCT00 provide the positive 2-input NAND function.

SYMBOL AND PARAMETER		CONDITIONS	TYPICAL		UNIT
			HC	HCT	
t_{PHL} t_{PLH}	Propagation delay nA, nB to nY	$C_L = 15pF$	8	8	ns
C_I	Input capacitance		3.5	3.5	pF
$C_{PD}{}^1$	Power dissipation capacitance per gate	See note 2	22	22	pF

NOTES:
1. C_{PD} is used to determine the dynamic power consumption -
$$P_D = C_{PD} \cdot V_{CC}{}^2 \cdot f_i + \Sigma\, C_L \cdot V_{CC}{}^2 \cdot f_0 \text{ where:}$$
f_i = input frequency; f_0 = output frequency
C_L = output load capacitance; V_{CC} = supply voltage
2. For HC, condition is V_I = GND to V_{CC}
For HCT, condtion is V_I = GND to $V_{CC} - 1.5V$

ORDERING CODE

PACKAGES	COMMERCIAL RANGES $T_A = -40°C \text{ to } +85°C$	MILITARY RANGES $T_A = -55°C \text{ to } +125°C$
Plastic DIP	N74HC00N, N74HCT00N	
Ceramic DIP		
Plastic SO	N74HC00D, N74HCT00D	

PIN DESCRIPTION

PIN NO.	SYMBOL	NAME AND FUNCTION
1,4,9,12	1A to 4A	Data inputs
2,5,10,13	1B to 4B	Data inputs
3,6,8,11	1Y to 4Y	Data outputs
7	GND	Ground (OV)
14	V_{CC}	Positive supply voltage

PIN CONFIGURATION

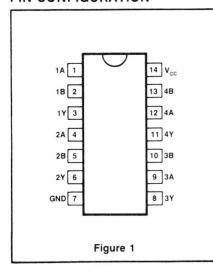

Figure 1

LOGIC SYMBOL

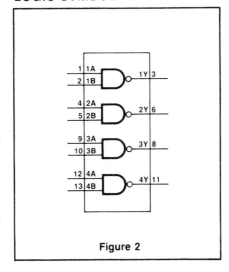

Figure 2

LOGIC SYMBOL (IEEE/IEC)

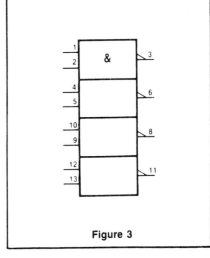

Figure 3

ABSOLUTE MAXIMUM RATINGS

Limiting values in accordance with the Absolute Maximum System (IEC134)

SYMBOL AND PARAMETER		RATING	UNIT
V_{CC}	Supply voltage	-0.5 to $+7.0$	V
$\pm I_{IK}$	Input diode current $V_I \leq -0.5V$ or $V_I \geq V_{CC} + 0.5V$	20	mA
$+ I_{DK}$	Output diode current $V_O \leq -0.5V$ or $V_O \geq V_{CC} + 0.5V$	20	mA
$\pm I_{OH}$, $\pm I_{OL}$	Output source or sink current $-0.5V \leq V_O \leq V_{CC} + 0.5V$	Standard output	25	mA
$\pm I_{CC}$, $\pm I_{GND}$	V_{CC} or GND current	Standard outputs	50	mA
T_{stg}	Storage temperature range	-65 to $+150$	°C
P_{tot}	Power dissipation per package Plastic and Ceramic (Cerdip) DIL (above $+60°C$; derate linearly with 8mW/K)	Standard temp -40 to $+85°C$	500	mW
	Power dissipation per package Plastic minipack (SO) (above $+70°C$; derate linearly with 5mW/K)	Standard temp -40 to $+85°C$	200	mW
	Power dissipation per package Ceramic (Cerdip) DIL (above $+100°C$; derate linearly with 8mW/K)	Extended temp -55 to $+125°C$	500	mW

Voltages are referenced to GND (ground = OV)

RECOMMENDED OPERATION CONDITIONS

SYMBOL AND PARAMETER			54/74 HC			54/74 HCT			UNIT
			Min	Typ	Max	Min	Typ	Max	
V_{CC}	Supply voltage		2.0	5.0	6.0	4.5	5.0	5.5	V
V_I	Input voltage range		0		V_{CC}	0		V_{CC}	V
V_O	Output voltage range		0		V_{CC}	0		V_{CC}	V
T_{amb}	Operating ambient temperature range	54	-55		$+125$	-55		$+125$	°C
		74	-40		$+85$	-40		$+85$	
t_r t_f	Input rise and fall times	$V_{CC} = 2.0V$			1000				ns
		$V_{CC} = 4.5V$		6.0	500		6.0	500	
		$V_{CC} = 6.0V$			400				

DC ELECTRICAL CHARACTERISTICS: 54/74HC

SYMBOL AND PARAMETER			T_{amb} (°C)						UNIT	TEST CONDITIONS[1]		
			54/74HC +25			74HC −40 to 85		54HC −55 to 125					
		Min	Typ	Max	Min	Max	Min	Max		V_{CC}	V_{IN}	OTHER
V_{IH}	HIGH-level input voltage	1.5 3.15 4.2			1.5 3.15 4.2		1.5 3.15 4.2		V	2V 4.5V 6V		
V_{IL}	LOW-level input voltage			0.3 0.9 1.2		0.3 0.9 1.2		0.3 0.9 1.2	V	2V 4.5V 6V		
V_{OH}	HIGH-level output voltage	1.9 4.4 5.9			1.9 4.4 5.9		1.9 4.4 5.9		V	2V 4.5V 6V	V_{IH} or V_{IL} V_{IH} or V_{IL} V_{IH} or V_{IL}	$-I_O = 20\mu A$ $-I_O = 20\mu A$ $-I_O = 20\mu A$
		3.98 5.48			3.84 5.84		3.7 5.2		V	4.5V 6V	V_{IH} or V_{IL} V_{IH} or V_{IL}	$-I_O = 4mA$ $-I_O = 5.2mA$
V_{OL}	LOW-level output voltage			0.1 0.1 0.1		0.1 0.1 0.1		0.1 0.1 0.1	V	2V 4.5V 6V	V_{IH} or V_{IL} V_{IH} or V_{IL} V_{IH} or V_{IL}	$I_O = 20\mu A$ $I_O = 20\mu A$ $I_O = 20\mu A$
				0.26 0.26		0.33 0.33		0.4 0.4	V	4.5V 6V	V_{IH} or V_{IL} V_{IH} or V_{IL}	$I_O = 4mA$ $I_O = 5.2mA$
$\pm I_I$	Input leakage current			0.1		1.0		1.0	μA	6V	V_{CC} or GND	
I_{CC}	Quiescent supply current SSI			2.0		20.0		40.0	μA	6V	V_{CC} or GND	$I_O = 0$

NOTE:
1. Voltages are referenced to GND (ground = OV).

AC ELECTRICAL CHARACTERISTICS: 54/74HC

GND = OV; $t_r = t_f = 6ns$; $C_L = 50pF$

SYMBOL AND PARAMETER			T_{amb}(°C)						UNIT	TEST CONDITIONS	
			54/74HC +25			74HC −40 to +85		54HC −55 to +125				
		Min	Typ	Max	Min	Max	Min	Max		V_{CC}	Figure
t_{PHL} t_{PLH}	Propagation delay nA, nB, to nY			100 20 17		125 25 21		150 30 26	ns	2V 4.5V 6V	4
t_{THL} t_{TLH}	Output transition time			75 15 13		95 19 16		112 22 19	ns	2V 4.5V 6V	4

DC ELECTRICAL CHARACTERISTICS: 54/74HCT

SYMBOL AND PARAMETER		T_{amb} (°C)						UNIT	TEST CONDITIONS[1]		
		54/74HCT +25			74HCT −40 to 85		54HCT −55 to 125					
		Min	Typ	Max	Min	Max	Min	Max		V_{CC}	V_{IN}	OTHER
V_{IH}	HIGH-level input voltage	2.0			2.0		2.0		V	4.5 to 5.5V		
V_{IL}	LOW-level input voltage			0.8		0.8		0.8	V	4.5 to 5.5V		
V_{OH}	HIGH-level output voltage	4.4			4.4		4.4		V	4.5	V_{IH} or V_{IL}	$-I_O = 20\mu A$
		3.98			3.84		3.7		V	4.5	V_{IH} or V_{IL}	$-I_O = 4mA$
V_{OL}	LOW-level output voltage			0.1		0.1		0.1	V	4.5	V_{IH} or V_{IL}	$I_O = 20\mu A$
				0.26		0.33		0.4	V	4.5	V_{IH} or V_{IL}	$I_O = 4mA$
$\pm I_I$	Input leakage current			0.1		0.1		0.1	μA	5.5V	V_{IH} or V_{IL}	
I_{CC}	Quiescent supply current	SSI		2.0		20.0		40.0	μA	5.5V	V_{CC} or GND	$I_O = 0$
I_C	Supply current								μA	5.5V	2.4V or 0.5V	$I_O = 0$[2]

NOTES:
1. Voltages are referenced to GND (ground = OV).
2. Per input-pin, other inputs at V_{CC} or GND.

AC ELECTRICAL CHARACTERISTICS: 54/74HCT

GND = OV; t_r = t_f = 6ns; C_L = 50pF

SYMBOL AND PARAMETER		T_{amb}(°C)						UNIT	TEST CONDITIONS	
		54/74HCT +25			74HCT −40 to +85		54HCT −55 to +125				
		Min	Typ	Max	Min	Max	Min	Max		V_{CC}	Figure
t_{PHL} t_{PLH}	Propagation delay nA, nB to nY			20		25		30	ns	4.5V	4
t_{THL} t_{TLH}	Output transition time			15		19		22	ns	4.5V	4

AC WAVEFORM

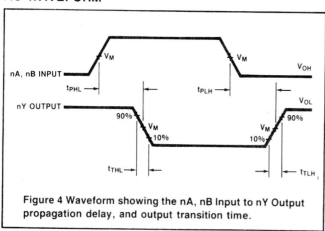

Figure 4 Waveform showing the nA, nB Input to nY Output propagation delay, and output transition time.

NOTE:
HC: V_I = GND to V_{CC}
    V_M = ½ V_{CC}
HCT: V_I = GND to 3.0V
    V_M = 1.3V

SN74LV00
QUADRUPLE 2-INPUT POSITIVE-NAND GATE

FEBRUARY 1993

- **Space-Saving Package Option:**
  Shrink Small-Outline Package (DB)
  Features EIAJ 0.65-mm Lead Pitch
- *EPIC* ™ **(Enhanced-Performance Implanted CMOS) 2-μm Process**
- **Typical V_{OLP} (Output Ground Bounce)**
  < 0.8 V at V_{CC} = 3.3 V, T_A = 25°C
- **Typical V_{OHV} (Output V_{OH} Undershoot)**
  > 2 V at V_{CC} = 3.3 V, T_A = 25°C
- **ESD Protection Exceeds 2000 V Per MIL-STD-883C, Method 3015; Exceeds 200 V Using Machine Model (C = 200 pF, R = 0)**
- **Latch-Up Performance Exceeds 250 mA Per JEDEC Standard JESD-17**
- **Package Options Include Plastic Small-Outline and Thin Shrink Small-Outline Packages**

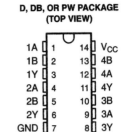

D, DB, OR PW PACKAGE
(TOP VIEW)

1A	1	14	V_{CC}
1B	2	13	4B
1Y	3	12	4A
2A	4	11	4Y
2B	5	10	3B
2Y	6	9	3A
GND	7	8	3Y

description

This quadruple 2-input positive-NAND gate is designed for 2.7-V to 3.6-V V_{CC} operation.

The SN74LV00 performs the Boolean functions $Y = \overline{A \cdot B}$ or $Y = \overline{A} + \overline{B}$ in positive logic.

The SN74LV00 is packaged in TI's shrink small-outline package (DB), which provides the same I/O pin count and functionality of standard small-outline packages in less than half the printed-circuit-board area.

The SN74LV00 is characterized for operation from −40°C to 85°C.

FUNCTION TABLE
(each gate)

INPUTS		OUTPUT
A	B	Y
H	H	L
L	X	H
X	L	H

EPIC is a trademark of Texas Instruments Incorporated.

TEXAS INSTRUMENTS

POST OFFICE BOX 655303 ● DALLAS, TEXAS 75265

Copyright © 1993, Texas Instruments Incorporated

SN74LV00
QUADRUPLE 2-INPUT POSITIVE-NAND GATE

FEBRUARY 1993

logic symbol†

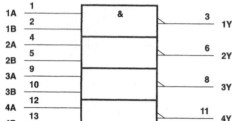

1A	1			
1B	2	&	3	1Y
2A	4			
2B	5		6	2Y
3A	9			
3B	10		8	3Y
4A	12			
4B	13		11	4Y

† This symbol is in accordance with ANSI/IEEE Std 91-1984 and
IEC Publication 617-12.

logic diagram, each gate (positive logic)

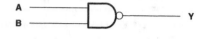

A
B ———— Y

absolute maximum ratings over operating free-air temperature range (unless otherwise noted)‡

Supply voltage range, V_{CC} .... −0.5 V to 4.6 V
Input voltage range, V_I (see Note 1) .... −0.5 V to V_{CC} + 0.5 V
Output voltage range, V_O (see Notes 1 and 2) .... −0.5 V to V_{CC} + 0.5 V
Input clamp current, I_{IK} (V_I < 0 or V_I > V_{CC}) .... ±20 mA
Output clamp current, I_{OK} (V_O < 0 or V_O > V_{CC}) .... ±50 mA
Continuous output current, I_O (V_O = 0 to V_{CC}) .... ±25 mA
Continuous current through V_{CC} or GND pins .... ±50 mA
Maximum power dissipation at T_A = 55°C (in still air): D package .... 0.7 W
DB package .... 0.4 W
PW package .... 0.4 W
Storage temperature range .... −65°C to 150°C

‡ Stresses beyond those listed under "absolute maximum ratings" may cause permanent damage to the device. These are stress ratings only and
functional operation of the device at these or any other conditions beyond those indicated under "recommended operating conditions" is not
implied. Exposure to absolute-maximum-rated conditions for extended periods may affect device reliability.
NOTES: 1. The input and output voltage ratings may be exceeded if the input and output clamp-current ratings are observed.
2. This value is limited to 4.6 V maximum.

recommended operating conditions (see Note 3)

			MIN	NOM	MAX	UNIT
V_{CC}	Supply voltage		2.7	3.3	3.6	V
V_{IH}	High-level input voltage	V_{CC} = 2.7 V to 3.6 V	2			V
V_{IL}	Low-level input voltage	V_{CC} = 2.7 V to 3.6 V			0.8	V
V_I	Input voltage		0		V_{CC}	V
V_O	Output voltage		0		V_{CC}	V
I_{OH}	High-level output current				−6	mA
I_{OL}	Low-level output current				6	mA
Δt/Δv	Input transition rise or fall rate		0		100	ns/V
T_A	Operating free-air temperature		−40		85	°C

NOTE 3: Unused or floating inputs must be held high or low.

TEXAS
INSTRUMENTS
POST OFFICE BOX 655303 ● DALLAS, TEXAS 75265

electrical characteristics over recommended operating free-air temperature range (unless otherwise noted)

PARAMETER	TEST CONDITIONS		V_{CC}†	MIN	TYP	MAX	UNIT
V_{IK}	$I_I = -18$ mA		2.7 V			−1.5	V
V_{OH}	$I_{OH} = -100$ µA		MIN to MAX	V_{CC}−0.2			V
	$I_{OH} = -6$ mA		3 V	2.4			
V_{OL}	$I_{OL} = 100$ µA		MIN to MAX			0.2	V
	$I_{OL} = 6$ mA		3 V			0.4	
I_I	$V_I = V_{CC}$ or GND		3.6 V			±1	µA
I_{OZ}	$V_O = V_{CC}$ or GND		3.6 V			±5	µA
I_{CC}	$V_I = V_{CC}$ or GND,	$I_O = 0$	3.6 V			20	µA
ΔI_{CC}	$V_{CC} = 3$ V to 3.6 V, Other inputs at V_{CC} or GND	One input at $V_{CC} - 0.6$ V,				500	µA
C_i	$V_I = V_{CC}$ or GND		3.3 V		TBD		pF
C_o	$V_O = V_{CC}$ or GND		3.3 V		TBD		pF

† For conditions shown as MIN or MAX, use the appropriate values under recommended operating conditions.

PRODUCT PREVIEW

TEXAS
INSTRUMENTS
POST OFFICE BOX 655303 ● DALLAS, TEXAS 75265

Philips Components—Signetics

Document No.	
ECN No.	853-1444 00227
Date of Issue	August 20, 1990
Status	Product Specification
Advanced BiCMOS Products	

74ABT244
Octal buffer/line driver
(3-State)

FEATURES

- Octal bus interface
- 3-State buffers
- Output capability: +64 mA/-32mA
- Latch-up protection exceeds 500mA per Jedec JC40.2 Std 17
- ESD protection exceeds 2000 V per MIL STD 883C Method 3015.6 and 200 V per Machine Model

DESCRIPTION

The 74ABT244 high-performance BiCMOS device combines low static and dynamic power dissipation with high speed and high output drive.

The 74ABT244 device is an octal buffer that is ideal for driving bus lines or buffer memory address registers. The device features two Output Enables (1$\overline{OE}$, 2$\overline{OE}$), each controlling four of the 3-State outputs.

QUICK REFERENCE DATA

SYMBOL	PARAMETER	CONDITIONS $T_{amb} = 25°C$; GND = 0V	TYPICAL	UNIT
t_{PLH} t_{PHL}	Propagation delay A_n to Y_n	C_L = 50pF; V_{CC} = 5V	2.9	ns
C_{IN}	Input capacitance	V_I = 0V or V_{CC}	4	pF
C_{OUT}	Output capacitance	V_I = 0V or V_{CC}	7	pF
I_{CCZ}	Total supply current	Outputs Disabled; V_{CC} = 5.5V	500	nA

ORDERING INFORMATION

PACKAGES	TEMPERATURE RANGE	ORDER CODE
20-Pin Plastic DIP	-40°C to +85°C	74ABT244N
20-Pin Plastic SOL	-40°C to +85°C	74ABT244D

PIN CONFIGURATION

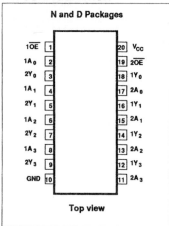

N and D Packages

Top view

LOGIC SYMBOL

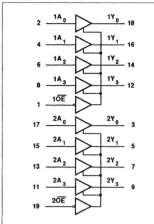

LOGIC SYMBOL (IEEE/IEC)

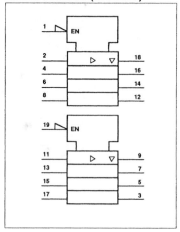

Octal buffer/line driver (3-State)

74ABT244

RECOMMENDED OPERATING CONDITIONS

SYMBOL	PARAMETER	LIMITS		UNIT
		Min	Max	
V_{CC}	DC supply voltage	4.5	5.5	V
V_I	Input voltage	0	V_{CC}	V
V_{IH}	High-level input voltage	2.0		V
V_{IL}	Input voltage		0.8	V
I_{OH}	High level output current		-32	mA
I_{OL}	Low level output current		64	mA
$\Delta t/\Delta v$	Input transition rise or fall rate	0	5	ns/V
T_{amb}	Operating free-air temperature range	-40	+85	°C

ABSOLUTE MAXIMUM RATINGS[1]

SYMBOL	PARAMETER	CONDITIONS	RATING	UNIT
V_{CC}	DC supply voltage		-0.5 to +7.0	V
I_{IK}	DC input diode current	$V_I < 0$	-18	mA
V_I	DC input voltage[2]		-1.2 to +7.0	V
I_{OK}	DC output diode current	$V_O < 0$	-50	mA
V_O	DC output voltage[2]	output in Off or High state	-0.5 to +5.5	V
I_O	DC output current	output in Low state	128	mA
T_{stg}	Storage temperature range		-65 to 150	°C

NOTES:
1. Stresses beyond those listed may cause permanent damage to the device. These are stress ratings only and functional operation of the device at these or any other conditions beyond those indicated under "recommended operating conditions" is not implied. Exposure to absolute-maximum-rated conditions for extended periods may affect device reliability.
2. The input and output voltage ratings may be exceeded if the input and output current ratings are observed.

FUNCTION TABLE

INPUTS				OUTPUT	
$1\overline{OE}$	$1A_n$	$2\overline{OE}$	$2A_n$	$1Y_n$	$2Y_n$
L	L	L	L	L	L
L	H	L	H	H	H
H	X	H	X	Z	Z

PIN DESCRIPTION

PIN NUMBER	SYMBOL	FUNCTION
2, 4, 6, 8	$1A_0 - 1A_3$	Data inputs
17, 15, 13, 11	$2A_0 - 2A_3$	Data inputs
18, 16, 14, 12	$1Y_0 - 1Y_3$	Data outputs
3, 5, 7, 9	$2Y_0 - 2Y_3$	Data outputs
1, 19	$1\overline{OE}, 2\overline{OE}$	Output enables
10	GND	Ground (0V)
20	V_{CC}	Positive supply voltage

Octal buffer/line driver (3-State)

DC ELECTRICAL CHARACTERISTICS

SYMBOL	PARAMETER	TEST CONDITIONS	LIMITS					UNIT
			T_{amb} = +25°C			T_{amb} = -40°C to +85°C		
			Min	Typ	Max	Min	Max	
V_{IK}	Input clamp voltage	V_{CC} = 4.5V; I_{IK} = -18mA			-1.2		-1.2	V
V_{OH}	High-level output voltage	V_{CC} = 4.5V; I_{OH} = -3mA; V_I = V_{IL} or V_{IH}	2.5			2.5		V
		V_{CC} = 5.0V; I_{OH} = -3mA; V_I = V_{IL} or V_{IH}	3.0			3.0		
		V_{CC} = 4.5V; I_{OH} = -32mA; V_I = V_{IL} or V_{IH}	2.0	2.4		2.0		
V_{OL}	Low-level output voltage	V_{CC} = 4.5V; I_{OL} = 64mA; V_I = V_{IL} or V_{IH}		0.42	0.55		0.55	V
I_I	Input leakage current	V_{CC} = 5.5V; V_I = GND or 5.5V		±0.01	±1.0		±1.0	μA
I_{OZH}	3-State output High current	V_{CC} = 5.5V; V_O = 2.7V; V_I = V_{IL} or V_{IH}		5.0	50		50	μA
I_{OZL}	3-State output Low current	V_{CC} = 5.5V; V_O = 0.5V; V_I = V_{IL} or V_{IH}		-5.0	-50		-50	μA
I_O	Short-circuit output current[1]	V_{CC} = 5.5V; V_O = 2.5V	-50	-100	-180	-50	-180	mA
I_{CCH}	Quiescent supply current	V_{CC} = 5.5V; Outputs High; V_I = GND or V_{CC}		0.5	50		50	μA
I_{CCL}		V_{CC} = 5.5V; Outputs Low; V_I = GND or V_{CC}		24	30		30	mA
I_{CCZ}		V_{CC} = 5.5V; Outputs 3-State; V_I = GND or V_{CC}		0.5	50		50	μA
ΔI_{CC}	Additional supply current per input pin[2]	Outputs enabled, one input at 3.4V, other inputs at V_{CC} or GND; V_{CC} = 5.5V		0.5	1.5		1.5	mA
		Outputs 3-State, one data input at 3.4V, other inputs at V_{CC} or GND; V_{CC} = 5.5V		0.5	50		50	μA
		Outputs 3-State, one enable input at 3.4V, other inputs at V_{CC} or GND; V_{CC} = 5.5V		0.5	1.5		1.5	mA

NOTES:
1. Not more than one output should be tested at a time, and the duration of the test should not exceed one second.
2. This is the increase in supply current for each input at 3.4V.

Octal buffer/line driver (3-State)

74ABT244

AC CHARACTERISTICS

GND = 0V; $t_R = t_F = 2.5ns$; $C_L = 50pF$, $R_L = 500\Omega$

SYMBOL	PARAMETER	WAVEFORM	LIMITS					UNIT
			$T_{amb} = +25°C$ $V_{CC} = +5.0V$			$T_{amb} = -40°C$ to $+85°C$ $V_{CC} = +5.0V \pm0.5V$		
			Min	Typ	Max	Min	Max	
t_{PLH} t_{PHL}	Propagation delay A_n to Y_n	1	1.0 1.0	2.6 2.9	4.1 4.2	1.0 1.0	4.6 4.6	ns
t_{PZH} t_{PZL}	Output enable time to High and Low level	2	1.1 2.1	3.1 4.1	4.6 5.6	1.1 2.1	5.1 6.1	ns
t_{PHZ} t_{PLZ}	Output disable time from High and Low level	2	2.1 1.7	4.1 3.7	5.6 5.2	2.1 1.7	6.6 5.7	ns

AC WAVEFORMS

($V_M = 1.5V$, V_{IN} = GND to 3.0V)

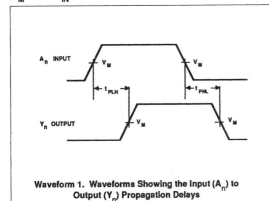

Waveform 1. Waveforms Showing the Input (A_n) to Output (Y_n) Propagation Delays

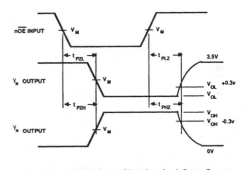

Waveform 2. Waveforms Showing the 3-State Output Enable and Disable Times

TEST CIRCUIT AND WAVEFORMS

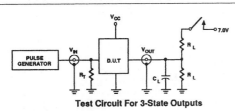

Test Circuit For 3-State Outputs

SWITCH POSITION

TEST	SWITCH
t_{PLZ}	closed
t_{PZL}	closed
All other	open

DEFINITIONS

R_L = Load resistor; see AC CHARACTERISTICS for value.

C_L = Load capacitance includes jig and probe capacitance; see AC CHARACTERISTICS for value.

R_T = Termination resistance should be equal to Z_{OUT} of pulse generators.

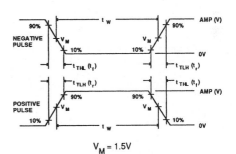

$V_M = 1.5V$

Input Pulse Definition

FAMILY	INPUT PULSE REQUIREMENTS				
	Amplitude	Rep. Rate	t_W	t_R	t_F
74ABT	3.0V	1MHz	500ns	2.5ns	2.5ns

Octal buffer/line driver (3-State)

74ABT244

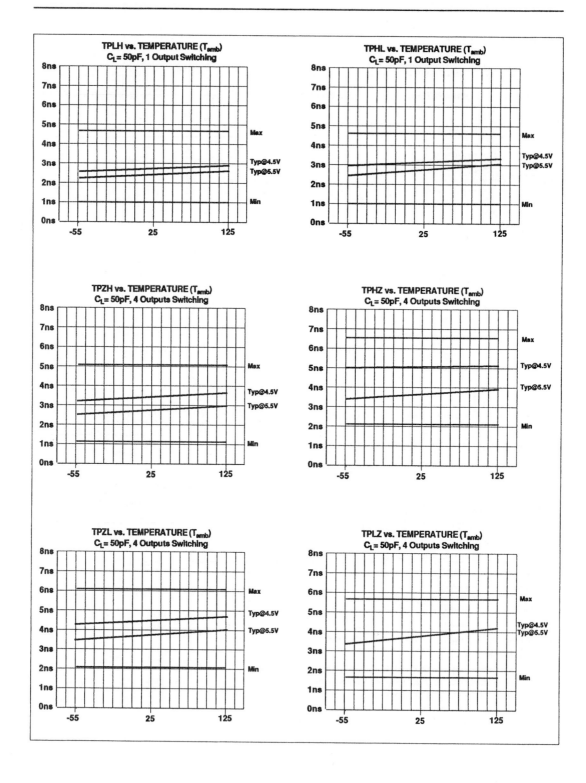

Octal buffer/line driver (3-State)

74ABT244

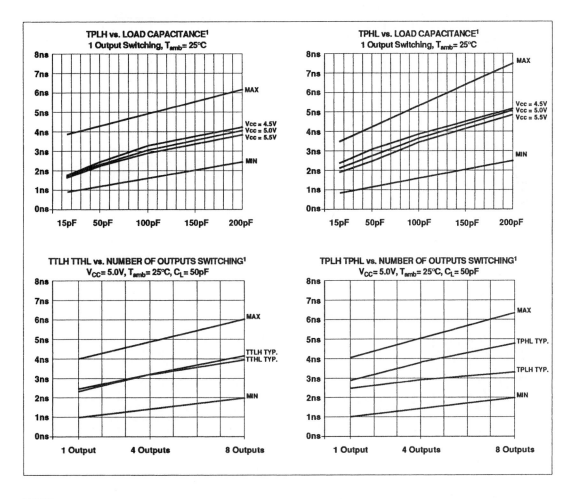

NOTES:
1. MIN and MAX lines are design characteristics and are not necessarily guaranteed by test.

Octal buffer/line driver (3-State)

74ABT244

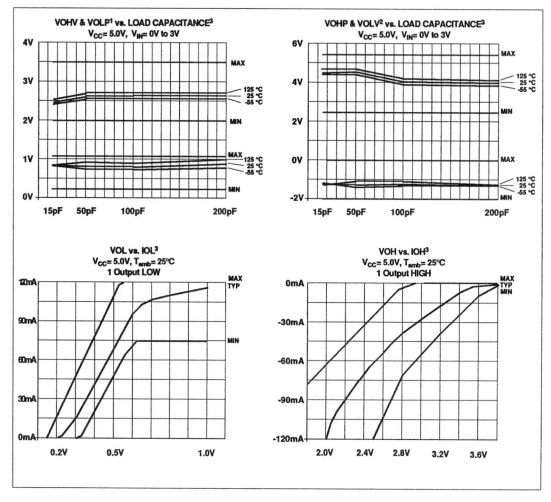

NOTES:
1. VOHV is defined as the minimum (valley) voltage induced on a quiescent high-level output during switching of other outputs. VOLP is defined as the maximum (peak) voltage induced on a quiescent low-level output during switching of other outputs.
2. VOHP is defined as the maximum (peak) voltage induced on a quiescent high-level output during switching of other outputs. VOLV is defined as the minimum (valley) voltage induced on a quiescent low-level output during switching of other outputs.
3. MIN and MAX lines are design and process characteristics. They are not necessarily guaranteed by test.

Quad Two-Input NAND Gate

TYPICAL PROPAGATION DELAY / TYPICAL SUPPLY CURRENT

TYPE	TYPICAL PROPAGATION DELAY	TYPICAL SUPPLY CURRENT (Total)
7400	9ns	8mA
74LS00	9.5ns	1.6mA
74S00	3ns	15mA

ORDERING CODE

PACKAGES	COMMERCIAL RANGES $V_{CC} = 5V \pm 5\%; T_A = 0°C$ to $+70°C$	MILITARY RANGES $V_{CC} = 5V \pm 10\%; T_A = -55°C$ to $+125°C$
Plastic DIP	N7400N • N74LS00N N74S00N	
Plastic SO	N74LS00D N74S00D	
Ceramic DIP	S5400F • S54S00F	S54LS00F
Flatpack	S5400W • S54S00W	S54LS00W
LLCC		S54LS00G

INPUT AND OUTPUT LOADING AND FAN-OUT TABLE

PINS	DESCRIPTION	54/74	54/74S	54/74LS
A, B	Inputs	1ul	1Sul	1LSul
Y	Output	10ul	10Sul	10LSul

NOTE
Where a 54/74 unit load (ul) is understood to be 40µA I_{IH} and -1.6mA I_{IL}, a 54/74S unit load (Sul) is 50µA I_{IH} and -2.0mA I_{IL}, and 54/74LS unit load (LSul) is 20µA I_{IH} and -0.4mA I_{IL}.

LOGIC SYMBOL

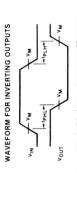

LOGIC SYMBOL (IEEE/IEC)

PIN CONFIGURATION

FUNCTION TABLE

INPUTS		OUTPUT
A	B	Y
L	L	H
L	H	H
H	L	H
H	H	L

H = HIGH voltage level
L = LOW voltage level

DC ELECTRICAL CHARACTERISTICS (Over recommended operating free-air temperature range unless otherwise noted.)

PARAMETER		TEST CONDITIONS[1]		54/7400 Min	54/7400 Typ[2]	54/7400 Max	54/74LS00 Min	54/74LS00 Typ[2]	54/74LS00 Max	54/74S00 Min	54/74S00 Typ[2]	54/74S00 Max	UNIT
V_{OH} HIGH-level output voltage	Mil	$V_{CC} = MIN, V_{IH} = MIN,$ $V_{IL} = MAX, I_{OH} = MAX$		2.4	3.4		2.5	3.4		2.5	3.4		V
	Com'l			2.4	3.4		2.7	3.4		2.7	3.4		V
V_{OL} LOW-level output voltage	Mil	$V_{CC} = MIN,$ $V_{IH} = MIN$	$I_{OL} = MAX$		0.2	0.4		0.25	0.4			0.5[4]	V
	Com'l		$I_{OL} = MAX$		0.2	0.4		0.35	0.5			0.5	V
	74LS		$I_{OL} = 4mA$					0.25	0.4				V
V_{IK} Input clamp voltage		$V_{CC} = MIN, I_I = I_{IK}$				-1.5			-1.5			-1.2	V
I_I Input current at maximum input voltage		$V_{CC} = MAX$	$V_I = 5.5V$			1.0			0.1			1.0	mA
			$V_I = 7.0V$										mA
I_{IH} HIGH-level input current		$V_{CC} = MAX$	$V_I = 2.4V$			40			20			50	µA
			$V_I = 2.7V$										µA
I_{IL} LOW-level input current		$V_{CC} = MAX$	$V_I = 0.4V$			-1.6			-0.4			-2.0	mA
			$V_I = 0.5V$										mA
I_{OS} Short-circuit output current[3]	Mil	$V_{CC} = MAX$		-20		-55	-20		-100	-40		-100	mA
	Com'l			-18		-55	-20		-100	-40		-100	mA
I_{CC} Supply current (total)		$V_{CC} = MAX$	Outputs HIGH		4	8		0.8	1.6		10	16	mA
			Outputs LOW		12	22		2.4	4.4		20	36	mA

NOTES
1. For conditions shown as MIN or MAX, use the appropriate value specified under recommended operating conditions for the applicable type.
2. All typical values are at $V_{CC} = 5V$, $T_A = 25°C$.
3. I_{OS} is tested with $V_{OUT} = +0.5V$ and $V_{CC} = V_{CC}$ MAX $+0.5V$. Not more than one output should be shorted at a time and duration of the short circuit should not exceed one second.
4. $V_{OL} = +0.45V$ MAX for 54S at $T_A = +125°C$ only.

AC WAVEFORM

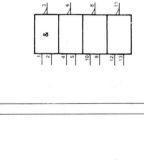

WAVEFORM FOR INVERTING OUTPUTS

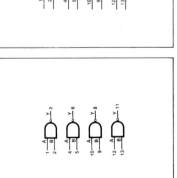

Waveform 1

$V_M = 1.3V$ for 54LS/74LS, $V_M = 1.5V$ for all other TTL families.

AC CHARACTERISTICS $T_A = 25°C$, $V_{CC} = 5.0V$

PARAMETER		TEST CONDITIONS	54/74 $C_L = 15pF, R_L = 400\Omega$ Min	54/74 Max	54/74LS $C_L = 15pF, R_L = 2k\Omega$ Min	54/74LS Max	54/74S $C_L = 15pF, R_L = 280\Omega$ Min	54/74S Max	UNIT
t_{PLH}	Propagation delay	Waveform 1		22		15		4.5	ns
t_{PHL}				15		15		5.0	ns

SCHMITT TRIGGERS

Hex Inverter Schmitt Trigger

TYPE	TYPICAL PROPAGATION DELAY	TYPICAL SUPPLY CURRENT (Total)
7414	15ns	31mA
74LS14	15ns	10mA

DESCRIPTION

The '14 contains six logic inverters which accept standard TTL input signals and provide standard TTL output levels. They are capable of transforming slowly changing input signals into sharply defined, jitter-free output signals. In addition, they have greater noise margin than conventional inverters.

Each circuit contains a Schmitt trigger followed by a Darlington level shifter and a phase splitter driving a TTL totem-pole output. The Schmitt trigger uses positive feedback to effectively speed-up slow input transition, and provide different input threshold voltages for positive and negative-going transitions. This hysteresis between the positive-going and negative-going input thresholds (typically 800mV) is determined internally by resistor ratios and is essentially insensitive to temperature and supply voltage variations.

FUNCTION TABLE

INPUT	OUTPUT
A	Y
0	1
1	0

ORDERING CODE

PACKAGES	COMMERCIAL RANGES $V_{CC} = 5V \pm 5\%$; $T_A = 0°C$ to $+70°C$		MILITARY RANGES $V_{CC} = 5V \pm 10\%$; $T_A = -55°C$ to $+125°C$	
Plastic DIP	N7414N	•		
Plastic SO	N74LS14D			
Ceramic DIP			S5414F	S54LS14F
Flatpack			S5414W	S54LS14W

INPUT AND OUTPUT LOADING AND FAN-OUT TABLE

PINS	DESCRIPTION	54/74	54/74LS
A	Inputs	1ul	1LSul
Y	Output	10ul	10LSul

NOTE
Where a 54/74 unit load (ul) is understood to be 40µA I_{IH} and − 1.6mA I_{IL} and a 54/74LS unit load (LSul) is 20µA I_{IH} and − 0.4mA I_{IL}.

LOGIC SYMBOL

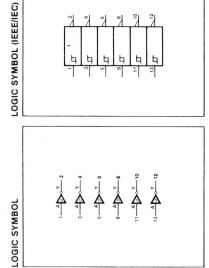

LOGIC SYMBOL (IEEE/IEC)

PIN CONFIGURATION

ABSOLUTE MAXIMUM RATINGS (Over operating free-air temperature range unless otherwise noted.)

	PARAMETER	54	54LS	54S	74	74LS	74S	UNIT
V_{CC}	Supply voltage	7.0	7.0	7.0	7.0	7.0	7.0	V
V_{IN}	Input voltage	−0.5 to +5.5	−0.5 to +7.0	−0.5 to +5.5	−0.5 to +5.5	−0.5 to +7.0	−0.5 to +5.5	V
I_{IN}	Input current	−30 to +5	−30 to +1	−30 to +5	−30 to +5	−30 to +1	−30 to +5	mA
V_{OUT}	Voltage applied to output in HIGH output state	−0.5 to +V_{CC}	−0.5 to +V_{CC}	−0.5 to +V_{CC}	−0.5 to +V_{CC}	−0.5 to +V_{CC}	−0.5 to +V_{CC}	V
T_A	Operating free-air temperature range	−55 to +125			0 to 70			°C

NOTE
$V_{IL} = 0.7V$ MAX for 54S at $T_A = +125°C$ only.

RECOMMENDED OPERATING CONDITIONS

	PARAMETER		54/74			54/74LS			54/74S			UNIT
			Min	Nom	Max	Min	Nom	Max	Min	Nom	Max	
V_{CC}	Supply voltage	Mil	4.5	5.0	5.5	4.5	5.0	5.5	4.5	5.0	5.5	V
		Com'l	4.75	5.0	5.25	4.75	5.0	5.25	4.75	5.0	5.25	V
V_{IH}	HIGH-level input voltage		2.0			2.0			2.0			V
V_{IL}	LOW-level input voltage	Mil			+0.8			+0.7			+0.8	V
		Com'l			+0.8			+0.8			+0.8	V
I_{IK}	Input clamp current				−12			−18			−18	mA
I_{OH}	HIGH-level output current	Mil			−400			−400			−1000	µA
		Com'l			−400			−400			−1000	µA
I_{OL}	LOW-level output current	Mil			16			4			20	mA
		Com'l			16			8			20	mA
T_A	Operating free-air temperature	Mil	−55		+125	−55		+125	−55		+125	°C
		Com'l	0		70	0		70	0		70	°C

TEST CIRCUITS AND WAVEFORMS

TEST CIRCUIT FOR 54/74 TOTEM-POLE OUTPUTS

INPUT PULSE DEFINITIONS

$V_M = 1.3V$ for 54LS/74LS; $V_M = 1.5V$ for all other TTL families

FAMILY	INPUT PULSE REQUIREMENTS				
	Amplitude	Rep. Rate	Pulse Width	t_{TLH}	t_{THL}
54/74	3.0V	1MHz	500ns	7ns	7ns
54LS/74LS	3.0V	1MHz	500ns	15ns	6ns
54S/74S	3.0V	1MHz	500ns	2.5ns	2.5ns

DEFINITIONS
R_L = Load resistor to V_{CC}; see AC CHARACTERISTICS for value.
C_L = Load capacitance includes jig and probe capacitance; see AC CHARACTERISTICS for value.
R_T = Termination resistance should be equal to Z_{OUT} of Pulse Generators.
D = Diodes are 1N916, 1N3064, or equivalent.
t_{TLH}, t_{THL} Values should be less than or equal to the table entries.

ABSOLUTE MAXIMUM RATINGS (Over operating free-air temperature range unless otherwise noted.)

	PARAMETER	54	74	54LS	74LS	UNIT
V_{CC}	Supply voltage	7.0	7.0	7.0	7.0	V
V_{IN}	Input voltage	−0.5 to +5.5	−0.5 to +5.5	−0.5 to +7.0	−0.5 to +7.0	V
I_{IN}	Input current	−30 to +5	−30 to +5	−30 to +1	−30 to +1	mA
V_{OUT}	Voltage applied to output in HIGH output state	−0.5 to +V_{CC}	−0.5 to +V_{CC}	−0.5 to +V_{CC}	−0.5 to +V_{CC}	V
T_A	Operating free-air temperature range	−55 to +125			0 to 70	°C

RECOMMENDED OPERATING CONDITIONS

	PARAMETER		54/74 Min	54/74 Nom	54/74 Max	54/74LS Min	54/74LS Nom	54/74LS Max	UNIT
V_{CC}	Supply voltage	Mil	4.5	5.0	5.5	4.5	5.0	5.5	V
		Com'l	4.75	5.0	5.25	4.75	5.0	5.25	V
I_{IK}	Input clamp current				−12			−18	mA
I_{OH}	HIGH-level output current				−800			−400	µA
I_{OL}	LOW-level output current	Mil			16			4	mA
		Com'l			16			8	mA
T_A	Operating free-air temperature	Mil	−55		+125	−55		+125	°C
		Com'l	0		70	0		70	°C

TEST CIRCUITS AND WAVEFORMS

TEST CIRCUIT FOR 54/74 TOTEM-POLE OUTPUTS

INPUT PULSE DEFINITIONS

FAMILY	Amplitude	Rep. Rate	Pulse Width	t_{TLH}	t_{THL}
54/74	3.0V	1MHz	500ns	7ns	7ns
54LS/74LS	3.0V	1MHz	500ns	15ns	6ns
54S/74S	3.0V	1MHz	500ns	2.5ns	2.5ns

INPUT PULSE REQUIREMENTS

$V_M=1.3V$ for 54LS/74LS; $V_M=1.5V$ for all other TTL families

DEFINITIONS:
R_L = Load resistor to V_{CC}, see AC CHARACTERISTICS for value
C_L = Load capacitance includes jig and probe capacitance, see AC CHARACTERISTICS for value
R_T = Termination resistance should be equal to Z_{OUT} of Pulse Generators
D = Diodes are 1N916, 1N3064, or equivalent
t_{TLH}, t_{THL} Values should be less than or equal to the table entries

---

DC ELECTRICAL CHARACTERISTICS (Over recommended operating free-air temperature range unless otherwise noted.)

	PARAMETER	TEST CONDITIONS[1]		54/7414 Min	54/7414 Typ[2]	54/7414 Max	54/74LS14 Min	54/74LS14 Typ[2]	54/74LS14 Max	UNIT
V_{T+}	Positive-going threshold	$V_{CC}=5.0V$		1.5	1.7	2.0	1.4	1.6	1.9	V
V_{T-}	Negative-going threshold	$V_{CC}=5.0V$		0.6	0.9	1.1	0.5	0.8	1.0	V
ΔV_T	Hysteresis ($V_{T+}-V_{T-}$)	$V_{CC}=5.0V$		0.4	0.8		0.4	0.8		V
V_{OH}	HIGH-level output voltage	$V_{CC}=MIN, V_I=V_{T-MIN}, I_{OH}=MAX$	Mil	2.4	3.4		2.5	3.4		V
			Com'l	2.4	3.4		2.7	3.4		V
V_{OL}	LOW-level output voltage	$V_{CC}=MIN, I_{OL}=MAX, V_I=V_{T+MAX}$	Mil		0.2	0.4		0.25	0.4	V
			Com'l		0.2	0.4		0.35	0.5	V
		$I_{OL}=4mA$	74LS					0.25	0.4	V
V_{IK}	Input clamp voltage	$V_{CC}=MIN, I_I=I_{IK}$				−1.5			−1.5	V
I_{T+}	Input current at positive-going threshold	$V_{CC}=5.0V, V_I=V_{T+}$			−0.43			−0.14		mA
I_{T-}	Input current at negative-going threshold	$V_{CC}=5.0V, V_I=V_{T-}$			−0.56			−0.18		mA
I_I	Input current at maximum input voltage	$V_{CC}=MAX$, $V_I=5.5V$ / $V_I=7.0V$				1.0			0.1	mA
I_{IH}	HIGH-level input current	$V_{CC}=MAX$, $V_I=2.4V$ / $V_I=2.7V$				40			20	µA
I_{IL}	LOW-level input current	$V_{CC}=MAX, V_I=0.4V$				−1.2			−0.4	mA
I_{OS}	Short-circuit output current[3]	$V_{CC}=MAX$	Mil	−20		−55	−20		−100	mA
			Com'l	−18		−55	−20		−100	mA
I_{CC}	Supply current (total)	I_{CCH} Outputs HIGH			22	36		8.6	16	mA
		I_{CCL} Outputs LOW			39	60		12	21	mA

NOTES:
1. For conditions shown as MIN or MAX, use the appropriate value specified under recommended operating conditions for the applicable type.
2. All typical values are at $V_{CC}=5V$, $T_A=25°C$.
3. I_{OS} is tested with $V_{OUT}=+0.5V$ and $V_{CC}=V_{CC}$ MAX $+0.5V$. Not more than one output should be shorted at a time and duration of the short circuit should not exceed one second.

AC WAVEFORMS

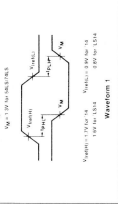

$V_M=1.5V$ for 54/74
$V_M=1.3V$ for 54LS/74LS

$V_{iref(HI)}=1.7V$ for 14 $V_{iref(LI)}=0.9V$ for 14
$V_{iref(HI)}=1.6V$ for LS14 $V_{iref(LI)}=0.8V$ for LS14

Waveform 1

AC CHARACTERISTICS $T_A=25°C$, $V_{CC}=5.0V$

	PARAMETER	TEST CONDITIONS	54/74 $C_L=15pF, R_L=400\Omega$ Min	54/74 Max	54LS/74LS $C_L=15pF, R_L=2k\Omega$ Min	54LS/74LS Max	UNIT
t_{PLH}	Propagation delay	Waveform 1		22		22	ns
t_{PHL}				22		22	ns

Monostable Multivibrator

- Very good pulse width stability
- Virtually immune to temperature and voltage variations
- Schmitt trigger input for slow input transitions
- Internal timing resistor provided

ORDERING CODE

		TYPICAL PROPAGATION DELAY	TYPICAL SUPPLY CURRENT (Total)†
TYPE			
74121		43ns	18mA

PACKAGES	COMMERCIAL RANGES $V_{CC} = 5V \pm 5\%$; $T_A = 0°C$ to $+70°C$	MILITARY RANGES $V_{CC} = 5V \pm 10\%$; $T_A = -55°C$ to $+125°C$
Plastic DIP	N74121N	
Plastic SO	N74121D	
Ceramic DIP		S54121F
Flatpack		S54121W

FUNCTION TABLE

INPUTS			OUTPUTS	
$\overline{A}_1$	$\overline{A}_2$	B	Q	$\overline{Q}$
L	X	H	H	H
X	L	H	H	H
X	X	L	H	H
H	H	X	H	H
H	↓	H	⊓	⊔
↓	H	H	⊓	⊔
↓	↓	H	⊓	⊔
L	X	↑	⊓	⊔
X	L	↑	⊓	⊔

H = HIGH voltage level
L = LOW voltage level
X = Don't care
↓ = LOW-to-HIGH transition
↑ = HIGH-to-LOW transition

DESCRIPTION

These multivibrators feature dual active LOW going edge inputs and a single active HIGH going edge input which can be used as an active HIGH enable input. Complementary output pulses are provided.

Pulse triggering occurs at a particular voltage level and is not directly related to the transition time of the input pulse. Schmitt trigger input circuitry (TTL hysteresis) for the B input allows jitter-free triggering from inputs with transition rates as slow as 1 volt/second, providing the circuit with an excellent noise immunity of typically 1.2 volts. A high immunity to V_{CC} noise of typically 1.5 volts is also provided by internal latching circuitry. Once fired, the outputs are independent of further transitions of the inputs and are a function only of the timing components. Input pulses may be of any duration relative to the output pulse. Output pulse length may be varied from 20 nanoseconds to 28 seconds by choosing appropriate timing components. With no external timing components (i.e., R_{int} connected to V_{CC}, C_{ext} and R_{ext}/C_{ext} open), an output pulse of typically 30 or 35 nanoseconds is achieved which may be used as a dc triggered reset signal. Output rise and fall times are TTL compatible and independent of pulse length.

Pulse width stability is achieved through internal compensation and is virtually independent of V_{CC} and temperature. In most applications, pulse stability will only be limited by the accuracy of external timing components.

Jitter-free operation is maintained over the full temperature and V_{CC} ranges for more than six decades of timing capacitance (10pF to 10µF) and more than one decade of timing resistance (2KΩ to 30KΩ)

INPUT AND OUTPUT LOADING AND FAN-OUT TABLE

PINS	DESCRIPTION	54/74
$\overline{A}_1$, $\overline{A}_2$	Inputs	1ul
B	Input	2ul
Q, $\overline{Q}$	Outputs	10ul

NOTE
A 54/74 unit load (ul) is understood to be 40µA I_{IH} and
-1.6mA I_{IL}.

for the 54121 and 2KΩ to 40KΩ for the 74121). Throughout these ranges, pulse width is defined by the relationship: (see Figure 1)

$$t_W(out) = C_{ext} R_{ext} \ln 2$$
$$t_W(out) = 0.7 C_{ext} R_{ext}$$

In circuits where pulse cutoff is not critical, timing capacitance up to 1000µF and timing resistance as low as 1.4kΩ may be used.

PIN CONFIGURATION

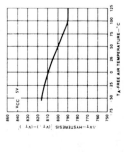

LOGIC SYMBOL

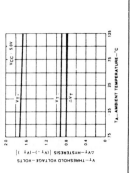

LOGIC SYMBOL (IEEE/IEC)

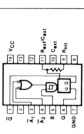
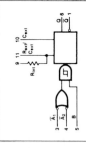
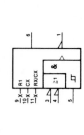

TYPICAL CHARACTERISTICS

(54/74, 54LS/74LS)
VIN vs VOUT
TRANSFER FUNCTION

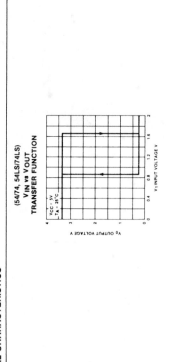

(54/74)
THRESHOLD VOLTAGE AND HYSTERESIS vs POWER SUPPLY VOLTAGE

(54/74)
HYSTERESIS vs TEMPERATURE

(54LS/74LS)
THRESHOLD VOLTAGE AND HYSTERESIS vs POWER SUPPLY VOLTAGE

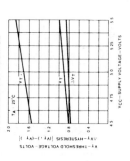

(54LS/74LS)
THRESHOLD VOLTAGE AND HYSTERESIS vs AMBIENT TEMPERATURE

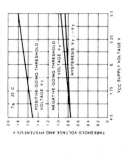

intel®

2716*
16K (2K × 8) UV ERASABLE PROM

- **Fast Access Time**
  - 350 ns Max. 2716-1
  - 390 ns Max. 2716-2
  - 450 ns Max. 2716
  - 490 ns Max. 2716-5
  - 650 ns Max. 2716-6

- **Single +5V Power Supply**

- **Low Power Dissipation**
  - 525 mW Max. Active Power
  - 132 mW Max. Standby Power

- **Pin Compatible to Intel® 2732 EPROM**

- **Simple Programming Requirements**
  - Single Location Programming
  - Programs with One 50 ms Pulse

- **Inputs and Outputs TTL Compatible during Read and Program**

- **Completely Static**

The Intel® 2716 is a 16,384 bit ultraviolet erasable and electrically programmable read-only memory (EPROM). The 2716 operates from a single 5-volt power supply, has a static standby mode, and features fast single address location programming. It makes designing with EPROMs faster, easier and more economical.

The 2716, with its single 5-volt supply and with an access time up to 350 ns, is ideal for use with the newer high performance +5V microprocessors such as Intel's 8085 and 8086. A selected 2716-5 and 2716-6 is available for slower speed applications. The 2716 is also the first EPROM with a static standby mode which reduces the maximum standby power dissipation is 525 mW while the maximum active power dissipation is only 132 mW, a 75% savings.

The 2716 has the simplest and fastest method yet devised for programming EPROMs — single pulse TTL level programming. No need for high voltage pulsing because all programming controls are handled by TTL signals. Program any location at any time—either individually, sequentially or at random, with the 2716's single address location programming. Total programming time for all 16,384 bits is only 100 seconds.

PIN CONFIGURATION

2732†

2716 16K

† Refer to 2732 data sheet for specifications

PIN NAMES

A0–A10	ADDRESSES
CE/PGM	CHIP ENABLE PROGRAM
OE	OUTPUT ENABLE
O0–O7	OUTPUTS

MODE SELECTION

PINS / MODE	CE/PGM (18)	OE (20)	Vpp (21)	Vcc (24)	OUTPUTS (9-11, 13-17)
Read	VIL	VIL	+5	+5	DOUT
Standby	VIH	Don't Care	+5	+5	High Z
Program	Pulsed VIL to VIH	VIH	+25	+5	DIN
Program Verify	VIL	VIL	+25	+5	DOUT
Program Inhibit	VIL	VIH	+25	+5	High Z

BLOCK DIAGRAM

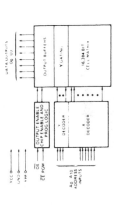

PROGRAMMING

The programming specifications are described in the Data Catalog PROM/ROM Programming Instructions Section.

Absolute Maximum Ratings*

Temperature Under Bias	−10°C to +80°C
Storage Temperature	−65°C to +125°C
All Input or Output Voltages with Respect to Ground	+6V to −0.3V
Vpp Supply Voltage with Respect to Ground During Program	+26.5V to −0.3V

*COMMENT: Stresses above those listed under "Absolute Maximum Ratings" may cause permanent damage to the device. This is a stress rating only and functional operation of the device at these or any other conditions above those indicated in the operational sections of this specification is not implied. Exposure to absolute maximum rating conditions for extended periods may affect device reliability.

DC and AC Operating Conditions During Read

	2716	2716-1	2716-2	2716-5	2716-6
Temperature Range	0°C – 70°C	0°C – 70°C	0°C – 70°C	0°C – 70°C	0°C – 70°C
Vcc Power Supply[1,2]	5V ±5%	5V ±10%	5V ±5%	5V ±5%	5V ±5%
Vpp Power Supply[2]	Vcc	Vcc	Vcc	Vcc	Vcc

READ OPERATION
D.C. and Operating Characteristics

Symbol	Parameter	Limits			Unit	Conditions
		Min.	Typ.[3]	Max.		
ILI	Input Load Current			10	μA	VIN = 5.25V
ILO	Output Leakage Current			10	μA	VOUT = 5.25V
IPP1[2]	Vpp Current			5	mA	Vpp = 5.25V
ICC1[2]	Vcc Current (Standby)		10	25	mA	CE = VIH, OE = VIL
ICC2[2]	Vcc Current (Active)		57	100	mA	OE = CE = VIL
VIL	Input Low Voltage	−0.1		0.8	V	
VIH	Input High Voltage	2.0		Vcc+1	V	
VOL	Output Low Voltage			0.45	V	IOL = 2.1 mA
VOH	Output High Voltage	2.4			V	IOH = −400 μA

NOTES:
1. Vcc must be applied simultaneously or before Vpp and removed simultaneously or after Vpp
2. Vpp may be connected directly to Vcc except during programming. The supply current would then be the sum of ICC and Ipp1.
3. Typical values are for TA = 25°C and nominal supply voltages
4. This parameter is only sampled and is not 100% tested.

Typical Characteristics

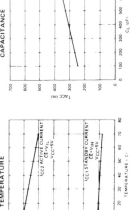

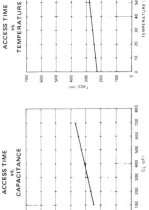

ERASURE CHARACTERISTICS

The erasure characteristics of the 2716 are such that erasure begins to occur when exposed to light with wavelengths shorter than approximately 4000 Angstroms (Å). It should be noted that sunlight and certain types of fluorescent lamps have wavelengths in the 3000–4000Å range. Data show that constant exposure to room level fluorescent lighting could erase the typical 2716 in approximately 3 years, while it would take approximately 1 week to cause erasure when exposed to direct sunlight. If the 2716 is to be exposed to these types of lighting conditions for extended periods of time, opaque labels are available from Intel which should be placed over the 2716 window to prevent unintentional erasure.

The recommended erasure procedure (see Data Catalog PROM/ROM Programming Instruction Section) for the 2716 is exposure to shortwave ultraviolet light which has a wavelength of 2537 Angstroms (Å). The integrated dose (i.e., UV intensity X exposure time) for erasure should be a minimum of 15 W-sec/cm². The erasure time with this dosage is approximately 15 to 20 minutes using an ultra-violet lamp with a 12000 μW/cm² power rating. The 2716 should be placed within 1 inch of the lamp tubes during erasure. Some lamps have a filter on their tubes which should be removed before erasure.

DEVICE OPERATION

The five modes of operation of the 2716 are listed in Table I. It should be noted that all inputs for the five modes are at TTL levels. The power supplies required are a +5V V_{CC} and a V_{PP}. The V_{PP} power supply must be at 25V during the three programming modes, and must be at 5V in the other two modes

OUTPUT OR-TIEING

Because 2716's are usually used in larger memory arrays, Intel has provided a 2 line control function that accomodates this use of multiple memory connections. The two line control function allows for:

a) the lowest possible memory power dissipation, and
b) complete assurance that output bus contention will not occur.

To most efficiently use these two control lines, it is recommended that $\overline{CE}$ (pin 18) be decoded and used as the primary device selecting function, while $\overline{OE}$ (pin 20) be made a common connection to all devices in the array and connected to the READ line from the system control bus. This assures that all deselected memory devices are in their low power standby mode and that the output pins are only active when data is desired from a particular memory device.

PROGRAMMING (See Programming Instruction Section for Waveforms.)

Initially, and after each erasure, all bits of the 2716 are in the "1" state. Data is introduced by selectively programming "0's" into the desired bit locations. Although only "0's" will be programmed, both "1's" and "0's" can be presented in the data word. The only way to change a "0" to a "1" is by ultraviolet light erasure.

The 2716 is in the programming mode when the V_{PP} power supply is at 25V and $\overline{OE}$ is at V_{IH}. The data to be programmed is applied 8 bits in parallel to the data output pins. The levels required for the address and data inputs are TTL.

When the address and data are stable, a 50 msec, active high, TTL program pulse is applied to the $\overline{CE}$/PGM input. A program pulse must be applied at each address location to be programmed. You can program any location at any time – either individually, sequentially, or at random. The program pulse has a maximum width of 55 msec. The 2716 must not be programmed with a DC signal applied to the $\overline{CE}$/PGM input.

Programming of multiple 2716s in parallel with the same data can be easily accomplished due to the simplicity of the programming requirements. Like inputs of the paralleled 2716s may be connected together when they are programmed with the same data. A high level TTL pulse applied to the $\overline{CE}$/PGM input programs the paralleled 2716s.

PROGRAM INHIBIT

Programming of multiple 2716s in parallel with different data is also easily accomplished. Except for $\overline{CE}$/PGM, all like inputs (including $\overline{OE}$) of the parallel 2716s may be common. A TTL level program pulse applied to a 2716's $\overline{CE}$/PGM input with V_{PP} at 25V will program that 2716. A low level $\overline{CE}$/PGM input inhibits the other 2716 from being programmed.

PROGRAM VERIFY

A verify should be performed on the programmed bits to determine that they were correctly programmed. The verify may be performed with V_{PP} at 25V. Except during programming and program verify, V_{PP} must be at 5V.

READ MODE

The 2716 has two control functions, both of which must be logically satisfied in order to obtain data at the outputs. Chip Enable ($\overline{CE}$) is the power control and should be used for device selection. Output Enable ($\overline{OE}$) is the output control and should be used to gate data to the output pins, independent of device selection. Assuming that addresses are stable, address access time (t_{ACC}) is equal to the delay from $\overline{CE}$ to output (t_{CE}). Data is available at the outputs 120 ns (t_{OE}) after the falling edge of $\overline{OE}$, assuming that $\overline{CE}$ has been low and addresses have been stable for at least $t_{ACC} - t_{OE}$.

STANDBY MODE

The 2716 has a standby mode which reduces the active power dissipation by 75%, from 525 mW to 132 mW. The 2716 is placed in the standby mode by applying a TTL high signal to the $\overline{CE}$ input. When in standby mode, the outputs are in a high impedence state, independent of the $\overline{OE}$ input.

TABLE I. MODE SELECTION

PINS MODE	$\overline{CE}$/PGM (18)	$\overline{OE}$ (20)	V_{PP} (21)	V_{CC} (24)	OUTPUTS (9-11, 13-17)
Read	V_{IL}	V_{IL}	+5	+5	D_{OUT}
Standby	V_{IH}	Don't Care	+5	+5	High Z
Program	Pulsed V_{IL} to V_{IH}	V_{IH}	+25	+5	D_{IN}
Program Verify	V_{IL}	V_{IL}	+25	+5	D_{OUT}
Program Inhibit	V_{IL}	V_{IH}	+25	+5	High Z

A.C. Characteristics

Symbol	Parameter	2716 Min.	2716 Max.	2716-1 Min.	2716-1 Max.	2716-2 Min.	2716-2 Max.	2716-5 Min.	2716-5 Max.	2716-6 Min.	2716-6 Max.	Test Conditions
t_{ACC}	Address to Output Delay		450		350		390		450		450	$\overline{CE} = \overline{OE} = V_{IL}$
t_{CE}	$\overline{CE}$ to Output Delay		450		350		390		490		650	$\overline{OE} = V_{IL}$
t_{OE}	Output Enable to Output Delay		120		120		120		160		200	$\overline{CE} = V_{IL}$
t_{DF}	Output Enable High to Output Float	0	100	0	100	0	100	0	100	0	100	$\overline{CE} = V_{IL}$
t_{OH}	Output Hold from Addresses, $\overline{CE}$ or $\overline{OE}$ Whichever Occurred First	0		0		0		0		0		$\overline{CE} = \overline{OE} = V_{IL}$

Limits (ns)

Capacitance [4] $T_A = 25°C$, $f = 1$ MHz

Symbol	Parameter	Typ.	Max.	Unit	Conditions
C_{IN}	Input Capacitance	4	6	pF	$V_{IN} = 0V$
C_{OUT}	Output Capacitance	8	12	pF	$V_{OUT} = 0V$

A.C. Test Conditions:

Output Load: 1 TTL gate and $C_L = 100$ pF
Input Rise and Fall Times: ≤20 ns
Input Pulse Levels: 0.8V to 2.2V
Timing Measurement Reference Level:
Inputs   1V and 2V
Outputs   0.8V and 2V

A.C. Waveforms [1]

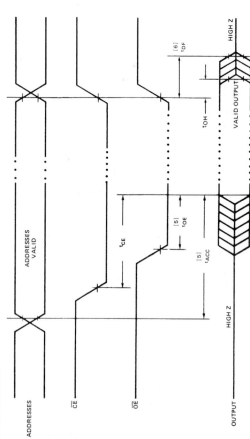

NOTE:
1. V_{CC} must be applied simultaneously or before V_{PP} and removed simultaneously or after V_{PP}.
2. V_{PP} may be connected directly to V_{CC} except during programming. The supply current would then be the sum of I_{CC} and I_{PP}.
3. Typical values are for $T_A = 25°C$ and nominal supply voltages.
4. This parameter is only sampled and is not 100% tested.
5. This parameter is only sampled and is not 100% tested.
6. $\overline{OE}$ may be delayed up to $t_{ACC} - t_{OE}$ after the falling edge of $\overline{CE}$ without impact on t_{ACC}.
7. t_{DF} is specified from $\overline{OE}$ or $\overline{CE}$, whichever occurs first.

CMOS 8-BIT A/D CONVERTERS

Preliminary

DESCRIPTION

The ADC0801 family is a series of five CMOS 8-bit successive approximation A/D converters using a resistive ladder and capacitive array together with an auto-zero comparator. These converters are designed to operate with microprocessor controlled buses using a minimum of external circuitry. The three-state output data lines can be connected directly to the data bus.

The differential analog voltage input allows for increased common-mode rejection and provides a means to adjust the zero scale offset. Additionally, the voltage reference input provides a means of encoding small analog voltages to the full 8 bits of resolution.

FEATURES

- Compatible with most microprocessors
- Differential inputs
- Three-state outputs
- Logic levels TTL and MOS compatible
- Can be used with internal or external clock
- Analog input range 0V to V_{CC}
- Single 5V supply
- Guaranteed specification with 1MHz clock

APPLICATIONS

- Transducer to microprocessor interface
- Digitally-controlled thermostat
- Digital thermometer
- Microprocessor-based monitoring and control systems

PIN CONFIGURATION

F,N PACKAGE

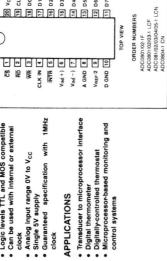

TOP VIEW

ORDER NUMBERS

ADC0801/02-1 F
ADC0801/02/03-1 LCF
ADC081/02/03/04/05-1 LCN
ADC0804-1 CN

ABSOLUTE MAXIMUM RATINGS

	SYMBOL & PARAMETER	RATING	UNIT
V_{CC}	Supply Voltage	6.5	V
	Logic Control Input Voltages	−0.3 to +16	V
	All Other Input Voltages	−0.3 to (V_{CC} +0.3)	V
T_A	Operating Temperature Range		
	ADC0801/02-1 F	−55 to +125	°C
	ADC0801/02/03-1 LCF	−40 to +85	°C
	ADC0801/02/03/04/05-1 LCN	−40 to +85	°C
	ADC0804-1 CN	0 to +70	°C
T_{STG}	Storage Temperature	−65 to +150	°C
T_{SOLD}	Lead Soldering Temperature (10 seconds)	300	°C
P_D	Package Power Dissipation at T_A = 25°C	875	mW

Preliminary

BLOCK DIAGRAMS

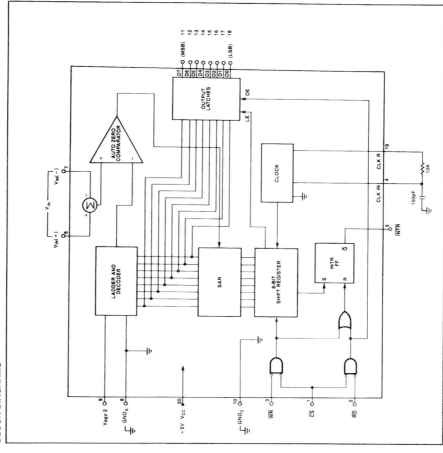

CMOS 8-BIT A/D CONVERTERS   ADC0801/2/3/4/5

Preliminary

AC ELECTRICAL CHARACTERISTICS

SYMBOL & PARAMETER	TO	FROM	TEST CONDITIONS	ADC0801/2/3/4/5 Min	Typ	Max	UNIT
Conversion Time			f_{CLK} = 1MHz[1]	66		73	µS
f_{CLK} Clock Frequency			See Note 1.	0.1	1.0	3.0	MHz
Clock Duty Cycle			See Note 1.	40		60	%
CR Free-Running Conversion Rate			$\overline{CS}$ = 0, f_{CLK} = 1MHz $\overline{INTR}$ Tied To $\overline{WR}$			13690	conv's
$t_{W(\overline{WR})L}$ Start Pulse Width			$\overline{CS}$ = 0	30			ns
t_{ACC} Access Time	Output	$\overline{RD}$	$\overline{CS}$ = 0, C_L = 100 pF		75	100	ns
t_{1H}, t_{0H} Three-State Control	Output	$\overline{RD}$	CL = 10 pF, RL = 10K See Three-State Test Circuit		70	100	ns
t_{W1}, t_{R1} $\overline{INTR}$ Delay	$\overline{INTR}$	$\overline{WD}$ or $\overline{RD}$			100	150	ns
C_{IN} Logic Input =Capacitance					5	7.5	pF
C_{OUT} Three-State Output Capacitance					5	7.5	pF

NOTE:
1. Accuracy is guaranteed at f_{CLK} = 1MHz. Accuracy may degrade at higher clock frequencies.

CMOS 8-BIT A/D CONVERTERS   ADC0801/2/3/4/5-1

Preliminary

DC ELECTRICAL CHARACTERISTICS V_{CC} = 5.0V, f_{CLK} = 1MHz, $T_{MIN} \le T_A \le T_{MAX}$, unless otherwise specified.

SYMBOL & PARAMETER	TEST CONDITIONS	ADC0801/2/3/4/5 Min	Typ	Max	UNIT
ADC0801 Relative Accuracy Error (Adjusted)	Full Scale Adjusted			0.25	LSB
ADC0802 Relative Accuracy Error (Unadjusted)	$\frac{V_{REF}}{2}$ = 2.500 V_{DC}			0.50	LSB
ADC0803 Relative Accuracy Error (Adjusted)	Full Scale Adjusted			0.50	LSB
ADC0804 Relative Accuracy Error (Unadjusted)	$\frac{V_{REF}}{2}$ = 2.500 V_{DC}			1	LSB
ADC0805 Relative Accuracy Error (Unadjusted)	$\frac{V_{REF}}{2}$ = has no connection			1	LSB
$\frac{V_{REF}}{2}$ Input Resistance		400	640		Ω
Analog Input Voltage Range	Over Analog Input Voltage Range[1]	-0.05		V_{CC} +0.05	V
DC Common Mode Error			1/16	1/8	LSB
Power Supply Sensitivity	V_{CC} = 5V ± 10%[1]				

CONTROL INPUTS

SYMBOL & PARAMETER	TEST CONDITIONS	Min	Typ	Max	UNIT
V_{IH} Logical "1" Input Voltage	V_{CC} = 5.25V_{DC}	2.0		15	V_{DC}
V_{IL} Logical "0" Input Voltage	V_{CC} = 4.75V_{DC}			0.8	V_{DC}
I_{IH} Logical "1" Input Current	V_{IN} = 5V_{DC}		0.005	1	μA_{DC}
I_{IL} Logical "0" Input Current	V_{IN} = 0V_{DC}	-1	-0.005		μA_{DC}

CLOCK IN AND CLOCK R

SYMBOL & PARAMETER	TEST CONDITIONS	Min	Typ	Max	UNIT
V_{T+} Clk In Positive-Going Threshold Voltage		2.7	3.1	3.5	V_{DC}
V_{T-} Clk In Negative-Going Threshold Voltage		1.5	1.8	2.1	V_{DC}
V_H Clk In Hysteresis (V_{T+}) – (V_{T-})		0.6	1.3	2.0	V_{DC}
V_{OL} Logical "0" Clk R Output Voltage	I_{OL} = 360µA, V_{CC} = 4.75 V_{DC}			0.4	V_{DC}
V_{OH} Logical "1" Clk R Output Voltage	I_{OH} = -360µA, V_{CC} = 4.75 V_{DC}	2.4			V_{DC}

DATA OUTPUT AND $\overline{INTR}$

SYMBOL & PARAMETER	TEST CONDITIONS	Min	Typ	Max	UNIT
V_{OL} Logical "0" Output Voltage					
Data Outputs	I_{OL} = 1.6mA, V_{CC} = 4.75 V_{DC}			0.4	V_{DC}
$\overline{INTR}$ Outputs	I_{OL} = 1.0mA, V_{CC} = 4.75 V_{DC}			0.4	V_{DC}
V_{OH} Logical "1" Output Voltage	I_{OH} = -360µA, V_{CC} = 4.75 V_{DC}	2.4			V_{DC}
	I_{OH} = -10µA, V_{CC} = 4.75 V_{DC}	4.5			V_{DC}
I_{OZL} 3-State Output Leakage	V_{OUT} = 0V_{DC}, $\overline{CS}$ = Logical "1"	-3			μA_{DC}
I_{OZH} 3-State Output Leakage	V_{OUT} = 5V_{DC}, $\overline{CS}$ = Logical "1"			3	μA_{DC}
I_{SC} + Output Short Circuit Current	V_{OUT} = 0V, T_A = 25°C	4.5	6		mA_{DC}
I_{SC} – Output Short Circuit Current	V_{OUT} = V_{CC}, T_A = 25°C	9.0	16		mA_{DC}
I_{CC} Power Supply Current	f_{CLK} = 1MHz, V_{REF} 2 = Open $\overline{CS}$ = Logical "1", T_A = 25°C		3.0	3.5	mA

NOTE:
1. Analog inputs must remain within the range: -0.05 ≤ V_{IN} ≤ V_{CC} + 0.05V.

Preliminary

FUNCTIONAL DESCRIPTION

The ADC0801 through ADC0805 series of A/D converters are successive approximation devices with 8-bit resolution and no missing codes. The most significant bit is tested first and after 64 clock cycles a digital 8-bit binary word is transferred to an output latch and the INTR pin goes low, indicating that conversion is complete. A conversion in progress can be interrupted by issuing another start command. The device may be operated in a continuous conversion mode by connecting the INTR and WR pins together and holding the CS pin low. To insure start-up when connected this way, an external WR pulse is required at power-up.

As the WR input goes low, when CS is low, the SAR is cleared and remains so as long as these two inputs are low. Conversion begins between 1 and 8 clock periods after at least one of these inputs goes high. As the conversion begins, the INTR line goes high. Note that the INTR line will remain low until 1 to 8 clock cycles after either the WR or the CS input (or both) goes high.

When the CS and RD inputs are both brought low to read the data, the INTR line will go low and the three-state output latches are enabled.

The digital control lines (CS, RD, and WR) operate with standard TTL levels and have been renamed when compared with standard A/D Start and Output Enable labels. For non-microprocessor based applications, the CS pin can be grounded, the WR pin can be interpreted as a START pulse pin, and the RD pin performs the OE (Output Enable) function.

The $V_{IN(-)}$ input can be used to subtract a fixed voltage from the input voltage. Because there is a time interval between sampling the $V_{IN(+)}$ and the $V_{I(-)}$ inputs, it is important that these inputs remain constant, during the entire conversion cycle.

THREE-STATE TEST CIRCUITS AND WAVEFORMS

Preliminary

TIMING DIAGRAMS (All timing is measured from the 50% voltage points)

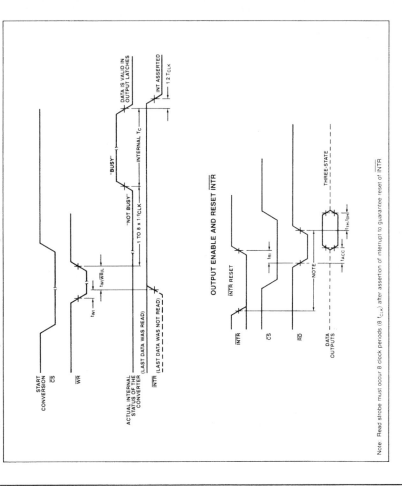

Note: Read strobe must occur 8 clock periods ($8\ t_{CLK}$) after assertion of interrupt to guarantee reset of INTR

8-BIT MULTIPLYING D/A CONVERTER   MC1508-8/1408-8/1408-7

DC ELECTRICAL CHARACTERISTICS[1]

$V_{CC} = +5.0$Vdc, $V_{EE} = -15$Vdc, $\frac{V_{ref}}{R_{14}} = 2.0$mA unless otherwise specified. MC1508: $T_A = -55°C$ to $125°C$. MC1408: $T_A = 0°C$ to $75°C$ unless otherwise noted.

Pin 3 must be 3V more negative than the potential to which R_{15} is returned

	PARAMETER	TEST CONDITIONS	MC1508-8 Min	MC1508-8 Typ	MC1508-8 Max	MC1408-8 Min	MC1408-8 Typ	MC1408-8 Max	MC1408-7 Min	MC1408-7 Typ	MC1408-7 Max	UNIT
E_r	Relative accuracy	Error relative to full scale I_o, Figure 3			±0.19			±0.19			±0.39	%
t_s	Settling time[1]	To within ½ LSB includes t_{PLH}, $T_A = +25°C$, Figure 4		70			70			70		ns
t_{PLH} t_{PHL}	Propagation delay time Low-to-high High-to-low	$T_A = +25°C$, Figure 4		35	100		35	100		35	100	ns
TCI_o	Output full scale current drift			-20			-20			-20		PPM/°C
V_{IH} V_{IL}	Digital input logic level (MSB) High Low	Figure 5	2.0		0.8	2.0		0.8	2.0		0.8	Vdc
I_{IH} I_{IL}	Digital input current (MSB) High Low	Figure 5 $V_{IH} = 5.0$V $V_{IL} = 0.8$V		0 -0.4	0.04 -0.8		0 -0.4	0.04 -0.8		0 -0.4	0.04 -0.8	µA
I_{15}	Reference input bias current	Pin 15, Figure 5		-1.0	-5.0		-1.0	-5.0		-1.0	-5.0	µA
I_{OR}	Output current range	$V_{EE} = -5.0$V $V_{EE} = -7.0$V to -15V	0 0	2.0 2.0	2.1 4.2	0 0	2.0 2.0	2.1 4.2	0 0	2.0 2.0	2.1 4.2	mA
I_O	Output current	Figure 5 $V_{ref} = 2.000$V, R14 = 1000Ω All bits low	1.9	1.99	2.1	1.9	1.99	2.1	1.9	1.99	2.1	mA
$I_{o(min)}$	Off-state	$E \leq 0.19\%$ at $T_A = +25°C$, Figure 5			4.0			4.0			4.0	µA
V_o	Output voltage compliance	$V_{EE} = -5$V V_{EE} below -10V		-0.6, +10 -5.5, +10	-0.55, +0.5 -5.0, +0.5		-0.6, +10 -5.5, +10	-0.55, +0.5 -5.0, +0.5		-0.6, +10 -5.5, +10	-0.55, +0.5 -5.0, +0.5	Vdc
SRI_{ref}	Reference current slew rate	Figure 6		8.0			8.0			8.0		mA/µs
$PSRR_{(-)}$	Output current power supply sensitivity	$I_{ref} = 1$mA		0.5	2.7		0.5	2.7		0.5	2.7	µA/V
I_{CC} I_{EE}	Power supply current Positive Negative	All bits low, Figure 5		+2.5 -6.5	+22 -13		+2.5 -6.5	+22 -13		+2.5 -6.5	+22 -13	mA
V_{CCR} V_{EER}	Power supply voltage range Positive Negative	$T_A = +25°C$, Figure 5	+4.5 -4.5	+5.0 -15	+5.5 -16.5	+4.5 -4.5	+5.0 -15	+5.5 -16.5	+4.5 -4.5	+5.0 -15	+5.5 -16.5	Vdc
P_D	Power dissipation	All bits low, Figure 5 $V_{EE} = -5.0$Vdc $V_{EE} = -15$Vdc		34 110	170 305		34 110	170 305		34 110	170 305	mW

NOTES:
1. All bits switched.

DESCRIPTION

The MC1508/MC1408 series of 8-bit monolithic digital-to-analog converters provide high speed performance with low cost. They are designed for use where the output current is a linear product of an 8-bit digital word and an analog reference voltage.

FEATURES

- Fast settling time—70ns (typ)
- Relative accuracy ±0.19% (max error)
- Non-inverting digital inputs are TTL and CMOS compatible
- High speed multiplying rate 4.0mA/µs (input slew)
- Output voltage swing -.5V to -5.0V
- Standard supply voltages +5.0V and -5.0V to -15V
- Military qualifications pending

APPLICATIONS

- Tracking A-to-D converters
- 2½-digit panel meters and DVM's
- Waveform synthesis
- Sample and hold
- Peak detector
- Programmable gain and attenuation
- CRT character generation
- Audio digitizing and decoding
- Programmable power supplies
- Analog-digital multiplication
- Digital-digital multiplication
- Analog-digital division
- Digital addition and subtraction
- Speech compression and expansion
- Stepping motor drive
- Modems
- Servo motor and pen drivers

CIRCUIT DESCRIPTION

The MC1508/MC1408 consists of a reference current amplifier, an R-2R ladder, and 8 high speed current switches. For many applications, only a reference resistor and reference voltage need be added.

The switches are non-inverting in operation, therefore, a high state on the input turns on the specified output current component.

The switch uses current steering for high speed, and a termination amplifier consisting of an active load gain stage with unity gain feedback. The termination amplifier holds the parasitic capacitance of the ladder at a constant voltage during switching, and provides a low impedance termination of equal voltage for all legs of the ladder.

The R-2R ladder divides the reference amplifier current into binarily-related components, which are fed to the switches. Note that there is always a remainder current which is equal to the least significant bit. This current is shunted to ground, and the maximum output current is 255/256 of the reference amplifier current, or 1.992mA for a 2.0mA reference amplifier current if the NPN current source pair is perfectly matched.

PIN CONFIGURATION

F,N PACKAGE

TOP VIEW

1	NC
2	GND
3	V_{EE}
4	I_o
5	A_1 (MSB)
6	A_2
7	A_3
8	A_4
9	A_5
10	A_6
11	A_7
12	A_8 (LSB)
13	V_{CC}
14	$V_{REF}(+)$
15	$V_{REF}(-)$
16	COMPEN

ORDER NUMBERS
MC1508-8F   MC1408-7N
MC1408-8F   MC1408-7F

D³ PACKAGE

TOP VIEW

1	V+
2	$V_{REF}(+)$
3	$V_{REF}(-)$
4	COMPEN
5	NC
6	GND
7	V-
8	I_o
9	A_1 (MSB)
10	A_2
11	A_3
12	A_4
13	A_5
14	A_6
15	A_7
16	A_8 (LSB)

ORDER NUMBER
MC1408D

NOTES:
1. SOL Released in Large SO package only.
2. SOL and non-standard pinout.
3. SO and non-standard pinouts.

BLOCK DIAGRAM

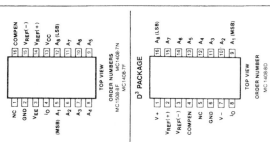

ABSOLUTE MAXIMUM RATINGS $T_A = +25°C$ unless otherwise specified

	PARAMETER	RATING	UNIT
V_{CC}	Power Supply Voltage Positive	+5.5	V
V_{EE}	Negative	-16.5	V
V_{S1-12}	Digital Input Voltage	0 to V_{CC}	V
V_o	Applied Output Voltage	-5.2 to +18	V
I_{14}	Reference Current	5.0	mA
V_{14}, V_{15}	Reference Amplifier Inputs	V_{EE} to V_{CC}	
P_D	Power Dissipation (Package Limitation) Ceramic Package	1000	mW
	Plastic Package	800	mW
	Lead Soldering Temperature (60 sec)	300	°C
T_A	Operating Temperature Range MC1508	-55 to +125	°C
	MC1408	0 to +75	°C
T_{STG}	Storage Temperature Range	-65 to +150	°C

TEST CIRCUITS

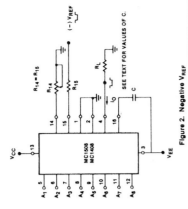

Figure 1. Positive V$_{REF}$

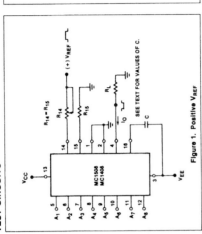

Figure 2. Negative V$_{REF}$

TYPICAL PERFORMANCE CHARACTERISTICS

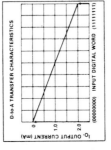

D-to-A TRANSFER CHARACTERISTICS

FUNCTIONAL DESCRIPTION

Reference Amplifier Drive and Compensation

The reference amplifier input current must always flow into pin 14 regardless of the setup method or reference supply voltage polarity.

Connections for a positive reference voltage are shown in Figure 1. The reference voltage source supplies the full reference current. For bipolar reference signals, as in the multiplying mode, R$_{15}$ can be tied to a negative voltage corresponding to the minimum input level. R$_{15}$ may be eliminated and pin 15 grounded, with only a small sacrifice in accuracy and temperature drift.

The compensation capacitor value must be increased with increasing values of R$_{14}$ to maintain proper phase margin. For R$_{14}$ values of 1.0, 2.5, and 5.0K ohms, minimum capacitor values are 15, 37, and 75pF. The capacitor may be tied to either V$_{EE}$ or ground, but using V$_{EE}$ increases negative supply rejection. (Fluctuations in the negative supply have more effect on accuracy than do any changes in the positive supply).

A negative reference voltage may be used if R$_{14}$ is grounded and the reference voltage is applied to R$_{15}$, as shown in Figure 2. A high input impedance is the main advantage of this method. The negative reference voltage must be at least 3.0V above the V$_{EE}$ supply. Bipolar input signals may be handled by connecting R$_{14}$ to a positive reference voltage equal to the peak positive input level at pin 15.

Capacitive bypass to ground is recommended when a DC reference voltage is used. The 5.0V logic supply is not recommended as a reference voltage, but if a

well regulated 5.0V supply which drives logic is to be used as the reference, R$_{14}$ should be formed of two series resistors and the junction of the two resistors bypassed with 0.1μF to ground. For reference voltages greater than 5.0V, a clamp diode is recommended between pin 14 and ground.

If pin 14 is driven by a high impedance such as a transistor current source, none of the above compensation methods apply and the amplifier must be heavily compensated, decreasing the overall bandwidth.

Output Voltage Range

The voltage at pin 4 must always be at least 4.5 volts more positive than the voltage of the negative supply (pin 3) when the reference current is 2mA or less, and at least 8 volts more positive than the negative supply when the reference current is between 2mA and 4mA. This is necessary to avoid saturation of the output transistors, which would cause serious degradation of accuracy.

Signetics MC1508/MC1408 does not need a range control because the design extends the compliance range down to 4.5 volts (or 8 volts—see above) above the negative supply voltage without significant degradation of accuracy. Signetics' MC1508/MC1408 can be used in sockets designed for other manufacturers' MC1508/MC1408 without circuit modification.

Output Current Range

Any time the full scale current exceeds 2mA, the negative supply must be at least 8 volts more negative than the output voltage. This is due to the increased internal voltage drops between the negative supply and the outputs with higher reference currents.

Accuracy

Absolute accuracy is the measure of each output current level with respect to its intended value, and is dependent upon relative accuracy and full scale current drift. Relative accuracy is the measure of each output current level as a fraction of the full scale current after zero scale current has been nulled out. The relative accuracy of the MC1508/MC1408 is essentially constant over the operating temperature range because of the excellent temperature tracking of the monolithic resistor ladder. The reference current may drift with temperature, causing a change in the absolute accuracy of output current; however, the MC1508/MC1408 has a very low full scale current drift over the operating temperature range

The MC1508/MC1408 series is guaranteed accurate to within ±1/2 LSB at +25°C at a full scale output current of 1.99mA. The relative accuracy test circuit is shown in Figure 3. The 12-bit converter is calibrated to a full scale output current of 1.99219mA; then the MC1508/MC1408's full scale current is trimmed to the same value with R$_{14}$ so that a zero value appears at the error amplifier output. The counter is activated and the error band may be displayed on the oscilloscope, detected by comparators, or stored in a peak detector.

Two 8-bit D-to-A converters may not be used to construct a 16-bit accurate D-to-A converter. Sixteen-bit accuracy implies a total of ±1/2 part in 65,536, or ±0.00076%, which is much more accurate than the ±0.19% specification of the MC1508/MC1408.

Monotonicity

A monotonic converter is one which always provides an analog output greater than or equal to the preceding value for a corresponding increment in the digital input code. The MC1508/MC1408 is monotonic for all values of reference current above 0.5mA. The recommended range for operation is a DC reference current between 0.5mA and 4.0mA.

Settling Time

The worst case switching condition occurs when all bits are switched on, which corresponds to a low-to-high transition for all input bits. This time is typically 70ns for settling to within 1/2 LSB for 8-bit accuracy. This time applies when RL < 500 ohms and C$_O$ < 25pF. The slowest single switch is the least significant bit, which typically turns on and settles in 65ns. In applications where the D-to-A converter functions in a positive going ramp mode, the worst case condition does not occur and settling times less than 70ns may be realized.

Extra care must be taken in board layout since this usually is the dominant factor in satisfactory test results when measuring settling time. Short leads, 100μF supply bypassing for low frequencies, minimum scope lead length, good ground planes, and avoidance of ground loops are all mandatory.

TEST CIRCUITS (Cont'd)

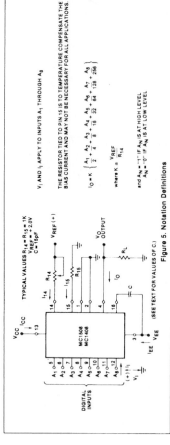

TYPICAL VALUES $R_{14} = R_{15} = 1K$
$V_{REF} = +2.0V$
$C = 15pF$

(SEE TEXT FOR VALUES OF C.)

THE RESISTOR TIED TO PIN 15 IS TO TEMPERATURE COMPENSATE THE BIAS CURRENT AND MAY NOT BE NECESSARY FOR ALL APPLICATIONS.

V_I AND I_I APPLY TO INPUTS A_1 THROUGH A_8

$$I_O = K \left\{ \frac{A_1}{2} + \frac{A_2}{4} + \frac{A_3}{8} + \frac{A_4}{16} + \frac{A_5}{32} + \frac{A_6}{64} + \frac{A_7}{128} + \frac{A_8}{256} \right\}$$

where $K = \dfrac{V_{REF}}{R_{14}}$

and $A_n = $ "1" IF A_n IS AT HIGH LEVEL
$A_n = $ "0" IF A_n IS AT LOW LEVEL

Figure 5. Notation Definitions

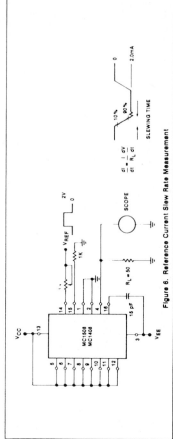

$\dfrac{di}{dt} = \dfrac{1}{R_L} \dfrac{dV}{dt}$

Figure 6. Reference Current Slew Rate Measurement

TEST CIRCUITS (Cont'd)

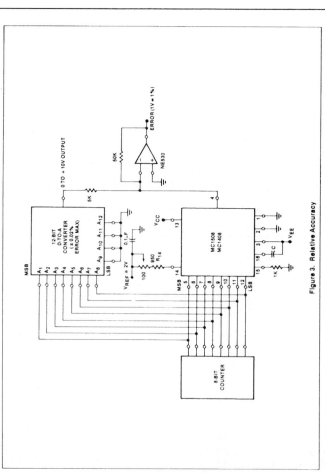

Figure 3. Relative Accuracy

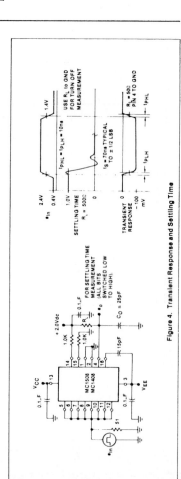

Figure 4. Transient Response and Settling Time

Signetics

Linear Products

µA741/µA741C/SA741C
General Purpose Operational Amplifier

Product Specification

DESCRIPTION

The µA741 is a high performance operational amplifier with high open-loop gain, internal compensation, high common mode range and exceptional temperature stability. The µA741 is short-circuit-protected and allows for nulling of offset voltage.

FEATURES

- **Internal frequency compensation**
- **Short circuit protection**
- **Excellent temperature stability**
- **High input voltage range**

PIN CONFIGURATION

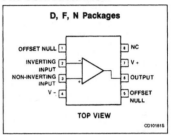

D, F, N Packages

OFFSET NULL	1	8 NC
INVERTING INPUT	2	7 V +
NON-INVERTING INPUT	3	6 OUTPUT
V −	4	5 OFFSET NULL

TOP VIEW

CD10181S

ORDERING INFORMATION

DESCRIPTION	TEMPERATURE RANGE	ORDER CODE
8-Pin Plastic DIP	−55°C to +125°C	µA741N
8-Pin Plastic DIP	0 to +70°C	µA741CN
8-Pin Plastic DIP	−40°C to +85°C	SA741CN
8-Pin Cerdip	−55°C to +125°C	µA741F
8-Pin Cerdip	0 to +70°C	µA741CF
8-Pin SO	0 to +70°C	µA741CD

EQUIVALENT SCHEMATIC

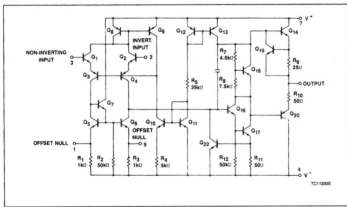

General Purpose Operational Amplifier

μA741/μA741C/SA741C

ABSOLUTE MAXIMUM RATINGS

SYMBOL	PARAMETER	RATING	UNIT
V_S	Supply voltage μA741C μA741	$\pm$18 $\pm$22	V V
P_D	Internal power dissipation D package N package F package	500 1000 1000	mW mW mW
V_{IN}	Differential input voltage	$\pm$30	V
V_{IN}	Input voltage[1]	$\pm$15	V
I_{SC}	Output short-circuit duration	Continuous	
T_A	Operating temperature range μA741C SA741C μA741	0 to +70 −40 to +85 −55 to +125	°C °C °C
T_{STG}	Storage temperature range	−65 to +150	°C
T_{SOLD}	Lead soldering temperature (10sec max)	300	°C

NOTE:
1. For supply voltages less than $\pm$15V, the absolute maximum input voltage is equal to the supply voltage.

DC ELECTRICAL CHARACTERISTICS (μA741, μA741C) $T_A = 25°C$, $V_S = \pm15V$, unless otherwise specified.

SYMBOL	PARAMETER	TEST CONDITIONS	μA741 Min	μA741 Typ	μA741 Max	μA741C Min	μA741C Typ	μA741C Max	UNIT
V_{OS} $\Delta V_{OS}/\Delta T$	Offset voltage	$R_S = 10k\Omega$ $R_S = 10k\Omega$, over temp.		1.0 1.0 10	5.0 6.0		2.0 10	6.0 7.5	mV mV μV/°C
I_{OS} $\Delta I_{OS}/\Delta T$	Offset current	 Over temp. $T_A = +125°C$ $T_A = −55°C$		20 7.0 20 200	200 200 500		20 300 200	200	nA nA nA nA pA/°C
I_{BIAS} $\Delta I_B/\Delta T$	Input bias current	 Over temp. $T_A = +125°C$ $T_A = −55°C$		80 30 300 1	500 500 1500		80 800 1	500	nA nA nA nA nA/°C
V_{OUT}	Output voltage swing	$R_L = 10k\Omega$ $R_L = 2k\Omega$, over temp.	$\pm$12 $\pm$10	$\pm$14 $\pm$13		$\pm$12 $\pm$10	$\pm$14 $\pm$13		V V
A_{VOL}	Large-signal voltage gain	$R_L = 2k\Omega$, $V_O = \pm10V$ $R_L = 2k\Omega$, $V_O = \pm10V$, over temp.	50 25	200		20 15	200		V/mV V/mV
	Offset voltage adjustment range			$\pm$30			$\pm$30		mV
PSRR	Supply voltage rejection ratio	$R_S \leqslant 10k\Omega$ $R_S \leqslant 10k\Omega$, over temp.		10	150		10	150	μV/V μV/V
CMRR	Common-mode rejection ratio	Over temp.	70	90					dB dB
I_{CC}	Supply current	$T_A = +125°C$ $T_A = −55°C$		1.4 1.5 2.0	2.8 2.5 3.3		1.4	2.8	mA mA mA

General Purpose Operational Amplifier

μA741/μA741C/SA741C

DC ELECTRICAL CHARACTERISTICS (Continued) (μA741, μA741C) $T_A = 25°C$, $V_S = \pm15V$, unless otherwise specified.

SYMBOL	PARAMETER	TEST CONDITIONS	μA741			μA741C			UNIT
			Min	Typ	Max	Min	Typ	Max	
V_{IN}	Input voltage range	(μA741, over temp.)	±12	±13		±12	±13		V
R_{IN}	Input resistance		0.3	2.0		0.3	2.0		MΩ
P_D	Power consumption	$T_A = +125°C$ $T_A = -55°C$		50 45 45	80 75 100		50	85	mW mW mW
R_{OUT}	Output resistance			75			75		Ω
I_{SC}	Output short-circuit current		10	25	60	10	25	60	mA

DC ELECTRICAL CHARACTERISTICS (SA741C) $T_A = 25°C$, $V_S = \pm15V$, unless otherwise specified.

SYMBOL	PARAMETER	TEST CONDITIONS	SA741C			UNIT
			Min	Typ	Max	
V_{OS}	Offset voltage	$R_S = 10k\Omega$ $R_S = 10k\Omega$, over temp.		2.0	6.0 7.5	mV mV
$\Delta V_{OS}/\Delta T$				10		μV/°C
I_{OS}	Offset current	Over temp.		20	200 500	nA nA
$\Delta I_{OS}/\Delta T$				200		pA/°C
I_{BIAS}	Input bias current	Over temp.		80	500 1500	nA nA
$\Delta I_B/\Delta T$				1		nA/°C
V_{OUT}	Output voltage swing	$R_L = 10k\Omega$ $R_L = 2k\Omega$, over temp.	±12 ±10	±14 ±13		V V
A_{VOL}	Large-signal voltage gain	$R_L = 2k\Omega$, $V_O = \pm10V$ $R_L = 2k\Omega$, $V_O = \pm10V$, over temp.	20 15	200		V/mV V/mV
	Offset voltage adjustment range			±30		mV
PSRR	Supply voltage rejection ratio	$R_S \leqslant 10k\Omega$		10	150	μV/V
V_{IN}	Input voltage range	(μA741, over temp.)	±12	±13		V
R_{IN}	Input resistance		0.3	2.0		MΩ
P_d	Power consumption			50	85	mW
R_{OUT}	Output resistance			75		Ω
I_{SC}	Output short-circuit current			25		mA

AC ELECTRICAL CHARACTERISTICS $T_A = 25°C$, $V_S = \pm15V$, unless otherwise specified.

SYMBOL	PARAMETER	TEST CONDITIONS	μA741, μA741C			UNIT
			Min	Typ	Max	
C_{IN}	Parallel input capacitance	Open-loop, f = 20Hz		1.4		pF
	Unity gain crossover frequency	Open-loop		1.0		MHz
t_R	Transient response unity gain Rise time Overshoot	$V_{IN} = 20mV$, $R_L = 2k\Omega$, $C_L \leqslant 100pF$		0.3 5.0		μs %
SR	Slew rate	$C \leqslant 100pF$, $R_L \geqslant 2k\Omega$, $V_{IN} = \pm10V$		0.5		V/μs

General Purpose Operational Amplifier

µA741/µA741C/SA741C

TYPICAL PERFORMANCE CHARACTERISTICS

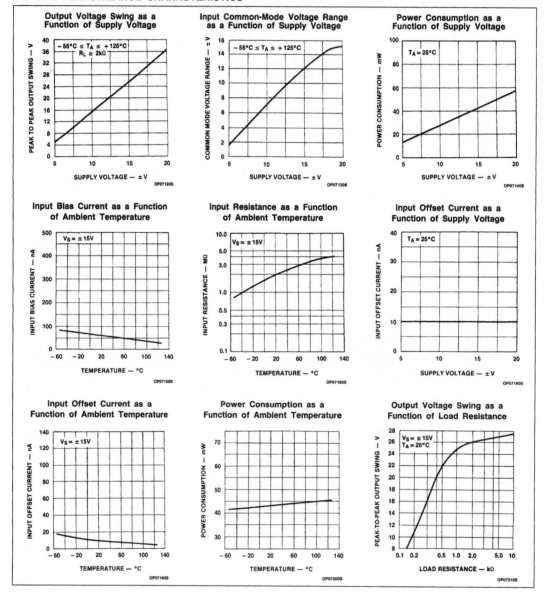

General Purpose Operational Amplifier

μA741/μA741C/SA741C

TYPICAL PERFORMANCE CHARACTERISTICS (Continued)

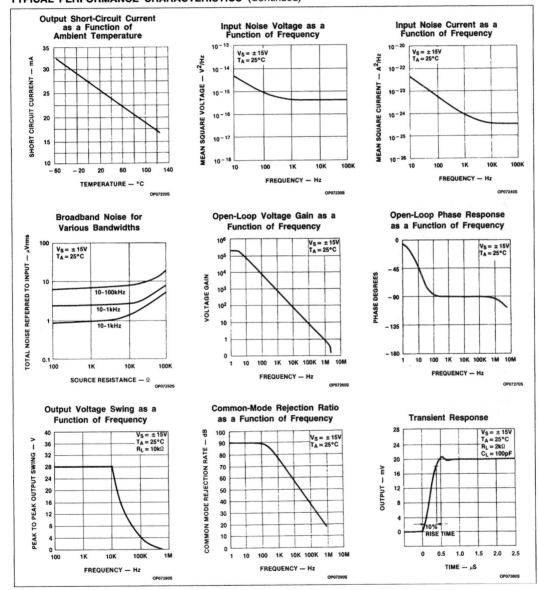

General Purpose Operational Amplifier

μA741/μA741C/SA741C

TYPICAL PERFORMANCE CHARACTERISTICS (Continued)

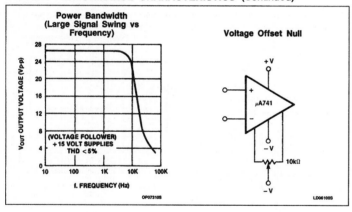

Signetics

NE/SE555/SE555C
Timer

Product Specification

Linear Products

DESCRIPTION

The 555 monolithic timing circuit is a highly stable controller capable of producing accurate time delays, or oscillation. In the time delay mode of operation, the time is precisely controlled by one external resistor and capacitor. For a stable operation as an oscillator, the free running frequency and the duty cycle are both accurately controlled with two external resistors and one capacitor. The circuit may be triggered and reset on falling waveforms, and the output structure can source or sink up to 200mA.

FEATURES
- Turn-off time less than 2µs
- Max. operating frequency greater than 500kHz
- Timing from microseconds to hours
- Operates in both astable and monostable modes
- High output current
- Adjustable duty cycle
- TTL compatible
- Temperature stability of 0.005% per °C

APPLICATIONS
- Precision timing
- Pulse generation
- Sequential timing
- Time delay generation
- Pulse width modulation
- Pulse position modulation
- Missing pulse detector

PIN CONFIGURATIONS

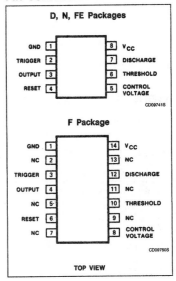

D, N, FE Packages

GND	1	8 VCC
TRIGGER	2	7 DISCHARGE
OUTPUT	3	6 THRESHOLD
RESET	4	5 CONTROL VOLTAGE

CD09741S

F Package

GND	1	14 VCC
NC	2	13 NC
TRIGGER	3	12 DISCHARGE
OUTPUT	4	11 NC
NC	5	10 THRESHOLD
RESET	6	9 NC
NC	7	8 CONTROL VOLTAGE

CD09750S

TOP VIEW

EQUIVALENT SCHEMATIC

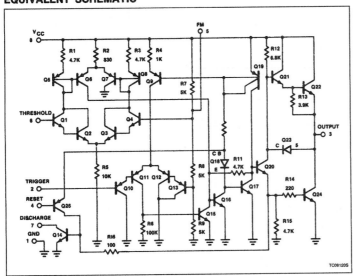

TC06120S

Timer

ORDERING INFORMATION

DESCRIPTION	TEMPERATURE RANGE	ORDER CODE
8-Pin Hermetic Cerdip	0 to +70°C	NE555FE
8-Pin Plastic SO	0 to +70°C	NE555D
8-Pin Plastic DIP	0 to +70°C	NE555N
8-Pin Hermetic Cerdip	−55°C to +125°C	SE555CFE
8-Pin Plastic DIP	−55°C to +125°C	SE555CN
14-Pin Plastic DIP	−55°C to +125°C	SE555N
8-Pin Hermetic Cerdip	−55°C to +125°C	SE555FE
14-Pin Ceramic DIP	0 to +70°C	NE555F
14-Pin Ceramic DIP	−55°C to +125°C	SE555F
14-Pin Ceramic DIP	−55°C to +125°C	SE555CF

ABSOLUTE MAXIMUM RATINGS

SYMBOL	PARAMETER	RATING	UNIT
V_{CC}	Supply voltage SE555 NE555, SE555C	+18 +16	V V
P_D	Maximum allowable power dissipation[1]	600	mW
T_A	Operating ambient temperature range NE555 SE555, SE555C	0 to +70 −55 to +125	°C °C
T_{STG}	Storage temperature range	−65 to +150	°C
T_{SOLD}	Lead soldering temperature (10sec max)	+300	°C

NOTE:
1. The junction temperature must be kept below 125°C for the D package and below 150°C for the FE, N and F packages. At ambient temperatures above 25°C, where this limit would be derated by the following factors:

    D package 160 °C/W
    FE package 150 °C/W
    N package 100 °C/W
    F package 105 °C/W

BLOCK DIAGRAM

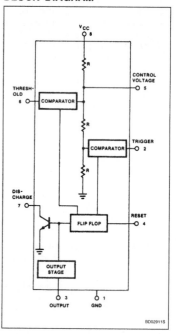

Timer

NE/SE555/SE555C

DC AND AC ELECTRICAL CHARACTERISTICS $T_A = 25°C$, $V_{CC} = +5V$ to $+15$ unless otherwise specified.

SYMBOL	PARAMETER	TEST CONDITIONS	SE555			NE555/SE555C			UNIT
			Min	Typ	Max	Min	Typ	Max	
V_{CC}	Supply voltage		4.5		18	4.5		16	V
I_{CC}	Supply current (low state)[1]	$V_{CC} = 5V$, $R_L = \infty$ $V_{CC} = 15V$, $R_L = \infty$		3 10	5 12		3 10	6 15	mA mA
t_M $\Delta t_M/\Delta T$ $\Delta t_M/\Delta V_S$	Timing error (monostable) Initial accuracy[2] Drift with temperature Drift with supply voltage	$R_A = 2k\Omega$ to $100k\Omega$ $C = 0.1\mu F$		0.5 30 0.05	2.0 100 0.2		1.0 50 0.1	3.0 150 0.5	% ppm/°C %/V
t_A $\Delta t_A/\Delta T$ $\Delta t_A/\Delta V_S$	Timing error (astable) Initial accuracy[2] Drift with temperature Drift with supply voltage	R_A, $R_B = 1k\Omega$ to $100k\Omega$ $C = 0.1\mu F$ $V_{CC} = 15V$		4 0.15	6 500 0.6		5 0.3	13 500 1	% ppm/°C %/V
V_C	Control voltage level	$V_{CC} = 15V$ $V_{CC} = 5V$	9.6 2.9	10.0 3.33	10.4 3.8	9.0 2.6	10.0 3.33	11.0 4.0	V V
V_{TH}	Threshold voltage	$V_{CC} = 15V$ $V_{CC} = 5V$	9.4 2.7	10.0 3.33	10.6 4.0	8.8 2.4	10.0 3.33	11.2 4.2	V V
I_{TH}	Threshold current[3]			0.1	0.25		0.1	0.25	μA
V_{TRIG}	Trigger voltage	$V_{CC} = 15V$ $V_{CC} = 5V$	4.8 1.45	5.0 1.67	5.2 1.9	4.5 1.1	5.0 1.67	5.6 2.2	V V
I_{TRIG}	Trigger current	$V_{TRIG} = 0V$		0.5	0.9		0.5	2.0	μA
V_{RESET}	Reset voltage[4]		0.3		1.0	0.3		1.0	V
I_{RESET}	Reset current Reset current	 $V_{RESET} = 0V$		0.1 0.4	0.4 1.0		0.1 0.4	0.4 1.5	mA mA
V_{OL}	Output voltage (low)	$V_{CC} = 15V$ $I_{SINK} = 10mA$ $I_{SINK} = 50mA$ $I_{SINK} = 100mA$ $I_{SINK} = 200mA$ $V_{CC} = 5V$ $I_{SINK} = 8mA$ $I_{SINK} = 5mA$		 0.1 0.4 2.0 2.5 0.1 0.05	 0.15 0.5 2.2 0.25 0.2		 0.1 0.4 2.0 2.5 0.3 0.25	 0.25 0.75 2.5 0.4 0.35	 V V V V V V
V_{OH}	Output voltage (high)	$V_{CC} = 15V$ $I_{SOURCE} = 200mA$ $I_{SOURCE} = 100mA$ $V_{CC} = 5V$ $I_{SOURCE} = 100mA$	 13.0 3.0	 12.5 13.3 3.3		 12.75 2.75	 12.5 13.3 3.3		 V V V
t_{OFF}	Turn-off time[5]	$V_{RESET} = V_{CC}$		0.5	2.0		0.5	2.0	μs
t_R	Rise time of output			100	200		100	300	ns
t_F	Fall time of output			100	200		100	300	ns
	Discharge leakage current			20	100		20	100	ns

NOTES:
1. Supply current when output high typically 1mA less.
2. Tested at $V_{CC} = 5V$ and $V_{CC} = 15V$.
3. This will determine the max value of $R_A + R_B$, for 15V operation, the max total $R = 10M\Omega$, and for 5V operation, the max. total $R = 3.4M\Omega$.
4. Specified with trigger input high.
5. Time measured from a positive going input pulse from 0 to $0.8 \times V_{CC}$ into the threshold to the drop from high to low of the output. Trigger is tied to threshold.

Timer

NE/SE555/SE555C

TYPICAL PERFORMANCE CHARACTERISTICS

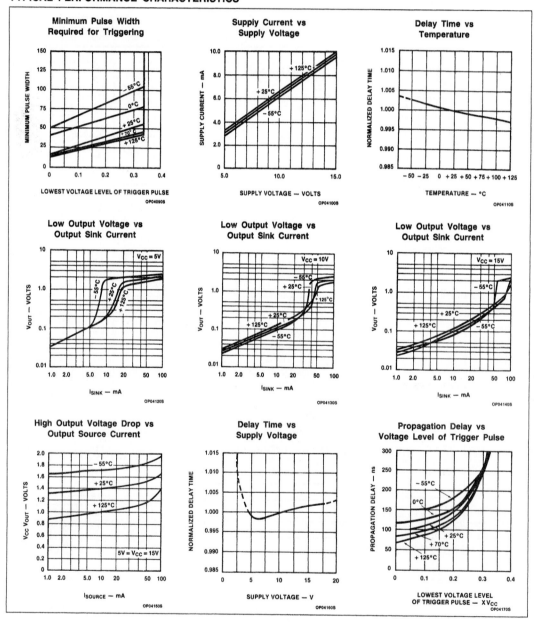

Timer

TYPICAL APPLICATIONS

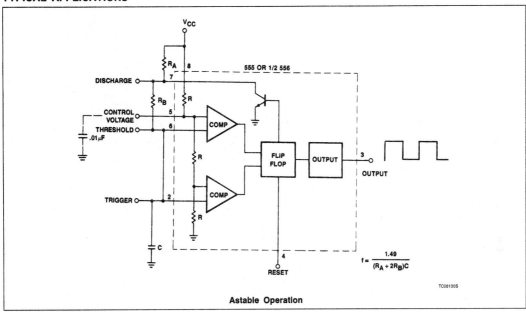

Astable Operation

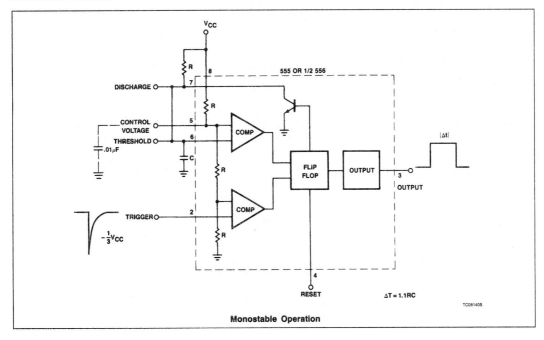

Monostable Operation

Timer

TYPICAL APPLICATIONS

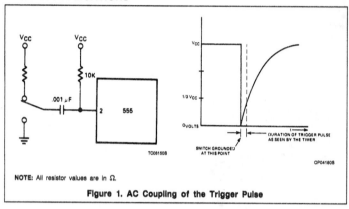

NOTE: All resistor values are in Ω.

Figure 1. AC Coupling of the Trigger Pulse

Trigger Pulse Width Requirements and Time Delays

Due to the nature of the trigger circuitry, the timer will trigger on the negative going edge of the input pulse. For the device to time out properly, it is necessary that the trigger voltage level be returned to some voltage greater than one third of the supply before the time out period. This can be achieved by making either the trigger pulse sufficiently short or by

AC coupling into the trigger. By AC coupling the trigger, see Figure 1, a short negative going pulse is achieved when the trigger signal goes to ground. AC coupling is most frequently used in conjunction with a switch or a signal that goes to ground which initiates the timing cycle. Should the trigger be held low, without AC coupling, for a longer duration than the timing cycle the output will remain in a high state for the duration of the low trigger signal, without regard to the

threshold comparator state. This is due to the predominance of Q_{15} on the base of Q_{16}, controlling the state of the bistable flip-flop. When the trigger signal then returns to a high level, the output will fall immediately. Thus, the output signal will follow the trigger signal in this case.

Another consideration is the "turn-off time". This is the measurement of the amount of time required after the threshold reaches 2/3 V_{CC} to turn the output low. To explain further, Q_1 at the threshold input turns on after reaching 2/3 V_{CC}, which then turns on Q_5, which turns on Q_6. Current from Q_6 turns on Q_{16} which turns Q_{17} off. This allows current from Q_{19} to turn on Q_{20} and Q_{24} to given an output low. These steps cause the $2\mu s$ max. delay as stated in the data sheet.

Also, a delay comparable to the turn-off time is the trigger release time. When the trigger is low, Q_{10} is on and turns on Q_{11} which turns on Q_{15}. Q_{15} turns off Q_{16} and allows Q_{17} to turn on. This turns off current to Q_{20} and Q_{24}, which results in output high. When the trigger is released, Q_{10} and Q_{11} shut off, Q_{15} turns off, Q_{16} turns on and the circuit then follows the same path and time delay explained as "turn off time". This trigger release time is very important in designing the trigger pulse width so as not to interfere with the output signal as explained previously.

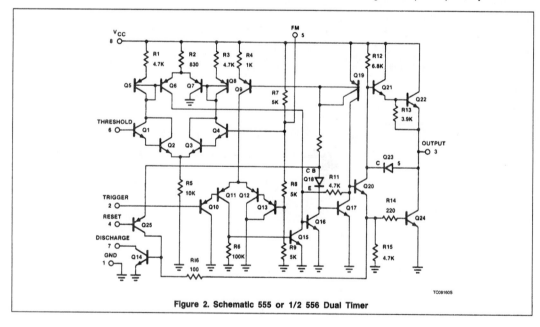

Figure 2. Schematic 555 or 1/2 556 Dual Timer

C

Explanation of the IEEE/IEC Standard for Logic Symbols (Dependency Notation)*

The IEEE/IEC standard for logic symbols introduces a method of determining the complete logical operation of a given device just by interpreting the notations on the symbol for the device. At the heart of the standard is *dependency notation,* which provides a means of denoting the relationship between inputs and outputs without actually showing all of the internal elements and interconnections involved. The information that follows briefly explains the standards publication IEEE Std. 91–1984 and is intended to help in the understanding of these new symbols.

A complete explanation of the logic symbols is available from Texas Instruments, Inc., in Publication SDYZ001, "Overview of IEEE Std 91–1984." They also have a publication (SZZZ003) entitled "Using Functional Logic Symbols," which explains several applications of the symbols.

Explanation of Logic Symbols

F. A. Mann
Texas Instruments Incorporated

Contents

1.0   INTRODUCTION

2.0   SYMBOL COMPOSITION

3.0   QUALIFYING SYMBOLS

---

*Courtesy of Texas Instruments, Inc.

IEEE standards may be purchased from:

Institute of Electrical and Electronics Engineers, Inc.
IEEE Standards Office
345 East 47th Street
New York, N.Y. 10017

International Electrotechnical Commission (IEC) publications may be purchased from:

American National Standards Institute, Inc.
1430 Broadway
New York, N.Y. 10018

1.0 INTRODUCTION

The International Electrotechnical Commission (IEC) has been developing a very powerful symbolic language that can show the relationship of each input of a digital logic circuit to each output without showing explicitly the internal logic. At the heart of the system is dependency notation, which will be explained in Section 4.

The system was introduced in the USA in a rudimentary form in IEEE/ANSI Standard Y32.14-1973. Lacking at that time a complete development of dependency notation, it offered little more than a substitution of rectangular shapes for the familiar distinctive shapes for representing the basic functions of AND, OR, negation, etc. This is no longer the case.

Internationally, Working Group 2 of IEC Technical Committee TC-3 has prepared a new document (Publication 617-12) that consolidates the original work started in the mid 1960's and published in 1972 (Publication 117-15) and the amendments and supplements that have followed. Similarly for the USA, IEEE Committee SCC 11.9 has revised the publication IEEE Std 91/ANSI Y32.14. Now numbered simply IEEE Std 91-1984, the IEEE standard contains all of the IEC work that has been approved, and also a small amount of material still under international consideration. Texas Instruments is participating in the work of both organizations and this document introduces new logic symbols in accordance with the new standards. When changes are made as the standards develop, future editions will take those changes into account.

The following explanation of the new symbolic language is necessarily brief and greatly condensed from what the standards publications will contain. This is not intended to be sufficient for those people who will be developing symbols for new devices. It is primarily intended to make possible the understanding of the symbols used in various data books and the comparison of the symbols with logic diagrams, functional block diagrams, and/or function tables will further help that understanding.

2.0 SYMBOL COMPOSITION

A symbol comprises an outline or a combination of outlines together with one or more qualifying symbols. The shape of the symbols is not significant. As shown in Figure 1, general qualifying symbols are used to tell exactly what logical operation is performed by the elements. Table I shows general qualifying symbols defined in the new standards. Input lines are placed on the left and output lines are placed on the right. When an exception is made to that convention, the direction of signal flow is indicated by an arrow as shown in Figure 11.

All outputs of a single, unsubdivided element always have identical internal logic states determined by the function of the element except when otherwise indicated by an associated qualifying symbol or label inside the element.

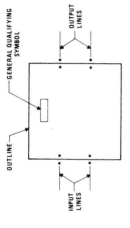

*Possible positions for qualifying symbols relating to inputs and outputs

Figure 1. Symbol Composition

The outlines of elements may be abutted or embedded in which case the following conventions apply. There is no logic connection between the elements when the line common to their outlines is in the direction of signal flow. There is at least one logic connection between the elements when the line common to their outlines is perpendicular to the direction of signal flow. The number of logic connections between elements will be clarified by the use of qualifying symbols and this is discussed further under that topic. If no indications are shown on either side of the common line, it is assumed there is only one connection.

When a circuit has one or more inputs that are common to more than one element of the circuit, the common-control block may be used. This is the only distinctively shaped outline used in the IEC system. Figure 2 shows that unless otherwise qualified by dependency notation, an input to the common-control block is an input to each of the elements below the common-control block.

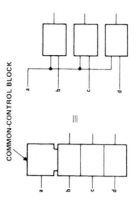

Figure 2. Common-Control Block

A common output depending on all elements of the array can be shown as the output of a common-output element. Its distinctive visual feature is the double line at its top. In addition the common-output element may have other inputs as shown in Figure 3. The function of the common-output element must be shown by use of a general qualifying symbol.

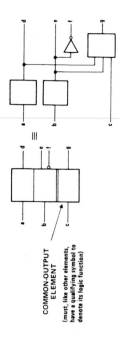

COMMON-OUTPUT ELEMENT

(must, like other elements, have a qualifying symbol to denote its logic function)

Figure 3. Common-Output Element

3.0 QUALIFYING SYMBOLS

3.1 General Qualifying Symbols

Table I shows general qualifying symbols defined by IEEE Standard 91. These characters are placed near the top center or the geometric center of a symbol or symbol element to define the basic function of the device represented by the symbol or of the element.

3.2 Qualifying Symbols for Inputs and Outputs

Qualifying symbols for inputs and outputs are shown in Table II and will be familiar to most users with the possible exception of the logic polarity and analog signal indicators. The older logic negation indicator means that the external O state produces the internal 1 state. The internal 1 state means the active state. Logic negation may be used in pure logic diagrams; in order to tie the external 1 and O logic states to the levels H (high) and L (low), a statement of whether positive logic (1 = H, O = L) or negative logic (1 = L, O = H) is being used is required or must be assumed. Logic polarity indicators eliminate the need for calling out the logic convention and are used in various data books in the symbology for actual devices. The presence of the triangular polarity indicator indicates that the L logic level will produce the internal 1 state (the active state) or that, in the case of an output, the internal 1 state will produce the external L level. Note how the active direction of transition for a dynamic input is indicated in positive logic, negative logic, and with polarity indication.

The internal connections between logic elements abutted together in a symbol may be indicated by the symbols shown in Table II. Each logic connection may be shown by the presence of qualifying symbols at one or both sides of the common line and if confusion can arise about the numbers of connections, use can be made of one of the internal connection symbols.

Table I. General Qualifying Symbols

SYMBOL	DESCRIPTION	CMOS EXAMPLE	TTL EXAMPLE
&	AND gate or function.	'HC00	SN7400
≥1	OR gate or function. The symbol was chosen to indicate that at least one active input is needed to activate the output.	'HC02	SN7402
=1	Exclusive OR. One and only one input must be active to activate the output.	'HC86	SN7486
=	Logic identity. All inputs must stand at the same state.	'HC86	SN74180
2k	An even number of inputs must be active.	'HC280	SN74180
2k+1	An odd number of inputs must be active.	'HC86	SN74LS86
1	The one input must be active.	'HC04	SN7404
△ or ▽	A buffer or element with more than usual output capability (symbol is oriented in the direction of signal flow).	'HC240	SN74S436
⎍	Schmitt trigger; element with hysteresis.	'HC132	SN74LS18
X/Y	Coder; code converter (DEC/BCD, BIN/OUT, BIN/7-SEG, etc.).	'HC42	SN74LS347
MUX	Multiplexer/data selector.	'HC151	SN74150
DMUX or DX	Demultiplexer.	'HC138	SN74138
Σ	Adder.	'HC283	SN74LS385
P-Q	Subtracter.		SN74LS385
CPG	Look-ahead carry generator	'HC182	SN74182
π	Multiplier.		SN74LS384
COMP	Magnitude comparator.	'HC85	SN74LS682
ALU	Arithmetic logic unit.	'HC181	SN74LS381
⊓	Retriggerable monostable.	'HC123	SN74LS422
1⊓	Nonretriggerable monostable (one shot)	'HC221	SN74121
G⎍	Astable element. Showing waveform is optional.		SN74LS320
!G⎍	Synchronously starting astable.		SN74LS624
G!⎍	Astable element that stops with a completed pulse.		
SRGm	Shift register. m = number of bits.	'HC164	SN74LS595
CTRm	Counter. m = number of bits; cycle length = 2m.	'HC590	SN54LS590
CTR DIVm	Counter with cycle length = m.	'HC160	SN74LS668
RCTRm	Asynchronous (ripple-carry) counter; cycle length = 2m	'HC4020	
ROM	Read-only memory.		SN74187
RAM	Random-access read/write memory.	'HC189	SN74170
FIFO	First-in, first-out memory.		SN74LS222
1=0	Element powers up cleared to 0 state.		SN74AS877
1=1	Element powers up set to 1 state.	'HC7022	SN74AS877
Φ	Highly complex function; "gray box" symbol with limited detail shown under special rules.		SN74LS608

*Not all of the general qualifying symbols have been used in TI's CMOS and TTL data books, but they are included here for the sake of completeness.

Table II. Qualifying Symbols for Inputs and Outputs

Logic negation at input. External 0 produces internal 1.

Logic negation at output. Internal 1 produces external 0.

Active-low input. Equivalent to ⊸ in positive logic.

Active-low output. Equivalent to ⊳ in positive logic.

Active-low input in the case of right-to-left signal flow.

Active-low output in the case of right-to-left signal flow.

Signal flow from right to left. If not otherwise indicated, signal flow is from left to right.

Bidirectional signal flow.

	POSITIVE LOGIC	NEGATIVE LOGIC	POLARITY INDICATION
Dynamic inputs active on indicated transition	1 / 0	0 / 1	H
	not used	not used	not used
	1	0	H / L
	0	1	L / H

Nonlogic connection. A label inside the symbol will usually define the nature of this pin.

Input for analog signals (on an analog symbol) (see Figure 14).

Input for digital signals (on an analog symbol) (see Figure 14).

Internal connection. 1 state on left produces 1 state on right.

Negated internal connection. 1 state on left produces 0 state on right.

Dynamic internal connection. Transition from 0 to 1 on left produces transitory 1 state on right.

Internal input (virtual input). It always stands at its internal 1 state unless affected by an overriding dependency relationship.

Internal output (virtual output). Its effect on an internal input to which it is connected is indicated by dependency notation.

The internal (virtual) input is an input originating somewhere else in the circuit and is not connected directly to a terminal. The internal (virtual) output is likewise not connected directly to a terminal. The application of internal inputs and outputs requires an understanding of dependency notation, which is explained in Section 4.

Table III. Symbols Inside the Outline

Postponed output (of a pulse-triggered flip-flop). The output changes when input initiating change (e.g., a C input) returns to its initial external state or level. See § 5.

Bi-threshold input (input with hysteresis).

N-P-N open-collector or similar output that can supply a relatively low-impedance L level when not turned off. Requires external pull-up. Capable of positive-logic wired-AND connection.

Passive-pull-up output is similar to N-P-N open-collector output but is supplemented with a built-in passive pull-up.

N-P-N open-emitter or similar output that can supply a relatively low-impedance H level when not turned off. Requires external pull-down. Capable of positive-logic wired-OR connection.

Passive-pull-down output is similar to N-P-N open-emitter output but is supplemented with a built-in passive pull-down.

3-state output.

Output with more than usual output capability (symbol is oriented in the direction of signal flow).

Enable input
When at its internal 1-state, all outputs are enabled.
When at its internal 0-state, open-collector and open-emitter outputs are off, three-state outputs are in the high-impedance state, and all other outputs (e.g., totem-poles) are at the internal 0-state.

J, K, R, S, T — Usual meanings associated with flip-flops (e.g., R = reset, T = toggle)

Data input to a storage element equivalent to:

Shift right (left) inputs, m = 1, 2, 3, etc. If m = 1, it is usually not shown.

Counting up (down) inputs, m = 1, 2, 3, etc. If m = 1, it is usually not shown.

Binary grouping. m is highest power of 2.

CT = 15 / CT = 9 — The contents-setting input, when active, causes the content of a register to take on the indicated value.

The content output is active if the content of the register is as indicated.

Input line grouping . . . indicates two or more terminals used to implement a single logic input. e.g., The paired expander inputs of SN7450.

Fixed-state output always stands at its internal 1 state. For example, see SN74185.

D

Answers to Odd-Numbered Problems

Chapter 1

1–1. (a) 6_{10} (b) 11_{10} (c) 9_{10} (d) 7_{10} (e) 12_{10} (f) 75_{10} (g) 55_{10} (h) 181_{10} (i) 167_{10} (j) 118_{10}

1–3. (a) 31_8 (b) 35_8 (c) 134_8 (d) 131_8 (e) 155_8

1–5. (a) 23_{10} (b) 31_{10} (c) 12_{10} (d) 58_{10} (e) 41_{10}

1–7. (a) $B9_{16}$ (b) DC_{16} (c) 74_{16} (d) FB_{16} (e) $C6_{16}$

1–9. (a) 134_{10} (b) 244_{10} (c) 146_{10} (d) 171_{10} (e) 965_{10}

1–11. (a) 98_{10} (b) 69_{10} (c) 74_{10} (d) 36_{10} (e) 81_{10}

1–13. (a) 010 0101 (b) 0100100 0110001 0110100 (c) 1001110 0101101 0110110 (d) 1000011 1010000 1010101 (e) 1010000 1100111

1–15. (a) Tank A, temperature high; tank C, pressure high
(b) Tank D, temperature and pressure high
(c) Tanks B and D, pressure high
(d) Tanks B and C, temperature high
(e) Tank C, temperature and pressure high

1–17. (a) sku43 (b) $534B553433_{16}$

1–19. 16-MAR 1995 Revision A

Chapter 2

2–1. (a) $0.5\,\mu s$ (b) $2\,\mu s$ (c) $0.234\,\mu s$ (d) 58.8 ns (e) 500 kHz (f) 10 kHz (g) 1.33 kHz (h) 0.667 MHz

2–3. (a) $2.16\,\mu s$ (b) LOW

2–5.

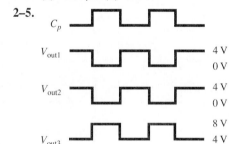

2–7. $V_1 = 0$ V $V_5 = 4.3$ V
$V_2 = 4.3$ V $V_6 = 5.0$ V
$V_3 = 4.3$ V $V_7 = 0$ V
$V_4 = 0$ V

2–9. That diode will conduct raising V_7 to 4.3 V ("OR").

2–11.

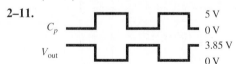

2–13. $V_{out} = 4.998$ V

2–15. Because, when the transistor is turned on (saturated), the collector current will be excessive ($I_C = 5$ V/R_C).

2–17. The totem-pole output replaces R_C with a transistor that acts like a variable resistor. The transistor prevents excessive collector current when it is cut off and provides a high-level output when turned on.

2–19. **(a)** 8.0 MHZ **(b)** 125 nS

2–21. P3 parallel, P2 serial

2–23. A HIGH on pin 2 will turn Q1 on, making RESET_B approximately zero.

Chapter 3

3–1.

A	B	C	X
0	0	0	0
0	0	1	0
0	1	0	0
0	1	1	0
1	0	0	0
1	0	1	0
1	1	0	0
1	1	1	1

3–3. 256

3–5. The output is HIGH whenever any input is HIGH; otherwise, the output is LOW.

3–7.

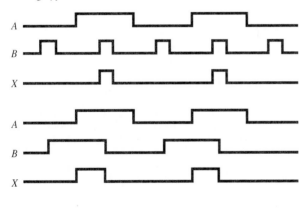

3–9.

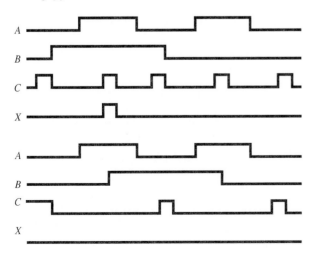

3–11.

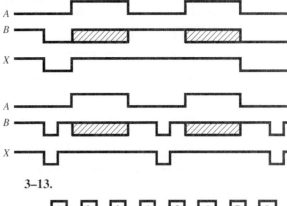

3–13.

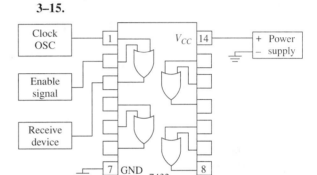

3–15.

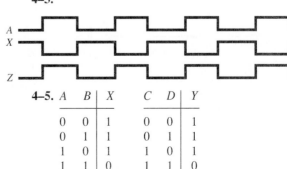

3–17. Two

3–19. To provide pulses to a digital circuit for troubleshooting purposes.

3–21. HIGH, to enable the output to change with pulser (if gate is good).

3–23. Pin 2 should be ON; the Enable switch is bad, or bad Enable connection.

3–25. AND-74HC08; U3:A = location C2, U3:B = location D2 OR-74HC32; location B7

3–27. Pin 20 = LOW (GND), pin 40 HIGH (+5)

Chapter 4

4–1. $X = \overline{A}, X = 0$

4–3.

4–5.

A	B	X		C	D	Y
0	0	1		0	0	1
0	1	1		0	1	1
1	0	1		1	0	1
1	1	0		1	1	0

4–7.

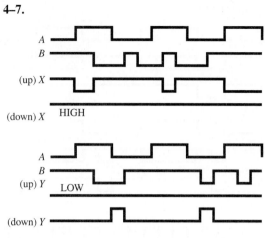

4–9. $X = \overline{A + B + C}$ $Y = \overline{D + E + F}$

4–11.

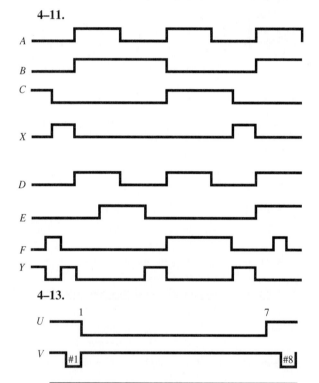

4–13.

4–15. $U = C_p AB$ $W = BC$
$\quad\quad V = \overline{C}\,\overline{D}$ $\quad X = C_p CD$

4–17.

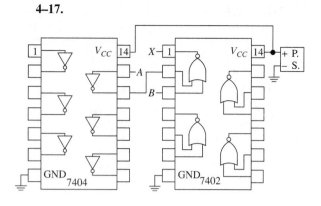

4–19. HIGH; to see inverted output pulses (otherwise, output would always be HIGH).

4–21. There is no problem.

4–23. With all inputs HIGH, pin 8 should be LOW. Next try making each of the 8 inputs LOW, one at a time, while checking for a HIGH at pin 8.

4–25. Because they are all part of one IC package.

4–27. all HIGH

Chapter 5

5–1. (a) $W = (A + B)(C + D)$
$\quad\quad\quad X = AB + BC$
$\quad\quad\quad Y = (AB + B)C$
$\quad\quad\quad Z = (AB + B + (B + C))D$

5–3.

A	B	C	D	M	N	Q	R	S		A	B	C	P
0	0	0	0	0	0	0	0	0		0	0	0	0
0	0	0	1	1	0	0	1	1		0	0	1	0
0	0	1	0	1	0	0	0	0		0	1	0	0
0	0	1	1	1	1	0	1	1		0	1	1	1
0	1	0	0	0	0	0	0	0		1	0	0	0
0	1	0	1	1	1	0	1	1		1	0	1	1
0	1	1	0	1	0	0	1	1		1	1	0	0
0	1	1	1	1	1	1	1	1		1	1	1	1
1	0	0	0	0	0	0	0	0					
1	0	0	1	1	1	0	1	1					
1	0	1	0	1	0	0	0	1					
1	0	1	1	1	1	0	1	1					
1	1	0	0	1	0	0	0	1					
1	1	0	1	1	1	0	1	1					
1	1	1	0	1	0	0	1	1					
1	1	1	1	1	1	1	1	1					

5–5. (a) Commutative law **(b)** Associative law
(c) Distributive law

5–7.

$W = (A + B)BC$
$W = BC$

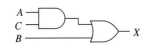

$X = (A + B)(B + C)$
$X = B + AC$

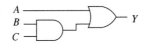

$Y = A + (A + B)BC$
$Y = A + BC$

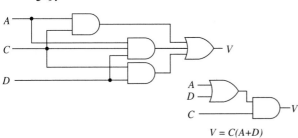

$Z = AB + B + BC$
$Z = B$

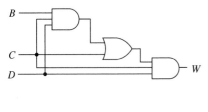

5–9.

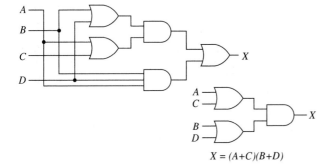

$V = C(A+D)$

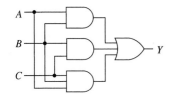

$W = CD$

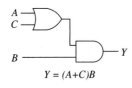

$X = (A+C)(B+D)$

$Y = (A+C)B$

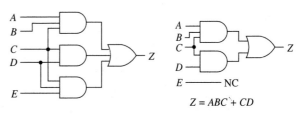

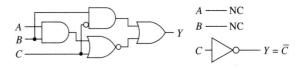

$Z = ABC + CD$

5–11. $X = (A + B)(D + C)$

5–13. Break the long bar and change the AND to an OR, or the OR to an AND.

5–15. Y and Z are both ORs

5–17.

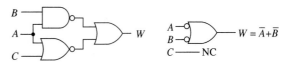

$W = \overline{A} + \overline{B}$

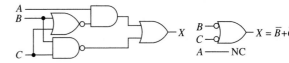

$X = \overline{B} + \overline{C}$

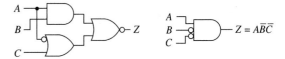

$Y = \overline{C}$

$Z = A\overline{B}\,\overline{C}$

5–19.

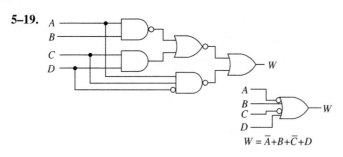

$$W = \bar{A} + B + \bar{C} + D$$

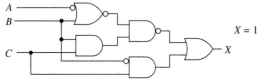

$X = 1$

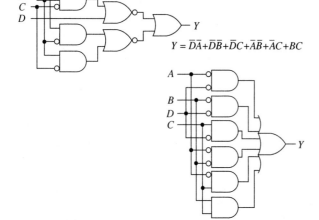

$$Y = \bar{D}\bar{A} + \bar{D}\bar{B} + \bar{D}C + \bar{A}\bar{B} + \bar{A}C + BC$$

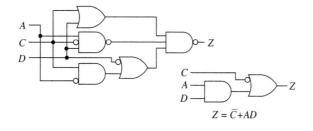

$$Z = \bar{C} + AD$$

5–21.

A
B
C
D

5–23.

A
B
C
D

5–25.

(2^3) A ————
(2^2) B ———— $ABCD > 11$
(2^1) C —— NC
(2^0) D —— NC

5–27.

A	B	C	W	X
0	0	0	0	1
0	0	1	1	1
0	1	0	1	1
0	1	1	1	0
1	0	0	1	0
1	0	1	1	1
1	1	0	0	1
1	1	1	0	0

A	B	C	D	Y	Z
0	0	0	0	1	1
0	0	0	1	1	1
0	0	1	0	1	1
0	0	1	1	0	1
0	1	0	0	0	0
0	1	0	1	1	0
0	1	1	0	1	1
0	1	1	1	1	1
1	0	0	0	0	1
1	0	0	1	1	1
1	0	1	0	0	1
1	0	1	1	0	0
1	1	0	0	0	0
1	1	0	1	1	0
1	1	1	0	0	1
1	1	1	1	1	0

5–29. (a)

(b)

(c)

(d)

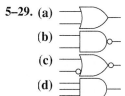

5–31. (a)

$$X = \overline{A}$$

(b)

$$X = \overline{A}$$

5–33.

(a)

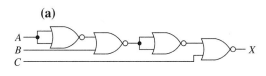

(b)

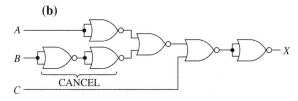

CANCEL

(c)

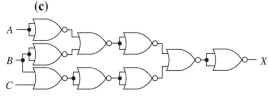

5–35. u. SOP   x. SOP
v. POS   y. POS
w. POS   z. POS, SOP

5–37. $X = \overline{A} + B\overline{C}$
$Y = B + \overline{A}C$
$Z = A\overline{C} + AB + \overline{A}\overline{B}C$

5–39. (a) $X = \overline{C}D + AC + B$ **(b)** $Y = 1$

5–41.

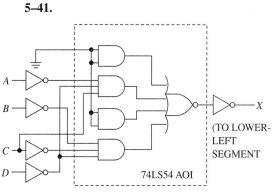

$X = \overline{A}C\overline{D} + \overline{B}\overline{C}D$ where A = MSB

5–43. The IC checks out OK. The problem is that pin 9 should be connected to pin 10 (not 9 to GND).

5–45. $\overline{WATCHDOG\_EN \cdot Qa}$

5–47. (a) pin $6 = \overline{\overline{P1.0} + \overline{A15}}$ **(b)** AND **(c)** quad 2 input AND **(d)** $\overline{RD}$ is LOW or $\overline{WR}$ is LOW

Chapter 6

6–1. (a) Exclusive-OR produces a HIGH output for one or the other input HIGH, but not both.
(b) Exclusive-NOR produces a HIGH output for both inputs HIGH or both inputs LOW.

6–3.

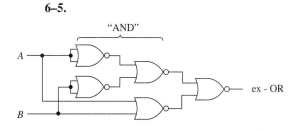

X   ex - OR
Y   ex - NOR

6–5.

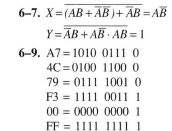

"AND"

ex - OR

6–7. $X = \overline{(AB + \overline{A}\overline{B}) + \overline{A}B} = A\overline{B}$

$Y = \overline{\overline{A}B} + A\overline{B} \cdot AB = 1$

6–9. A7 = 1010 0111   0
4C = 0100 1100   0
79 = 0111 1001   0
F3 = 1111 0011   1
00 = 0000 0000   1
FF = 1111 1111   1

6–11.

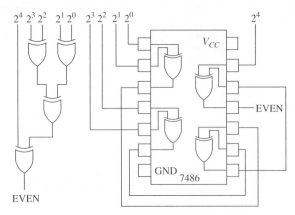

6–13.

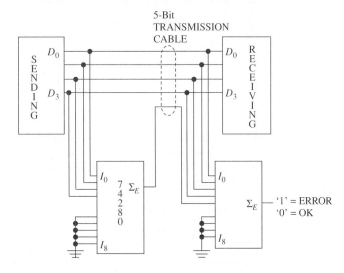

6–15. Yes; LOW

6–17.

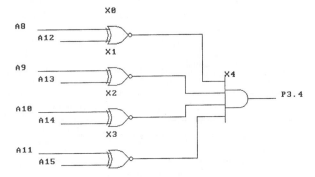

Chapter 7

7–1. (a) 1001 (b) 1111 (c) 1 1100 (d) 100 0010 (e) 1100 1000 (f) 10010 0010 (g) 10100 1111 (h) 10110 0000

7–3. (a) 1 0101 (b) 10 1010 (c) 11 1100 (d) 1 0001 0001 (e) 1 1110 1100 0011 (f) 111 0111 0001 (g) 1 1001 0011 (h) 111 1110 1000 0001

7–5.

+15	0000 1111	−1	1111 1111
+14	0000 1110	−2	1111 1110
+13	0000 1101	−3	1111 1101
+12	0000 1100	−4	1111 1100
+11	0000 1011	−5	1111 1011
+10	0000 1010	−6	1111 1010
+9	0000 1001	−7	1111 1001
+8	0000 1000	−8	1111 1000
+7	0000 0111	−9	1111 0111
+6	0000 0110	−10	1111 0110
+5	0000 0101	−11	1111 0101
+4	0000 0100	−12	1111 0100
+3	0000 0011	−13	1111 0011
+2	0000 0010	−14	1111 0010
+1	0000 0001	−15	1111 0001
0	0000 0000		

7–7. (a) 0001 0110 = +22 (b) 0000 1111 = +15 (c) 0101 1100 = +92 (d) 1000 0110 = −122 (e) 1110 1110 = −18 (f) 1000 0001 = −127 (g) 0111 1111 = +127 (h) 1111 1111 = −1

7–9. (a) 0000 1100 (b) 0000 0110 (c) 0011 0010 (d) 0000 1110 (e) 0000 1010 (f) 0011 1011 (g) 1111 0100 (h) 1010 1100

7–11. (a) E (b) D (c) 21 (d) CA (e) 10C (f) 162 (g) AB45 (h) A000

7–13. 2CF0H

7–15. b, c, e

7–17. For the LSB addition of two binary numbers.

7–19. Σ_o = OK
C_o = NO

7–21.

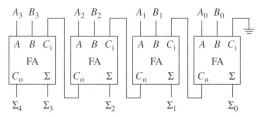

7–23.

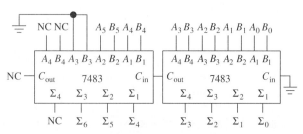

7–25.

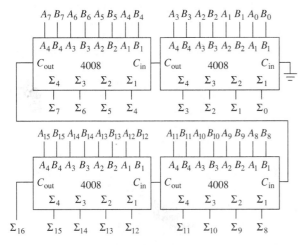

7–27. The Ex-OR gate third from right is bad. Also, the full-adder fourth from right is bad.

7–29. $S_3 - S_0 = 0100$.
 (a) 1110 (b) 0001

7–31.

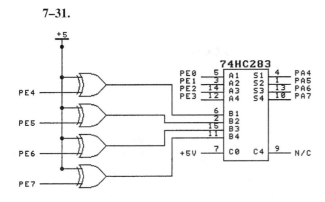

8–5.

2^3	2^2	2^1	2^0	$\bar{0}$	$\bar{1}$	$\bar{2}$	$\bar{3}$	$\bar{4}$	$\bar{5}$	$\bar{6}$	$\bar{7}$	$\bar{8}$	$\bar{9}$
0	0	0	0	0	1	1	1	1	1	1	1	1	1
0	0	0	1	1	0	1	1	1	1	1	1	1	1
0	0	1	0	1	1	0	1	1	1	1	1	1	1
0	0	1	1	1	1	1	0	1	1	1	1	1	1
0	1	0	0	1	1	1	1	0	1	1	1	1	1
0	1	0	1	1	1	1	1	1	0	1	1	1	1
0	1	1	0	1	1	1	1	1	1	0	1	1	1
0	1	1	1	1	1	1	1	1	1	1	0	1	1
1	0	0	0	1	1	1	1	1	1	1	1	0	1
1	0	0	1	1	1	1	1	1	1	1	1	1	0

8–7. That input is a "don't care" and will have no effect on the output for that particular table entry.

8–9.

Time interval	Low output pulse at:
t_0–t_1	None (E_3 disabled)
t_1–t_2	None (E_3 disabled)
t_2–t_3	$\bar{5}$
t_3–t_4	$\bar{4}$
t_4–t_5	$\bar{3}$
t_5–t_6	$\bar{2}$
t_6–t_7	$\bar{1}$
t_7–t_8	$\bar{0}$
t_8–t_9	$\bar{7}$
t_9–t_{10}	$\bar{6}$
t_{10}–t_{11}	$\bar{5}$
t_{11}–t_{12}	None (E_3 disabled)
t_{12}–t_{13}	None (E_3 disabled)

8–11. All HIGH

8–13. The higher number

Chapter 8

8–1.

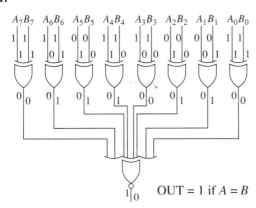

OUT = 1 if $A = B$

8–3. See #1 (lower numbers)

8–15.

8-17.

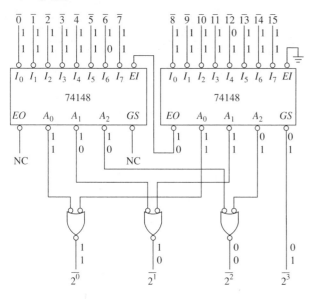

8-19. (a) 100000_2 **(b)** 101110_2 **(c)** 110111_2
(d) 1000100_2

8-21. See #20 (lower numbers).

8-23. (a) $1010_2 = 1111$ **(b)** $1111_2 = 1000$
(c) $0011_2 = 0010$ **(d)** $0001_2 = 0001$

8-25.

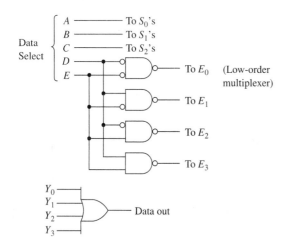

8-27.

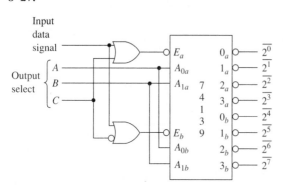

8-29. (a) The 74150 is not working. The data select is set for input D_7, which is 0. Therefore, $\overline{Y}$ should be 1 but it is not.
(b) The 74151 is OK. The data select is set for input I_o, which is 1. Y should equal 1 and $\overline{Y} = 0$, which they do.
(c) The 74139 has two bad decoders. Decoder A is enabled and should output 1011 but does not. Decoder B is disabled and should output 1111 but does not.
(d) The 74154 is OK. The chip is disabled, so all outputs should be HIGH, which they are.

8-31. (a) Write to memory bank 5, location 3H.
(b) Read from memory bank 6, location C7H.

8-33. (a) 05H **(b)** 07H **(c)** 82H **(d)** Impossible

8-35. $\overline{0} = 0, \overline{1} = 1, \overline{2} = 1, \overline{3} = 1$

8-37. The D input of U1:B

8-39.

	LCD_SL	KEY_SL
AD3	0	1
AD4	0	0
AD5	0	0
AD11	1	1
AD12	1	1
AD13	0	0
AD14	0	0
AD15	0	0
AS	0	0

8-41.

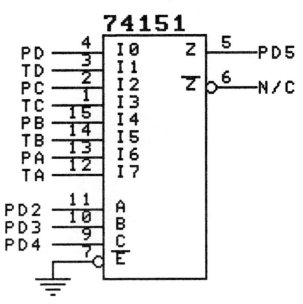

Chapter 9

9-1. D_1 and D_2 provide some protection against negative input voltages.

9–3. From Figure 9–2(a) there is = 0.2 V dropped across the 1.6 k, 0.7 V across V_{BE_3}, and 0.7 V across D_3, leaving ~ 3.4 V at the output terminal.

9–5. Negative sign signifies current *leaving* the input or output of the gate.

9–7a. [STD]
(a) $V_a = 3.4$ V (typ)
$I_a = 120\ \mu A$
(b) $V_a = 4.6$ V
@V_{in} = LOW, $I_a = 20\ \mu A$ (typ)
@V_{in} = HIGH, $I_a = 340\ \mu A$ (typ)
(c) $V_a = V_{OL} = 0.2$ V (typ)
$I_a = 3.2$ mA
(d) $V_a = V_{OH} = 3.4$ V (typ)
$I_a = 380\ \mu A$

9–7b. [LS]
(a) $V_a = V_{OH} = 3.4$ V (typ)
$I_a = 60\ \mu A$
(b) $V_a = 4.8$ V
@V_{in} = LOW, $I_a = 35\ \mu A$ (typ)
@V_{in} = HIGH, $I_a = 340\ \mu A$ (typ)
(c) $V_a = V_{OL} = 0.35$ V (typ)
$I_a = 0.8$ mA
(d) $V_a = V_{OH} = 3.4$ V (typ)
$I_a = 360\ \mu A$

9–9.

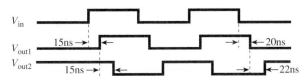

9–11. 7400: $P_D = 75$ mW (max)
74LS00: $P_D = 15$ mW (max)

9–13. (a)   7400: HIGH state (min levels) = 0.4 V
LOW state (max levels) = 0.4 V
74LS00: HIGH state (min levels) = 0.7 V
LOW state (max levels) = 0.3 V
(b) The 74LS00 has a wider margin for the HIGH state. The 7400 has a wider margin for the LOW state.

9–15. The open-collector FLOAT level is made a HIGH level by using a pull-up resistor.

9–17. The 7400 series is faster than the 4000B series but dissipates more power.

9–19. Because MOS ICs are prone to electrostatic burnout.

9–21. Where speed is most important, ECL is faster but uses more power.

9–23. The 74HC family

9–25. Interfacing (c) and (e) will require a pull-up resistor to "pull up" the TTL HIGH-level output to meet the minimum HIGH-level input specifications of the CMOS gates.

9–27. (a) 10 (b) 400

9–29.

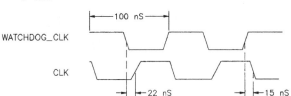

Chapter 10

10–1.

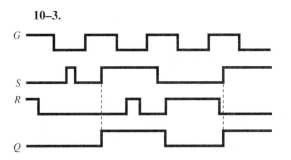

10–3.

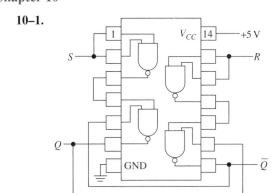

10–5.

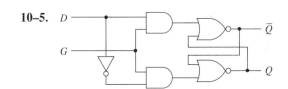

10–7.

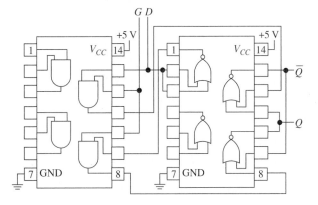

10–9.

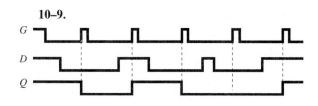

10–11.

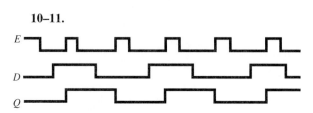

10–13. HIGH, LOW

10–15.

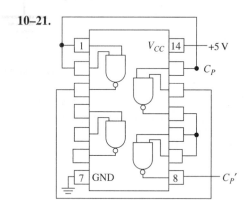

10–17. The 7474 is edge-triggered; the 7475 is pulse-triggered. The 7474 has asynchronous inputs at $\overline{S_D}$ and $\overline{R_D}$.

10–19. The triangle indicates that it is an edge-triggered device as opposed to being pulse-triggered.

10–21.

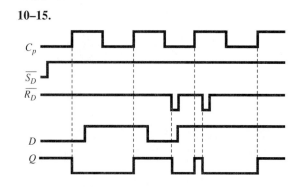

10–23. The toggle mode

10–25. The 7476 accepts J and K data during the entire positive level of CP, whereas the 74LS76 only looks at J and K at the negative edge of CP.

10–27.

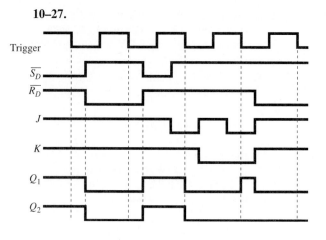

10–29.

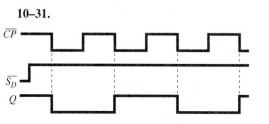

10–31.

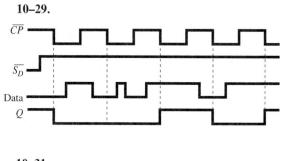

10–33. The '373 is a transparent latch. If the timing pulses are connected to E, the BCD will pass through to Q while E is HIGH and latch on to the data when E goes LOW. As long as the positive timing pulses are very narrow (<10 ms), the display will not flicker. The '273 is the preferred device, however, because it is edge triggered.

10–35. **(a)** HIGH **(b)** no **(c)** Qa must go HIGH while WATCHDOG_EN is HIGH. (Qa will go HIGH after Qb of U1:B goes HIGH.)

10–37. WATCHDOG_SEL is pulsed

Chapter 11

11–1.

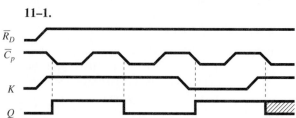

11–3. $t_a = 20$ ns $\qquad$ $t_d = 30$ ns
$t_b = 30$ ns $\qquad$ $t_e = 20$ ns
$t_c = 20$ ns $\qquad$ $t_f = 30$ ns

11–5. Proper circuit operation depends on t_p of the 7432 being ≥10 ns. The worst-case t_p is specified as 15 ns but the actual t_p may be less. If it's actually less than 10 ns, the circuit won't operate properly.

11–7.

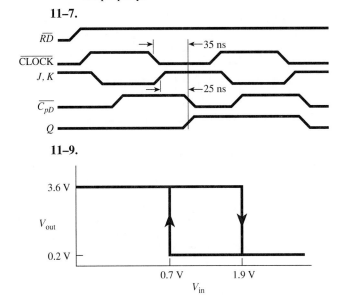

11–9.

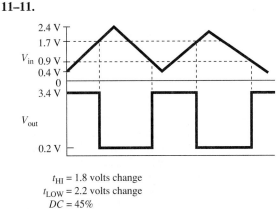

11–11.

$t_{HI} = 1.8$ volts change
$t_{LOW} = 2.2$ volts change
$DC = 45\%$

11–13. It is caused by switch bounce. If the switch bounces an even number of times, the LED will be off. A debounce circuit would correct the problem.

11–15. Check the output of the 7805 for +5 V dc. Check the fuse. If the fuse is OK, you should check for approximately 12.6 V ac at the transformer secondary, 20 V dc at the +/– output of the diode bridge and the input to the 7805.

11–17. Light: $V_A = 0.0495$ V
Dark: $V_A = 4.55$ V

11–19. The 7474 can sink 16 mA. Connect the positive lead of the buzzer to +5 V and the negative lead to $\overline{Q}$. When Q is HIGH, the buzzer is energized via the LOW $\overline{Q}$ output.

11–21. $I_{coil} = 240$ mA

11–23.

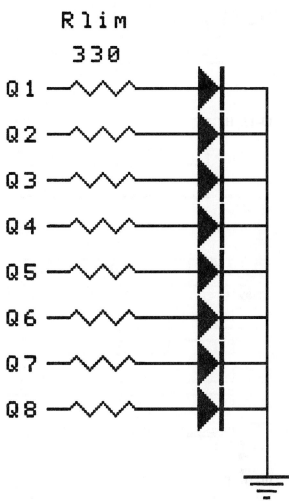

11–25. The switches of U12 and the pull-up resistors are used to place either a HIGH or LOW on the MODA and MODB lines. To place a HIGH on one of these lines the corresponding switch must be open. A closed switch pulls the line to ground.

Chapter 12

12–1. Sequential circuits follow a predetermined sequence of digital states triggered by a timing pulse or clock. Combination logic circuits operate almost instantaneously based on the levels placed at its inputs.

12–3.

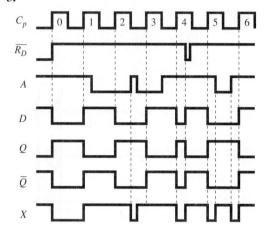

12–5.

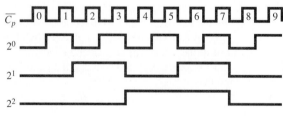

12–7. (a) 3 (b) 3 (c) 1 (d) 5 (e) 6 (f) 4

12–9.

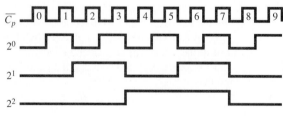

12–11. (a) 3 (b) 15 (c) 127 (d) 1

12–13. (a) 2 (b) 4 (c) 4 (d) 5

12–15.

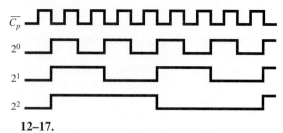

12–17.

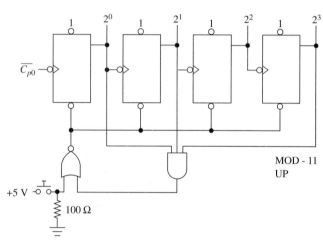

12–19.

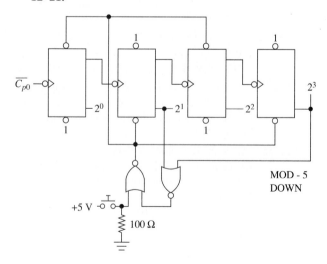

12–21.

12–23.

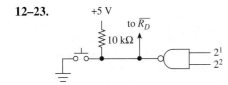

12–25. Yes, it will work. The inputs to the AND are normally 1–1 until the push button is pressed, or the NAND output goes LOW. If either AND input goes LOW, its output goes LOW, resetting the counter. This is an improvement over Figure 12–19, because the 10 kΩ draws less current than the 100-Ω resistor when the push button is depressed.

12–27.

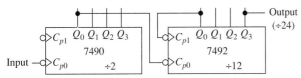

12–29.

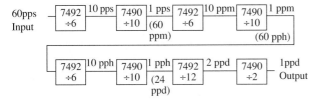

12–31.

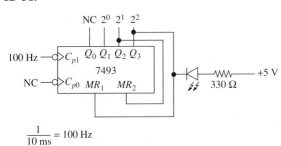

$$\frac{1}{10 \text{ ms}} = 100 \text{ Hz}$$

12–33. Connect a RESET push button across the 0.001-μF capacitor. When momentarily pressed, it will RESET the circuit to its initial condition.

12–35. With all the current going through the same resistor, as more segments are turned ON, the voltage that reaches the segments is reduced, making them dimmer. The displayed ⊟ is much dimmer than the ⌐.

12–37.

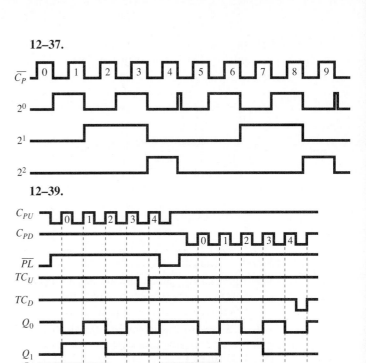

12–39.

12–41.

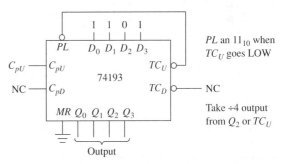

PL an 11_{10} when TC_U goes LOW

NC

Take ÷4 output from Q_2 or TC_U

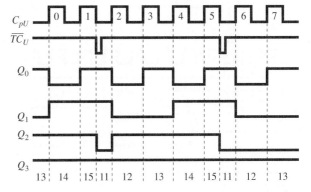

12–43. (a) HIGH order U10, LOW order U9 (b) no
(c) by a LOW at the output of U3:B

12–45.

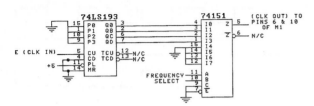

Chapter 13

13–1. Right, negative

13–3. 1110, 1111

13–5. Apply a LOW pulse to $\overline{R_D}$ to RESET all Q outputs to zero. Next, apply a LOW pulse to the active-LOW D_3, D_1, D_0 inputs.

13–7. J_3, K_3 are the data input lines. Q_3, Q_2, Q_1, Q_0 are the data output lines. (D_3, D_2, D_1, D_0 are held HIGH.)

13–9. Three

13–11. The Q_0 flip-flop and the Q_3 flip-flop

13–13.

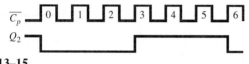

13–15.

13–17. Put a switch in series with the phototransistor's collector. With the switch open, the input to inverter 1 will be HIGH, simulating nighttime conditions, causing the yellow light to flash, day or night.

13–19.

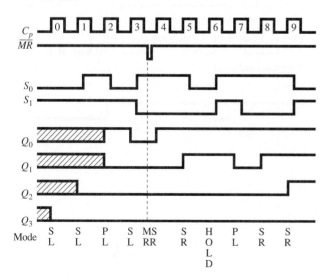

13–21.

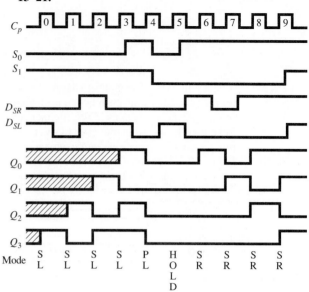

13–23.

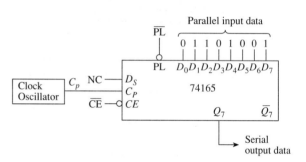

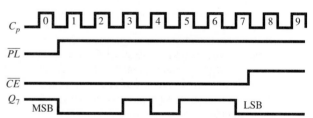

13–25. The 74164 is a serial-in, parallel-out, whereas the 74165 is a serial or parallel-in, serial-out shift register. The 74165 provides a clock-enable input, $\overline{CE}$. The serial input to the 74164 is the logical AND of two data inputs ($D_{sa} \cdot D_{sb}$).

13–27.

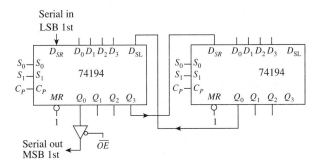

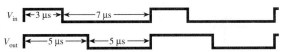

13–29.

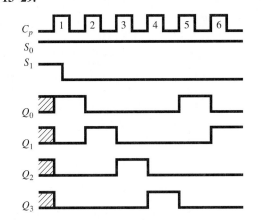

13–31. A buffer is a transparent device that connects two digital circuits; a latch is a storage device that can hold data. A buffer allows data to flow in only one direction; a transceiver is bidirectional.

13–33. (a) U4, U12, (b) U5, U30, U31, U32, U36, U37, U38 (c) U3, U6 (d) U11, U33

13–35. To load IA0-IA7 of U30, valid data must be placed on BD0-BD7 by U33, then a LOW to HIGH pulse must be applied to the CLK input of U30. When pin 3 of U13:A is LOW, a HIGH will appear at LE of U33. This allows the valid data at D0-D7 to pass through to BD0-BD7. Next, pin 3 of U13:A goes HIGH making LE LOW which causes the data outputs of U33 to remain latched. The HIGH at pin 3 of U13:A also enables the decoder U23. Just before pin 3 went HIGH A0-A1-A2 are set to 1-0-1 to provide an active output at /IOADDR. This is the clock input for U30 which passes the valid data at BD0-BD7 of U30 out to IA0-IA7. To load ID0-ID7 of U32 the same process is followed

except A0-A1-A2 are made 0-1-1 before pin 3 of U13:A is made HIGH.

Chapter 14

14–1. (a) Monostable (b) Bistable (c) Astable

14–3. 75.6 μs

14–5. 32.6 μs

14–7. Because a Schmitt device has two distinct switching thresholds, V_{T+} and V_{T-}; a regular Inverter does not. The capacitor voltage charges and discharges between those two levels.

14–9.

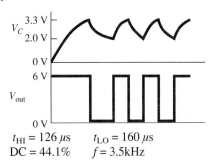

$t_{HI} = 126 \, \mu$s $\quad t_{LO} = 160 \, \mu$s
DC = 44.1% $\quad f = 3.5$kHz

14–11.

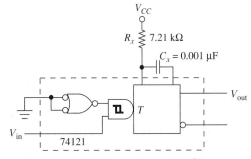

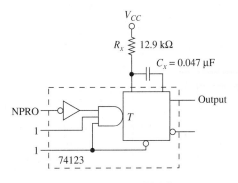

Pick $C_x = 0.001 \, \mu$F; $R_x = 7.21$ kΩ

14–13. Choose an output pulse width that is longer than 150 μs, let's say, 170 μs. That way, if a pulse is missing after 170 μs (the O.S. is not retriggered), the output will go LOW.

Pick $C_x = 0.047 \, \mu$F; $R_x = 12.9$ kΩ

14–15.

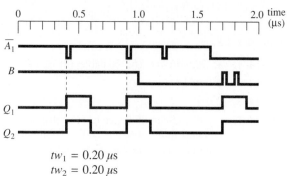

$$tw_1 = 0.20 \ \mu s$$
$$tw_2 = 0.20 \ \mu s$$

14–17. @ 0 Ω, f = 89.0 kHz
DC = 71.0%
@ 10 kΩ, f = 39.8 kHz
DC = 59.4%

14–19.

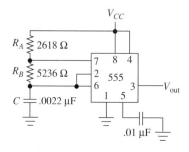

14–21.

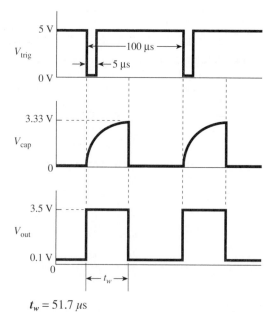

$$t_w = 51.7 \ \mu s$$

14–23.

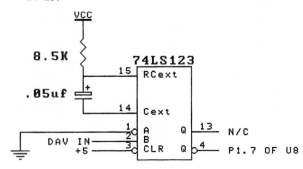

Chapter 15

15–1. Converts physical quantities into electrical quantities.

15–3. Very high input impedance, very high voltage gain, and very low output impedance.

15–5. The (−) input is at 0-V potential.

15–7. **(a)** 20 kΩ, 40kΩ, 80 kΩ

(b)

D_3	D_2	D_1	D_0	V_{out}
0	0	0	0	0.00
0	0	0	1	−1.25
0	0	1	0	−2.50
0	0	1	1	−3.75
0	1	0	0	−5.00
0	1	0	1	−6.25
0	1	1	0	−7.50
0	1	1	1	−8.75
1	0	0	0	−10.00
1	0	0	1	−11.25
1	0	1	0	−12.50
1	0	1	1	−13.75
1	1	0	0	−15.00
1	1	0	1	−16.25
1	1	1	0	−17.50
1	1	1	1	−18.75

15–9. All values of V_{out} would become positive.

15–11.

D_3	D_2	D_1	D_0	V_{out}	D_3	D_2	D_1	D_0	V_{out}
0	0	0	0	0.00	1	0	0	0	−2.00
0	0	0	1	−0.25	1	0	0	1	−2.25
0	0	1	0	−0.50	1	0	1	0	−2.50
0	0	1	1	−0.75	1	0	1	1	−2.75
0	1	0	0	−1.00	1	1	0	0	−3.00
0	1	0	1	−1.25	1	1	0	1	−3.25
0	1	1	0	−1.50	1	1	1	0	−3.50
0	1	1	1	−1.75	1	1	1	1	−3.75

15–13. To convert the analog output current (I_{out}) of the MC1408 to a voltage

15–15. **(a)** 2.5 V **(b)** 2.11 V **(c)** 1.25 V **(d)** 1.17 V

15–17. By making V_{REF} = 7.5 V. The range of I_{out} would then be 0 to 1.5 mA. The range of V_{out} would be 0 to 7.47 V

15–19.

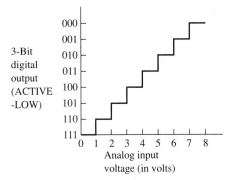

15–21. (a) 0 V (b) They are equal

15–23. $t_{\text{tot}} = 8 \times \dfrac{1}{50 \text{ kHz}} = 0.16$ ms

15–25. % error = −0.197%

15–27. $\overline{CS}$ (chip select) and $\overline{RD}$ (READ); active-LOW

15–29. (a) The three-state output latches (D_0 to D_7) would be in the float condition.
(b) The outputs would float and $\overline{WR}$ (start conversion) would be disabled.
(c) It issues a LOW at power-up to start the first conversion.
(d) 1.0 V

15–31. The temperature transducer is selected by setting up the appropriate code on the *ABC* multiplexer select inputs. The voltage level passes to the LF 198, which takes a sample at some precise time and holds the level on the hold capacitor. The LH0084 adjusts the voltage to an appropriate level to pass into the ADC. The microprocessor issues $\overline{CS}_1$, $\overline{WR}_1$, then waits for $\overline{INTR}$ to go LOW. It then issues $\overline{CS}_1$, $\overline{RD}_1$ to transfer the converted data to the data bus, then $\overline{CS}_2$, $\overline{WR}_2$, to transfer the data to RAM.

15–33. 2.439 to 2.564 V

15–35. 9.60 kg

15–37.

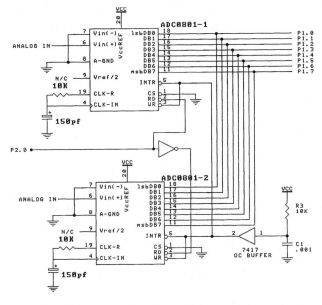

Chapter 16

16–1. Bipolar, faster; MOS, more dense.

16–3. Data are being written into memory *from* the data bus. Bus contention occurs only when two or more devices are writing *to* the data bus at the same time.

16–5.

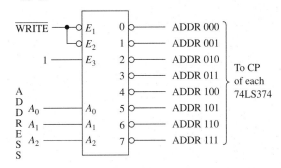

16–7. $\overline{CS}$ = LOW, $\overline{WE}$ = LOW

16–9. 10, 12, 13

16–11. (a) 8192 (b) 16,384 (c) 65,536 (d) 16,384

16–13. $\overline{CS}$ is not held LOW long enough. The access time (t_{acs}) for the 2147H is given in Figure 16–6(a) as 35 ns minimum. To correct, increase the LOW $\overline{CS}$ pulse to 35 ns or more.

16–15. The high- and low-order address lines are multiplexed to minimize the IC pin count.

16–17. (a)

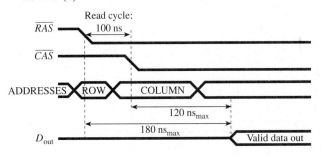

16–17. (a) Write cycle:

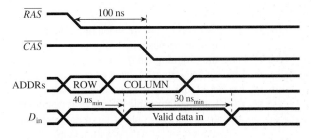

(b) 220 ns max
(c) 60 ns max

16–19. Address line multiplexing and memory refresh cycling

16–21. 2816

16–23.

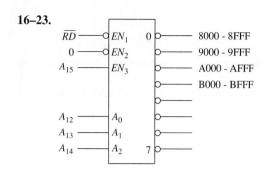

16–25. Address line A_{15} is stuck LOW.

16–27. The circuit design will be similar except the 2716 address inputs are A_0 to A_{10}. The new 74LS138 connections will be as follows:

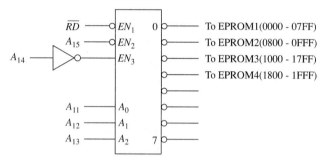

16–29.

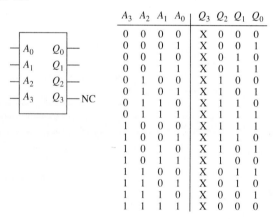

A_3	A_2	A_1	A_0	Q_3	Q_2	Q_1	Q_0
0	0	0	0	X	0	0	0
0	0	0	1	X	0	0	1
0	0	1	0	X	0	1	0
0	0	1	1	X	0	1	1
0	1	0	0	X	1	0	0
0	1	0	1	X	1	0	1
0	1	1	0	X	1	1	0
0	1	1	1	X	1	1	1
1	0	0	0	X	1	1	1
1	0	0	1	X	1	1	0
1	0	1	0	X	1	0	1
1	0	1	1	X	1	0	0
1	1	0	0	X	0	1	1
1	1	0	1	X	0	1	0
1	1	1	0	X	0	0	1
1	1	1	1	X	0	0	0

16–31. The PAL and FPLA are much more flexible than an AOI because their internal gate connections are programmed by the user, and they provide several outputs instead of just one. The basic PAL has fixed OR gates at its output. The PLA has programmable OR gates at its output. This gives the PLA greater flexibility, but adds to its complexity and propagation delay.

16–33. All equations can be implemented. It can have a maximum of 16 inputs and 48 product terms.

16–35. $32k \times 8$

16–37. (a) $8k \times 8$ (b) SMN_SEL, MON_SEL (c) A000H–BFFFH (d) E000H–FFFFH

Chapter 17

17–1. A microprocessor-based system would be used whenever calculations are to be made, decisions based on inputs are to be made, a memory of events is needed, or a modifiable system is needed.

17–3. The address bus is used to select a particular location or device within the system.

17–5.

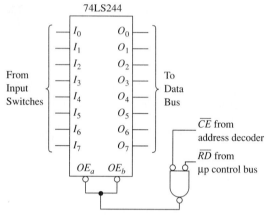

17–7. The input port has three-stated outputs so that it can be disabled when it is not being read.

17–9. 2^{16} (65,536)

17–11. It stores the contents of the accumulator out to address 6000H.

17–13. Instruction decoder and register: register and circuitry inside the microprocessor that receives the machine language code and produces the internal control signals required to execute the instruction.

17–15. (1) Pulse: read memory location 2000 (LDA) (2) and (3) Pulse: read address bytes at 2001, 2002 (4000H) (4) Pulse: read data at address 4000H

17–17. A high-level language (FORTRAN, BASIC, etc.) has the advantage of being easier to write and understand. Its disadvantage is that the programs are not memory efficient.

17–19. (a) LOW (b) LOW (c) HIGH (d) HIGH

17–21. U6a and U6b are drawn as inverted-input NAND gates to make the logical flow of the schematic easier to understand.

17–23. (a) IN instruction, $\overline{RD}$ is pulsed LOW.
(b) OUT instruction, $\overline{WR}$ is pulsed LOW.

17–25. (1) $A8$ to $A15$ = FEH
(2) $IO/\overline{M}$ = HIGH
(3) $\overline{WR}$ is pulsed LOW/HIGH

17–27. MVI A, 4FH

17–29.

Year	Value
2010	3E
2011	04
2012	3D
2013	CA
2014	10
2015	20
2016	C3
2017	12
2018	20

17–31.

	LCD_SL	KEY_SL
AD3	0	1
AD4	0	0
AD5	0	0
AD11	1	1
AD12	1	1
AD13	0	0
AD14	0	0
AD15	0	0

E

Designing with PLD Software

The basics of PLDs were presented in Section 16–6. A PLD solution to a BCD-to-seven-segment decoder using a PLS100 PLA is presented in Application 16–3. In that application you are shown the step-by-step procedure involved in manually building a programming table based on the FPLA logic diagram and the decoder function table. This is a good exercise to illustrate the theory behind PLD programming, but practically all new PLD designs are solved using PLD computer software. The manufacturers of PLDs offer this programming software at a slight fee or free of charge to encourage designers to use their PLDs.

The software prompts the user to input the logic operations to be performed and automatically determines which fuses are to be blown. The logic operations are usually input as Boolean equations or truth tables, but in some cases they can even be entered as timing waveforms or schematic diagrams. The software then reduces the input logic to its simplest form and determines if the resultant equations can be implemented on the PLD that you have chosen. It then determines which fuses are to be blown and produces a *fuse map,* which is output to a computer file that conforms to the JEDEC format. JEDEC stands for Joint Electronic Device Engineering Council. The council developed the standard so that the hardware that programs the PLD ICs (blows fuses) will use the same fuse-map configuration regardless of the PLD vendor or software.

After the JEDEC file has been created, the designer uses a hardware device called a PLD programmer to program the actual PLD IC. The device programmer is usually connected to a personal computer so that it can access the JEDEC file created by the PLD design software. The device programmer also has software associated with it that runs on the PC. This software basically asks the user what the name of the JEDEC file is and what the target PLD IC is.

Before you actually program the PLD IC, some PLD design software packages allow you to *simulate* the operation of the logic on a model of the PLD that you have chosen. To perform a simulation, you tell the simulator the timing and levels that you want to inject into the PLD model. The simulator output shows the resultant truth table and timing waveforms so that you can see if the PLD is responding as you wanted. Figure E–1 shows the flow of the development cycle for PLD programming.

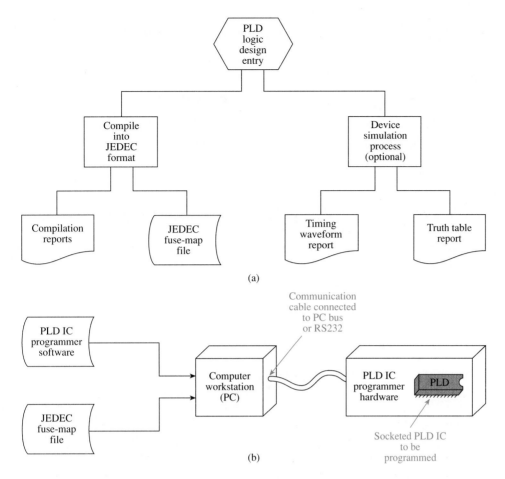

Figure E–1   The PLD programming cycle: (a) design and simulation operations; (b) hardware for programming the PLD IC.

Several versions of PLD design software are marketed today. Most PLD IC manufacturers want to encourage colleges and industries to use their product, so they offer evaluation versions of their design software free or at a reduced charge. Table E–1 lists the names of companies that provide logic design software.

Table E–1   Sample of Companies That Produce PLD ICs and/or Software

Company	Phone Number
Advanced Micro Devices	(512) 462-4360
Altera	(800) 925-8372
Cypress	(408) 943-2600
Intel	(800) 468-8118
Lattice Semiconductor	(800) 327-8425
National Semiconductor	(800) 272-9959
Orcad	(800) 671-9505
Philips (Signetics)	(408) 991-2000
Texas Instruments	(800) 477-8924
Xilinx	(800) 291-3381

The software comes on computer disks, which usually have to be loaded onto a hard disk drive before using. The documentation included gives step-by-step procedures leading you through several software design examples.

Table E–2	Sample of Companies That Provide PLD Programming Hardware
Company	Phone Number
Actel	(800) 228-3532
Advin Systems	(800) 627-2456
Altera	(408) 894-7000
Data I/O	(800) 426-1045
Emulation Technology	(408) 982-0660
Logical Devices	(800) 315-7766
Needhams Electronics	(916) 924-8037

The software by itself gives you a good experience in PLD design, but to actually program a PLD you need to purchase PLD programming hardware. Depending on the level of sophistication and features that you need, the cost of this hardware ranges from $200 to $2000, or more. Table E–2 lists some of the companies commonly recommended by PLD IC manufacturers. (Usually, these companies will also offer a PLD design software package for sale.)

Using Advanced Micro Device's PALASM 4 Software

Programmable array logic was invented at Monolithic Memories, Inc., now a part of Advanced Micro Devices (AMD), in the late 1970s. AMD's line of programmable logic ICs, design software, and programming hardware is commonly used in electronic designs. In the following pages we will use AMD's design software PALASM 4 to solve three problems previously covered in the text. The problems to be solved and the PLD used to implement the design are given in Table E–3.

Table E–3 Circuit Designs to Be Solved with PALASM 4	
Circuit Design	PLD Used
Example 5–15 (combinational logic circuit)	PAL16H2
Figure 8–7 (74138 octal decoder)	PAL16L8
Figure 13–6 (Johnson shift counter)	PAL16R8

The Users Manual provided with PALASM 4 describes everything that you need to know to design a PLD solution. It tells you to first run the INSTALL program, which copies the PALASM programs from the floppy disks provided over to the hard drive on your PC. After installation, you run the PALASM program. The program is completely menu driven. This means that there will be prompts at the top of the computer screen that you respond to as you proceed through the design stages. There are several good examples in the documentation that lead you through the design process. The design solutions that follow are not intended to explain all the details of running the software, but instead they show you some of the capabilities that are most pertinent to our basic understanding of PLD software.

Each of the following design solutions will illustrate the three most important files associated with the design process. The file names always start with the name that you have chosen and have one of the following extensions: .PDS, .XPT, or .HST (*Example:* PROBLEM1.PDS).

The .PDS file is created by the user. It declares which PLD IC is to be used, defines its pins, defines the logic to be implemented, and gives the simulation criteria.

The .XPT file shows each node within the PLD as having the fuse blown or left intact. By comparing this output with the logic diagram, you can see exactly what logic function has been created by the software.

The .HST file is the output results of the simulation. This is in the form of a truth table and/or timing waveforms. By studying it you can determine if the output response to your predefined input criteria is what it is supposed to be.

PAL Solution to Example 5–15 (Combinational Logic)

The goal of this design is to solve the equation given in Example 5–15 using a PAL. This is a combinational logic problem that can be implemented with one of the basic PAL devices like the 16H2 shown in Figure E–2.

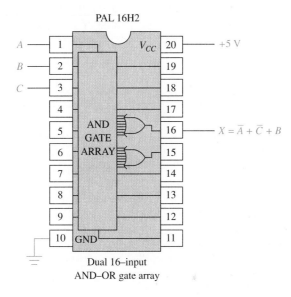

Figure E–2   Connections to a PAL16H2 to solve the equation in Example 5–15.

Figure E–3 shows the file EX5-15.PDS as it would be defined by the user. In the *Declaration Segment,* you name the design and PLD to be used. In the *PIN Declarations Segment,* you assign the inputs and outputs to specific pins. In the *Boolean Equation Segment,* you enter the original equation to be simplified and mapped into the PLD. In the *Simulation Segment,* you list the simulation activity that you want to occur.

Figure E–4 shows the file EX5-15.XPT as it would be output by the PALASM software. It is called a fuse map because it shows which fuses are to be blown (−) and which are to be left intact (X). This file is used by the PALASM compiler to create the JEDEC file.

To interpret the XPT file, you need to compare it to a logic diagram for a 16H2 (see Figure E–5). The 16H2 has 16 inputs (pins 1 to 9, 11 to 14, and 17 to 19) and two outputs (pins 15 and 16). Both true and complement inputs are connected to the 32 vertical lines. Each OR gate output can have up to eight product terms (AND gates), with each product term having up to 32 input variables (chosen from the vertical lines). In our design we have the A-variable input on pin 1. This makes the vertical line labeled 2 connect to A and 3 is connected to $\bar{A}$. B is input on pin 2, so vertical line 0 is B and line 1 is $\bar{B}$. C is input on pin 3, so vertical line 4 is C and line 5 is $\bar{C}$.

Now, by comparing Figure E–5 to Figure E–4, you should be able to determine that the output at X (pin 16) is the correctly *simplified equation* that we expected, $X = \bar{A} + \bar{C} + B$.

```
;PALASM Design Description


;------------------------------------------- Declaration Segment -------------


TITLE      EXAMPLE 5-15 (COMBINATIONAL LOGIC)
PATTERN    00
REVISION   00
AUTHOR     KLEITZ
COMPANY    TCCC
DATE       03/15/95
CHIP       AMD PAL16H2

;------------------------------------------- PIN Declarations ----------------

PIN 1         A                COMBINATORIAL                            ;INPUT
PIN 2         B                COMBINATORIAL                            ;INPUT
PIN 3         C                COMBINATORIAL                            ;INPUT
PIN 4         NC                                                        ;
PIN 5         NC                                                        ;
PIN 6         NC                                                        ;
PIN 7         NC                                                        ;
PIN 8         NC                                                        ;
PIN 9         NC                                                        ;
PIN 10        GND                                                       ;
PIN 11        NC                                                        ;
PIN 12        NC                                                        ;
PIN 13        NC                                                        ;
PIN 14        NC                                                        ;
PIN 15        NC                                                        ;
PIN 16        X                COMBINATORIAL                            ;OUTPUT
PIN 17        NC                                                        ;
PIN 18        NC                                                        ;
PIN 19        NC                                                        ;
PIN 20        VCC                                                       ;
```

;-- Boolean Equation Segment ----

```
EQUATIONS                        ———— Original equation to be simplified
X=/(A*/B)+/(A*(/A+C))
```
$$X = \overline{A\overline{B}} + \overline{A(\overline{A}+C)}$$

;-- Simulation Segment -----------

```
SIMULATION
TRACE_ON A B C X ◄——— Trace activity on A, B, C, and X
FOR I := 1 TO 3 DO ◄— Repeat the following block three times
  BEGIN
SETF /A /B /C ◄———————— Set A = 0, B = 0, C = 0
SETF /A /B C ◄———————— Set A = 0, B = 0, C = 1
SETF /A B /C              and so on
SETF /A B C
SETF A /B /C
SETF A /B C
SETF A B /C
SETF A B C
  END
TRACE_OFF
```

;---

Figure E–3 EX5-15.PDS file (design definitions). (Copyright © Advanced Micro Devices, Inc. 1991. Reprinted with permission of copyright owner. All rights reserved.)

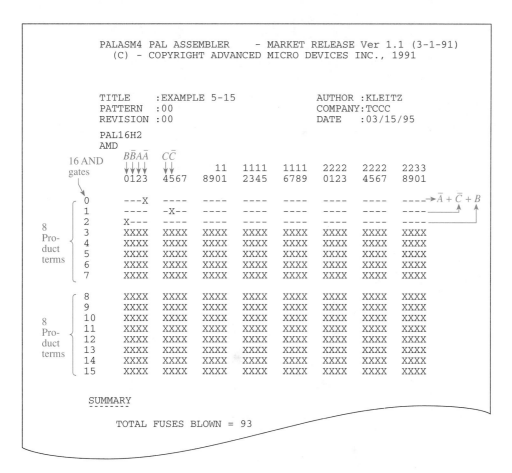

Figure E–4 EX5-15.XPT file (fuse map). (Copyright © Advanced Micro Devices, Inc. 1991. Reprinted with permission of copyright owner. All rights reserved.)

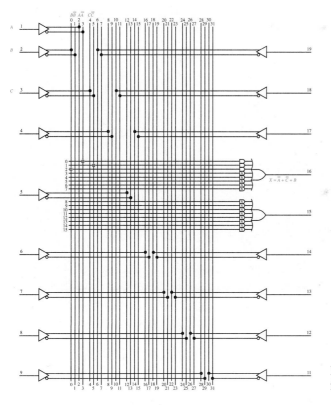

Figure E–5 Logic diagram for a PAL16H2.

695

Figure E–6 is the simulator output showing the output at X for the eight possible input conditions at $A, B,$ and C repeated three times. There are two ways to view the simulation. Figure E–6(a) shows the simulation in the form of a truth table, and Figure E–6(b) shows a photograph of the waveform output as it appears on a computer display.

```
PALASM4 PLDSIM    - MARKET RELEASE Ver 1.1 (3-1-91)
  (C) - COPYRIGHT ADVANCED MICRO DEVICES INC., 1991

PALASM SIMULATION HISTORY LISTING

TITLE    :EXAMPLE 5-15               AUTHOR :KLEITZ
PATTERN  :00                         COMPANY:TCCC
REVISION :00                         DATE   :03/15/95

PAL16H2
Page: 1

        ggggggggggggggggggggggggg  ──→ g represents SETF
    A   LLLLHHHHLLLLHHHHLLLLHHHH ⎤
    B   LLHHLLHHLLHHLLHHLLHHLLHH ⎬  The eight combinations of A, B, C
    C   LHLHLHLHLHLHLHLHLHLHLHLH ⎦  repeated three times
  GND   LLLLLLLLLLLLLLLLLLLLLLLL
    X   HHHHHLHHHHHHHLHHHHHHHLHH  ──→ X = HIGH for Ā+C̄+B
  VCC   HHHHHHHHHHHHHHHHHHHHHHHH
```

(a)

(b)

Figure E–6   PALASM simulator output: (a) EX5-15.HST file (truth table format); (b) computer display showing waveform simulation at $A, B, C,$ and X. (Copyright © Advanced Micro Devices, Inc. 1991. Reprinted with permission of copyright owner. All rights reserved.)

Designing an Octal Decoder (74138) Using a PAL16L8

A common use of programmable logic devices is to implement customized data conversion functions like decoding and encoding. To illustrate this application, we will design an octal decoder to function exactly like the 74138 presented in Figure 8–7.

By reviewing Figure 8–7, you can see that we need to provide eight active-LOW outputs (O_7 to O_0). The output to be made active is selected by a 3-bit binary code (A_0 to A_2) as long as the three enable inputs (E_1 to E_3) are satisfied.

A good choice to implement the circuit is the 16L8 PAL IC. It has eight active-LOW outputs (that is what the L8 stands for) and 16 true and complemented inputs available as product terms. Figure E–7 shows the inputs and outputs that we will define on the PAL16L8.

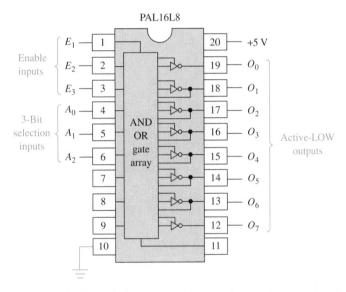

Figure E–7   Pin connections to a 16L8 PAL to function as a 74138 octal decoder.

Figure E–8 shows the FIG8–7.PDS design definition file. The pin declaration variable names are all listed in their true (not complemented) form. The equations section forms an active-LOW input or output by using a slash (/) in front of the variable name.

The simulation segment is set up to trace the activity on all inputs and outputs for 12 different input conditions. The first SETF sets E_1 HIGH. Because the other five inputs are not specified, they will be don't-cares. The first column in the simulation listing (Figure E–9) shows this condition. Notice that all outputs are HIGH as they should be, because E_1 must be LOW to enable the decoder.

The next SETF sets E_1 LOW and E_2 HIGH. This is also a disable condition, because E_1 and E_2 must both be LOW. The second column in Figure E–9 shows all outputs are still HIGH.

The third SETF is still a disable condition, and the outputs are still HIGH. The fourth SETF satisfies all three enables (E_1 = LOW, E_2 = LOW, E_3 = HIGH). Now the eight outputs are undetermined (X) because the 3-bit binary input has not been specified yet.

The next eight SETFs specify a 3-bit binary count on A_0 to A_2. The last specification given for E_1 to E_3 remains in effect unless specified otherwise, so the decoder remains enabled. Notice that as the inputs are counting, the last eight columns in the simulation listing (Figure E–9) show the active-LOW output moving from O_0 to O_1 to O_2, and so on, just as it's supposed to!

```
;PALASM Design Description

;-------------------------------------------------- Declaration Segment -------------
TITLE     Fig. 8-7 (74138 Octal Decoder)
PATTERN   00
REVISION  00
AUTHOR    Kleitz
COMPANY   TCCC
DATE      03/15/95

CHIP      AMD PAL16L8
;-------------------------------------------------- PIN Declarations ----------------

PIN 1          E1             COMBINATORIAL                          ;ENABLE
PIN 2          E2             COMBINATORIAL                          ;ENABLE
PIN 3          E3             COMBINATORIAL                          ;ENABLE
PIN 4          A0             COMBINATORIAL                          ;INPUT
PIN 5          A1             COMBINATORIAL                          ;INPUT
PIN 6          A2             COMBINATORIAL                          ;INPUT
PIN 7          NC                                                    ;
PIN 8          NC                                                    ;
PIN 9          NC                                                    ;
PIN 10         GND                                                   ;
PIN 11         NC                                                    ;
PIN 12         O7             COMBINATORIAL                          ;OUTPUT
PIN 13         O6             COMBINATORIAL                          ;OUTPUT
PIN 14         O5             COMBINATORIAL                          ;OUTPUT
PIN 15         O4             COMBINATORIAL                          ;OUTPUT
PIN 16         O3             COMBINATORIAL                          ;OUTPUT
PIN 17         O2             COMBINATORIAL                          ;OUTPUT
PIN 18         O1             COMBINATORIAL                          ;OUTPUT
PIN 19         O0             COMBINATORIAL                          ;OUTPUT
PIN 20         VCC                                                   ;
;-------------------------------------------------- Boolean Equation Segment ----
EQUATIONS
/O0=/E1*/E2*E3*/A0*/A1*/A2  ◀── Read as O0 is LOW for 001000
/O1=/E1*/E2*E3* A0*/A1*/A2  ◀── Read as O1 is LOW for 001100
/O2=/E1*/E2*E3*/A0* A1*/A2        and so on
/O3=/E1*/E2*E3* A0* A1*/A2
/O4=/E1*/E2*E3*/A0*/A1* A2
/O5=/E1*/E2*E3* A0*/A1* A2
/O6=/E1*/E2*E3*/A0* A1* A2
/O7=/E1*/E2*E3* A0* A1* A2
;-------------------------------------------------- Simulation Segment -----------
SIMULATION
TRACE_ON E1 E2 E3 A0 A1 A2 O1 O2 O3 O4 O5 O6 O7  ◀── Trace activity on
SETF  E1                                             all inputs and outputs
SETF /E1   E2
SETF /E1  /E2  /E3
SETF /E1  /E2   E3
SETF /A0  /A1  /A2
SETF  A0  /A1  /A2        Simulate outputs for
SETF /A0   A1  /A2        these twelve input conditions
SETF  A0   A1  /A2
SETF /A0  /A1   A2
SETF  A0  /A1   A2
SETF /A0   A1   A2
SETF  A0   A1   A2
TRACE_OFF
;--------------------------------------------------------------------------
```

Figure E–8   FIG8–7.PDS design definition file. (Copyright © Advanced Micro Devices, Inc. 1991. Reprinted with permission of copyright owner. All rights reserved.)

```
PALASM4 PLDSIM    - MARKET RELEASE Ver 1.1 (3-1-91)
  (C) - COPYRIGHT ADVANCED MICRO DEVICES INC., 1991

PALASM SIMULATION HISTORY LISTING

Title    :FIG. 8-7 (74138)          Author :KLEITZ
Pattern  :00                        Company:TCCC
Revision :00                        Date   :03/30/95

PAL16L8
Page: 1

        gggggggggggg
  E1    HLLLLLLLLLLL
  E2    XHLLLLLLLLLL  } 0 0 1 Enables decoder
  E3    XXLHHHHHHHHH  |
  A0    XXXXLHLHLHLH  |
  A1    XXXXLLHHLLHH  } Output selection bits
  A2    XXXXLLLLHHHH  |
  GND   LLLLLLLLLLLL
  O7    HHHXHHHHHHHL  |
  O6    HHHXHHHHHHLH  |
  O5    HHHXHHHHHLHH  |
  O4    HHHXHHHLHHHH  |
  O3    HHHXHHHLHHHH  } Output goes LOW when selected by A0 to A2
  O2    HHHXHHLHHHHH  |
  O1    HHHXHLHHHHHH  |
  O0    HHHXLHHHHHHH  |
  VCC   HHHHHHHHHHHH

          |← Enabled →|
       →| |← Disabled |
```

Figure E–9   FIG8–7.HST simulation truth table listing. (Copyright © Advanced Micro Devices, Inc. 1991. Reprinted with permission of copyright owner. All rights reserved.)

Implementing the Johnson Shift Counter of Figure 13–6 Using the Registered Outputs of the PAL16R8

The two previous PAL solutions were strictly combinational logic circuits. Sequential logic circuits, such as counters and shift registers, can be implemented only by using PAL devices that have *registered* outputs. The registered outputs are usually D flip-flops.

The PAL that we will use to solve the sequential logic requirements of a Johnson shift counter is the 16R8. The R8 in the part number signifies that it has 8 registered outputs. The abbreviated logic diagram in Figure E–10 shows that the eight D flip-flops are driven by a common buffered clock on pin 1 (CLK). The Q outputs are available on pins 12 through 19 via three-stated inverting buffers controlled by an active-LOW output enable on pin 11 *(OE)*.

The design description file in Figure E–11 defines the eight pins used on the chip to implement the function. As before, all variable names are given in their true form. The Qs are declared as "registered" because they are flip-flop outputs. The common clock (CLK) and output enable (OE) are declared as "combinatorial."

The equations section defines the necessary connections from flip-flop to flip-flop. The cross connection from the last flip-flop back to the first (which is required for the Johnson configuration) is defined by the first equation: /Q0 = Q3. The next three equations define the other three flip-flop connections. Because the Qs are registered outputs, they change states only after a LOW-to-HIGH clock transition. Therefore, the first equation should *not* be interpreted as "/Q0 *equals* Q3," but instead it should be interpreted as "/Q0 receives the level of Q3 after the active clock transition."

To simulate the circuit operation, we need to make *OE* LOW (SETF /OE) and apply clock pulses to the CLK input (CLOCKF CLK). The simulation history listing

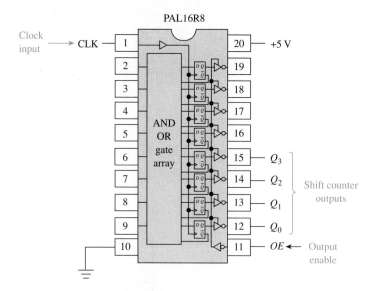

Figure E–10   Circuit connections to a registered-output PAL16R8 to be used as the Johnson shift counter of Figure 13-6.

is given in Figure E–12. The letter c is used across the top of the listing to indicate each occurrence of a LOW-to-HIGH transition of CLK. By scanning across the output listing, you can see that the Q outputs are performing the shift counter operation correctly.

```
;PALASM Design Description


;-------------------------------------- Declaration Segment -------------
TITLE     Fig. 13-6 Johnson Shift Counter
PATTERN   00
REVISION  00
AUTHOR    Kleitz
COMPANY   TCCC
DATE      04/13/95

CHIP      AMD        PAL16R8
;-------------------------------------- PIN Declarations ---------------

PIN 1          CLK           COMBINATORIAL                  ;CLOCK
PIN 10         GND                                          ;
PIN 11         OE            COMBINATORIAL                  ;ENABLE
PIN 12         Q0            REGISTERED                     ;OUTPUT
PIN 13         Q1            REGISTERED                     ;OUTPUT
PIN 14         Q2            REGISTERED                     ;OUTPUT
PIN 15         Q3            REGISTERED                     ;OUTPUT
PIN 20         VCC                                          ;

;---------------------------------------Boolean Equation Segment----

EQUATIONS


/Q0=Q3            The variable on the left
/Q1=/Q0           side of the equation assumes
/Q2=/Q1           the level of the variable on
/Q3=/Q2           the right at the LOW-to-HIGH
                  clock transition

;---------------------------------------Simulation Segment-----------

SIMULATION

TRACE_ON CLK Q0 Q1 Q2 Q3
SETF /OE            ◄────── Make OE LOW
FOR I := 1 TO 20 DO
         BEGIN
                   CLOCKF CLK       Apply 20 clock pulses
         END                        to input CLK (pin 1)
TRACE_OFF

;---------------------------------------------------------------------
```

Figure E–11   FIG13–6.PDS design description file for a PAL16R8. (Copyright © Advanced Micro Devices, Inc. 1991. Reprinted with permission of copyright owner. All rights reserved.)

701

```
PALASM4 PLDSIM    - MARKET RELEASE Ver 1.1 (3-1-91)
  (C) - COPYRIGHT ADVANCED MICRO DEVICES INC., 1991

PALASM SIMULATION HISTORY LISTING

Title    :FIG. 13-6 Johnson Shift         Author :KLEITZ
Pattern  :00                              Company:TCCC
Revision :00                              Date   :04/13/95

PAL16L8                                        ───── c represents LOW-to-HIGH
Page: 1                                                clock edge
                                                 ↓   ↓   ↓
      gc c c c c c c c c c c c c c c c c c c c c c
CLK   LHHLHHLHHLHHLHHLHHLHHLHHLHHLHHLHHLHHLHHLHHLHHLHHLHHLHH
GND   LLLLLLLLLLLLLLLLLLLLLLLLLLLLLLLLLLLLLLLLLLLLLLLLLLLLL
OE    LLLLLLLLLLLLLLLLLLLLLLLLLLLLLLLLLLLLLLLLLLLLLLLLLLLLL
Q0    HHLLLLLLLLLLLLHHHHHHHHHHHHHLLLLLLLLLLLLLHHHHHHHHHHHHHHLLLLLLLLLLL ⎤ Shift
Q1    HHHHHLLLLLLLLLLLLHHHHHHHHHHHHHLLLLLLLLLLLLLHHHHHHHHHHHHHHLLLLLLL ⎬ counter
Q2    HHHHHHHHLLLLLLLLLLLLLHHHHHHHHHHHHHLLLLLLLLLLLLLHHHHHHHHHHHHHHLLLL ⎬ output
Q3    HHHHHHHHHHHLLLLLLLLLLLLLHHHHHHHHHHHHHLLLLLLLLLLLLLHHHHHHHHHHHHHHL ⎦ pattern
VCC   HHHHHHHHHHHHHHHHHHHHHHHHHHHHHHHHHHHHHHHHHHHHHHHHHHHHH
```

Figure E–12   FIG13–6.HST simulator truth table listing for a Johnson shift counter. (Copyright © Advanced Micro Devices, Inc. 1991. Reprinted with permission of copyright owner. All rights reserved.)

Review of Basic Electricity Principles

Definitions for Figure F–1

$V \equiv$ voltage source that pushes the current *(I)* through the circuit, like water through a pipe
$I \equiv$ current that flows through the circuit
$R \equiv$ resistance to the flow of current

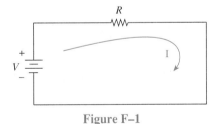

Figure F–1

Units

voltage = volts (V), for example, 12 V, 6 mV
current = amperes (A), for example, 2 A, 2.5 mA
resistance = ohms (Ω), for example, 100 Ω, 4.7 kΩ

Ohms Law

The current *(I)* in a complete circuit is proportional to the applied voltage *(V)* and inversely proportional to the resistance *(R)* of the circuit.

Formulas:

$$I = \frac{V}{R}$$

$$V = I \times R$$

$$R = \frac{V}{I}$$

EXAMPLE F–1

Determine the current *(I)* in the circuit of Figure F–1 if $V = 10$ V and $R = 1\ k\Omega$.

Solution:

$$I = \frac{V}{R}$$

$$= \frac{10\ V}{1\ k\Omega}$$

$$= 10 \times 10^{-3}\ A$$

$$= 10\ mA$$

EXAMPLE F–2

A *series circuit* has two or more resistors end to end. The total resistance is equal to the sum of the individual resistances $(R_T = R_1 + R_2)$. Also, the sum of the voltage drops across all resistors will equal the total applied voltage $(V_T = V_{R1} + V_{R2})$.

Find the current in the circuit *(I)*, the voltage across R_1 (V_{R1}), and the voltage across R_2 (V_{R2}) in Figure F–2.

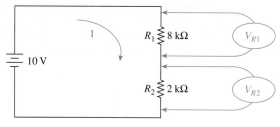

Figure F–2

Solution:

$$R_T = 8\ k\Omega + 2\ k\Omega = 10\ k\Omega$$

$$I = \frac{10\ V}{10\ k\Omega} = 1\ mA$$

$$V_{R1} = 1\ mA \times 8\ k\Omega = 8V$$

$$V_{R2} = 1\ mA \times 2\ k\Omega = 2V$$

Check:

$$V_T = 8\ V + 2\ V = 10\ V$$

Notice that the voltage across any resistor in the series circuit is proportional to the size of the resistor. That fact is used in developing the *voltage-divider equation:*

$$V_{R1} = V_T \times \frac{R_1}{R_1 + R_2}$$

$$= 10\ V \times \frac{8\ k\Omega}{2\ k\Omega + 8\ k\Omega}$$

$$= 8\ V$$

EXAMPLE F–3

Use the voltage-divider equation to find V_{out} in Figure F–3. (V_{out} is the voltage from the point labeled V_{out} to the ground symbol.)

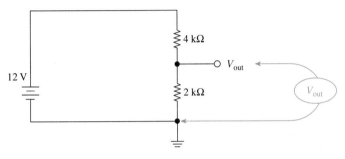

Figure F–3

Solution:

$$V_{out} = 12\ V \times \frac{2\ k\Omega}{2\ k\Omega + 4\ k\Omega}$$

$$= 4V$$

EXAMPLE F–4

A *short circuit* occurs when an electrical conductor is purposely or inadvertently placed across a circuit component. The short causes the current to bypass the shorted component. Calculate V_{out} in Figure F–4.

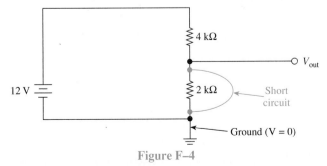

Figure F–4

Solution: V_{out} is connected directly to ground; therefore, $V_{out} = 0\ V$.

Find V_{out} in Figure F–5.

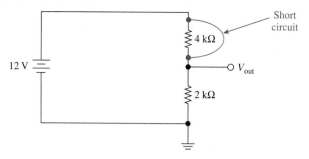

Figure F–5

Solution: V_{out} is connected directly to the top of the 12-V source battery. Therefore, V_{out} = 12 V.

An *open circuit* is a break in a circuit. This break will cause the current to stop flowing to all components fed from that point. Calculate V_{out} in Figure F–6.

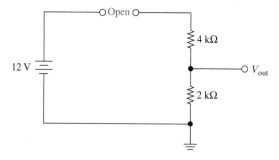

Figure F–6

Solution: Because $I = 0$ A,

$$V_{2 k\Omega} = 0 \text{ A} \times 2 \text{ k}\Omega = 0 \text{ V}$$
$$V_{out} = V_{2 k\Omega} = 0 \text{ V}$$

Calculate V_{out} in Figure F–7.

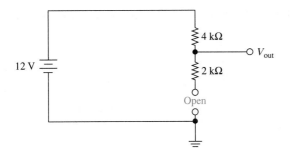

Figure F–7

Solution: Because $I = 0$ A,

$$V_{\text{drop (4 k}\Omega)} = 0 \text{ A} \times 4 \text{ k}\Omega = 0 \text{ V}$$
$$V_{\text{out}} = 12 \text{ V} - V_{\text{drop}} = 12 \text{ V}$$

EXAMPLE F–8

The symbol for a battery is seldom drawn in schematic diagrams. Figure F–8 is an alternative schematic for a series circuit. Solve for V_{out}.

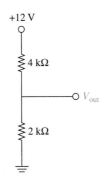

Figure F–8

Solution:

$$V_{\text{out}} = 12 \text{ V} \times \frac{2 \text{ k}\Omega}{2 \text{ k}\Omega + 4 \text{ k}\Omega}$$
$$= 4 \text{ V}$$

EXAMPLE F–9

A relay's contacts or a transistor's collector–emitter can be used to create opens and shorts. Figure F–9(a) uses a relay to short one resistor in a series circuit (relay operation is described in Section 2–6). Sketch the waveform at V_{out} in Figure F–9(a).

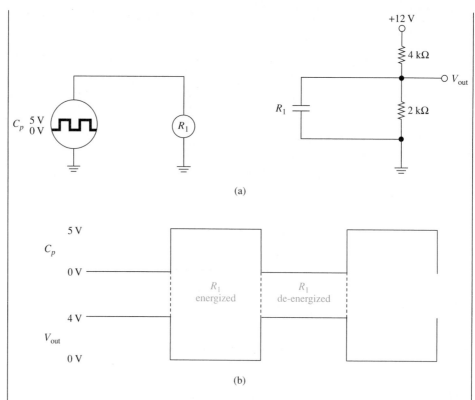

(a)

(b)

Figure F–9

Solution: When the R_1 coil energizes, the R_1 contacts close, shorting the 2-kΩ resistor and making $V_{out} = 0$ V. When the coil is deenergized, the contacts are open and V_{out} is found using the voltage-divider equation.

$$V_{out} = 12 \text{ V} \times \frac{2 \text{ k}\Omega}{2 \text{ k}\Omega + 4 \text{ k}\Omega}$$

$$= 4 \text{ V}$$

The clock oscillator (C_p) and V_{out} waveforms are given in Figure F–9(b).

Review Questions

F–1. What value of voltage will cause 6 mA to flow in Figure F–1 if $R = 2 \text{ k}\Omega$ (3 V, 0.333 V, or 12 V)?

F–2. To increase the current in Figure F–1, the resistor value should be _____ (increased, decreased).

F–3. In a series voltage-divider circuit like Figure F–2, the larger resistor will have the larger voltage across it. True or false?

F–4. In Figure F–2, if R_1 is changed to 8 MΩ and R_2 is changed to 2 Ω, V_{R2} will be close to _____ (0 V, 10 V).

F–5. If the supply voltage in Figure F–3 is increased to 18 V, V_{out} becomes _____ (6 V, 12 V).

F–6. A short circuit causes current to stop flowing in the part of the circuit not being shorted. True or false?

F–7. The current leaving the battery in Figure F–4 _____ (increases, decreases) if the short circuit is removed.

F–8. The short circuit in Figure F–5 causes V_{out} to become 12 V because the current through the 2-kΩ resistor becomes 0 A. True or false?

F–9. In Figure F–6, the voltage across the 2-kΩ resistor is the same as that across the 4-kΩ resistor. True or false?

F–10. If the 4-kΩ resistor in Figure F–7 is doubled, V_{out} will _____ (increase, decrease, remain the same)?

Answers to Review Questions

F–1.	12 V	**F–6.**	False
F–2.	Decreased	**F–7.**	Decreases
F–3.	True	**F–8.**	False
F–4.	0 V	**F–9.**	True
F–5.	6 V	**F–10.**	Remain the same

G

Schematic Diagrams

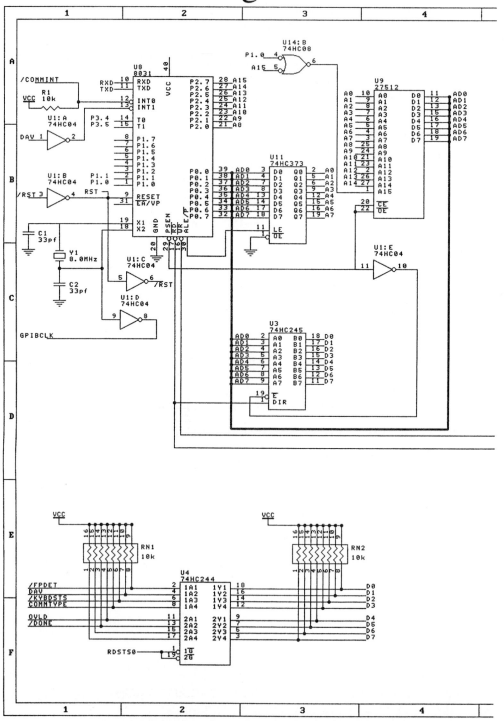

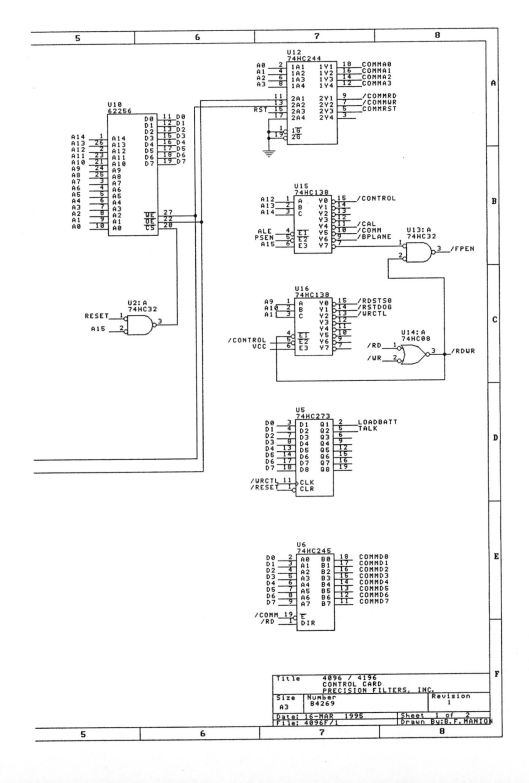

Title	4096 / 4196 CONTROL CARD PRECISION FILTERS, INC.		
Size A3	Number B4269		Revision 1
Date: 16-MAR 1995		Sheet 1 of 2	
File: 4096F/1		Drawn By:B.F.MANION	

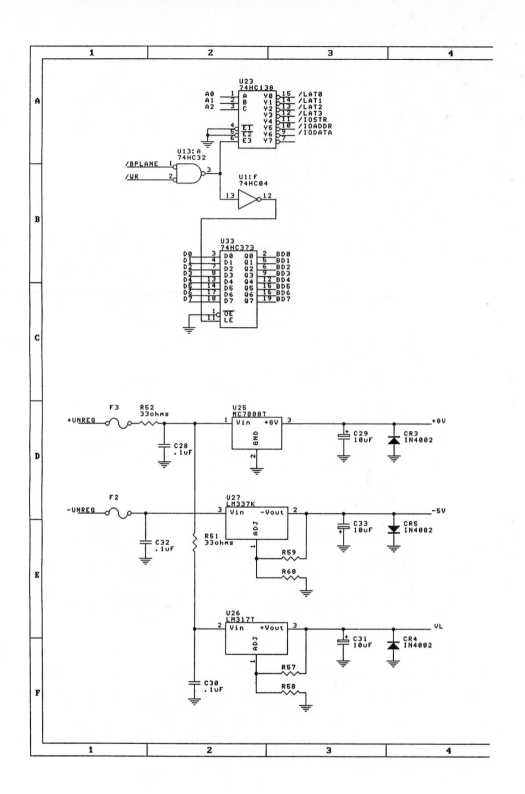

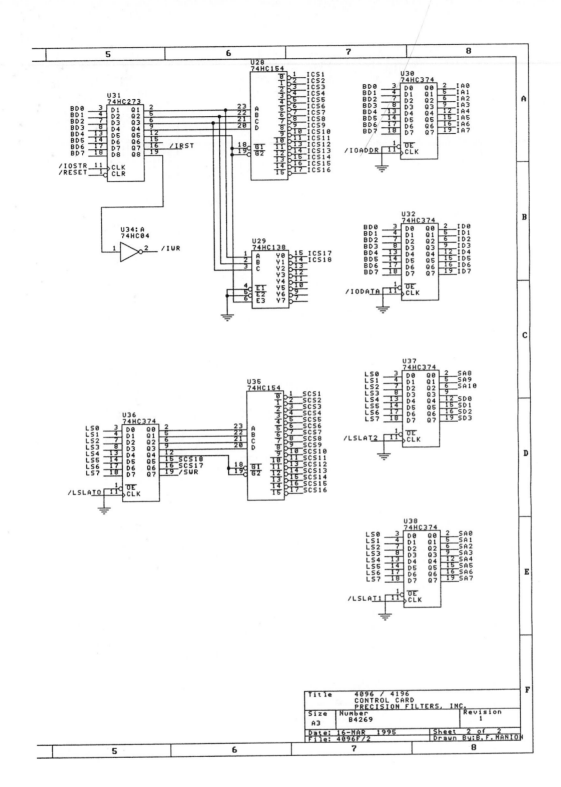

Title	4096 / 4196		
	CONTROL CARD		
	PRECISION FILTERS, INC.		
Size	Number		Revision
A3	B4269		1
Date: 16-MAR 1995		Sheet 2 of 2	
File: 4096F/2		Drawn By:B.F.MANION	

713

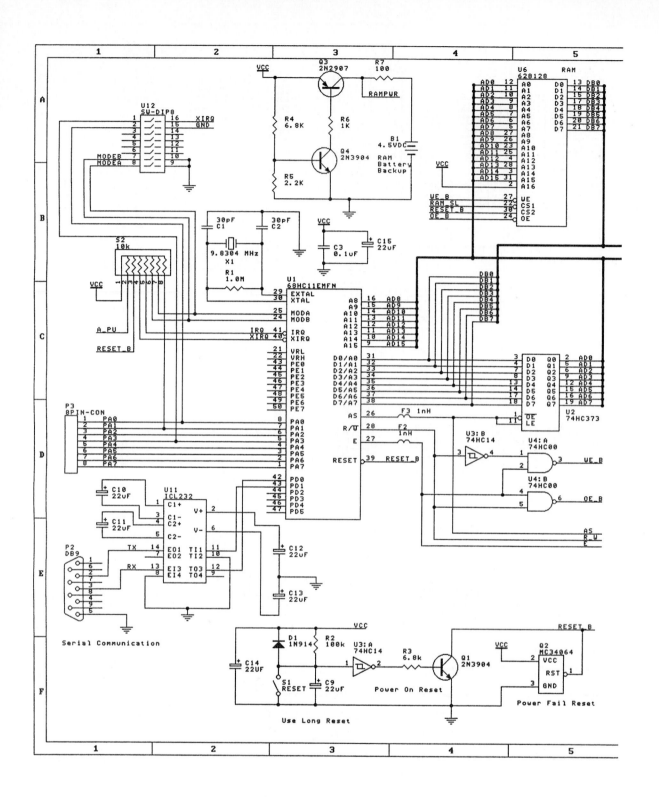

714

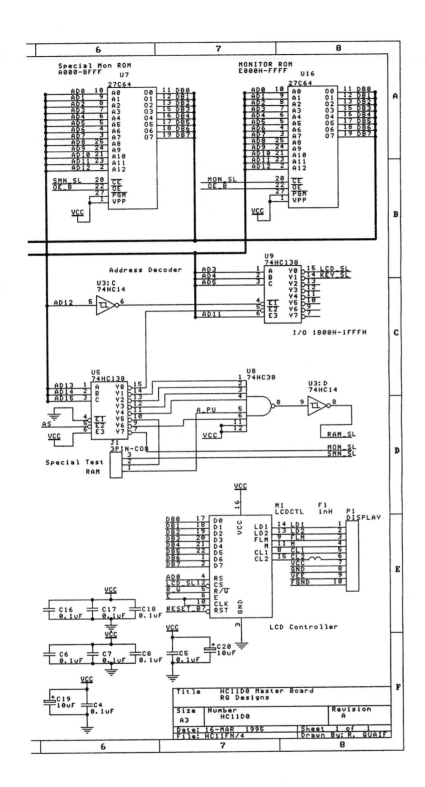

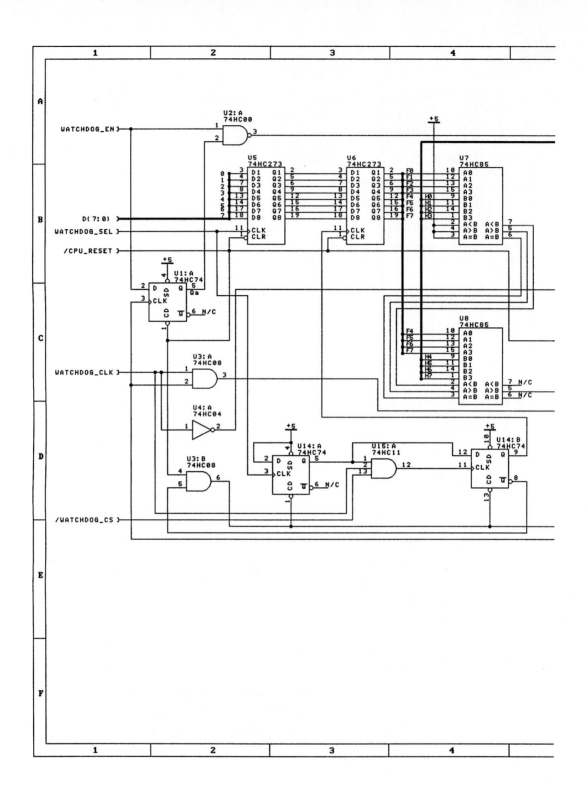

716

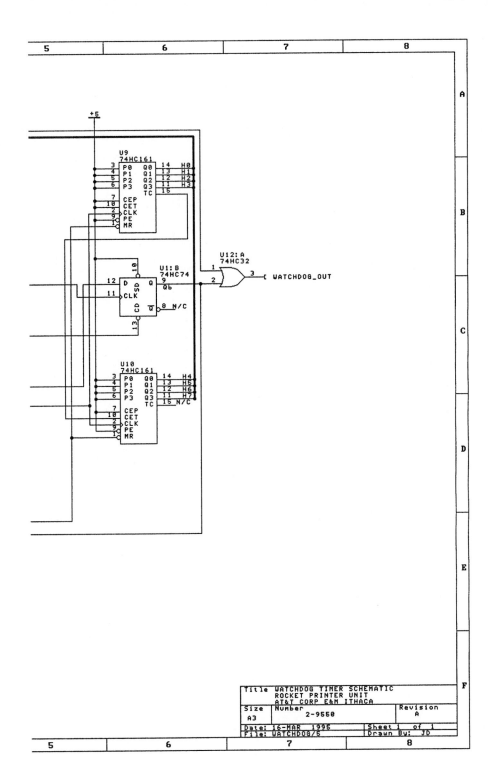

Title	WATCHDOG TIMER SCHEMATIC ROCKET PRINTER UNIT AT&T CORP E&M ITHACA		
Size A3	Number 2-9550		Revision A
Date: 16-MAR 1995		Sheet 1 of 1	
File: WATCHDOG/5		Drawn By: JD	

Index

Ac waveforms, 350, **382**
Access time, 570, 575
Accumulator, 607–12, **620**
Active clock edge, 350–51, **382**
Active-HIGH, 223–24, 317, 419
Active-LOW, 223–24, 228, 320, **337**, 419
Adder ICs, 201–06
Address bus, 582–83, 605, **620**
Address decoding, 252, 582–83, **596**, 605
Alphanumeric, 15, **20**
ALS TTL (*see* TTL)
Analog, 2–4, **20**, 527–28
Analog multiplexer, 549
Analog multiplexer/demultiplexer, 250
Analog-to-digital converter (ADC), 528, 537–47, **555**
  ADC ICs, 542–49, 626
  counter-ramp ADC, 539
  parallel-encoded ADC, 537–38
  specifications, 536–37
  successive-approximation ADC, 540
AND gate, 58–60
AND-OR-INVERT gate, 135–39
Anode, 37, 378
Arithmetic circuits, 197–203
  adder IC, 201–03
  ALU, 207–10, **211**
  BCD adder, 205–06
  full adder, 198–99
  half adder, 198
  two's complement adder/subtractor, 204
Arithmetic/logic unit (ALU), 207–10, **211**

ASCII, 15–16
Assembly language, 607, 612, 614–16, **620**
Astable (free-running) multivibrator, 494, 498, 510–11, **521**
Asynchronous, 314, 319–21, **337**
Asynchronous counter, 397
Automatic reset, 365–67, **382**

BCD adder circuit, 205–06
BCD arithmetic, 195–96
BCD-to-seven-segment decoder, 420–21
Biasing diodes and transistors, 37–40
BiCMOS, 289
Bidirectional, 250, **258**
Binary arithmetic, 182–87
  addition, 182–83
  BCD addition, 195–96, 205–06
  division, 187–88
  hexadecimal addition, 192–94
  hexadecimal subtraction, 194–95
  multiplication, 185–86
  subtraction, 183–85
  two's complement, 188–92, 204–05
Binary conversion (*see* Conversion)
Binary-coded-decimal (BCD), 14, **20**
Binary numbering system, 5–6, **20**
Binary string, 174, **176**, 220
Binary weighted DAC, 530
Binary word, 201, **211**, 221
Binary-to-Gray code conversion, 241–42
Bipolar memory, 563
Bipolar transistor (*see* Transistor)

NOTE: Page number in **boldface** type indicates end-of-chapter glossary definition for the term.

Bistable multivibrator (S-R flip-flop), 494
Bit, 8, **21**
Block diagram, 200, **211**
Boolean algebra, 110–19
  associative law, 110
  commutative law, 110
  distributive law, 110
  reduction, 115–19
  rules, 112–14
  simplification, 115–19
Boolean equation defined, 58, **73**
Bubble pushing, 128–29
Buffer, 295, **301**, 476–78, **482**, 585, **596**
Burn-in, 518, **520**
Bus, 256–**58**, 543, **556** 565–66
Bus contention, 566, **596**
Byte, 564, **596**

CAD (Computer-Aided Design), 96
Capacitor charging/discharging formula, 494–95
Carry, 182, 197–200, **211**
CAS (Column Address Strobe for dynamic RAMs), 574–77, **596**
Cascading flip-flops, 396, **441**
Cathode, 37, 378
Central Processing Unit (CPU), 609–10, **620**
Chip, 44, **49**
Clock enable, 462, 482
Clock signal, 318, **337**
Clock waveform, 26–28
CMOS (Complementary metal-oxide semiconductor), 285–89
  comparison table, 292, 296
  handling MOS devices, 287
  high-speed CMOS, 288
  interfacing, 293–300
  Low Voltage (LV), 289
  MOSFET, 286
  NMOS, 286
  PMOS, 286
  silicon dioxide layer, 286
  substrate, 286
Code converters, 235–42
  BCD-to-binary, 237
  binary-to-BCD, 238
  binary-to-Gray code, 240–42
  Gray code-to-binary, 240–42
  (*see also* Encoder and Decoder)
Combinational logic, 108–10, **149**
Common-anode LED, 418–19, **441**
Comparator, 174, 220–21, **259**, 510–11, **520**, 538
Complement, 82, **98** (*see also* Controlled inverter)
Contact bounce, 372–75
Controlled inverter, 175
Conversion (number systems):
  BCD-to-decimal, 14
  binary-to-decimal, 5–6
  binary-to-hexadecimal, 12
  binary-to-octal, 9
  decimal-to-BCD, 14

Conversion (number systems), *continued.*
  decimal-to-binary, 7–8
  decimal-to-hexadecimal, 13
  decimal-to-octal, 10
  hexadecimal-to-binary, 11–12
  hexadecimal-to-decimal, 11–12
  octal-to-binary, 10
  octal-to-decimal, 10
  successive division, 6, 8, 10
Conversion time (of an ADC), 539–41, **556**
Counter (*see* Ripple counter and Synchronous counter)
Counter-ramp ADC, 539
Crystal, 518–19, **521**

Data acquisition system, 547–50
Data bit, 451–54, **482**
Data bus, 565, 571–72, 605, **620**
Data distributor (*see* Demultiplexer)
Data selector (*see* Multiplexer)
Data transmission, 29, 453–54
Debouncing a switch, 372–75
Decimal conversion (*see* Conversion)
Decimal numbering system, 4, **21**
Decoder, 222–30, **259**
  BCD decoder IC, 227–28
  BCD-to-seven segment, 420–21
  binary-to-octal, 223–25
  hex decoder IC, 228–29
  octal decoder IC, 225–27
Delay gate, 361–63
Delay line (for DRAMs), 577, **596**
DeMorgan's theorem, 119–27, **149**
Demultiplexer, 248–50
Dependency notation, 665–69
Differential measurement, 544, 546, **556**
Digital, 2–4, **21**
Digital clock, 415–16
Digital-to-analog converter (DAC), 530–36
  binary weighted DAC, 530–31
  DAC ICs, 533–36
  R-2R DAC, 531–33
  specifications, 536–37
Diode, 36–39, **49**
DIP (Dual In-Line Package), 45, **49**
Disable, 64–65, **73**, 226
Don't care condition, 64, 127, **149**
Down-counter, 398–99, 429
Duty cycle, 369–70, **382**, 504, 514
Dynamic RAM, 574–78, **596**
  controller, 575–76
  RAS and CAS, 574–75, **596**
  refresh, 575
Dynamic RAM controller, 575–76

ECL (Emitter-coupled logic), 290–91, 297–98
Edge-detection circuit, 318–19
Edge-triggered flip-flop, 326–30, **338**
EEPROM, 579
Egg timer circuit, 416–18
Electrostatic noise, **176, 338**

EMI (electromagnetic interference), **301**
Enable, 64–65, **73,** 226
Encoder, 230–35
   decimal-to-BCD, 231–33
   octal-to-binary, 233–35
EPROM, 579–81
Error detection (*see* Parity)
Even parity, 170–74
Exclusive-NOR gate, 167–169, **176**
Exclusive-OR gate, 166–169, **176**
Exponential charge/discharge, 494–95, **521**

Fall time, 278–79, **301**
Fan-out, 272–73, **301**
FAST TTL (*see* TTL)
Fast-look-ahead carry, 202, **211**
555 IC timer, 510–18
   astable (oscillator) connections, 510–15
   monostable (one-shot) connections, 500–10
   pin definitions, 510
Flash converter ADC, 537
Flip-flop, 310–336, **338**
   cross-NAND S-R flip-flop, 311
   cross-NOR S-R flip-flop, 310
   D flip-flop, 316, 318–22
   J-K flip-flop, 322–27
   S-R flip-flop, 310–311
Float, **73, 382, 482**
Flowchart, 615, **620**
FPLA (Field-programmable logic array), 589–94
Frequency, 26–28, **49**
Full adder, 198–200, **211**
Function table, 310, **338**
Fusible link, 538–39, **555** (*see also* EPROM, EEPROM, and
   Programmable array)

Gate, **73**
Gate loading, 273–75
Glitch, 401, **442**
Gray code, 240–42, **259**
   binary-to-Gray code conversion, 241–42
   Gray code-to-binary conversion, 241–42

Half adder, 198–200, **211**
Handshaking, **482, 556**
Hardware, 613, **620**
Hertz, 27
Hexadecimal arithmetic, 192–94
Hexadecimal conversion (*see* Conversion)
Hexadecimal numbering system, 11, **21**
High impedance state, 466, 543 (*see also* Float)
Hold time, 351, **382**
Hysteresis, 367, **382**

IEEE/IEC standard for logic symbols, 97–98, 665–69
Integrated circuit (IC), 44–46 (*see also* Supplementary Index
   of ICs)
Inversion bar, 82
Inversion bubble, 82
Inverter gate, 82

Johnson shift counter, 87–93, 456–59

Karnaugh map, 139–48, **150**

Latch, 316–17, **338**
LCD (Liquid Crystal Display), 420, **442,** 586, **597**
Least significant bit (LSB), 8, **21**
LED (Light-emitting-diode), 378
Level shifter, 297, **302**
Linear temperature sensor, 551–52
Linearity (of a DAC or ADC), 536, **556**
Logic analyzer, 88
Logic probe, 68–**73**
Logic pulser, 68–**73**
Logic state, 31, **49**
Look-up table, 585–86, **597**
Low Voltage CMOS (LV), 289
LS TTL (*see* TTL)

Magnetic memory, 563, **597**
Magnitude comparator (*see* Comparator)
Mask ROM, 578, **597**
Master-slave flip-flop, 322–25, **338**
Memory, 564–87
Memory address (location), 564, **597**
Memory cell, 568, **597**
Memory contents, 564, **597**
Memory expansion, 570, 582–85
Memory mapping, 582, 612
Memory timing, 566, 569–70 (*see also* Read cycle and Write
   cycle)
Microcontroller, 253–55, 336, 423
Microprocessor, 604–617, **621**
Mnemonic, 607, **621**
Modulus, 396, **442**
Monostable (one-shot) multivibrator, 500–10
   555 monostable, 515–18
   IC monostable, 503–07
   retriggerable monostable, 507–10, **521**
Monotonicity (of a DAC), 537, **556**
MOSFET (Metal-oxide-semiconductor field-effect transis-
   tor), 285–87, **302**
   N-channel, 285–87
   P-channel, 285–87
   power MOSFET, 380
Most significant bit (MSB), 8, **21**
Multiplexed address bus, 574
Multiplexed display, 257, 423–24
Multiplexer, 243–48, **259**
Multivibrator, 498–518
   astable (free-running), 498–500
   bistable (S-R flip-flop), 494
   555 astable, 510–15
   555 monostable, 515–18
   IC monostable, 503–07
   monostable (one-shot), 500–10
   retriggerable monostable, 507–10, **521**

NAND gate, 82–85
Negative-edge, 326, **338**

NMOS (N-channel MOSFET) (*see* MOSFET)
Noise, 176, 324, 338
Noise margin, 275–76, **302**
NOR gate, 85–87

Octal conversion (*see* Conversion)
Octal numbering system, 10
Odd parity, 170–74
Ohm's law, 703
Ones catching, 324, **338**
One's complement, 189
One-shot (monostable) multivibrator, 500–510
Open-collector output, 280–82, **302**
Operational amplifier, 529–30, **556**
Optocoupler, 379, **382**
OR gate, 60–61
Oscillator, 498–518
Oscilloscope, 26, **49**, 278–79, 333

PAL (Programmable array logic), 587, 690
Parallel representation, 29–30, **49**
Parallel-encoded ADC, 537–38
Parity, 170–74, **176**
Period, 26–28
Phototransistor, 378–79, **382**
PLA (Programmable logic array), 588–94
PLD (Programmable logic device), 587–95, 690
PMOS (P-channel MOSFET), 285–87
Positive edge, 326, **338**
Power dissipation, 280, **302**, 375
Power supply circuit, 376–77
Power supply decoupling, 283, 301
Power-up reset, 365–66, **382**
Priority encoder, 231–33, **259**, 538
Product-of-sums (POS) expression, 135–137, **150**
Programmable array (*see* PAL, PLA, and PLD)
Programmable read-only memory (*see* PROM)
Programmable-gain amplifier, 549, **556**
PROM (Fusible-link), 578, **596** (*see also* EPROM and EEPROM)
Propagation delay, 278–79, **302**, 352–54, 397
Pull-down resistor, 375
Pull-up resistor, 294, **302**, 375
Pulse-triggered flip-flop, 322

Quartz crystal, 518

R-2R DAC, 531–33
Race condition, 350, 383
RAM (Random-access memory), 567–77
  dynamic RAM, 573–77, **597**
  static RAM, 567–72, **597**
RAS (Row Address Strobe for dynamic RAMs), 574–77, **597**
Read cycle, 569
Read/write memory (RWM) (*see* RAM)
Recirculating shift register, 454, **482**
Rectifier, 376, **383**
Register, 335, **338**
Relay, 32–36, **49**
Reset, 310–11, **338**

Resolution (of a DAC or ADC), 528, **556**
Ring shift counter, 456–59
Ripple blanking a display, 420, **442**
Ripple counter, 396–424
  digital clock, 415–16
  divide-by-N counter, 400–406
  down-counter, 398
  egg timer circuit, 416–18
  ICs, 406–11
  system design applications, 412–18
  three digit counter, 414–15
Rise time, 278–79, **302**
ROM (Read-only memory), 578–81
  EEPROM, 579
  EPROM, 579–81
  fusible-link PROM, 578, **596**
  mask ROM, 578, **597**

Sample and hold circuit, 549, **556**
Schematic diagrams, 710
Schmitt trigger, 367–71, **383**
Schottky TTL (*see* TTL)
Scientific notation, 27
Semiconductor, 36
Semiconductor memory, 564–87, **597**
  memory concepts, 564
  memory expansion, 570, 582–85
  RAM (Random-access memory), 567–77
  ROM (Read-only memory), 578–81
Sequential logic circuits, 392–95, **442**
Serial representation, 28–29, **50**
Setup time, 310–11, **338**, 350–51, **383**
Seven-segment display, 418–24, **442** (*see also* LED and LCD)
Shift counter, 456–59
Shift register, 451–81, 483
  basics, 452–54
  ICs, 460–68
  Johnson shift counter, 87–93, 456–59
  parallel-to-serial conversion, 454
  recirculating, 454
  ring shift counter, 456–59
  serial-to-parallel conversion, 456
  system design applications, 468–72
  universal shift register, 462–64
Sign bit, 188, **212**
Sink current, 273, **302**
Software, 607, **621**
Source current, 273, **302**
Speed-power product, 284, 292
Static RAM, 567–72, **597**
Stepper motor, 473–76
Storage (*see* Memory)
Storage register, 335, **338**
Strain gage, 553–54
Strobe, 313, **339**, 456–57, **483**
Substrate, 286, **302**
Successive division (*see* Conversion)
Successive-approximation ADC, 540–42, **556**
Sum-of-products (SOP) expression, 124, 135–39, **150**

Surface-mount devices (SMDs), 47–48, **50**
Switch bounce, 372–75, **383**
Synchronous counter, 424–40, **442**
   divide-by-200, 439
   down counter, 429–33
   system design applications, 427–28, 437–40
   two digit decimal counter, 432–33
   up counter, 426
   up/down counter IC, 429–33
Synchronous flip-flop operations, 314, **339**

Thermistor, 550–51, **556**
Thermometer, 551–52, 586–87
Three-state output (tri-state), 465–66, **483,** 476–77
Timer IC (*see* 555 IC timer)
Timing diagram, 62–63
Toggle 322–23, **339**
Totem-pole output, 44, **50,** 270, 280, **302**
Transceiver, 479–80, **483**
Transducer, 527
Transfer function, 367–68, **383**
Transistor:
   as an inverter, 43–44, **49**
   basic switch, 39–42, **50**
   bipolar (NPN and PNP), 285, **301**
   MOSFET, 285–87, **302**
Transmission, 29–30, 170, **176**
Transparent latch, 316–17, **338**
Troubleshooting techniques, 67–72, **73**
Truth table defined, 58–59, **73**
TTL (Transistor-transistor logic), 270–85
   advanced low-power Schottky TTL (ALS), 284
   basic inverter, 43–44, **49**
   circuit operation, 270–72

TTL (Transistor-transistor logic), *continued.*
   comparison table 292, 296
   current ratings, 272–77
   disposition of unused inputs and gates, 283
   FAST TTL, 285
   interfacing, 293–300
   low-power Schottky TTL (LS), 284
   multiemitter transistor, 270
   open collector, 280–82, **302**
   power dissipation, 280
   Schottky TTL (S), 284
   time parameters, 278–79
   totem-pole output, 44, 50, 270, 280, **302**
   voltage ratings, 272–77
   wired output, 281–82, **302**
Two's complement, 188–92, **212**
   adder/subtractor circuit, 204
   arithmetic, 191
   conversion, 189–90
   representation, 188–89

Undetermined state, 350, 363
Up/down counter IC, 429–33

Virtual ground, 529, **556**
Volatile memory, 578
Voltage regulator, 376, **383**

Waveform generation, 87, **98** (*see also* Oscillator)
Wired-AND, 281–82, **302**
Write cycle, 570

Zener diode, 377, **383**
Zero suppression, 420–22

Supplementary Index of ICs

This is an index of the integrated circuits (ICs) discussed in this book. The page numbers indicate where the IC is first discussed. Page numbers in **boldface** type indicate pages containing a data sheet for the device.

7400 Series ICs

7400	Quad 2-input NAND gate, 94, **641**	
74HC00	Quad 2-input NAND gate, **627**	
74LV00	Quad 2-input NAND gate, 290, **631**	
7402	Quad 2-input NOR gate, 94	
7404	Hex inverter, 94	
74HC04	Hex inverter, 519	
7408	Quad 2-input AND gate, 67	
74HC08	Quad 2-input AND gate, 67	
74HCT08	Quad 2-input AND gate, 154	
7411	Triple 3-input AND gate, 67	
74HC11	Triple 3-input AND gate, 67	
7414	Hex inverter Schmitt trigger, 368, **643**	
74HC14	Hex inverter Schmitt trigger, 498	
7421	Dual 4-input AND gate, 67	
74HC21	Dual 4-input AND gate, 67	
7427	Triple 3-input NOR gate, 105	
7430	8-input NAND gate, 104	
7432	Quad 2-input OR gate, 67	
74HC32	Quad 2-input OR gate, 67	
74HCT32	Quad 2-input OR gate, 154	
7442	BCD-to-decimal decoder (1-of-10), 228	
7447	BCD-to-seven-segment decoder/driver, 420	
7454	4-wide 2- and 3-input AND-OR-invert gate (AOI), 137	
7474	Dual D-type flip-flop, 319	
7475	Quad bistable latch, 317	

7476	Dual J-K flip-flop, 328	
74LS76	Dual J-K flip-flop, 328	
7483	4-bit full adder, 201	
7485	4-bit magnitude comparator, 221	
74HC86	Quad 2-input exclusive-OR gate, 178	
7490	Decade counter, 409	
7492	Divide-by-twelve counter, 409	
7493	4-bit binary ripple counter, 407	
74H106	Dual J-K negative edge-triggered flip-flop, 352	
74109	Dual J-K positive edge-triggered flip-flop, 331	
74LS112	Dual J-K negative edge-triggered flip-flop, 363	
74121	Monostable multivibrator, 503, **644**	
74123	Dual retriggerable monostable multivibrator, 507	
74S124	Voltage-controlled oscillator, 519	
74132	Quad 2-input NAND Schmitt trigger, 369	
74138	1-of-8 decoder/demultiplexer, 225	
74LS138	1-of-8 decoder/demultiplexer, 225	
74HCT138	1-of-8 decoder/demultiplexer, 252	
74139	Dual 1-of-4 decoder/demultiplexer, 249	
74147	10-line-to-4-line priority encoder, 232	
74148	8-input priority encoder, 234	
74150	16-input multiplexer, 248	
74151	8-input multiplexer, 245	
74154	1-of-16 decoder/demultiplexer, 229	
74LS154	1-of-16 decoder/demultiplexer, 565	
74160	4-bit binary counter, 435	
74164	8-bit serial-in parallel-out shift register, 460	

74165	8-bit serial/parallel-in, serial-out shift register, 461
74181	4-bit arithmetic logic unit, 207
74184	BCD-to-binary converter, 237
74185	Binary-to-BCD converter, 238
74190	Presettable BCD decade up/down counter, 434
74191	Presettable 4-bit binary up/down counter, 434
74192	Presettable BCD decade up/down counter, 429
74193	Presettable 4-bit binary up/down counter, 429
74194	4-bit bidirectional universal shift register, 462
74HCT238	1-of-8 demultiplexer, 254
74241	Octal buffer (3-state), 585
74LS244	Octal buffer (3-state), 478
74ABT244	Octal buffer (3-state), 289, **634**
74LS245	Octal transceiver (3-state), 480
74HCT273	Octal D flip-flop, 336
74280	9-bit odd/even parity generator/checker, 173
74HC283	4-bit full adder with fast carry, 203
74LS373	Octal latch with 3-state outputs, 479
74LS374	Octal D flip-flop with 3-state outputs, 479
74395A	4-bit cascadable shift-register with 3-state outputs, 466
74HC583	4-bit BCD adder, 206
74LS640	Inverting octal bus transceiver, 585
74HCT4543	BCD-to-seven-segment latch/decoder/driver, 586

Other ICs

4001	Quad 2-input NOR gate, 94
4008	4-bit binary full adder, 204
4011	Quad 2-input NAND, 94
4049	Hex inverting buffer, 94
4050	Hex non-inverting buffer, 295
4051	8-channel analog multiplexer/demultiplexer, 251

4069B	Hex inverter, 294
4504B	Level-shifting buffer, 297
4543	BCD-to-seven-segment latch/decoder/driver, 586
2118	16K × 1 dynamic RAM, **573**
2147H	4K × 1 static RAM, **568**
2716	2K × 8 EPROM, 579, 645
2732	4K × 8 EPROM, 583
3242	Dynamic RAM controller, 576
LF198	Sample and hold circuit, 549
LH0084	Programmable-gain instrumentation amplifier, 549
ADC0801	Analog-to-digital converter, 544, **647**
DAC0808	Digital-to-analog converter, 535
MC1408	Digital-to-analog converter, 534, **650**
AM3705	Analog multiplexer switch, 549
NE5034	Analog-to-digital converter, 543
10124	ECL level shifter, 298
10125	ECL level shifter, 298
82S100 (PLS100)	Field programmable logic array, 594
PAL16H2	Programmable array logic, 690
PAL16L8	Programmable array logic, 690
PAL16R8	Programmable array logic, 690
8085A	8-bit microprocessor, 608
8051	8-bit microcontroller, 254
68HC11	8-bit microcontroller, 336
7805	Voltage regulator, 376
555	Timer, 511, **659**
741	Operational amplifier, 534, **653**
4N35	Opto-coupler, 379
IRF130	Power MOSFET, 380
TIL601	Phototransistor, 379
LM35	Linear temperature sensor, 552
LM185	Precision reference diode, 552
1N749	Zener diode, 377